BIRKHÄUSER

E.C.G. Stueckelberg, An Unconventional Figure of Twentieth Century Physics

Selected Scientific Papers with Commentaries

Jan Lacki
Henri Ruegg
Gérard Wanders
Editors

Birkhäuser
Basel · Boston · Berlin

Editors:
Jan Lacki
REHSEIS, UMR 7596, CNS, and
Université de Genève
Unité „Histoire et Philosophie des Sciences“
Faculté des Sciences
24, Quai E. Ansermet
1211 Genève 4
Switzerland
e-mail: Jan.Lacki@physics.unige.ch

Henri Ruegg
6, Chemin du Claiset
1256 Troinex
Switzerland

Gérard Wanders
13, Chemin de la Cure
1012 Lausanne
Switzerland

2000 Mathematical Subject Classification: 81-03, 00B60, 01A75

Library of Congress Control Number: 2008935062

Bibliographic information published by Die Deutsche Bibliothek. Die Deutsche Bibliothek lists this publication in the Deutsche Nationalbibliografie; detailed bibliographic data is available in the Internet at http://dnb.ddb.de

ISBN 978-3-7643-8877-5 Birkhäuser Verlag AG, Basel - Boston - Berlin

Basel · Boston · Berlin
P.O. Box 133, CH-4010 Basel, Switzerland
Part of Springer Science+Business Media
Printed on acid-free paper produced from chlorine-free pulp. TCF ∞
Cover figures: Institute of Physics, Geneva
Printed in Germany

ISBN 978-3-7643-8877-5
9 8 7 6 5 4 3 2 1

e-ISBN 978-3-7643-8878-2
www.birkhauser.ch

List of Contributors

Olivier Darrigol
REHSEIS, UMR 7596, CNRS
Paris
France
darrigol@paris7.jussieu.fr

Werner Israel
Department of Physics and Astronomy
University of Victoria
Victoria BC, V8W 3P6
Canada
israel@uvic.ca

Jan Lacki
REHSEIS, UMR 7596, CNRS, and
Unité "Histoire et Philosophie des Sciences"
Faculté des Sciences
Université de Genève
24, quai E. Ansermet
CH-1211 Genève 4
Switzerland
jan.lacki@unige.ch

Henri Ruegg
Professor emeritus
Section de Physique
Université de Genève
6, chemin du Claiset
CH-1256 Troinex
Switzerland
Henri.Ruegg@unige.ch

Dr. Marti Ruiz-Altaba
Collège de Staël, Genève
25 route de St. Julien
CH-1227 Carouge
Switzerland
Marti.Ruiz-Altaba@cern.ch

Gérard Wanders
Professor emeritus
University of Lausanne
Chemin de la Cure 13
CH-1012 Lausanne
Switzerland
gbwanders@bluewin.ch

Preface

Born in 1905, Ernst C. G. Stueckelberg was one of the most eminent Swiss physicists of the 20th century. He spent most of his career as professor of theoretical physics at the Universities of Geneva and Lausanne, in the years 1930–70. He took part in the development of modern theoretical physics through his work on molecular physics, on the theory of nuclear forces, and his breakthroughs in elementary particle physics, the S-matrix and the renormalization group. In spite of their relevance and originality, his results were often unnoticed and did not attract the attention they deserved. In some way, Stueckelberg became a kind of legend: his name is famous but one doesn't really know why.

An International Symposium for the centenary of Ernst C. G. Stueckelberg's birth took place at the University of Geneva on 2nd and 3rd December 2005. The aim of this Symposium was to go beyond the legend, to explore Stueckelberg's scientific work in its context, and to assess its significance and relevance for contemporary physics. The present book is a continuation of that Symposium and is based, in part, on talks delivered there.

Stueckelberg published most of his articles in local journals of little international audience. Some of them ceased publication. Therefore the access to Stueckelberg's papers can be tedious for international scholars. To facilitate this access a selection of Stueckelberg's most significant papers is reproduced in this volume. As a matter of fact, reading these papers is not always an easy task. They are often plagued with idiosyncratic notations and obscured by enigmatic statements. This led us to include a set of articles explaining, analysing and commenting on important topics of Stueckelberg's work. We hope that this publication will provide a starting point for scholars wishing to dwelve into Stueckelberg's physics.

Valentine L. Telegdi, who died this year, was a great admirer of Ernst C. G. Stueckelberg. He was a fervent promoter of the disclosure of his work. At the Symposium, he payed a moving tribute to Stueckelberg, an outstanding man who devoted his life to science in spite of a terrible mental disease.

The editors thank the Swiss National Science Foundation, The University of Geneva, especially its Physics Institute, and the Swiss Federal Institute of Technology Lausanne (EPFL) for their support of the Symposium and of this publication.

The helpful assistance of the librarians of the Physics Institute, Claire-Lise Held and Jocelyne Favre is also acknowledged.

Geneva, July 2008

J. Lacki
H. Ruegg
G. Wanders

Conventions and Notations

In order to avoid repetition, all references to Stueckelberg's papers that are enclosed in square brackets (i.e [...]), refer to the complete list of Stueckelberg's publications at the end of this volume. The squared numbers in bold (i.e. [**81**]) signal that the corresponding article is reprinted in this volume.

Contents

I

Ernst C. G. Stueckelberg: Biography and Personal Recollections

CHAPTER 1

Overview of Stueckelberg's Life as a Scientist

Gérard Wanders

Ernst Carl Gerlach Stueckelberg was born in Basel on February 1, 1905. His full name was: Johann Melchior Ernst Karl Gerlach Stueckelberg, Freiherr von Breidenbach zu Breidenstein und Melsbach. He inherited his German title from his mother's family. His father was a lawyer and his paternal grandfather was a well-known swiss painter.

At secondary school, the young Stueckelberg was a very gifted student. He obtained his "Matura" in 1923 and started to study physics at the University of Basel.

As president of the Students' Union, he invited Arnold Sommerfeld, the eminent theoretical physicist, to give a talk in Basel. Sommerfeld accepted the invitation and was so impressed by Stueckelberg that he invited him to spend the academic year 1924–25 in Munich. The resulting stay in Munich was a crucial experience in Stueckelberg's life. He attended Sommerfeld's famous lectures in theoretical physics and discovered the early stages of quantum mechanics. He became also acquainted with Werner Heisenberg.

Back home in Basel, Stueckelberg prepared his Ph. D. thesis under A. Hagenbach's supervision. It is a rather unexceptional work on cathodes temperatures, both theoretical and experimental. He got his Ph. D. in 1927 and also succeeded in becoming an officer in the Swiss army.

At the end of 1927, Stueckelberg left for the United States. On Sommerfeld's recommendation he got a position as research associate at the prestigious Palmer Physical Laboratory in Princeton, led by K. T. Compton. There he worked on the quantum properties of molecules, a new subject, at the forefront of research at that time. He was very efficient, collaborated with J. G. Winans and P. Morse and became assistant professor in 1930.

In January 1932 Stueckelberg suffered from the first attack of a serious mental illness (manic-depressive psychosis) which would severely handicap him for the rest of his life. He felt unable to carry on with a scientific career in the United States, left Princeton and returned to Switzerland.

A. Hagenbach managed to provide him with an assistant post at the University of Basel and, by the end of 1933, Stueckelberg was offered a position as "Privatdozent" at the University of Zürich. This put him in contact with two prominent leaders in theoretical physics, Wolfgang Pauli, at the ETHZ, and Gregor Wentzel, at Zürich University. He abandoned once and for all his previous research field and embarked on a completely new subject, the theory of elementary particles.

As a mere "Privatdozent", Stueckelberg was faced with serious financial difficulties and he even contemplated a career in the army. But his situation improved at the end of 1934. After Arthur Schidlof's death, the University of Geneva asked Stueckelberg to take over the teaching of theoretical physics. Stueckelberg replaced Schidlof ad interim, before being appointed as "professeur extraordinaire" in Spring 1935, which provided him financial security. In 1942, he also started to teach theoretical physics at the University of Lausanne as "chargé de cours".

Details on Stueckelberg's life after his nomination in Geneva can be found in the next contribution "Stueckelberg in Geneva and Lausanne".

Ever since Stueckelberg started working at the University of Zürich, his research was mainly focused on the theory of elementary particles. He took part, as an outsider in some ways, in the development of relativistic quantum field theory. His contributions range from an explanation of the nuclear forces, in the 1930s, to construction of the renormalized S-matrix – a tool providing the description of elementary particle collisions – with Dominique Rivier, in the late 1940s. He invented also the renormalization group with André Petermann, in the early 1950s. Stueckelberg had a very original approach and disregarded conventional methods. He carried out his research in his own way, always working in parallel with outstanding physicists of the 20th century, such as Hideki Yukawa and Richard P. Feynman.

Besides elementary particles, thermodynamics was another recurrent topic of interest to Stueckelberg. This led to a formulation of relativistic thermodynamics in 1953, and, subsequently, to various studies on non-relativistic thermodynamics in the 1960s. The outcome was a book written by Stueckelberg and Paul B. Scheurer [116].

Apart from his relationship with Wolfgang Pauli, Stueckelberg had little contact with the scientific community. This was partly due to his mental illness. But he nevertheless attended an International Conference on Fundamental Particles and Low Temperature held at the Cavendish Laboratory (Cambridge, UK) in 1946. He also went to Copenhagen where he met Niels Bohr in 1947, and he was in Basel in 1949 for the Basel-Como Konferenz über Kernphysik und Quantenelektrodynamik.

Teaching theoretical physics was an important and gratifying mission in Stueckelberg's life. His classes were meticulously designed, detailed "talk and chalk" lectures, modelled on Arnold Sommerfeld's teaching. Although classical in style, his lectures were original in content, particularly in the unconventional way Stueckelberg introduced new subjects. Anyone who attended one of Stueckelberg's lectures realized immediately that he was in the presence of an exceptional character.

During Wolfgang Pauli's stay in the United States, Stueckelberg took over his teaching at the ETH, with Markus Fierz, in the year 1943. Stueckelberg was also invited to lecture at the University of Bern during the academic year 1960–61.

With regard to honours, the doctorate honoris causa was conferred on Stueckelberg by the Universities of Neuchâtel and Bern in 1962. He was elected correspondent member of the Academy of Coïmbra (Portugal) in 1944 and of the Academy of Chieti (Italy) in 1965. Stueckelberg was awarded the "Prix des Sciences de la Ville de Genève" in 1971 and the Max-Planck Medal of the German Physical Society in 1976.

In 1975, Stueckelberg retired from the Universities of Geneva and Lausanne, which both awarded him the title of "professeur honoraire". He died in Geneva nine years later, on September 4, 1984.

Ernst Carl Gerlach Stueckelberg is an unusual figure of 20th century theoretical physics. Despite his severe mental illness and his countless stays in psychiatric hospitals, he was able to pursue a scientific career, always at the forefront of research. Due to his lack of contact with the scientific community after the Princeton period, his contributions were poorly recognized or simply ignored. So he did not participate in a clearly visible way in the advancement of the theory of elementary particles, in spite of his exceptionally original and sound ideas.

A detailed account of Stueckelberg's life is given in: Wenger, R. *Ernest C. G. Stueckelberg von Breidenbach – étude biographique*. Genève: Université de Genève, Bibliothèque de l'Ecole de Physique, 1986, p. 42.

An interview of Stueckelberg in his old age gives a pleasant idea of his personality: Crease, R. P. and Mann, Ch.C. "The physicist that physics forgot: Baron Stueckelberg's brilliantly obscure career"; in *The Science*, July-Aug. 1985, vol. 25, no 4, pp. 18–23.

Stueckelberg's achievements in quantum field theory are acknowledged in: Schweber, S. S. *Q. E. D. and the men who made it.* Princeton: Princeton University Press, 1994, pp. 576–582.

A chapter has been devoted recently to Stueckelberg in a book by Peter Freund, *A Passion for Discovery*. Singapore: World Scientific Pub. Co., 2007, pp. 14-19.

CHAPTER 2

Stueckelberg in Geneva and Lausanne

Gérard Wanders

I met Stueckelberg as a student in Lausanne more than fifty years ago. I was captivated by his lectures and became his assistant in 1953 until 1957. We wrote a few papers together. This talk is based on my recollections of that time and on what I have heard and read about Stueckelberg (see Wenger [1986]). My aim is to give you an idea of the context of Stueckelberg's work and life in Geneva and Lausanne.[1]

When we try to picture what the universities of the French speaking part of Switzerland were like in the 1930s we discover a vanished world which is very far from what we have nowadays. Until the first years after the war, physics in particular was very provincial, provincial to a point that is hard to believe. Each university had only one Professor of Experimental Physics, assisted by very few subordinates. Geneva University was an exception, since it also had a Professor of Theoretical Physics. The main task was training secondary school teachers and research was not a high priority. The Federal Technical High School (EPFL, Ecole Polytechnique Fédérale de Lausanne) as we know it now in Lausanne didn't exist. The University of Lausanne

[1] Talk given at the Symposium "E. C. G. Stueckelberg (1905–1984), Symposium for the Centenary of his Birth", Geneva University, 2nd–3rd December 2005.

had an Engineering School (EPUL, Ecole Polytechnique de l'Université de Lausanne), which was in the Science Faculty, but it didn't offer a physics curriculum. Some of the subjects belonging nowadays to theoretical physics were taught by mathematicians: the so-called "rational mechanics", analytical mechanics and mathematical methods of physics. Even astronomy could be in the hands of mathematicians. One was suspicious about quantum mechanics, at least in Lausanne where it wasn't taught at all until the 1950s.

Stueckelberg was appointed as "professeur extraordinaire" of Theoretical Physics in Geneva in 1935, at the instigation of Jean Weigle, who was Professor of Experimental Physics. After the war, Weigle pioneered biophysics in Switzerland by initiating the use of electron microscopes for the study of biological organisms. He left Geneva for CalTech in 1948; this was a great loss for Stueckelberg. As Stueckelberg had no previous contacts in Geneva, his rapid appointment is quite surprising. He started at the end of 1934 with a temporary job and was fully appointed at the beginning of 1935, at the age of 30. As Weigle and Stueckelberg were both acquainted with Sommerfeld, one may suspect that Sommerfeld played some role in Stueckelberg's appointment. Stueckelberg was promoted to "professeur ordinaire" in 1939.

In 1942 Stueckelberg started teaching also at Lausanne University, as a "chargé de cours" at the Science Faculty. His lectures in Theoretical Physics were the first in Lausanne to be given by a true theoretical physicist. In 1957 he became "professeur extraordinaire" in Lausanne and "professeur ordinaire" in 1967.

Thinking about the circumstances of the beginning of Stueckelberg's career in Geneva, one realizes that he was very isolated. This lasted until the end of the war. He was the only theoretical physicist in Geneva. He published 42 articles between 1935 and 1943, 40 as sole author. This shows that he was quite productive, but it also indicates that he had no assistant during this period. This is surprising according to modern standards, but the same situation prevailed at Zürich University. As far as I know Gregor Wentzel had no assistant at that time.

Stueckelberg had practically no contact with the international physics community. He did not go to international conferences, although he attended regularily the meetings of the Swiss Physical Society. The recurrent attacks of his very severe mental illness (maniac-depressive psychosis) explain partly this situation. Only when he went to Zürich could he have discussions with colleagues such as Wolfgang Pauli and Gregor Wentzel.

This situation changed radically and in an astonishing way after the end of the war. Suddenly, Stueckelberg started to have assistants. Between 1945 and 1951 he had simultaneously up to three collaborators, Dominique Rivier and André Petermann among them. He went regularly to Zürich with his assistants to attend the joint theoretical physics seminar of the EPFZ and the University of Zürich, which took place on Monday, every other week. At that time, it was the only such meeting in the whole country. Stueckelberg met Niels Bohr in Copenhagen and attended an international conference at Cavendish Laboratory in 1947. This was certainly one of the most fruitfull and rewarding periods in Stueckelberg's life in Geneva. Sometimes he even went horse-riding with his assistants.

The sudden appearence of the atomic bomb raised the question of the use of atomic energy in Switzerland. At the end of 1945, the Swiss Federal Government created a Commision to answer that question, only four months after the explosion of the bomb. It was called the Swiss Atomic Energy Commission and was presided over by Paul Scherrer, the most prominent figure of Swiss physics at the time (see Kupper [2003]). Surprisingly, Stueckelberg was a member of this Commission. He seemed pleased and proud to belong to it, while, to my knowledge, he didn't show much interest in University or Faculty business. A huge amount of money was available and, besides the commitment with nuclear energy, the Commission took the initiative to subsidize fundamental research. We can therefore consider that the Swiss Atomic Energy Commission was the true predecessor of the Swiss National Science Foundation (which was created in 1954).

It is interesting to notice that all Stueckelberg's assistants from 1945 to 1960, including myself, got their main financial support from this Commission. Funnily enough, money which was originally intended for atomic energy was thus partly diverted to esoteric topics like the S-matrix, causality and the renormalization group.

An unexpected and dramatic development took place in 1950. As a result of an attack of his disease, Stueckelberg handed in his resignation. He had done this several times previously, when attacks occurred, but the authorities had always ignored his resignations. This time, it was accepted and Stueckelberg became "professeur honoraire" at Geneva University, at the age of 45. He couldn't have an office in the Ecole de Physique, a building which was inaugurated in 1952, and his lectures had to be held privately. This sad situation prevailed until the arrival of Josef Maria Jauch, who created the Department of Theoretical Physics and succeeded in getting Stueckelberg reinstated as "professeur ordinaire" in 1961 .

The 1950s saw an extraordinary and rapid mutation in the French-speaking Swiss universities. New posts were created, research groups established and the equipment was improved and up-dated. New courses were introduced and the curricula were modernized The provincial character of the 1930s disappeared once and for all. In physics, the EPUL introduced an "Ingénieur-Physicien" degree and Stueckelberg took a keen interest in this novelty. A more fundamental development resulted from the creation of CERN in 1954 and its settling in Geneva. The first Genevan CERN activities took place in the Ecole de Physique, including those of the Theoretical Division. Stueckelberg welcomed this arrival, established some contacts and attended the CERN seminars and colloquia, but no effective collaboration came out of these contacts. In a way, it was too late and Stueckelberg's interest in particle physics was fading.

If one considers Stueckelberg's career in Geneva, one may say that his appointment was in itself providential. His previous financial situation as an unpaid "Privat Dozent" in Zürich was rather precarious. The number of academic positions in theoretical physics in Switzerland was very restricted and he was even led to contemplate the possibility of a military career. These sources of worry disappeared with his appointment. Despite his prominent scientific achievements and the occasionally spectacular outbreaks of his illness, one may say that his academic career was rather discreet. He didn't attempt to improve the situation of theoretical physics in Geneva. On the basis of his experience in Princeton, he could have tried to create a permanent and well-structured research group. He didn't follow the example of his colleague Weigle and his biophysics project. Obiously he had no interest in research management. Research itself and teaching were his only passion. His lectures and his discussions with his Ph. D. students were his daily scientific contacts and he didn't want more.

References

Kupper, Patrick, Sonderfall Atomenergie, Die bundesstaatliche Atompolitik 1945–1970, *Schweizerische Zeitschrift für Geschichte*, vol. 53 (2003), pp. 87–95.

Wenger, Ruth, *Ernst C. G. Stückelberg von Breidenbach, Etude Biographique*, Bibliothèque de l'Ecole de Physique, Faculté des Sciences de l'Université de Genève, 1986.

II

Scientific Work

CHAPTER 3

Stueckelberg and Molecular Physics

Jan Lacki

The first period of E. C. G. Stueckelberg's scientific career was marked by important contributions he made to molecular physics.[1] After publishing his thesis in 1927 in Basel [1] Stueckelberg joined the prestigious Palmer Physical Laboratory in Princeton where he worked under the guidance of Karl Taylor Compton, brother of Arthur Holly Compton. Stueckelberg owed this position to a recommendation by A. Sommerfeld.[2] In this stimulating environment, he

[1] The content of this paper is partly based on a talk given by Majed Chergui (Ecole Polytechnique Fédérale de Lausanne, Laboratoire de Spectroscopie Ultrarapide) at the Symposium "E. C. G. Stueckelberg (1905–1984), Symposium for the Centenary of his Birth", Geneva University, 2nd–3rd December 2005.

[2] Arnold Sommerfeld was instrumental in the career of Stueckelberg. In a letter of 30 March 1935, Stueckelberg informed Sommerfeld of his nomination in Geneva and continued: "I owe my knowledge in the domain of theoretical physics mainly to the two years which I could spend in your Institute, once as a student and later as a National Research Fellow. Furthermore I received my appointment in Princeton as well as my habilitation in Zurich thanks to your recommendations. Thus I know that I owe it mainly to you, highly esteemed Herr Geheimrat, to have obtained this position in Geneva.".

Until Sommerfeld's death, Stueckelberg kept contact with his teacher and mentor. From 7 to 11 March 1937, A. Sommerfeld was hosted by Stueckelberg in his house in Geneva following a visit of his son Johann Wolfgang in June 1936 (Dr. Georges Stueckelberg, private communication). In a letter of 31 March 1949 (two years before Sommerfeld's death), Stueckelberg thanks Sommerfeld for sending him a copy of his textbook on electrodynamics and discusses the substraction method

devoted several papers to problems of molecular physics. Stueckelberg had the benefit at Princeton of exchanges with other gifted members of the Palmer Physical Laboratory, Philip M. Morse and E. U. Condon among others.[3]

The Princeton stay was a quite a productive period in Stueckelberg's life. From 1928 till his return to Switzerland in 1932, he authored 18 papers both in English and German [2–**19**], dealing with both theoretical [3, 4, 7, 8, 9, 12–**19**] and experimental [2, 5, 6, 10, 11] problems of molecular physics. His research interests were mostly oriented to the issues of ion-molecule, electron-molecule and molecule-molecule collisions, to the interpretation of the continuum emission and absorption spectra, and to calculations of the energy levels of molecules such as H_2^+ and NH_3. Some of Stueckelberg's works represented, at the time, genuine achievements, and will be commented upon now.

3.1. Stueckelberg's Work on Spectra of Molecules

In 1927, Heitler and London [1927] and slightly later Sugiura [1927] published the first calculations of the potential energy curves of H_2 opening the era of quantum chemistry.[4] The novelty was the presence, besides the bound $1\,{}^1S$ state, of the purely repulsive state $1\,{}^3S$: their potential curves are shown in Figure 3.1.[5]

At the time, one of the issues raised by the H_2 molecule was the continuum of its discharge spectrum ($\sim$ 160 nm to $\sim$ 360 nm): the best hypothesis was that it originated in the dissociation of a hydrogen molecule with electronic (and perhaps vibrational) energy into two atoms with various amounts of relative kinetic energy. The new state $1\,{}^3S$ suggested to Stueckelberg and Winans another explanation. In their paper [4] entitled "The Origin of the Continuous Spectrum of the Hydrogen Molecule", they argued that this spectrum originated actually from the fact that the upper $P(\Pi)$-type triplet states (Figure 3.2) can only emit to the lowest triplet state $1\,{}^3S$, that is purely repulsive. More precisely, Winans and Stueckelberg considered the transition of the lowest vibrational level of $2\,{}^3S$ to $1\,{}^3S$ (using their notation): to compute the corresponding matrix element of the transition probability they proceeded as follows (see Figure 3.1). Considering the vibratory motion as harmonic, they

of Dirac, his work with Rivier and Schwinger's theory (the 1935 and 1949 letters have been kindly communicated by K. von Meyenn).

3 Morse is best known today for proposing a model of the potential energy of a diatomic molecule which has been named after him Morse [1929]. We owe Condon, among many other contributions, the quantum mechanical interpretation of the classical Franck principle for the intensity of the simultaneous changes in electronic and vibrational energy levels of a molecule, see Condon [1926].

4 For a history of the early developments in quantum chemistry, see Ballhausen [1979], Gavroglu and Simões [1994].

5 In modern notation, these states are written today as $1\,{}^1\Sigma$ and $1\,{}^3\Sigma$. At the time, the atomic notation was still used for the total angular momentum.

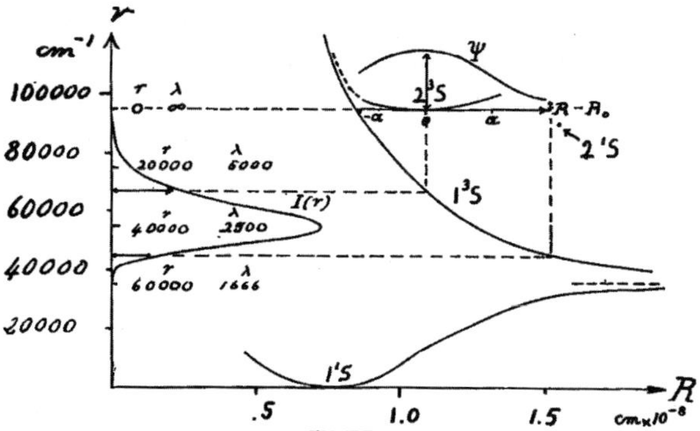

FIGURE 1

Figure 3.1: Potential energy curves of the lowest electronic states of the H_2 molecule as obtained in Sugiura [1927]. *R* is the internuclear separation. The upper 2 ^{3}S state is approximated as a harmonic potential (reproduced with permission from [3]).

took the corresponding wave function as gaussian; then, to estimate the matrix element of the transition, because of the continuous set of nuclear motion energies of 1 ^{3}S, they "reflected" this wave function on the potential energy curve of 1 ^{3}S. This was sufficient to determine the band width and energy of its emission band to the 1 ^{3}S state. The whole procedure was based on Condon's quantum mechanical interpretation of the classical Franck principle for the intensity of the simultaneous changes in electronic and vibrational energy levels of a molecule (Condon [1926, 1927], Franck [1926]); Winans and Stueckelberg had also access to the manuscript of an extensive work of Condon exposing the underlying ideas (among others the reflection recipe) that appeared shortly after (Condon [1928], in particular, pp. 861–864). As it turns out, the energy extension of the continuum stems from the fact that not only the 2 ^{3}S state, but the whole series of 2 ^{3}S$-$m ^{3}P states contributes to the emission (see Figure 3.2). The above procedure is what has become known since as the reflection method in molecular physics and is widely used as a qualitative approach to estimate continuum spectra including in condensed phases (see Schinke [1993]).

A few years later, Stueckelberg interpreted in a more formal way the continuum absorption of molecular oxygen between 130 and 175 nm, which had been published the same year by Ladenburg et al. [1932]. In this publication [18], he modelled again the ground state with a harmonic potential but this time calculated explicitly the wave functions for the purely repulsive $^3\Sigma$ upper state.

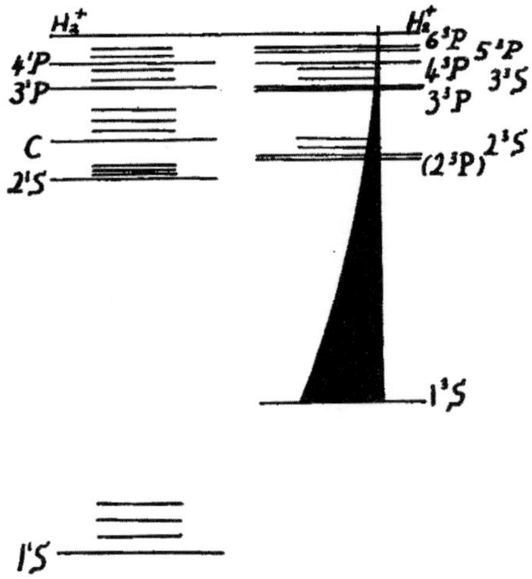

Figure 3.2: Term arrangements of H_2. Electronic levels are represented by long lines, while vibrational ones, by short lines. The non-quantized levels of the lowest purely repulsive $1\,^3S$ state are represented by the black triangle (reproduced with permission from [3]).

Another object of interest was at the time the H_2^+ molecule.[6] In a paper with Morse [4], Stueckelberg studied, using wave mechanics separated in elliptical coordinates, the electronic energies of H_2^+ as a function of nuclear separation 2ρ for several values of the quantum numbers. They performed this investigation by cleverly taking advantage of the information provided by the nodal surfaces (see Figure 3 for an example). When $\rho = \infty$, the resulting system amounts to a hydrogen atom and a separated nucleus; when $\rho = 0$, the system is instead that of a helium ion He^+. The geometry of the nodal surfaces changes between the two cases: these are paraboloids and planes for $\rho = \infty$; and surfaces with spherical symmetry for $\rho = 0$. Using the fact that the number of these surfaces in any coordinate equals the quantum number in that coordinate, Morse and Stueckelberg were able to relate the quantum numbers in both extreme cases to the quantum numbers in elliptical coordinates. This enabled them in particular to express the electronic energies at $\rho = 0$ in terms of the quantum numbers at $\rho = \infty$. Moreover, this rule of correspondence provided them an efficient way of controlling the validity of the approximations used in their calculations, where they studied the H_2^+ molecule perturbing the He^+ ion (at $\rho = 0$) for small separation of nuclei,

---

[6] Today, this molecule still remains the model system for testing computational methods of energy levels of a molecule.

and, alternately, perturbing the H atom and nucleus (at $\rho = \infty$), for a diminishing ρ. Morse and Stueckelberg found that the perturbation methods on both sides were not covering the range $a/2 < \rho < 3a/2$ (a is the radius of the first Bohr orbit of Hydrogen), but they managed to extrapolate this gap using again their nodal reasoning. Morse and Stueckelberg's work, based on the use of the recently proposed Born–Oppenheimer approximation which separates the nuclear from the electronic motion (Born and Oppenheimer [1927]), was one of the first that calculated molecular energies as a function of internuclear separation.

During his few month stay in Cambridge (and then in Munich with Sommerfeld) in 1931, Stueckelberg published two more papers with Morse. In one of them, they confronted the problem of the energy of a double-minimum potential [16]. This problem had been suggested to them by Dennison who used perturbation theory and applied it to the case of the infrared spectrum of the NH_3 molecule (Dennison [1931]). In their paper, Morse and Stueckelberg were able to propose an exact solution at the price of choosing a peculiar double-well potential. However, their choice was not appropriate to describe the interactions in NH_3 because it deviates from a harmonic potential already near the minimum.[7] Retrospectively, one witnesses in this paper a certain tension, to be found later in Stueckelberg's other contributions in quantum physics, between the rigour of approach and the practical value of the result.

3.2. Collisional Problems

As important as may be his contributions commented on above, a survey of Stueckelberg's scientific activity in the years 1928–1932 should however mostly put emphasis on his work on collision problems (ion-molecule, atom-molecule, electron-atom). This is indeed where Stueckelberg put the greatest part of his energy, judging by the fact that the 11 articles that he devoted to this topic make up more than the half of his total scientific production during this period [2, 5, 8–13, 15, 17, **19**]. This is also where he made presumably his most celebrated contribution to molecular physics [**19**]. In it, he provided his theoretical description of inelastic collisions between atoms. Thanks to this contribution, Stueckelberg's name is attached nowadays to a formalism which is still commonly used, known as the "Landau–Zener–Stueckelberg" formalism. At the beginning of the 1930s the problem of inelastic collisions saw important contributions, as is witnessed by the contemporary review papers by Condon and Morse [1931], Morse [1932]. It has many aspects, some tech-

---

7 It is certainly possible to find better potential forms but these do not enable exact solutions (Herzberg [1989])

nically quite challenging. The Landau–Zener–Stueckelberg theory emerged in relation to the following problem.

Consider two atomic systems approaching (colliding). Suppose that there are two diabatic potential energy curves available, functions of the inter-system distance R (curves A_0A_1 and B_0B_1 on Figure 3.3). They will give rise to a so-called "avoided crossing" near the point of crossing (dashed lines): whenever the systems approach at the distance where crossing occurs, there is a probability that, instead of following on the given potential curve, say A_0A_1, a (non-adiabatic) hoping on the other curve occurs. The resulting curves A_0B_2 and B_0A_1 define adiabatic curves of the potential energy of interaction. This non-adiabatic mechanism was shown to be fundamental to many state and phase changes in physics, chemistry and biology.[8] The Landau–Zener–Stueckelberg theory enables one to calculate this hoping probability.

Let us now introduce some concepts and notations.[9] Consider two atoms at a fixed inter-nuclear distance R and assume that the motion of the electrons has been solved and potential energy curves obtained as functions of R. Because the motion of the nuclei is supposed slow with respect to that of the electrons, R can play the role of an adiabatic parameter. Assume further that the potential curves for two states, say $V_1(R)$ and $V_2(R)$, cross at some value $R = R_c$; one can then define corresponding adiabatic states $E_\pm(R)$ as:

$$E_\pm(R) = \frac{1}{2}\left\{V_1(R) + V_2(R) \pm \sqrt{\big(V_1(R) - V_2(R)\big)^2 + 4v(R)^2}\right\}$$

where $v(R)$ is the coupling potential between the two states. For a non-vanishing v at the distance $R = R_c$, the adiabatic states can thus come close together but cannot cross, hence the situation of avoided crossing. The Figure 3.3 illustrates the situation.

If the nuclear motion in now taken into account, and hence energy transfers between nuclear and electronic degrees of freedeom, this can induce transitions between the adiabatic states, most likely in the crossing region where the gap is minimal. In the so-called Landau–Zener situation, the curve crossing is characterized by the fact that the two curves have slopes of the same sign as in Figure 3.3.

The problem of nonadiabatic dynamics for a two-state one-dimensional system in atomic collisions was treated in [**19**], Landau [1932], Zener [1932] under the following assumptions: the potentials $V_1(R)$ and $V_2(R)$ are linear functions of the distance R with slopes F_1 and F_2, and the kinetic energy of the colliding nuclei at the crossing distance R_c, corresponding to the velocity

[8] See for instance Nakamura [2002], in particular Chapter 3 for a historical survey.

[9] The following paragraphs are inspired by the accounts in Nakamura [2002] and Di Giacomo and Nikitin [2005].

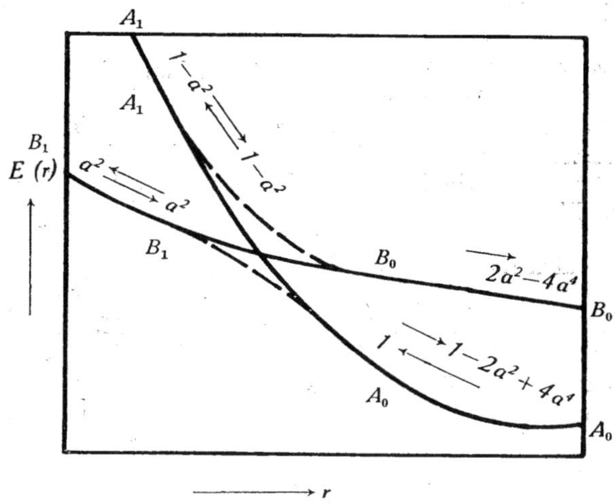

Figure 3.3: Extracted from [**19**, p. 371], showing the crossing potential energy curves and the adiabatic ones.

ν_c, is much larger than the energy difference between the two adiabatic states given by $2a$. The potential energy 2×2 matrix for electronic degrees of freedom in a region of non-adiabatic coupling is then of the form

$$V(R) = \begin{bmatrix} E_c - F_2(R - R_c) & a \\ a & E_c - F_1(R - R_c) \end{bmatrix} \tag{3.1}$$

where E_c, R_c are the energy and interatomic distance at the point of crossing, F_1, F_2 are the slopes of the two diabatic curves at the point of crossing, and $2a$ is the energy separation of the adiabatic curves (dashed curves) at the point of crossing.[10]

The first instance of the LZS mechanism is to be found in a 1932 paper of Lev Landau devoted to the theory of energy transfer in collisions Landau [1932]. Landau obtained his expression for the transition probability using first-order perturbation theory and the WKB theory for the relevant wave functions. He also used analytical continuation of adiabatic potential energy surface (PES) into the complex region of inter-nuclear separation R. His expression for the transition probability reads

$$p^L_{WKB} = C \exp(-\zeta) \tag{3.2}$$

[10] The eigenvalues of this matrix correspond to the adiabatic potential curves in the avoided crossing region.

in the regime where the dimensionless ratio ζ, expressed with the characteristic parameters of the problem

$$\zeta = \frac{2\pi a^2}{\hbar\,|F_1 - F_2|\,v_c} \tag{3.3}$$

was supposed large $\zeta \gg 1$. The prefactor C was, according to Landau, of the order of unity.

Clarence Zener showed that Landau's result was in fact correct in a simplified case where an exact treatment was possible (Zener [1932]). Zener considered that the inter-nuclear motion was uniform $R - R_c = v_c t$ (semi-classical (SC) approximation) in (3.1) which enabled him to set up two time-dependent coupled equations. He solved the problem by transforming the two first-order coupled equations into a second-order one, which he eventually solved in terms of parabolic cylinder functions. The asymptotic analysis of these functions enabled him to find the accurate transition probability

$$p_{SC}^{LZ} = \exp(-\zeta)\,, \tag{3.4}$$

which shows that Landau's prefactor C in Eq. (3.2) should be unity and that the exponential form of the transition probability in formula (3.2) is actually valid for any values of ζ, and not just for $\zeta \gg 1$. Eq. (3.4) is known as the Landau–Zener formula.

Landau's paper was submitted in December 1931 and appeared the next year. In November 1932, Stueckelberg published in turn a sophisticated formulation of the theory of inelastic collisions between atoms going beyond his predecessors' work [**19**]. Stueckelberg was then back in Basel and this work was presented as his Habilitationsschrift.

Stueckelberg [**19**] considered two coupled coordinate-dependent second-order wave equations with the Hamiltonian

$$H(t) = H_N(R) + U(R)$$

where $H_N(R)$ is the operator of the kinetic energy of the nuclei, and $U(R)$ is an arbitrary matrix potential that is approximated by $V(R)$ within the avoided crossing region. He applied then the approximate WKB analysis to the fourth order differential equation obtained from the original second order coupled Schrödinger equations. In using the WKB approximation, Stueckelberg in particular analyzed with much care the associated problem of the analytical continuation of the WKB solution across the Stokes lines. His involved solution to this problem enabled him to write down the following expression for the total inelastic transition probability:

$$P_{WKB}^{St} = 4\exp(-\zeta)\big[1 - \exp(-\zeta)\big]\sin^2(\Phi^{St})\,. \tag{3.5}$$

Here, Φ^{St} is the so-called "Stueckelberg phase" depending on adiabatic potentials in the region between the avoided crossing point and points where the kinetic nuclear energy vanishes (turning points). This expression (3.5) when averaged over the phase shows then, using the relation between the phase-averaged total transition probability $\bar{P}$ and the "single-passage" one p, $\bar{P} = 2p(1-p)$, that Stueckelberg could recover in turn the Landau–Zener formula

$$p^{St} = \exp(-\zeta)\,. \tag{3.6}$$

It is interesting to report here that a recent review by Di Giacomo and Nikitin [2005] proposes a fourth name to be associated to the Landau–Zener–Stueckelberg trio. As the authors point out, the great Italian physicist Ettore Majorana, in a work devoted to the problem of the probability of spin-flipping in a time-dependent magnetic field, arrived at a mathematical situation similar to the LZS one. He derived a formula describing the probability of the spin $\frac{1}{2}$ in a magnetic field with one component changing sign linearly with time. Majorana's work [1932] was actually published the same year as Stueckelberg's. The substantial similarity between Stueckelberg's and Majorana's solutions justifies, in the opinion of Nikitin and Di Giacomo, that the LZS formalism be renamed as the Landau–Zener–Stueckelberg–Majorana formalism. At the end of the Nikitin–Di Giacomo article, one finds this concise appreciation of the various contributions to the theory of nonadiabatic transitions. It captures something of Stueckelberg's style with respect to his peers.

> [Majorana's] elegant solution nicely complements the artistic derivation by Landau, straightforward solution by Zener, and sophisticated treatment by Stueckelberg. One may say that five papers of 1932 laid the foundation of different methods in the theory of nonadiabatic transitions: expressing the solution as an integral representation, using the analytical continuation of classical dynamical quantities into a complex plane, resorting to well-documented higher transcendental functions, and understanding the role of the Stokes phenomenon in quasiclassical analysis of coupled wave equations.

References

Ballhausen, C. J., Quantum mechanics and chemical bonding in inorganic complexes. I, *Journal of Chemical Education*, vol. 56 (1979), pp. 215–218.

Born, M. and Oppenheimer, R., Quantum theory of molecules, *Annalen der Physik*, vol. 84 (20) (1927), pp. 457–484.

Condon, E., A theory of intensity distribution in band systems, *Physical Review*, vol. 28 (6) (1926), pp. 1182–1201.

Condon, E. U., Coupling of electronic and nuclear motions in diatomic molecules, *Proceedings of the National Academy of Sciences of the United States of America*, vol. 13 (1927), pp. 462–466.

Condon, E. U., Nuclear motions associated with electron transitions in diatomic molecules, *Physical Review*, vol. 32 (6) (1928), pp. 858–872.

Condon, E. U. and Morse, P. M., Quantum mechanics of collision processes, *Reviews of Modern Physics*, vol. 3 (1931), pp. 43–88.

Dennison, D. M., The infrared spectra of polyatomic molecules. Part I, *Reviews of Modern Physics*, vol. 3 (2) (1931), pp. 280–345.

Di Giacomo, F. and Nikitin, E. E., The Majorana formula and the Landau–Zener–Stueckelberg treatment of the avoided crossing problem, *Physics-Uspekhi*, vol. 48 (5) (2005), pp. 515–517.

Franck, J., Elementary processes of photochemical reactions, *Transactions of the Faraday Society*, vol. 21 (3) (1926), pp. 536–542.

Gavroglu, K. and Simões, A., The Americans, the Germans and the beginnings of quantum chemistry: The confluence of diverging traditions, *Historical Studies in the Physical Sciences*, vol. 25 (1994), pp. 47–110.

Heitler, W. and London, F., Wechselwirkung neutraler Atome und homöopolare Bindung nach der Quantenmechanik, *Zeitschrift für Physik*, vol. 44 (1927), pp. 455–472.

Herzberg, G., *Molecular Spectra and Molecular Structure*, 2nd ed. Malabar, Fla.: R. E. Krieger Pub. Co., 1989.

Ladenburg, R., Van Voorhis, C. C. and Boyce, J. C., Absorption of oxygen in the region of short wave-lengths, *Physical Review*, vol. 40 (6) (1932), pp. 1018–1020.

Landau, L., Zur Theorie der Energieübertragung bei Stössen, *Physikalisches Zeitschrift der Sowjetunion*, vol. 1 (1932), 88–51.

Majorana, E., Atomi orientati in campo magnetico variabile, *Nuovo Cimento*, vol. 9 (1932), pp. 45–50.

Morse, P. M., Diatomic molecules according to the wave mechanics. II. Vibrational levels, *Physical Review*, vol. 34 (1929), pp. 57–64.

Morse, P. M., Quantum mechanics of collision processes. Part II, *Reviews of Modern Physics*, vol. 4 (1932), pp. 577–634.

Nakamura, H., *Nonadiabatic Transition. Concepts, Basic Theories and Applications.* Singapore: World Scientific, 2002.

Schinke, R., *Photodissociation dynamics: Spectroscopy and fragmentation of small polyatomic molecules*. Cambridge; New York: Cambridge University Press, 1993.

Sugiura, Y., Über die Eigenschaften des Wasserstoffmoleküls im Grundzustande, *Zeitschrift für Physik*, vol. 45 (1927), pp. 484–492.

Zener, C., Non-adiabatic crossing of energy levels, *Proceedings of the Royal Society of London* A, vol. 137 (833) (1932), pp. 696–702.

CHAPTER 4

Stueckelberg's Covariant Perturbation Theory

Jan Lacki

After a period of intensive research in molecular physics, Stueckelberg, back in Switzerland, became interested in 1934 in quantum electrodynamics.[1] He was then Privatdozent at the University of Zurich with Professor Gregor Wentzel. QED was at that time a prominent topic and many among the most renowned physicists were contributing.[2] In a letter to the president of the *Schulrat* of E. T. H. in Zürich (8 March 1934), W. Pauli writes:

> Dr. Stückelberg has stated his desire to get deeper involved with QED and agrees with the nomination of Mr. Weisskopf.[3]

Although new in the field, Stueckelberg managed nonetheless, only a couple of months later, to submit a paper full of original findings to the *Annalen*. As its title "Relativistic invariant perturbation theory of the Dirac electron; Part I: Radiative scattering and Bremsstrahlung" [**20**] shows[4], its aim is to

[1] For a biography see the contribution of Gérard Wanders in this volume and references therein. On Stueckelberg's molecular physics, see the contribution of J. Lacki in this volume.

[2] To witness, e.g the publications by Bethe and Fermi [1932], Bethe and Heitler [1934], Fermi [1932], Heisenberg [1934], Weisskopf [1934], Wentzel [1933, 1934a]. See also the contribution of Olivier Darrigol in this volume.

[3] Stueckelberg was himself considering this position, see [Enz, 1997, p. 57].

[4] Part II actually never appeared.

provide a unified covariant perturbative treatment of Compton scattering and Bremsstrahlung. Stueckelberg used also his method in several other papers [23–25, 29, 39, 40, see also 45]; in particular, he applied his scheme to the annihilation and creation of particle-antiparticle pairs in [24].[5]

The idea behind Stueckelberg's work appears to have been due to G. Wentzel who Stueckelberg thanks for the suggestion to work out an invariant perturbation theory using a four-dimensional Fourier transform.[6] In his publications from the period 1925–1930, Wentzel was mainly interested by the problem of light scattering off free and bound electrons and by the study of atomic spectra with the methods of quantum mechanics, including the relativistic H-atom (see e.g. Wentzel [1925, 1926, 1927, 1929]). Wentzel showed explicit interest in manifest Lorentz covariance when he later used in two instances the multi-time formalism of Dirac, Fock and Podolsky, first, in his tentative attempt to eliminate the self-energy divergences in classical and quantum electrodynamics (see Wentzel [1933, 1934a], see also [Schweber, 1994, pp. 577–578]), then, in his study of the spin 1 photon as a composite state of a spin 1/2 particle and its antiparticle, following an idea of de Broglie.[7]

4.1. Setting the Context: Quantum Electrodynamics in the Early 1930s

The road to a genuine quantum theory of electromagetic processes began with the quantization of the radiation field. In 1926, Dirac first showed how to quantize the radiation field as an assembly of non-interacting bosons and proposed a perturbation scheme enabling one to deal with its time-dependent coupling to matter Dirac [1927c]. His method of quantization made possible the first quantum treatment of emission-absorption processes and provided a decisive step towards a genuine quantum theory of photon-electron interaction. Until the advent of the modern S-matrix approach within quantum field theory at the end of the 1940s, the spirit of this method was very influential. Dirac's theory considered only the transverse degrees of the electromagnetic field in the absence of electric sources, but from 1928 on, Heisenberg and Pauli embarked on their study of a general theory of quantized fields Heisenberg and Pauli [1929, 1930] which aimed at covering, among others, all the aspects of the quantized electromagnetic field. Fermi, in 1932, proposed another approach to the quantization of the electromagnetic field which simplified the whole matter and offered a particularly simple scheme for calculations of concrete processes Fermi [1932].

---

5 See also the section devoted to covariant electrodynamics in O. Darrigol's contribution to this volume.

6 See Pauli's letter to Heisenberg quoted in section 4.4.

7 Wentzel [1934b], see Olivier Darrigol's contribution to this volume.

The main problem one was facing was a consistent formulation of a (relativistic) theory of particles interacting with and through the radiation field. Before the advent of quantum mechanics, the problem of charged systems (bound or free charges) interacting with an incident radiation field was tackled in a "correspondence principle" way as illustrated for instance by the work of Kramers and Heisenberg [1925]. The quantum evolution equations of matrix and wave mechanics provided next the foundations of "semi-classical" treatments, like those of Dirac [1926b, 1927a], of Gordon [1926], and of Klein and Nishina [1929] for the interaction of charged particles with a radiation field. In these works, only charged matter was (first)-quantized whereas the field was left classical. In the closely related approaches of Gordon and Klein–Nishina, one solves the first-quantized matter equation (respectively for spin 0, and spin 1/2 charges) in the presence of a classical incident (primary) field, then one computes the resulting currents and deduces the radiated (secondary) field. With the quantization of the radiation field according to Dirac's "second quantized procedure", one can define quantum states of the field, distinguished by their photon occupation numbers and related by creation and annihilation operators. The various processes of the interaction of charges with radiation can then be conceptualized in terms of transitions between states of the radiation field.[8]

Dirac brought further progress with his interaction picture and the multitime formalism (Dirac [1932], Dirac, Fock and Podolsky [1932]). However, the problem could not find a satisfying solution until it was realized that both radiation *and* matter had to be expressed as quantized fields. The point-particle singularities were then "dissolved" into field excitations and the problem of consistently coupling matter to radiation transformed into that of field-field coupling. This modern standpoint was initiated by the work of Heisenberg and Pauli [1929, 1930] and preceded by Jordan's insights [1927a, 1927b].

As the previous paragraph makes it already clear, the matrix elements for the electromagnetic processes of interest at the time (Compton scattering, Bremshstrahlung, etc.), were obtained using different and sometimes difficult to relate approaches and approximation schemes (see Lacki, Ruegg and Telegdi [1999]): a perturbative approach, easily adaptable for various processes, and where one could generate in a systematic way the successive terms of the perturbative series, was in retrospect certainly worth obtaining. This is precisely what Stueckelberg achieved in the paper that shall be examined here [**20**]: not only was his scheme at the time the first complete and easily generalizable instance of a perturbative calculus, it was moreover *manifestly* relativistically covariant. Furthermore, and most remarkably, Stueckelberg's perturbative calculus preserves the gauge invariance characteristic of the full

[8] For a detailed survey of methods applied to quantum electrodynamic processes, especially to Compton scattering, at the time of Stueckelberg's paper [**20**], see [Lacki, Ruegg and Telegdi, 1999, Sections 1–4]

equations. Because of these properties, in retrospect, Stueckelberg's scheme could have offered a suitable starting point to study the fundamental difficulties related to the infinities arising in higher-order processes, such as self-energy or vacuum polarization. The last section of my contribution makes it plain, relying on the testimonies of some of the key actors of the story of Quantum Electrodynamics.[9]

4.2. Time-Dependent Perturbation Theory: Problems and Issues

Accounting for the interaction of matter with electromagnetic radiation requires, because of the time-varying light waves, the use of a time-dependent perturbation theory scheme. After some early attempts to deal with time-dependent interaction (Dirac [1926c], Klein [1926], Schrödinger [1926d]), the time-dependent perturbation scheme based on Dirac's approach of expanding the perturbed solutions over a basis of non-perturbed ones became standard. However, this scheme has a number of drawbacks, the most notorious one being its lack of manifest covariance. Even if the latter was not at the time a widespread theoretical requirement, manifest covariance proved eventually to be of considerable advantage in getting a clearer picture in the analysis of theoretical situations, such as the proper analysis of divergences met in quantum electrodynamics. Retrospectively (but it took time to recognize it), one needed thus a reliable, practical perturbation scheme, with virtues going beyond mere computational adequacy. A more or less complete list of requirements that one could ask, possibly with hindsight, such a scheme to fulfill would be that:

- it should be "algorithmic", in the sense that at each new perturbation order the "same procedure" should be used as for lower orders, without ad hoc modifications;
- it should be compliant enough to enable handling various cases "at once"; this will be illustrated in what follows;
- it should be manifestly Lorentz covariant at each order;
- it should preserve any other symmetry of the exact equations: in the case of electromagnetism, it should preserve, most importantly, gauge symmetry.

As I shall explain below, Stueckelberg's scheme fulfills all these requirements, a remarkable achievement at the time.

[9] For historical accounts of the problem of infinities in the 1930s, see e. g. Pais [1986], chapter 16; Schweber [1994], Chapter 1, 2; Brown [1993]. For personal recollections of contributors see for example Wentzel [1960] and Weisskopf [1983].

In order to illustrate the importance and scope of the requirements mentioned, let us examine by way of contrast the working of the standard scheme.[10] The story starts with Dirac's paper *On the Theory of Quantum Mechanics* (Dirac [1926c], see the discussion of Bromberg [1977]), well known for its derivation from first principles of Einstein's B coefficient of stimulated emission. Dirac's perturbation theory for Hamiltonians with explicit time dependence was, as explained above, a key ingredient of his quantization of the radiation field (Dirac [1927c], see e.g. Bromberg [1977], Cao [1997], Darrigol [1982], Jost [1972]) and set a lasting standard in construing elementary processes as transitions between "free states".

Assume that we have a (complete) set of solutions φ_k of a problem associated with the Hamiltonian H_0 , typically a non-interacting atom or particle; the general solution then reads: $\varphi = \sum_i c_i \varphi_i$. Then consider a new Hamiltonian $H = H_0 + H_{\text{int}}$. To solve the corresponding problem, Dirac proposes to expand its solution ψ on the set of the free ones:

$$\psi = \sum_k a_k(t) \varphi_k \, . \tag{4.1}$$

The a_k 's are the expansion coefficients that remain to be determined.[11] Dirac remarks further that instead of $|a_k|^2$ expressing the probabilities of being in the state k, one can as well think of them here as giving the number of individual (disturbed) systems occupying the state k (Dirac [1926c], p. 674).[12] The *exact* evolution equations for the expansion coefficients a_k are

$$i\hbar \frac{d}{dt} a_k = \sum_n a_n \, [H_{\text{int}}]_{kn} \, , \tag{4.2}$$

where $[H_{\text{int}}]_{kn}$ are matrix elements of H_{int} (in the φ basis). All that remains to be done is to solve these equations, since the whole time-dependence of the solution is contained in the a 's. This must, in most cases, be done with the help of approximations. For instance, assuming that at $t = 0$, the system is entirely in the unperturbed state, say $a_0(0) = 1$, so that $a_k = 0; k \neq 0$, then one can get, inserting on the right-hand side of (4.2) the initial values, at first order

$$a_k = \frac{[H_{\text{int}}]_{k0} \, (e^{i(E_0 - E_k)t/\hbar} - 1)}{(E_0 - E_k)} \, . \tag{4.3}$$

---

[10] A good presentation can be found in Heitler's well-known text-book (Heitler [1936]).

[11] One should notice at this point that in his study of the relativistic scalar equation, Klein developed a perturbative scheme that used already the expansion on free solutions (Klein [1927]). His technique is however less general and can be seen as a special case of Dirac's. Klein's approach was derived for the classical approximation of the scalar relativistic equation, involving only a first order time derivative.

[12] This is precisely what opens the road to Dirac's later quantization of the electromagnetic field and to the expression of its interaction with charges in terms of varying occupation numbers of light quanta (Dirac [1927b]).

If the matrix elements $[H_{\text{int}}]_{k0}$ vanish (so that there is no direct contribution to a_k) then the method of approximation has to be refined.[13] If, for some indices n', $[H_{int}]_{kn'} \neq 0 \neq [H_{int}]_{n'0}$, then one can get the following second-order contribution passing through "intermediate states" n':

$$a_k = \sum_{n'} \frac{[H_{int}]_{kn'}\,[H_{int}]_{n'0}}{(E_0 - E_{n'})} \times \left[\frac{(e^{i(E_0 - E_k)t/\hbar} - 1)}{(E_0 - E_{n'})} - \frac{(e^{i(E_{n'} - E_k)t/\hbar} - 1)}{(E_{n'} - E_k)} \right]. \tag{4.4}$$

This brief reminder of the "standard scheme" makes immediately explicit its shortcomings. First, and most of all, because of the emphasis put on the time evolution, it obviously breaks the relativistic covariance. This is particularly unsatisfying when dealing with a relativistic wave equation. Indeed, in order to use the scheme, one has first to manipulate the equation to force it into an analogue of the Schrödinger form, distinguishing the energy variable to obtain the corresponding Hamiltonian: for instance, this is indeed what, Waller did with Dirac's equation in his analysis of Compton scattering (Waller [1930]). Such a move obviously breaks manifest covariance hence the appearance of the non-covariant energy denominators in the expression (4.4) above.

Next, we see that the method has to be adapted to the specific problem at hand, as the case of the vanishing of the direct matrix element $[H_{\text{int}}]_{k0}$ already illustrates quite well. Indeed, the method does not provide a unique guide to how to obtain higher-order terms: there is lack of an iterative algorithm and, as a result, the intermediate states must be "put by hand" to obtain (4.4). It is remarkable that Stueckelberg addressed successfully all these issues in his version of the perturbation theory.

4.3. Some Remarkable Features of Stueckelberg's Scheme

The main innovation of the 1934 paper is the introduction of a new perturbative scheme yielding *manifestly* covariant expressions for the matrix elements. This is achieved by performing a four-dimensional Fourier transformation of the wave-function, thus eliminating space *and* time variables.[14] Another orig-

---

[13] The formula (4.2) as well as (4.3) are not written or discussed in their general form in Dirac [1926c]. There Dirac considers specifically the first- and second- order solutions to the problem of dispersion of radiation by atoms, where the expansion (4.1) is for the wave function of the atom. The case of intermediate states will be discussed by Dirac later in his paper (Dirac [1927c]). I am following here Heitler [1936], p. 89.

[14] Møller [1931] uses also the four-dimensional Fourier transform but for the classical retarded potential in his correspondence – theoretic treatment of electron-electron scattering (see Kragh [1992], Roqué [1992]). See also Bethe and Fermi [1932].

inal feature of this paper is its use of integrations over the complex energy-plane to enforce the mass-shell relations. Stueckelberg's scheme relies here on the systematicity which, at each perturbative order, introduces a new singularity on the integration path over the energy variable.[15] This is what implements in his scheme the algorithmicity in passing from one perturbative order to the next one. Finally, because Stueckelberg does not commit himself to any specific choice of gauge, his matrix elements are manifestly gauge invariant which is rather unusual in his times.[16]

Let us now go into the details.[17]

4.3.1. Four-Dimensional Expansions and Contour Integrals

Stueckelberg's paper deals with charged spin-1/2 particles, but in order to present the essential points of his scheme as clearly as possible, avoiding the complication of spinor expressions, let us apply faithfully his procedure to spin-0 particles.[18] The total wave-function $\phi(x)$ of a spin-0 particle interacting with the electromagnetic field A_μ quantized according to Dirac [1927c] obeys the equation

$$\big(-i\partial_\mu + eA_\mu(x)\big)\big(-i\partial^\mu + eA^\mu(x)\big)\phi(x) + M^2\phi(x) = 0 \tag{4.5}$$

where $M = \frac{mc}{\hbar}$. This equation is written in the interaction picture of Dirac, Fock and Podolsky (see Dirac [1932], Dirac, Fock and Podolsky [1932]) which Stueckelberg used again in later publications [29, 39, 40].[19] In this picture, A_μ is a free quantized electromagnetic field with the Fourier expansion

$$A^\mu(x) = \sum_{\mathbf{k}} e^\mu_{\mathbf{k}} V_k \left[a_k e^{i(k\cdot x)} + a^\dagger_k e^{-i(k\cdot x)}\right]. \tag{4.6}$$

The sum over **k** refers to photons in a box of dimension G with $V_k^2 = \frac{2\pi}{Gk_0 c\hbar}$. The corresponding 4-momentum k and 4-vector of polarization e_k satisfy the mass-shell and transversality conditions $k^2 = 0$; $k \cdot e_k = 0$; $e_k^2 = 1$. The annihilation, resp. creation operators $a_k, a^\dagger_{k'}$ obey the usual commutation relations. The first step consists in expanding the total wave-function of the

---

[15] There is a mathematical inconsistency in Stueckelberg's approach, but it does not endanger his end results, see below.

[16] For instance Fermi in his review (Fermi [1932]), and after him Heitler [1936] work in the radiation (or Coulomb) gauge.

[17] The following sections follow closely a joint work with H. Ruegg and V. Telegdi [1999].

[18] However, for the sake of clarity, Stueckelberg's original notations and conventions have not been retained. A dictionary relating those used below to his own is provided in the Appendix.

[19] Stueckelberg discussed the multi-time formalism in his [25, 39, 40]

system on the photon number eigenstates $|N^j\rangle$ with coefficients $\varphi^j(x)$,

$$\phi(x) = \sum_j \varphi^j(x)|N^j\rangle \tag{4.7}$$

where N^j denotes all possible photon configurations $\{N_1, N_2, \ldots, N_k, \ldots\}$. Stueckelberg's (-Wentzel's) fundamental idea is now to consider the four-dimensional Fourier transform for the particle wave function $\varphi^j(x)$,

$$\varphi^j(x) = \int d^4 p e^{i(p\cdot x)} \chi^j(p)\,. \tag{4.8}$$

As the expansion coefficients $\chi^j(p)$ are by definition independent of space and time variables, they are relativistic invariants so that the resulting calculus will be manifestly covariant.[20] Introducing these expansions into the Klein–Gordon equation (4.5) yields

$$\begin{aligned}\int d^4 p e^{i(p\cdot x)}\Bigg[&\sum_j (p^2 + M^2)\chi^j(p)|N^j\rangle \\ &+ e\sum_j\sum_{\mathbf{k}} (2p-k)\cdot e_{\mathbf{k}} V_k (a_k + a^\dagger_{-k})\chi^j(p-k)|N^j\rangle \\ &+ e^2 \sum_j \sum_{\mathbf{k},\mathbf{k}'} e_k \cdot e_{k'} V_k V_{k'} (a_k + a^\dagger_{-k})(a_{k'} + a^\dagger_{-k'}) \\ &\times \chi^j(p-k-k')|N^j\rangle\Bigg] = 0\,.\end{aligned} \tag{4.9}$$

Following Stueckelberg, one considers now the perturbation expansion in powers of the charge e applied to the time-independent functions $\chi^j(p)$ written as

$$\chi^j(p) = \chi^{j(0)}(p) + \chi^{j(1)}(p) + \chi^{j(2)}(p) + \cdots \tag{4.10}$$

Substituting this expansion into (4.9) and matching terms of the same order yields successive equations expressing a given $\chi^{j(k)}(p)$ in terms of the lower-order ones. At zero-th order one has the free equation

$$\int d^4 p e^{i(p\cdot x)} (p^2 + M^2)\chi^{j(0)}(p) = 0\,. \tag{4.11}$$

[20] Notice that although the photon states are discretized (i.e. radiation is in a box), the Fourier expansion above is continuous, the states considered being not necessarily on mass-shell.

To get the first order one inserts $\chi^{j(0)}(p-k)$ into the second term of (4.9):

$$\int d^4pe^{i(p\cdot x)}\Bigg[\sum_j\Big[p^2+M^2\Big]\chi^{j(1)}(p)|N^j\rangle + e\sum_j\sum_{\mathbf{k}}(2p-k)\cdot e_{\mathbf{k}}V_k(a_k+a^\dagger_{-k}) \times \chi^{j(0)}(p-k)|N^j\rangle\Bigg]=0\,. \tag{4.12}$$

For the second order, $\chi^{j(1)}$ is introduced into the second term and $\chi^{j(0)}$ into the third one:

$$\begin{aligned}\sum_j\int d^4p'e^{i(p'\cdot x)}\Big[p'^2+M^2\Big]\chi^{j(2)}(p')|N^j\rangle \\ = -\sum_j\Bigg[\int d^4p'e^{i(p'\cdot x)}\Bigg\{e\sum_{\mathbf{k}'}(2p'-k') \\ \cdot e_{\mathbf{k}'}V_{k'}(a_{k'}+a^\dagger_{-k'})\chi^{j(1)}(p'-k') \\ + e^2\sum_{\mathbf{k},\mathbf{k}'}e_k\cdot e_{k'}V_kV_{k'}(a_k+a^\dagger_{-k})(a_{k'}+a^\dagger_{-k'}) \\ \times \chi^{j(0)}(p-k-k')\Bigg\}\Bigg]\Bigg|N^j\rangle=0\,.\end{aligned} \tag{4.13}$$

For the solution of the 0-th order equation (4.11), Stueckelberg introduces another peculiarity of his approach, the use of contour inegrals to implement the mass-shell conditions. It should be remembered that complex integration techniques were popularized by Sommerfeld (see for instance Sommerfeld [1916]). Since Stueckelberg followed Sommerfeld's lectures in Munich in 1924–1925 and later visited him, in 1930, as a National Research Fellow (see the contribution of G. Wanders in this volume), he probably owed to Sommerfeld the knowledge of this techniques.

Following Stueckelberg, we first introduce the mass-shell 4-momentum $\bar{p}$, defined by $\bar{p}=(\mathbf{p},p_0)$; with $\bar{p}_0^2=\mathbf{p}^2+M^2$. Stueckelberg now makes the surprising Ansatz

$$\chi^{(0)}(p)=\frac{1}{i\pi}\frac{\omega^{(0)}(p)}{p^2+M^2} \tag{4.14}$$

where $\omega^{(0)}(p)$ is a non-vanishing function on the mass-shell hyperboloid $\bar{p}_0^2=\mathbf{p}^2+M^2$ with additional properties to be specified. The reader might be puzzled by this Ansatz, at odds with modern treatment, which sets $\chi^{(0)}(p)$

proportional to the delta-function $\delta(p^2 + M^2)$. Indeed, the meaning of this Ansatz is ultimately (and solely) justified by the fact that after a suitable integration over the p_0 variable, the expression (4.8) should yield, for the 0-th order wave function $\varphi^0(x)$, a wave packet composed of on-shell plane waves. Hence, one should not think of it as satisfying directly (4.11), which it does not, as one easily sees: it has rather a formal value as the starting point of the iteration procedure generating higher orders. Be that as it may, Stueckelberg, to transform back to $\varphi^{(0)}$, integrates first over p_0 using contour integration in the complex p_0-plane in the following integral:

$$\int_{\mathcal{C}} dp_0 \chi^{(0)}(p) e^{ip_0 x_0} . \tag{4.15}$$

Notice that $p^2 + M^2 = \bar{p}_0^2 - p_0^2 = (\bar{p}_0 - p_0)(\bar{p}_0 + p_0)$ so that $\chi^{(0)}(p)$ has two singularities on the real-axis of the energy variable p_0. Stueckelberg claimed to have performed the integration over a contour from $-\infty$ to $+\infty$ which, for positive times ($x_0 > 0$), turns around the singularities $p_0 = -\bar{p}_0$ and $p_0 = \bar{p}_0$ respectively in the negative and positive sense, and is closed in the upper half-plane. Stueckelberg is quite short on the details [**20**, p. 371], but his result suggests he considers the contribution of the $p_0 = \bar{p}_0$ singularity since he claimed to have obtained

$$\begin{aligned} \int_{\mathcal{C}} dp_0 \chi^{(0)}(p) e^{ip_0 x_0} &= \int_{\mathcal{C}} dp_0 e^{ip_0 x_0} \frac{1}{i\pi} \frac{\omega^{(0)}(p)}{p^2 + M^2} \\ &= e^{i\bar{p}_0 x_0} \frac{\omega^{(0)}(\mathbf{p}, \bar{p}_0)}{2\bar{p}_0} . \end{aligned} \tag{4.16}$$

As we see, this contour integration amounts to picking-up, from $\chi^{(0)}(p)$, the correct physical on mass-shell behaviour which is then fed into the remaining integrations over the components of momentum variable $\mathbf{p}$:

$$\varphi^{(0)}(x) = \int d^3\mathbf{p} e^{i(\mathbf{px} - \bar{p}_0 x_0)} \frac{\omega^{(0)}(\mathbf{p}, \bar{p}_0)}{2\bar{p}_0} . \tag{4.17}$$

This has the form of a wave packet of plane waves with on-shell momenta, solving individually the free equation $(\partial^2 - M^2)\phi(x) = 0$ as it should. When introducing his Ansatz, Stueckelberg requires from the numerator $\omega^{(0)}(p)$ that it be non-vanishing on the mass-shell, and vanishing sufficiently fast outside it. Upon inspection, it appears that the use of the Cauchy formula is unwarranted in obtaining (4.16), but one can leave this inconsistency aside since all that eventually matters is just to secure (4.17). As to the form of (4.14) with its explicit singularity, it merely offers a suitable starting point for the iteration steps of the perturbative expansion (4.10) which are then easily monitored thanks to the explicit singularities appearing in the denominators. Indeed, each iteration adds a new pole and hence a new contribution after the integration over p_0. The kinematical information about the values of p where the non-trivial solution is valid is then retrieved from the appropriate singularity (be it through complex p_0-integration or something else). For instance, in the first-order term there are two poles, the one inherited from the 0-th order term in $\chi^{j(0)}(p-k)$, and another one coming from the first term in (4.9). The "final" first-order solution is thus given as

$$\chi^{f(1)}(p) = \frac{-e}{\left[p^2+M^2\right]} \sum_{\mathbf{k}} (2p-k) \cdot e_{\mathbf{k}} V_k \chi^{\alpha(0)}(p-k) \langle N^f | (a_k + a^{\dagger}_{-k}) | N^{\alpha} \rangle , \tag{4.18}$$

where it has been supposed that only one "initial" $\chi^{j(0)}$, say $\chi^{\alpha(0)}$, is different from 0.

At second order, specializing to the Compton scattering, where the initial particle and photon have momenta p and k (the initial photon state is $|N^{\alpha}\rangle = |\, 0, \ldots, 1_k, 0, 0, \ldots\rangle$), and projecting on the final state $|N^f\rangle = |\, 0, 0, \ldots, 1_{k'}, 0, \ldots\rangle$, one obtains, after taking care of the matrix elements between the relevant photon states,

$$\chi^{f(2)}(p') = \frac{e^2}{p'^2+M^2} V_k V_{k'} \Omega \chi^{\alpha(0)}(p) , \tag{4.19}$$

where

$$\Omega = \frac{(2p'+k')\cdot e_{\mathbf{k}'} \quad (2p+k)\cdot e_{\mathbf{k}}}{(p+k)^2+M^2} + \frac{(2p'-k)\cdot e_{\mathbf{k}} \quad (2p-k')\cdot e_{\mathbf{k}'}}{(p-k')^2+M^2} - 2e_{\mathbf{k}}\cdot e_{\mathbf{k}'} . \tag{4.20}$$

Here, the sign of k' has been changed, so that, with the final momenta p', k', the conservation law reads $p+k = p'+k'$; $e_{\mathbf{k}}$, $e_{\mathbf{k}'}$ are the initial, resp. final polarizations of the photon. The matrix element $\Omega(p)$ is obviously Lorentz invariant. In particular, notice the presence of invariant denominators which replace the non-covariant energy denominators of the standard scheme (see previous section). These "propagation" functions enclose the contributions of positive as well as negative energies, and correspond to what was much later called the Feynman propagator (see Feynman [1949a,b]) and, in the wording of Stueckelberg and Rivier, the causal function (see G. Wanders' contribution in this volume on Stueckelberg's S-matrix research). $\Omega(p)$ is also invariant under the gauge transformations $e_{\mathbf{k}} \to e_{\mathbf{k}} + \lambda k$, $e_{\mathbf{k}'} \to e_{\mathbf{k}'} + \lambda' k'$ if the external electrons are on the mass-shell. The first two terms with denominators contain the contribution of intermediate states, whereas the last term corresponds to the direct interaction corresponding to A^2 in (4.5). As one sees, the contribution of the intermediate states comes out automatically, contrary to what is needed in the conventional scheme where a refinement of the perturbative method has to be made (see Heitler [1936], p. 88).

Equation (4.20) is a witness to the modernity of Stueckelberg's approach, since one had to wait until 1948 to find similar expressions. Ω is for instance identical (except for normalizations) to the corresponding S-matrix term given by Bjorken and Drell [1964], eq. (9.30), using Feynman techniques.[21]

Now, what about the computation of cross-sections? In the conventional approach (see Tamm [1930], Waller [1930]), they are obtained directly from the value of the expansion coefficients (see (4.4)). Stueckelberg uses instead the conserved currents of free particles

$$j_\mu(x) = -i(\varphi^* \partial_\mu \varphi - \varphi \partial_\mu \varphi^*) \tag{4.21}$$

to define the number of incident charged particles n as

$$n = \int d^3\mathbf{x} j_0(x) = -i \int d^3\mathbf{x}(\varphi^* \partial_0 \varphi - \varphi \partial_0 \varphi^*) \,. \tag{4.22}$$

To obtain the flux of outgoing particles, Stueckelberg uses the same expression taking for the φ the perturbed solutions. At the second order, the computation of $n^{(2)}$ will in particular require us to evaluate the expectation value $\Omega^*\Omega$:

$$\overline{\Omega^*\Omega} \equiv \frac{(2\pi)^3}{n^{(0)}} \int d^3\mathbf{p} 2\bar{p}_0 \frac{\omega^{(0)}(p)^\dagger}{\bar{p}_0} \Omega^*(p)\Omega(p) \frac{\omega^{(0)}(p)}{\bar{p}_0} . \tag{4.23}$$

4.3.2. The Spinor Case: Further Niceties

To further illustrate the care Stueckelberg took to keep Lorentz and gauge invariance manifest, and how he elegantly solved some technical problems avoiding the more cumbersome methods of his time, let us take up now the spinor case, the one originally considered in [**20**]. The starting equation in [**20**] is, instead of (4.5), the Dirac equation (Dirac [1928a,b]) for the spinor wave-function $\Psi(x)$:

$$\left[(-i\partial^\mu + eA^\mu)\gamma_\mu + M\right]\Psi(x) = 0 \,, \tag{4.24}$$

where γ_μ are the Dirac matrices , $\mu = 0, 1, 2, 3$. [22] The analogue of the expansion (4.7) involves now spinor functions $\varphi^j(x)$ which are Fourier ex-

[21] Observe that (4.20) is greatly simplified in the rest system of the initial charged particle ($\mathbf{p} = 0$) and in the Coulomb or radiation gauge where $e_0 = e'_0 = 0$ and hence $p \cdot e_k = p \cdot e_{k'} = 0$. In this case, $\Omega_{\text{lab.}} = -2e_k \cdot e_{k'}$.

[22] The following convention for γ matrices is used: $\gamma_\mu \gamma_\nu + \gamma_\nu \gamma_\mu = -2g_{\mu\nu}$.

panded:

$$\Psi(x, N) = \sum_j \int d^4 p e^{i(p \cdot x)} u^j(p) |N^j\rangle \tag{4.25}$$

with $u^j(p)$ now spinors in momentum space.

After Fourier transforming the equation (4.24) and expanding $u^j(p)$ in perturbative series as in (4.10), one obtains at first order (compare with (4.18)):

$$u^{f(1)}(p) = -\frac{e}{p^2 + M^2}\{(\gamma \cdot p) - M\} \sum_{\mathbf{k}} V_k P_{kf\alpha}(e_k \cdot \gamma) u^{\alpha(0)}(p - k) \tag{4.26}$$

and, at the second order, one finds, in similarity with (4.19) and (4.20):

$$u^{f(2)}(p') = \frac{e^2}{p'^2 + M^2} \sum_{\mathbf{k}',\mathbf{k}} V_{k'} V_k (P_{k'})_{fi} (P_k)_{i\alpha} \Omega(p) u^{\alpha(0)}(p) \tag{4.27}$$

with $p = p' - k - k'$. Here, the P_{kji} are matrix elements of P_k which stands for one of the operators a_k, a^+_{-k} , or 1:

$$P_k |N^j\rangle = \sum_i |N^i\rangle P_{kij} \,. \tag{4.28}$$

Stueckelberg introduces this notation to allow a *unified* treatment of the various processes (see our second requirement in the list of the previous section): for Compton scattering P_k will stand for one of the a_k, or $a^\dagger_k$ and for Bremsstrahlung, one will instead set $P_k = 1$, see below. Ω, now a spinor operator, is

$$\Omega(p) = \big((\gamma \cdot p') - M\big) \left\{ \frac{(e_{k'} \cdot \gamma)\big((\gamma \cdot p' - k') - M\big)(e_k \cdot \gamma)}{(p' - k')^2 + M^2} + \frac{(e_k \cdot \gamma)\big((\gamma \cdot p' - k) - M\big)(e_{k'} \cdot \gamma)}{(p' - k)^2 + M^2} \right\} . \tag{4.29}$$

This expression is obviously Lorentz-invariant and *corresponds exactly to the expressions obtained by "modern" Feynman rules.*[23]

The spinor zero-order approximation $u^{(0)}(p)$ is still given by the Ansatz (see eq. (4.14)) but this time ω satisfies, on the hyperboloid $\bar{p}_0^2 = \mathbf{p}^2 + M^2$, the free Dirac equation

$$(\gamma \cdot \bar{p} + M)\omega^{(0)}(\bar{p}) = 0 \,. \tag{4.30}$$

[23] Notice that, contrary to the scalar case, one gets *only* the contributions corresponding to intermediate states as there is no direct transition because of the minimal coupling of the e.m. field to the spinors in (4.24).

The four linearly independent solutions are, for each sign of the energy, the two spin states $\omega^{(0)r}(\bar{p})$, indexed by r. Observe that in eq. (4.29), Stueckelberg gets automatically the projection operator

$$\Lambda^{+}(\bar{p}') = \frac{\gamma \cdot \bar{p}' - M}{2M} = \sum_{r=1}^{2} \omega^{(0)r}(\bar{p}')\bar{\omega}^{(0)r}(\bar{p}') .$$

This shows that the summation over final spin states of the electron is naturally built in.

Fourier-tranforming back to $\varphi^{(0)}(x)$ yields (cf. (4.16))

$$\begin{aligned}\varphi^{(0)}(x) &= \int d^4 p e^{i(p \cdot x)} u^{(0)}(p) \\ &= \int d^3 \mathbf{p} e^{i(\mathbf{px})} \frac{\omega^{(0)}(\bar{p})}{\bar{p}_0} e^{-i\bar{p}_0 x_0} .\end{aligned} \tag{4.31}$$

At second order, one finds accordingly [**20**, pp. 374]:

$$\frac{\omega^{(2)}(\bar{p}, x_0)}{\bar{p}_0} = \sum_{\mathbf{k}',\mathbf{k}} \frac{(1 - e^{is_0 x_0})}{s_0} V_k V_{k'} (P_k)_{ji} (P_{k'})_{i0} \frac{\Omega(\bar{p})\omega^{(0)}(\bar{p})}{\bar{p}'_0 \bar{p}_0}$$

plus higher-order terms in $s_0 = \bar{p}'_0 - \bar{p}_0 - k_0 - k'_0$.[24]

The scattering cross section is obtained following the same steps as in the scalar case provided that, for the particle number, one defines now

$$n = \int d^3 \mathbf{x} j_0(x) = \int d^3 \mathbf{x} \psi^{\dagger} \psi .$$

At second order, it leads, after the computation of $\overline{\Omega^{\dagger}\Omega}$, to the Klein–Nishina formula (Klein and Nishina [1929]).

Lorentz and Gauge Invariant Squared Matrix Element $\overline{\Omega^{\dagger}\Omega}$

$\overline{\Omega^{\dagger}\Omega}$ exhibits in a most striking way the manifest Lorentz and gauge invariance provided by the Stueckelberg scheme. In his calculation, Stueckelberg considers a final photon which is on mass-shell and transverse, i.e., $k'^2 = k' \cdot e_{k'} = 0$. On the other hand, he keeps terms proportional to k^2 and $k \cdot e_k$. This allows him to use his formulas also for Bremsstrahlung (see below). For Compton scattering he will of course put also $k^2 = k \cdot e_k = 0$.

---

[24] Later, when computing the scattering cross section, one will have to integrate over s_0 ; only the leading order in s_0 yields a number of scattered particles proportional to time.

In the calculation of $\overline{\Omega^\dagger\Omega}$ with the help of eq. (4.29), the sum over final spins is already taken care of. In order to obtain a simpler expression, what remains to be done is to apply the free Dirac equation for $\omega(\bar{p})$, the mass shell conditions for the initial and final electrons, and the energy momentum conservation. The general expression eventually found by Stueckelberg is the following [**20**, pp. 377–379]:

$$\begin{aligned}
\frac{p_0}{2p'_0}\overline{\Omega^\dagger\Omega} = {}& \frac{(e_{k'}\cdot p)^2}{(k'\cdot p)^2}\left\{2(e_k\cdot p')^2 + \frac{1}{2}e_k^2k^2 - 2(e_k\cdot p')(e_k\cdot k)\right\}\\
&+ \frac{(e_{k'}\cdot p')^2}{(k'\cdot p')^2}\left\{2(e_k\cdot p)^2 + \frac{1}{2}e_k^2k^2 + 2(e_k\cdot p)(e_k\cdot k)\right\}\\
&- 2\frac{(e_{k'}\cdot p)(e_{k'}\cdot p')}{(k'\cdot p)(k'\cdot p')}\left\{2(e_k\cdot p)(e_k\cdot p') + \frac{1}{2}e_k^2k^2\right.\\
&\left. - (e_k\cdot k')(e_k\cdot k)\right\}\\
&+ 2\frac{(e_{k'}\cdot p)}{(k'\cdot p)}(e_{k'}\cdot e_k)(e_k\cdot 2p' - k)\\
&- 2\frac{(e_{k'}\cdot p')}{(k'\cdot p')}(e_{k'}\cdot e_k)(e_k\cdot(2p + k)) + 2(e_k\cdot e_{k'})^2\\
&+ \frac{1}{2(k'\cdot p)(k'\cdot p')}\{e_k^2(k'\cdot k)^2 - 2(e_k\cdot k')(e_k\cdot k)(k'\cdot k)\\
&+ (e_k\cdot k')^2k^2\},
\end{aligned} \tag{4.32}$$

in which only Lorentz invariant scalar products appear.[25] In this expression p' and p satisfy the mass-shell conditions $(p')^2 + M^2 = 0$ and $(p)^2 + M^2 = 0$.[26] Equation (4.32) is again manifestly Lorentz invariant. It is also invariant under the two independent gauge transformations

$$\begin{aligned}
e_k &\to e_k + const.\cdot k\,,\\
e_{k'} &\to e_{k'} + const.\cdot k'\,.
\end{aligned} \tag{4.33}$$

Remarkably, Stueckelberg points out explicitly that checking this invariance helps to prevent algebraic errors [**20**, p. 379]. Stueckelberg's insistence on gauge invariance was rather unusual at the time of his paper. In most contemporary works, the gauge is fixed from the beginning[27], and it is not much

---

[25] In the case of Compton scattering, Stueckelberg's result was obtained later by Wannier taking traces of Dirac matrices, see Wannier [1935].

[26] In order not to overload the notations, the "bars" on p and p' have been skipped.

[27] For instance in Fermi's review [1932] and Heitler's book [1936], the Coulomb gauge $e_0 = 0$ is chosen.

commented on anyway. Again, the virtue of Eq. (4.32) is its generality. Since the signs of p_0 and p'_0 are not fixed, the formula can be used, besides for Compton scattering and Bremsstrahlung, to discuss pair production and annihilation.[28]

Klein–Nishina Formula for Moving Electrons

Stueckelberg could vindicate the value of his manifestly relativistic formalism on the occasion of his elegant answer to a question raised by Pauli. At the time, discrepancies were found between the Klein–Nishina formula for Compton scattering and experimental data for high-energy incoming photons (Schweber [1994], p.82). All Compton scattering calculations for spin 1/2 electrons done before Stueckelberg were performed in the rest system of the initial electron. Pauli consequently studied the question whether the K-N formula was still valid in the limit where the initial and final light frequencies ν and ν' go to infinity, their ratio being kept constant (Pauli [1933]). To this end he Lorentz-transformed the K-N formula from the rest system of the initial electron to the one with arbitrary velocity v. Taking the infinite frequency limit, Pauli found that the cross-section depended explicitly on v; for unpolarized photons it is (in modern notation)

$$\frac{d\sigma}{d\Omega} = \frac{\alpha^2}{2M^2}\left(\frac{mc^2}{E}\right)^2\left(\frac{\nu'}{\nu}\right)^2\frac{1}{D^2}\left[\frac{\nu D}{\nu' D'} + \frac{\nu' D'}{\nu D} - \sin^2\theta\right] \tag{4.34}$$

where θ is the scattering angle, ν and ν' the initial resp. final frequencies of the photon. D and D' are further the Doppler factors $D = 1 - \frac{v}{c}\cos\alpha$ and $D' = 1 - \frac{v}{c}\cos\alpha'$, where α resp. α' are the angles between the initial electron and the initial, resp. final photon.[29] Pauli's calculation, as clever as it was, nevertheless took five pages, whereas Stueckelberg could very easily specialize his general formula to this case [**20**, pp. 379–380]. He put in his eq. (4.32): $k^2 = k \cdot e_k = 0$ and $e_k^2 = 1$, which specializes to Compton scattering; he further chose a gauge for which $p \cdot e_k = p \cdot e_{k'} = 0$, corresponding to transversality in the initial electron rest system. Hence he obtained:

$$W \equiv \frac{p_0}{2p'_0}\overline{\Omega^\dagger\Omega} = 2(e_k \cdot e_{k'})^2 + \frac{(k \cdot k')^2}{2(k' \cdot p)(k' \cdot p')}\,. \tag{4.35}$$

---

[28] Franz [1938] gave the matrix elements for arbitrary polarizations of the photon *and* the electron, in the electron rest-system. See also Nishina [1929a,b].

[29] This formula differs from the Klein–Nishina one by the necessary changes when going from the initial electron rest system to an electron moving with velocity v, namely $mc^2 \to E$, $\frac{\nu'}{\nu} \to \frac{\nu'}{\nu}\frac{D'}{D}$.

Using energy-momentum conservation and mass-shell conditions for electrons, he recognized that

$$\frac{(k\cdot k')^2}{2(k'\cdot p)(k'\cdot p')} = \frac{1}{2}\left[\frac{\nu D}{\nu' D'} + \frac{\nu' D'}{\nu D}\right] - 1\,. \tag{4.36}$$

Stueckelberg then averaged over initial polarizations e_k and summed over final polarizations $e_{k'}$. He found for the first term in eq. (4.35):

$$\frac{1}{2}\sum_{e_k}\sum_{e_{k'}} 2(e_k\cdot e_{k'})^2 = 1+\cos^2\theta \tag{4.37}$$

$$\cos\theta = 1 - \frac{(k\cdot k')p^2}{(k'\cdot p)(k\cdot p)} \tag{4.38}$$

and for the second term he found it doubled. Therefore

$$\frac{1}{2}\sum_{e_k}\sum_{e_{k'}} W = \frac{\nu D}{\nu' D'} + \frac{\nu' D'}{\nu D} - \sin^2\theta \tag{4.39}$$

which is the bracket in eq. (4.34) now naturally written in the invariant form

$$\frac{(p\cdot k)}{(p\cdot k')} + \frac{(p\cdot k')}{(p\cdot k)} + \left(\frac{(k\cdot k')(p\cdot p)}{(k\cdot p)(k'\cdot p)}\right)^2 - 2\left(\frac{(k\cdot k')(p\cdot p)}{(k\cdot p)(k'\cdot p)}\right).$$

After the publication of Stueckelberg's paper, Pauli expressed his worry about Stueckelberg's procedure which consisted in first averaging over the photon polarization in the rest system of the electron, and *only then* Lorentz-transforming it back to the moving frame. Stueckelberg replied to Pauli's objection by remarking that, since unpolarized light is a Lorentz-invariant notion, the averaging is a Lorentz-invariant operation and can therefore be performed in any reference system. Nevertheless, in a subequent paper [23], Stueckelberg showed in a clever way that one can average in a way which exhibits manifest invariance from the beginning (see Lacki, Ruegg and Telegdi [1999], pp. 505–507).

Bremsstrahlung

Bremsstrahlung is the emission of a free photon by an electron interacting with the Coulomb field of a nucleus. It is thus analogous to Compton scattering, where the initial photon k is replaced by a classical electromagnetic field[30]

$$A_k = \frac{2\hbar c}{e^2} M_k e_k \cos(k\cdot x) \quad \text{with} \quad M_k = \frac{4\pi Z e^2}{k^2\hbar c}\,. \tag{4.40}$$

[30] This corresponds to putting P_k to 1 in the general formulas, see (4.28).

In a particular reference system where $k_0 = 0$, M_k is the Fourier transform of the Coulomb potential Ze^2/r of a nucleus. e_k is again a polarization vector.

Previously to Stueckelberg, the formula for Bremsstrahlung in the electrostatic field had been obtained by Bethe and Heitler [1934] and by Sauter [1934] in the second-order of perturbation theory. Using qualitative arguments, Williams and von Weizsäcker found, on the other hand, an approximate formula, which for large initial energy E_e of the electron and large energy compared with $mc^2$of the emitted photon, agrees with the Bethe–Heitler–Sauter formula (v. Weizsäcker [1934], Williams [1934]). The idea of von Weizsäcker and Williams, inspired by Fermi [1924], goes as follows:

In the rest system of the nucleus the Fourier transform of the static field is given by

$$V^L = \frac{1}{(2\pi)^3} \int d^4k\delta(k_0)M_k(e_k \cdot k)e^{i(k \cdot x)} . \tag{4.41}$$

Since $E_e \gg mc^2$ by assumption, the electron has a large velocity v. In the rest system of the electron, the partial waves of (4.41) move with velocity $V = -v$. $|V|$ is almost equal to the light velocity c, and the partial waves for small k^2 are almost transverse ($(e_k \cdot k) \cong 0$). To these quasi-lightwaves, v. Weizsäcker-Williams apply the Klein–Nishina formula for electrons at rest. The Bremsstrahlung, calculated in this frame, appears as an incoherent sum of the scattering amplitudes of the individual partial waves k. This can be justified by the fact that one averages over wave-packets large in comparison with the nuclear field.

To treat Bremsstrahlung, the strategy of Stueckelberg is to define what he calls the "generalized Klein–Nishina formula" [**20**, pp. 376–377]. To this end he replaces $V_k\sqrt{N_k}$ by M_k ($k^2 \neq 0, (e_k \cdot k) \neq 0$) everywhere in the calculation for Compton scattering leading to the K-N formula. Stueckelberg shows next that the Bethe–Heitler formula for Bremsstrahlung, exact to second order, can be deduced from his generalized K-N formula in the same way as the approximation for large velocities was deduced by v. Weizsäcker-Williams from the ordinary K-N formula [**20**, pp. 380–385].[31] Here too, Stueckelberg takes advantage of the manifest relativistic and gauge invariance of his formalism, which allows him to go back and forth between the nucleus rest system (B-H-S formula) and the electron rest system (v. W-W formula). For example, to go from the former to the latter, a gauge transformation is used from a polarization with only a time component, to one with only space components. Finally, Stueckelberg shows that the incoherent addition of the contributions of partial waves is rigorously justified.

---

[31] In Stueckelberg [24] the same idea was used to obtain the formula for pair creation by a fast electron in the field of a nucleus.

4.4. The Response to Stueckelberg's Paper

A survey of the literature after Stueckelberg's 1934 paper [**20**] shows that his covariant approach remained essentially ignored. The first paper which referred to it was Wannier's (Wannier [1935]) where the author mentioned that his result for the trace over spinor indices had also been obtained by Stueckelberg in his eq. (4.32). But Stueckelberg's scheme is for instance not mentioned by Heitler in his book (Heitler [1936]). This could be an illustration of the difference of "styles" of Heitler and Stueckelberg. While the first tried to get as simple expressions as possible, ready for applications, the second was more interested in the general structure of the theory with an emphasis on symmetries and other fundamental principles. Strangely enough, even Wentzel, who suggested the key idea of Stueckelberg's work, never quoted it.

Pauli had certainly noticed the 1934 paper, since, as discussed above, Stueckelberg wrote a sequel on a Lorentz-covariant polarization vector answering the criticism of Pauli [23]. Later on, in a letter to Heisenberg (5 February 1937), Pauli drew Heisenberg's attention to Stueckelberg's achievement:

> I would also like to draw your attention to a work of Stueckelberg concerning the formalism of perturbation theory. The paper is not well written, but the basic idea (due to Wentzel) seems reasonable to me; it consists in making evident the relativistic invariance by eliminating space and time completely from the theory and examining directly the coefficients of the *four*-dimensional Fourier expansion of the wave function (Pauli [1937]).

This lack of interest appears retrospectively as very unfortunate, as was recognized by Victor Weisskopf in his recollections of that period. Recalling the difficulties encountered in higher-order corrections to QED, Weisskopf remarks:

> Already in 1934 [...] it seemed that a systematic theory could be developed in which these infinities [divergent radiative corrections] were circumvented. At that time nobody attempted to formulate such a theory [...]. There was one tragic exception and that was Ernst C. G. Stueckelberg. He wrote several papers in which a manifestly invariant formulation of field theory was put forward. This could have been a perfect basis for developing the ideas of renormalization. Later on, he actually carried out a complete renormalization procedure in papers with D. Rivier, independently of the efforts of other authors.[32] Unfortunately, his writings and his talks were rather obscure, and it was very difficult

[32] See Gérard Wanders contribution to this volume on Stueckelberg's S-matrix research.

> to understand them or to make use of his methods. He came frequently to Zurich in the years 1934–6, when I was working with Pauli, but we could not follow his way of presentation. Had Pauli and I myself been capable of grasping his ideas, we might well have calculated the Lamb shift and the correction to the magnetic moment of the electron at the time. (see also Weisskopf [1983], p. 73–74; Weisskopf [1981], p. 78)

After the 1939–45 war, a whole new era started in quantum field theory with the renormalization program. In his report to the 1948 Solvay congress, J. R. Oppenheimer insisted on the necessity to preserve covariance in all steps of the calculation if one wants to eliminate the infinities. As an example of such a covariant theory he quotes Stueckelberg's paper.

> Now it is true that the fundamental equations of quantum-electrodynamics are gauge and Lorentz covariant. But they have in a strict sense no solutions expansible in powers of e. If one wishes to explore these solutions, bearing in mind that certain infinite terms will, in a later theory, no longer be infinite, one needs a covariant way of identifying these terms, and for that, not merely the field equations themselves, but the whole method of approximation and solution must at all stages preserve covariance. This means that the familiar Hamiltonian methods, which imply a fixed Lorentz frame t=constant, must be renounced; neither Lorentz frame nor gauge can be specified until after, in a given order in e, all terms have been identified, and those bearing on the definition of charge and mass recognized and relegated; then of course, in the actual calculation of transition probabilities and the reactive corrections to them, or in the determination of stationary states in fields which can be treated as static, and in the reactive corrections thereto, the introduction of a definite coordinate system and gauge for these no longer singular and completely well-defined terms can lead to no difficulty.
>
> It is probable that, at least to order e^2, more than one covariant formalism can be developed. Thus Stueckelberg's four-dimensional perturbation theory would seem to offer a suitable starting point, as also do the related algorithms of Feynman (see the reprint in Schwinger [1958], p. 150).

In the same spirit F. J. Dyson (1949) comments in the notes added in proof to his paper (Dyson [1949]: "A covariant perturbation theory similar to that of section III has previously been developed by E. C. G. Stueckelberg") referring to the 1934 paper [**20**] and to Stueckelberg's [**67**].

The importance of *manifest* Lorentz and gauge invariance for the development of the theory in these times has been also emphasized by Abraham

Pais. When commenting on Schwinger's Lorentz invariant 1948 calculation of various terms contributing to the Lamb shift, Pais writes: "Schwinger's direct calculation of the electric term produced a most unpleasant surprise, however: it was too small by a factor 1/3!" (Pais [1986], p. 457).[33] Pais continues by quoting Schwinger: "This difficulty is attributable to the incorrect transformation properties of the electron self-energy in the conventional Hamiltonian treatment and is completely removed in the covariant formalism now employed" (see Schwinger [1949]). Pais concludes with the following discussion:

> How can a fully covariant theory yield non-covariant results? Because during the calculation one has to subtract infinity from infinity, which in general is not a well-defined step. How can one hope to avoid non-covariant answers? By computing in such a way that covariance is manifest at every stage; and likewise for gauge invariance. Take, for example, Heisenberg and Pauli's treatment of quantum electrodynamics in the Coulomb gauge which [...] may not look covariant but is covariant nevertheless. It is not, however, *manifestly* covariant at every stage. Thus the Coulomb gauge does not lend itself (readily) to the evaluation of radiative corrections. Likewise the second-order perturbation formula [...] and its higher-order partners, though actually covariant, are not manifestly so. One can, however, cast them in an equivalent manifestly covariant form, as in fact Stueckelberg had already shown in 1934.[34]

Finally let us report this opinion of Gell-Mann:

> By about 1950 it was known that QED is renormalizable for charged spinor particles [...] The second-order renormalizability of the charge in QED had been established in 1934 by Dirac and Heisenberg, and that of the mass by a number of authors in 1948. Of those, the first ones to complete correct relativistic calculations of the Lamb shift (Willis Lamb and Norman Kroll and J. Bruce French and Victor Weisskopf) actually used the clumsy old non-covariant method. The place where the new covariant methods played a crucial role, particularly those of E. C. G. Stueckelberg and Richard P. Feynman, which are still used today, was in permitting calculations to be done quickly, especially to fourth and

---

[33] Pais' source is here J. Schwinger in Brown and Hoddeson [1983], p. 329.

[34] In his book, S. S. Schweber reports on the other hand this revealing reaction of Schwinger when asked about the disparities between his calculation of the Lamb shift and that of French and Weisskopf: "Well, if you do not keep the calculation explicitly covariant, anything can happen" (Schweber [1994], p. 244–245).

> higher orders (which would have been impractical with the old methods), and in making possible the proof of renormalizability to all orders. (Gell-Mann [1989], p. 702)

There is no single answer as to why Stueckelberg's contemporaries did not take advantage of his method. Various reasons could have been indeed at play. Stueckelberg's paper was certainly not written in a transparent style (remember Weisskopf's and Pauli's comments). His notations were rather clumsy. Also, some of the mathematical features of Stueckelberg's theory were less common (in particular his unconventional use of complex integration to put particles on mass-shell). More generally, the paper did not provide a new physical result but "only" a new method for deriving results already obtained; no one realised (see Weisskopf's reminiscences) that it could have been used to study in a more consistent way the unsolved problem of the divergences.

4.5. Appendix

The following conventions were used. Four-vectors are written in normal type, with their spatial part noted in boldface: and the time-component a_0; thus $a = (a_1, a_2, a_3, a_0) = (\mathbf{a}, a_0)$. In particular, the space-time location of an event is given by the four-vector $(\mathbf{x}, ct)$ where the time-component $x_0 = ct$ and c is the speed of light in vacuum. The relativistic metric is given by the matrix $g_{\mu\nu}$ with $g_{ii} = -g_{00} = 1$, $i = 1, 2, 3$. In the historical works mentioned in this commentary, the euclidean metric is used instead, introducing for each time-component a_0 its euclidean counterpart $a_4 = ia_0$. Thus, the space-time scalar product between two four-vectors a and b is $a \cdot b = \mathbf{ab} - a_0 b_0 \equiv a_\mu b^\mu$ where the sum convention has been used in the last expression, whereas, in the euclidean metric, $a \cdot b = \mathbf{ab} + a_4 b_4$.

An electromagnetic plane wave ψ of angular frequency $\omega = 2\pi\nu$, wavelength λ ,and direction of propagation given by the unit vector $\hat{\mathbf{k}}$ is given by

$$\psi = \exp\left(i\frac{\omega}{c}(\hat{\mathbf{k}}\mathbf{x} - ct)\right) = \exp(ik \cdot x)$$

where the four-vector $k = (\mathbf{k}, k_0) = \frac{\omega}{c}(\hat{\mathbf{k}}, 1)$ has been introduced. Similarly, using the Einstein–Planck-de Broglie relations: $\mathbf{p} = \hbar\mathbf{k}$, $E = \hbar\omega$, the matter-wave associated to a particle with four-momentum $p = (\mathbf{p}, \frac{E}{c})$ is

$$\exp\left(\frac{i}{\hbar}(\mathbf{px} - Et)\right).$$

The wave vectors of the initial and final radiation are denoted respectively by $\hat{\mathbf{k}}$ and $\hat{\mathbf{k}}'$, and in general, quantities related to the final state are primed.

The original notations of Stueckelberg appear quite unusual and are certainly not most transparent. A dictionary for going from his notations to the ones used here is the following.

- Space-time metric: $\delta_{\mu\nu} \to g_{\mu\nu}$
- Four momenta: initial electron $l^0 \to p$; final electron $l \to p'$, initial photon $k \to k$, final photon $-p$,then $m \to k'$
- Photon polarization: $\sigma^k \to e_k$
- Photon state: $u(N) \to |N\rangle$
- Photon annihilation and creation operators: $\Gamma_k, \Gamma_k^\dagger \to a_k, a_k^\dagger$
- Electron spinor: $A(l) \to u(p)$
- Electron spinor on mass-shell: $B(l) \to \omega(p)$
- Adjoint spinor: $B^\dagger = B^* \gamma_4 \to \bar{\omega} = \omega^\dagger \gamma_0$

References

Bethe, H. and Fermi, E., Über die Wechselwirkung von zwei Elektronen, *Zeit. f. Phys.*, vol. 77 (1932), pp. 296–306.

Bethe, H. and Heitler, W., On the stopping of fast particles and the creation of positive electrons, *Proc. Roy. Soc. A*, vol. 146 (1933), pp. 83–112.

Bjorken, J. D. and Drell, S. D., *Relativistic Quantum Mechanics*. New York: McGraw-Hill Book Company, 1964.

Bromberg, J., Dirac's quantum electrodynamics and the wave-particle equivalence, in *History of 20th Century Physics, Varenna 1977*, C. Weiner (ed.). New York: Academic Press 1977, pp. 147–157.

Brown, L. M., Introduction: Renormalization 1930–1950, in *Renormalization. From Lorentz to Landau (and beyond)*, L. M. Brown (ed.).New York: Springer 1993, pp. 1–27.

Brown, L. M. and Hoddeson, L. (eds), *The Birth of Particle Physics*. Cambridge: Cambridge Univ. Press, 1983.

Cao, T. Y., *Conceptual Developments of 20th Century Field Theories*. Cambridge: Cambridge University Press, 1997.

Darrigol, O., *Les débuts de la théorie quantique des champs*, thèse de 3ème cycle, Université de Paris 1, 1982.

Dirac, P. A. M., Relativity quantum mechanics with an application to Compton scattering, *Proc. Roy. Soc. A*, vol. 111 (1926), pp. 405–23.

Dirac, P. A. M., On the theory of quantum mechanics, *Proc. Roy. Soc. A*, vol. 112 (1926), pp. 661–77.

Dirac, P. A. M., The Compton effect in wave mechanics, *Proceedings of the Cambridge Philosophical Society*, vol. 23, Part V (1927), pp. 500–507.

Dirac, P. A. M., The physical interpretation of quantum dynamics, *Proc. Roy. Soc. A*, vol. 113, (1927), pp. 621–641.

Dirac, P. A. M., The quantum theory of emission and absorption of radiation, *Proc. Roy. Soc. A*, vol. 114 (1927), pp. 243–265.

Dirac, P. A. M., The quantum theory of the electron, *Proc. Roy. Soc. A*, vol. 117 (1928), pp. 610–624.

Dirac, P. A. M., The quantum theory of the electron, Part II, *Proc. Roy. Soc. A*, vol. 118 (1928), pp. 351–361.

Dirac, P. A. M., Relativistic quantum mechanics, *Proc. Roy. Soc. A*, vol. 136 (1932), pp. 453–464.

Dirac, P. A. M., Fock, V. A. and B. Podolsky, On quantum electrodynamics, *Physikalische Zeitschrift der Sowjetunion*, vol. 2 (6), (1932), pp. 468–79.

Dyson, F. J., The radiation theories of Tomonaga, Schwinger, and Feynman, *Phys. Rev.*, vol. 75 (1949), pp. 486–502.

Enz, C. P., Glaus, B. and Oberkofler, G. (eds), *Wolfgang Pauli und sein Wirken an der ETH Zürich*. Zürich: Vdf Hochschulverlag an der ETH Zürich, 1997.

Fermi, E., Über die Theorie des Stosses zwischen Atomen und elektrisch geladenen Teilchen, *Zeit. f. Phys.*, vol. 29 (1924), pp. 315–327.

Fermi, E., Quantum theory of radiation, *Rev. Mod. Phys.*, vol. 4 (1932), pp. 87–132.

Feynman, R. P., in "Conference on Physics-Pocono Manor, Pennsylvania, 30 March–1 April 1948, sponsored by the National Academy of Sciences". According to Schweber (1994) p. 631 n143, the Pocono conference notes were prepared informally by J. A. Wheeler. See also Schweber (1994) pp. 436–445.

Feynman, R. P., Space-time approach to quantum electrodynamics, *Phys. Rev.*, vol. 76 (1949), pp. 769–789.

Feynman, R. P., The theory of positrons, *Phys. Rev.*, vol. 76 (1949), pp. 749–759, reprinted in Schwinger (1958), pp. 225–235.

Franz, W., Die Streuung von Strahlung am magnetischen Elektron, *Ann. d. Phys.*, vol. 33 (1938), pp. 689–707.

Gell-Mann, M., Progress in elementary particle theory, 1950–1964, in: *Pions to Quarks*, Fermilab Symposium 1985, L. M. Brown, M. Dresden and L. Hoddeson (eds).Cambridge: Cambridge Univ. Press, 1989.

Gordon, W., Der Comptoneffekt nach der Schrödingerschen Theorie, *Zeit. f. Phys.*, vol. 40 (1926), pp. 117–133.

Heitler, W., *The Quantum Theory of Radiation*. Oxford: Clarendon Press, 1936.

Heisenberg, W., Bemerkung zur Diracschen Theorie des Positrons, *Zeit. f. Phys.*, vol. 90 (1934), pp. 209–231.

Heisenberg, W. and Pauli, W., Zur Quantentheorie der Wellenfelder I, *Zeit. f. Phys.*, vol. 56 (1929), pp. 1–61.

Heisenberg, W. and Pauli, W., Zur Quantentheorie der Wellenfelder II, *Zeit. f. Phys.*, vol. 59 (1930), pp. 168–190.

Jordan, P., Zur Quantenmechanik der Gasentartung, *Zeit. f. Phys.*, vol. 44 (1927), pp. 473–480.

Jordan, P., Philosophical foundations of quantum theory, *Nature*, vol. 119 (1927), pp. 566–569, 779.

Jost, R., Foundation of quantum field theory, in *Aspects of Quantum Theory*, A. Salam and E. Wigner (eds). Cambridge: Cambridge University Press 1972, pp. 61–77

Klein, O., Quantentheorie und fünfdimensionale Relativitätstheorie, *Zeit. f. Phys.*, vol. 37 (1926), pp. 895–906.

Klein, O., Elektrodynamik und Wellenmechanik vom Standpunkt des Korrespondenzprinzips, *Zeit. f. Phys.*, vol. 41 (1927), pp. 407–442.

Klein, O. and Nishina, Y., Über die Streuung von Strahlung durch freie Elektronen nach der neuen relativistischen Quantendynamik von Dirac, *Zeit. f. Phys.*, vol. 52 (1929), pp. 853–868.

Kragh, H., Relativistic collisions: The work of Christian Møller in the early 1930's, *AHES*, vol. 43 (1992), pp. 299–328

Kramers, H. A. and Heisenberg, W., Über die Streuung von Strahlung durch Atome, *Zeit. f. Phys.*, vol. 31 (1925), p. 681.

Lacki, J., Ruegg, H. and Telegdi, V., The road to Stueckelberg's covariant perturbation theory as illustrated by successive treatments of Compton scattering, *Studies in History and Philosophy of Modern Physics*, vol. 30 (1999), pp. 457–518.

Møller, Ch., Über den Stoss zweier Teilchen unter der Berücksichtigung der Retardation der Kräfte, *Zeit. f. Phys.*, vol. 70 (1931), pp. 786–795.

Nishina, Y., Die Polarisation der Comptonstreuung nach der Diracschen Theorie des Elektrons, *Zeit. f. Phys.*, vol. 52 (1929), pp. 869–877.

Nishina, Y., Polarisation of Compton scattering according to Dirac's new relativistic dynamics, *Nature*, vol. 123 (1929), p. 349.

Pais, A., *Inward Bound. Of Matter and Forces in the Physical World*. Oxford: Clarendon Press, 1986.

Pauli, W., Über die Intensität der Streustrahlung bewegter freier Elektronen, *Helv. Phys. Acta*, vol. 6 (1933) pp. 279–286.

Letter to Heisenberg, date: 5 Febr. 1937, in: *Wolfgang Pauli. Wissenschaftlicher Briefwechsel, vol. II, 1930–1939*, K. von Meyenn (ed.). Berlin: Springer Verlag, 1985, pp. 512–514.

Roqué, X., Møller scattering: A neglected application of early quantum electrodynamics, *AHES*, vol. 44 (1992), pp. 187–264

Sauter, F., Über die Bremsstrahlung schneller Elektronen, *Ann. d. Phys.*, vol. 20 (1934), pp. 404–412.

Schrödinger, E., Quantisierung als Eigenwertproblem (vierte Mitteilung), *Ann. d. Phys.*, vol. 81 (4) (1926), pp. 109–139.

Schweber, S. S., *QED and the Men Who Made It*. Princeton: Princeton Univ. Press, 1994.

Schwinger, J., On radiative corrections to electron scattering, *Phys. Rev.*, vol. 75 (1949), pp. 898–899, reprinted in Schwinger 1958, pp. 143–144.

Schwinger, J. (ed), *Quantum Electrodynamics*. New York: Dover, 1958.

Sommerfeld, A., Zur Theorie des Zeeman-Effekts der Wasserstofflinien mit einem Anhang über den Stark-Effekt, *Phys. Zs.*, vol. 17 (1916), pp. 309–325.

Tamm, Ig., Über die Wechselwirkung der freien Elektronen mit der Strahlung nach der Diracschen Theorie des Elektrons und nach der Quantenelektrodynamik, *Zeit. f. Phys.*, vol. 62 (1930), pp. 545–568.

Waller, I., Die Streuung von Strahlung durch gebundene und freie Elektronen nach der Diracschen relativistischen Mechanik, *Zeit. f. Phys.*, vol. 61 (1930), pp. 837–851.

Wannier, G., Eine vereinfachte Ableitung der Klein–Nishina-Formel, *Helv. Phys. Acta*, vol. 8 (1935), pp. 665–673.

Weisskopf, V., Über die Selbstenergie des Elektrons, *Zeit. f. Phys.*, vol. 89 (1934), pp. 27–39, and Berichtigung, *Zeit. f. Phys.*, vol. 90 (1934), pp. 817–818.

Weisskopf, V., The development of field theory in the last fifty years, *Physics Today*, Nov. 1981, pp. 69–85.

Weisskopf, V., Growing up with field theory, the development of quantum electrodynamics, in: *The Birth of Particle Physics* (International Symposium on the History of Physics, Fermilab 1980), L. M. Brown and L. Hoddeson (eds).Cambridge: Cambridge Univ. Press, 1983.

Weizsäcker, C. F. v., Austrahlung bei Stössen sehr schneller Elektronen, *Zeit. f. Phys.*, vol. 88 (1934), pp. 612–625.

Wentzel, G., Die Theorie des Compton-Effektes. I., *Phys. Zeitschrift*, vol. 26 (1925), pp. 436–454.

Wentzel, G., Die mehrfach periodischen Systeme in der Quantenmechanik, *Zeit. f. Phys.*, vol. 37 (1926), pp. 80–94.

Wentzel, G., Zur Theorie des Comptoneffekts, *Zeit. f. Phys.*, vol. 43 (1927), pp. 1–8 and 779–787.

Wentzel, G., Über den Rückstoss beim Comptoneffekt am Wasserstoffatom, *Zeit. f. Phys.*, vol. 58 (1929), pp. 348–367.

Wentzel, G., Über die Eigenkräfte der Elementarteilchen I und II, *Zeit. f. Phys.*, vol. 86 (1933), pp. 479–494 and 635–645.

Wentzel, G., Über die Eigenkräfte der Elementarteilchen III, *Zeit. f. Phys.*, vol. 87 (1934), pp. 726–733.

Wentzel, G., Zur Frage der Aequivalenz von Lichtquanten und Korpuskelpaaren, *Zeit. f. Phys.*, vol. 92 (1934), pp. 337–358.

Wentzel, G., Quantum theory of fields (until 1947), in *Theoretical Physics in the Twentieth Century*, M. Fierz and V. F. Weisskopf (eds). New York: Interscience Publishers Inc., 1960, pp. 48–77.

Williams, E. J., Nature of the high energy particles of penetrating radiation and status of ionization and radiation formula, *Phys. Rev.*, vol. 45 (1934), pp. 729–730.

CHAPTER 5

Stueckelberg's Unitary Field Theory of 1936–1939

Olivier Darrigol

The scientific production of Ernst Carl Gerlach Stueckelberg von Breidenbach reached a peak in the years 1936–39, in which he produced a series of deeply original papers on quantum field theory and its applications to nuclear matter. The only trace of this work in the memory of today's physicists is the Stueckelberg B-field, a trick to avoid troubles in quantizing massive vector-fields. Yet Stueckelberg's innovations went far beyond this formal contribution. For instance, he devised the first manifestly covariant perturbation theory, and he greatly contributed to the meson-field theory of nuclear interactions. Considerations of style, timing, and character explain the disparity between his achievements and their long-term appreciation.

5.1. From the Electric Arc to the Quantum Fields

Stueckelberg learned the older quantum theory under Arnold Sommerfeld, in Munich in the years 1924–25. His first use of it was an indirect one, in the experimental determination of the temperature of an electric arc through Planck's black-body law. This work was the basis of his dissertation [1] under August Hagenbach in his native city Basel. In the six following years, he worked as a house-theoretician at Palmer Laboratory in Princeton in collaboration with Henry DeWolf Smyth and Philip Morse. His main research topic was the theoretical interpretation of data on inelastic collisions, on the basis of the new quantum mechanics. His now best-remembered result was his solution of the "Landau–Zener–Stueckelberg problem" of inelastic collisions between two-level systems, in which he unveiled the "Stueckelberg oscillations" of the effective cross section in the case of crossing levels [**19**]. This and similar work established him as a competent and promising quantum physicist.[1]

In 1933–35, he came to the University of Zurich as a *Privatdozent*, which brought him in contact with Wolfgang Pauli and Gregor Wentzel. Two new particles, the positron and the neutron, had just been discovered. Hopes were rising to apply the young and troublesome quantum field theory to high-energy processes and to nuclear physics. Pauli was one of the main contributors to this theory and to this physics, as a lucid critic of Paul Dirac's and Pascual Jordan's foundations of 1926–28, as the author, with Werner Heisenberg, of two major memoirs on relativistic quantum field theory in 1929–30, and as the inventor of the neutrino in 1930. When Stueckelberg arrived in Zurich, Pauli was working with Victor Weisskopf on a new theory of charged scalar fields and on vacuum polarization effects in quantum electrodynamics; Wentzel was exploring a way to eliminate the infinities that result from the perturbative coupling of quantum fields.[2]

In March 1934, Pauli informed a local authority that "Dr. Stueckelberg ha[d] expressed a desire to get deeper involved with QED." In September of the same year, he wrote to Weisskopf:

> Perhaps we could invite Stueckelberg to collaborate with us. He is slow but reliable, and I would be pleased if we could bring him closer to the sort of scientific problems on which we have been working.

It was a suggestion by Wentzel, however, that prompted Stueckelberg's first important contribution to quantum field theory. Wentzel had been working on a new theory of point-like electrons based on the classical limit of a new version of quantum electrodynamics that Dirac had initiated in 1932. As this new quantum electrodynamics, unlike the earlier theories by Dirac, Heisenberg, and Pauli, was manifestly covariant, Wentzel suspected that it might lend itself to easier perturbative calculations of relativistic processes. He asked Stueckelberg to explore this possibility.[3]

---

1 On Stueckelberg's biography, cf. Wenger [1986]; Enz [1986] and Gérard Wanders' contribution to this volume. On Stueckelberg's works on molecular dynamics, cf. Jan Lacki's contribution to this volume.

2 On the early history of quantum field theory, cf. Schweber [1994], Darrigol [1984], Darrigol [1986].

3 Pauli to Weisskopf, 5 Sep 1934, in von Meyenn [1985], vol. 3, pp. 763–764. Wentzel's role is acknowledged in Stueckelberg's paper [**20**]. See the contribution of Jan Lacki to this volume.

5.2. Covariant Quantum Electrodynamics

With his theory of 1932, Dirac aimed to solve the difficulties of the former quantum electrodynamics in a manner similar to that through which Heisenberg had solved the difficulties of the old quantum theory. The idea was to focus on transition probabilities, between two quantum states of the same atom in the Heisenberg case, between two free-field quantum states in the case of quantum electrodynamics. Dirac injected the 4-potential of the free quantized electromagnetic field in the Dirac equation for each charged particle, and interpreted the solution of the resulting system as an operator connecting in- and out-states of the field. In the spinor case, this procedure yields the system subsequently established by Dirac and his Russian friends Vladimir Fock and Boris Podolsky:

$$\partial^2 A_\mu = 0\,,$$
$$\left(i\partial\!\!\!/_\alpha - e_\alpha A\!\!\!/(x_\alpha) - m_\alpha\right)\Psi(x_1,\ldots,x_\alpha,\ldots x_N) = 0\,,$$
$$\left(\partial\cdot A - \sum_\alpha e_\alpha \Delta(x-y)\right)\Psi = 0\,,$$

where A_μ is the electromagnetic potential obeying the commutation relations

$$\left[A_\mu(x), A_\nu(y)\right] = i\eta_{\mu\nu}\Delta(x-y)\,,$$

with $\eta_{\mu\nu}$ the Lorentz-metric tensor, Δ the invariant delta function of Jordan and Pauli, e_α and $m_\alpha$the charge and mass of a particle; the / sign denotes contraction with the Dirac matrices γ_μ and $\hbar$ has been set to 1. Worth noting is the occurrence of a different time for each charged particle, which permits the manifestly covariant form of the equations. (Dirac [1932]; Dirac, Podolsky and Fock [1932]. Cf. Schweber [1994], pp. 46–55; Darrigol [1982], pp. 106–109)

Although Dirac thus hoped to provide a new divergence-free quantum electrodynamics, Léon Rosenfeld soon showed that the new theory was nothing but the multi-time extension of ordinary quantum electrodynamics in a mixed interaction picture.

In the Schrödinger picture, the evolution of the state vector of the combined system of field and particles is given by the equation

$$\left(H_R + \sum_\alpha h_\alpha(A^S)\right) \mid \Psi\rangle^S = i\hbar\frac{\partial}{\partial t} \mid \Psi\rangle^S,$$

where H_R is the field Hamiltonian and h_α is the Dirac-electron Hamiltonian. The unitary transformation $U = \exp(iH_R t)$ yields the equation of evolution

$$\sum_\alpha h_\alpha\big[A(t)\big] \mid \Psi\rangle = i\hbar\frac{\partial}{\partial t} \mid \Psi\rangle \quad \text{or}$$

$$\sum_\alpha \left(i\gamma_0\frac{\partial}{\partial t} - i\gamma\nabla_\alpha - e_\alpha \not{A}(\mathbf{x}_\alpha, t) - m_\alpha \right) \mid \Psi\rangle = 0$$

which results from the Dirac–Fock–Podolsky system

$$(i\not{\partial}_\alpha - e_\alpha \not{A}(x_\alpha) - m_\alpha)\Psi(x_1, \ldots, x_\alpha, \ldots x_N) = 0$$

if the vector is related to the solution of this system by

$$\langle \mathbf{x}_1, \ldots, \mathbf{x}_\alpha, \ldots \mathbf{x}_N \mid \Psi\rangle = \Psi(\mathbf{x}_1, t; \ldots; \mathbf{x}_\alpha, t; \ldots \mathbf{x}_N, t)\,.$$

Consequently, the DFP matter field $\Psi(x_1, \ldots, x_\alpha, \ldots x_N)$ is a multi-time extension of the single-time evolution of the state-vector $\mid \Psi\rangle$ in the mixed picture for which the electromagnetic field evolves according to the free-field equations (Rosenfeld [1932]).

Although this equivalence disappointed Dirac, the DFP theory remained attractive because of its manifestly covariant character. Being based on equal-time commutators, the older theory of Heisenberg and Pauli required a forbiddingly complex proof of covariance of its physical results. The new theory needed no proof at all. Besides Wentzel, it seduced three budding geniuses of quantum field theory, Julian Schwinger, Sin-itiro Tomonaga, and Hideki Yukawa.

Following Wentzel's suggestion, Stueckelberg sought a perturbative solution of the DFP equations. For a single electron, the DFP interaction is ruled by

$$(i\not{\partial} - e\not{A} - m)\Psi(x) = 0\,,$$

wherein A_μ is the free quantized potential, and $\Psi(x)$ belongs to the Hilbert space of electromagnetic field states. Stueckelberg wrote

$$\Psi = \Psi_0 + \Psi_1 + \Psi_2 + \cdots$$

with

$$\Psi_0 = e^{ip\cdot x}u(p) \mid 1_{\mathbf{k}\varepsilon}\rangle \quad (\text{such that } (i\not{\partial} - m)\Psi_0 = 0)$$

$$(i\not{\partial} - m)\Psi_1 = e\not{A}\Psi_0\,,$$

$$(i\not{\partial} - m)\Psi_2 = e\not{A}\Psi_1\,,$$

$$\cdots$$

The Ψ_2 term has components proportional to $e^{i(p+k-k')\cdot x} \mid 0_{\mathbf{k}\varepsilon}, 1_{\mathbf{k}'\varepsilon'}\rangle$, to be interpreted as Compton scattering. The corresponding probability amplitude is identical to that given by the Feynman rules applied to the sum of the diagrams,

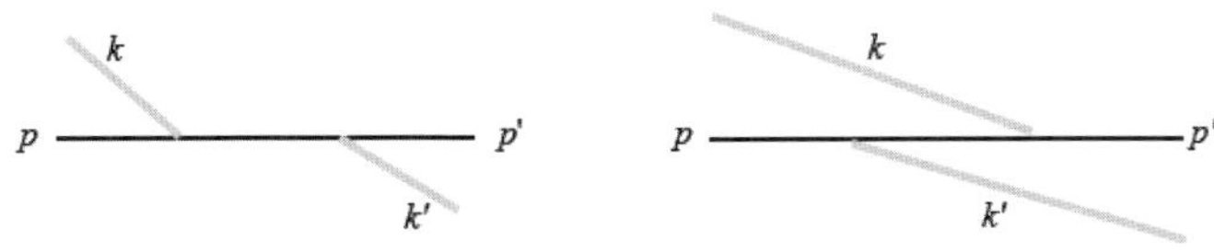

with the propagator $\frac{\not{p}+m}{p^2-m^2}$ (the $i\varepsilon$ is irrelevant at this order).[4]

Stueckelberg thus retrieved the Klein–Nishina formula and generalized it to a moving target electron. Whereas Oskar Klein and Yoshio Nishina's original derivation required several pages of difficult calculations, the new derivation took only a few lines. It was explicitly covariant at every step, and it did not involve any summation over intermediate energy states.[5]

Despite these evident qualities, Stueckelberg's paper went largely unnoticed. The only known comment is Pauli's, in a contemporary letter to Heisenberg:

> This paper is not very well written, but the basic idea (which came from *Wentzel*), seems reasonable to me; it consists in making the relativistic invariance evident by completely eliminating time and space from the theory and dealing directly with the *four*-dimensional Fourier development of the wave function.

Pauli's reaction gives a first hint for understanding why Stueckelberg's anticipation of Feynman-style perturbation theory did not strike his contemporaries as a major achievement: his idiosyncratic style, with personal notations and occasionally unexplained steps, made his writings difficult to follow. More fundamentally, manifest relativistic covariance and computational efficiency were not at the top of the agenda of contemporary field theories. They only became so after the advent of renormalization theory. In the 1930s and early 1940s the most urgent problems were the UV divergencies of the higher-order terms of perturbative expansions, and the compatibility of the lower-order (finite) terms with the known phenomenology of electromagnetic

---

[4] See [**20**]; concerning this paper, see the thorough study (Lacki, Ruegg and Telegdi [1999]); and Jan Lacki's contribution to this volume.

[5] Sin-itiro Tomonaga performed a similar calculation after seeing the DFP paper, but did not publish it, probably because he judged the result not to be sufficiently new: see Darrigol [1988], p. 16. In 1938, Heisenberg wrote the T-exponential formula for the S-matrix in the interaction picture, on which the modern relativistic perturbation theory is based (Heisenberg [1938]). He emphasized the explicit covariance of this formula, but only used it for the relativistic introduction of a fundamental length. Cf. Darrigol [1982], p. 175.

and nuclear forces. Few were those who believed that covariant methods were essential in solving these problems.[6]

In 1936, Stueckelberg applied his covariant perturbation technique (limited to tree-diagram processes) to a new field theory of his own. In 1938, he retrieved the Møller formula for electron-electron scattering with this method. In order to explain the agreement with Christian Møller's semi-classical derivation (based on retarded potentials), he appealed to the Heisenberg picture in which the electromagnetic field evolves according to the classical field equations. He did so firstly in the particle picture, and secondly in the quantized-wave picture for the electrons. In the multi-time extension of the latter case, he obtained an equation similar to the Tomonaga–Schwinger equation, with a state vector defined on a space-like surface. However, the operator in his version of this equation was not relativistically invariant, because he did not use the interaction picture of Tomonaga and Schwinger (in which the interaction Hamiltonian is invariant). The physical understanding of retardation, not manifest covariance, then motivated him.[7]

5.3. Nuclear Forces: 1930–1935

In the five years preceding Stueckelberg's arrival in Zurich, enormous progress had been made in understanding the physics of the nucleus. It should be remembered that until the early 1930s the nucleus was commonly regarded as made of protons and electrons, the only elementary particles of the time (besides the lightquantum). Around 1930, Niels Bohr and other quantum theorists noted that quantum mechanics did not allow the confinement of electrons in the tiny volume of a nucleus. Moreover, the process of β-decay was believed to violate three basic conservation laws regarding energy, spin, and statistics. Bohr argued that electrons did not truly exist as separate dynamical entities in the nucleus, that they were created in the β process. At the same time, he argued that quantum mechanics legitimately applied to heavy components of the nucleus such as the proton or α particles, thus justifying George Gamow's explanation of α-decay in terms of tunneling through a potential barrier. The alleged reason for this difference between electron and heavy particle was that the former's electromagnetic radius was of the same order as the diameter of the nucleus, whereas the latter's was a thousand times smaller. (Cf. Stuewer [1983]; Brown and Rechenberg [1996], Chapter 1; Darrigol [1988b], pp. 240–247)

---

[6] Pauli to Heisenberg, 5 Feb 1937, in von Meyenn [1985], vol. 2, p. 513.

[7] See [29] and [**39**], especially on p. 242. Stueckelberg's insistence on retardation in this paper may have come from Kemmer's criticism of his unified theory of 1936 (see below).

In the same year 1930, Pauli proposed to save the conservation of energy, spin, and statistics in β-decay by introducing a new, heretofore unobserved particle of vanishing mass, spin one-half, and Fermi statistics. In 1932, James Chadwick discovered the neutron in the radiation emitted by beryllium under α-particle bombardment. Heisenberg immediately realized that every nucleus could be built of neutrons and protons, whose interactions should obey quantum mechanics according to Bohr. For the interaction Hamiltonian, he used the form $-\frac{1}{2}J(r_{12})(\tau_1^+\tau_2^- + \tau_2^+\tau_1^-)$ wherein r_{12} is the distance between two nuclear particles, and $\tau^\pm$ an operator changing a neutron state into a proton state or reciprocally. This choice being guided by analogy with the matrix-mechanics treatment of the H_2^+ ion and the H_2 molecule, Heisenberg spoke metaphorically of the exchange of an electron between the two nuclear particles, even though he agreed with Bohr that quantum mechanics could not possibly apply to such a process. (Cf. Brown [1978]; Carson [1996]; Brown and Rechenberg [1996], pp. 33–36; Darrigol [1988b], pp. 247–254)

In the summer of the same year 1932, Carl Anderson identified a new light positive particle in cosmic-ray induced processes. A few months later, Pauli briefly contemplated the idea that this new particle may have Bose statistics and thus be the exchanged particle responsible for Heisenberg's nuclear force. The Leningrad field theorist Heinrich Mandel further suggested that this exchange could be described as a quantum-field-theoretic process, in analogy with photon exchange in quantum electrodynamics. By 1934, most physicists agreed with Patrick Blackett and Giuseppe Occhialini that Anderson's "positron" was nothing but a hole in the infinite sea of negative-energy solution of the Dirac equation for the electron, no matter how suspicious this idea had seemed before. Pauli and Mandel gave up the Bose-positron and derived speculations.[8]

In 1933–34, Enrico Fermi proposed a new theory of β-decay based on Bohr's idea that the electron was *created* in this process, on Pauli's idea that a neutrino was also created, and on Jordan's quantized field concept for representing the creation and annihilation of matter particles. His interaction Hamiltonian was the Hermitian part of $g\tau_-^t\,\psi_e^+\delta\psi_\nu^+$, where g is the coupling constant, τ_- is the operator that transforms a neutron state into a similar proton state, ψ_e and ψ_ν are the quantized spinor fields for the electron and the neutrino, and δ is a 4×4 matrix that makes ${}^t\psi_e^+\delta\psi_\nu^+$ the time component of a 4-vector. Analogy with photon emission in quantum electrodynamics inspired the latter choice. As long as it was limited to first-order processes, the theory yielded reasonable results. Heisenberg and Fermi concluded that the true source of the former troubles of nuclear theory was not the fundamental

8 Pauli to Heisenberg, 4 Jul 1933, in von Meyenn [1985], vol. 2, p. 314, Heinrich Mandel, "Bemerkung zur Heisenbergschen Theorie des Atomkernes", Cf. Darrigol [1988b], pp. 256–259.

failure of quantum mechanics at that scale, but rather the misidentification of basic building blocks and processes; once the neutron and the neutrino were available, and once the discovery of the positron made particle-creation more natural, quantum mechanics and quantum field theory (as far as it was well-defined) provided the desired dynamics. (Fermi [1934a,b]. Cf. Brown and Rechenberg [1996], pp. 41–44; Darrigol [1988b], pp. 263–266)

Heisenberg, Fermi, Igor Tamm, and Dmitri Iwanenko soon extended Fermi's quantum-electrodynamical analogy to the neutron-proton force. Somewhat like Mandel, who relied on the creation and absorption of virtual Bose-positrons, they explained this force through the creation and re-absorption of an electron-neutrino pair. This process derives from Fermi's theory at second order and may be anachronistically represented by the diagram.

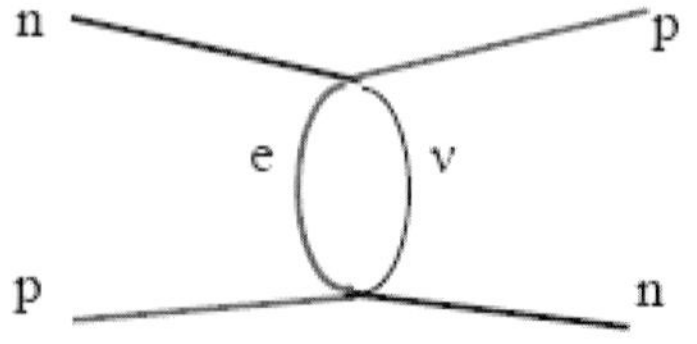

Among several difficulties, the "Fermi-field theory" involved an artificial elimination of infinities (yielding the potential g^2/hcr^5) and gave the wrong order of magnitude for the nuclear forces. Pauli severely criticized it, to the point of calling Fermi "a half-experimental opportunist." Other theorists admired its beautiful simplicity and unity of this scheme, and struggled to solve the order-of-magnitude difficulty by modifying the Fermi coupling.[9]

5.4. The Neutrino Theory of Light

Quite independently of nuclear developments, in 1932 Louis de Broglie proposed to regard the photon as a compound particle made of two elementary particles of spin one-half. In this way, he argued, the analogy between matter and light would be more perfect, since the elementary particles of matter (electron and proton) also had spin one-half. Moreover, simplicity seemed to require that every spin in nature should be obtained by composition of spins one-half. In 1934, as Pauli's neutrino had grown more popular, de Broglie built the photon out of two neutrinos. Pauli and Jordan applauded him, the former because he welcomed every new employment of his dear neutrino, the latter because he saw a new opportunity to apply his concept of quantized matter waves. In 1935, Jordan gave the quantum-field-theoretic form

[9] Pauli to Heisenberg, 24 Nov 1936, in von Meyenn [1985], vol. 2, p. 454. Cf. Brown and Rechenberg [1996], Chapter 3.

$e(\bar{\psi}_e \gamma_\mu \psi_e)(\bar{\psi}_\nu \gamma^\mu \psi_\nu)$ to the coupling between electron field and neutrino field, and noted the analogy with the relativistic extension of the Fermi coupling $g(\bar{\psi}_p \gamma_\mu \psi_n)(\bar{\psi}_e \gamma^\mu \psi_\nu)$. Around the same time, Iwanenko and Arsenij Sokolov detailed the analogy between neutron-proton interaction in Fermi's theory and electron-electron interaction in de Broglie's theory. In modern terms, the diagram (Cf. Brown and Rechenberg [1996], pp. 76–80; Darrigol [1988b], pp. 291–293).

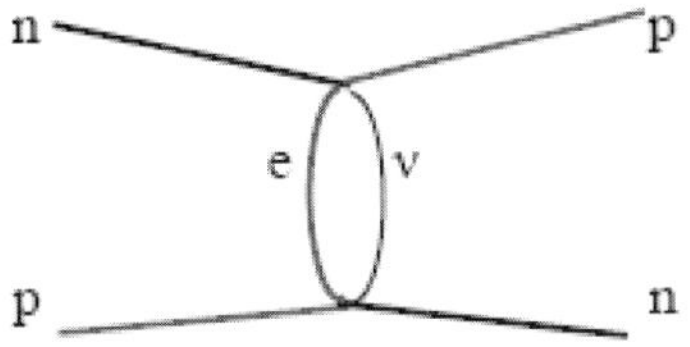

is similar to

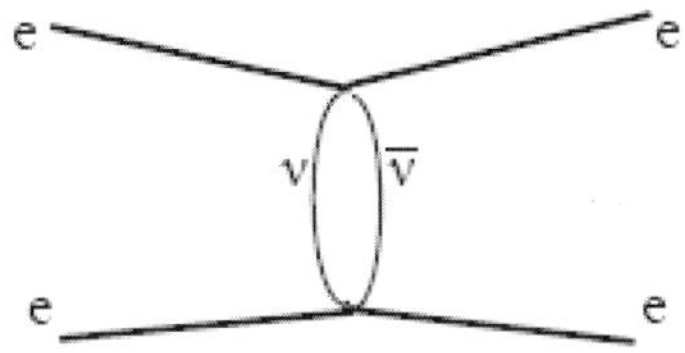

5.5. Stueckelberg's Spinor-Field World of 1936

Impressed by this analogy and by the unity brought by Fermi's and de Broglie's theories, Stueckelberg imagined a completely general field theory of matter and light based on quantized spinor fields. He ambitiously situated his attempt in the context of earlier unified theories by Gustav Mie, Hermann Weyl, and Albert Einstein (with Walther Mayer). Whereas the basic field of these theories always was a tensor field, Stueckelberg agreed with de Broglie that the fundamental field should be a spinor. He treated electron, proton, neutron, and neutrino as quantum states of the same particle, represented by a 16-component Ψ wave. (Stueckelberg [**26**, **27**]. Cf. Darrigol [1988b], pp. 293–295)

For the basic Lagrangian, Stueckelberg assumed the form

$$L = \bar{\Psi}(i\not{\partial} - M)\Psi - \int d^4 y K(x,y) J^\mu(x) J'_\mu(y) \,,$$

with

$$J^\mu(x) = \bar{\Psi}\gamma_\mu \Lambda \Psi \quad \text{and} \quad J^\mu(x) = \bar{\Psi}\gamma_\mu \Lambda' \Psi \,,$$

wherein the matrices Λ and Λ' act on the particle-kind index of the Ψ wave. He further required the conservation of the electric current $\bar{\Psi}_e \gamma_\mu \Psi_e + \bar{\Psi}_p \gamma_\mu \Psi_p$

and of the "neutrinic" or "dual" current $\bar{\Psi}_\nu \gamma_\mu \Psi_\nu + \bar{\Psi}_n \gamma_\mu \Psi_n$, and he excluded matter-annihilating processes. These constraints lead to the currents

$$J_\mu = \bar{\Psi}_e \gamma_\mu \Psi_e + \bar{\Psi}_p \gamma_\mu \Psi_p \,, \quad J'_\mu = \bar{\Psi}_\nu \gamma_\mu \Psi_\nu + \bar{\Psi}_n \gamma_\mu \Psi_n$$

for electromagnetic forces

and

$$J_\mu = \bar{\Psi}_e \gamma_\mu \Psi_\nu + \bar{\Psi}_p \gamma_\mu \Psi_n \,, \quad J'_\mu = J^*_\mu = \bar{\Psi}_\nu \gamma_\mu \Psi_e + \bar{\Psi}_n \gamma_\mu \Psi_p$$

for nuclear forces.

In a sequel, he introduced the fields $A^\mu = \bar{\Psi}_\nu \gamma^\mu \Psi_\nu$ and $B^\mu = \bar{\Psi}_p \gamma^\mu \Psi_n$ through which the interaction terms of the Lagrangian become more similar to the $e\bar{\Psi} \not{A} \Psi$ of quantum electrodynamics. The result should not be confused with a genuine mesic theory of nuclear forces, since the B_μ field here remains a compound field without specific quanta.[10]

Within this framework and with his covariant perturbation theory, Stueckelberg computed the probability of β-decay, and more innovatively, that of K-electron capture with and without emission of γ rays. In a lecture given in Basel in November of the same year 1936, he characterized the underlying philosophy of this theory as a positivism inherited from Ernst Mach and Pascual Jordan. The Machian idea of "a simpler and better description," not any suspicious metaphysics, was the alleged motor of his field-unifying enterprise. In the same lecture, Stueckelberg introduced the conservation of the "heavy charge" and the "light charge" as a replacement for his earlier conservation law. The following table of his lecture shows how he defined these new quantum numbers (see the columns for "leichte Ladung" and "schwere Ladung"), which anticipated our leptonic and hadronic numbers.[11]

---

[10] Stueckelberg [**26**, **27**, 29]. Note that Stueckelberg's field is not the Fermi field. In his view, the basic field process for nucleon-nucleon forces involves only the nucleon fields.

[11] Stueckelberg, ibid., [**30**, **33**].

Die Objekte der quantenmechanischen Beschreibung.

Feld	Symbol	Transformations-charakter	Partikel	Antipartikel	Masse	elektr.	neutr.	leichte	schwere	Geschwindigkeit	Drehmoment $\times 2\pi h^{-1}$	Statistik
						Ladung der Partikel						
Materie	ψ_α	Spinor	positives Elektron (+)	negatives Elektron (−)	1	+1	0	+1	0	$<c$	½	Fermi-Dirac
	φ_α		Neutrino (+)	Antineutrino (−)	0	0	+1	+1	0	$=c$	½	
	u_α		positives Proton (+)	negatives Proton (−)	1847	+1	0	0	+1	$<c$	½	
	v_α		Neutron (+)	Antineutron (−)	1848	0	+1	0	+1	$<c$	½	
Äther	A_i	Vektor	Photon	—	0	0	0	0	0	$=c$	0, 1	Einstein-Bose
	g_{ik}	Tensor	Graviton	—	0	0	0	0	0	$=c$	0, 1, 2	

— bedeutet: „keine Antipartikel"
$\alpha = 1, 2, 3, 4$
$i, k = 0, 1, 2, 3$

Die Antipartikel haben in diesen Kolonnen −1 statt +1 stehen

5.6. Stueckelberg's Mediating Field

As the kernel K in his interaction Lagrangian $-\int d^4yK(x,y)J^\mu(x)J'_\mu(y)$, Stueckelberg took $K(x,y) = C\exp((x-y)^2/\lambda^2)$ for the terms yielding nuclear forces, and $K(x,y) = C\delta((x-y)^2)$ for the terms yielding electromagnetic forces. The former choice was intended to reproduce the short range of nuclear forces, and the latter choice optimized the analogy with quantum electrodynamics. Nicholas Kemmer promptly noted that the choice $K(x,y) = C\delta((x-y)^2)$ was the only one compatible with the relativistic retardation of interactions. He also showed that the adoption of this kernel for nuclear forces would lead to catastrophic results, such as an infinite binding energy for the deuteron. (Stueckelberg [**26**, **27**]; Kemmer [1936], pp. 48–49)

In reaction to Kemmer's criticism, in 1937 Stueckelberg decided to imitate the field-mediated retardation found in usual electrodynamics, and replaced his biquadratic coupling with a field-mediated coupling, $\bar{\Psi}A\Psi$, with

$$\mathrm{A} = \Phi + a_\mu \gamma^\mu + b_{\mu\nu}\sigma^{\mu\nu} + \cdots$$

The mediating field A now is a dynamical field contributing to the kinetic part of the Lagrangian. For the simplest pertinent choice of this field,

$$a_\mu = QA_\mu\,, \quad \Phi = T\varphi + T^+\varphi^+\,,$$

where A_μ is the electromagnetic potential, Q the charge matrix, φ a charge scalar (Pauli–Weisskopf) field, and T is a matrix connecting electron states to neutrino states, and neutron states to proton states, with adequate coupling constants. (Stueckelberg [**38**, **39**]. Cf. Darrigol [1988b], pp. 295–296)

Stueckelberg identified the quanta of the scalar field φ with the "heavy electron" discovered by Carl Anderson, Seth Neddermeyer, Jabez Street, and Edward Stevenson in cosmic rays during the previous year. He was first to assert the spontaneous decay of this particle into an electron and an (anti)neutrino owing to the weak coupling between electron field, neutrino field, and φ field. In the short note announcing his main results, he emphasized the similarity of his theory with Yukawa's earlier theory of nuclear forces (Stueckelberg [**38**, p. 41]):

> The writer wishes to call attention to an explanation of the nuclear forces, given as early as 1934 [in fact 1935], by Yukawa, which predicts particles of this sort [mediators for the nuclear forces and for β-decay]. Independently of Yukawa, the writer has arrived at the same conclusion.

There is no reason to doubt Stueckelberg's sincerity regarding the independence of his invention. Indeed his new unified theory naturally came out of Kemmer's criticism of his earlier unified theory with interaction kernels. Its formal expression and its higher generality confirm this genesis. He may nonetheless have been aware of Yukawa's theory before he himself introduced new mediating fields, because Yukawa probably sent his paper to European authorities with whom Stueckelberg was in contact.

5.7. Yukawa's U-Field

In a talk given at Sendai in April 1933, the young Yukawa sketched a new theory of nuclear forces based on Bohr's idea that electrons were *created* in β-decay and on a literal understanding of Heisenberg's idea that electron exchange caused the force between two nucleons. The corresponding diagram would be:

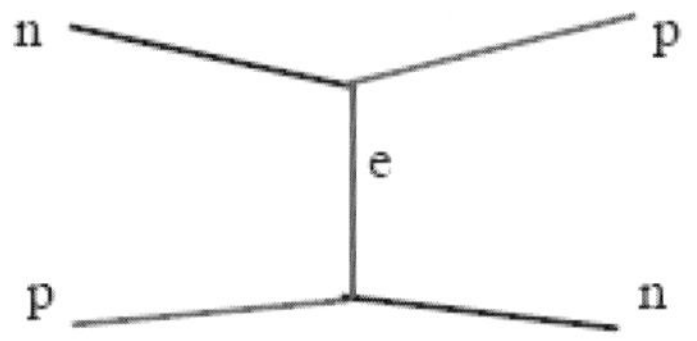

In analogy with quantum electrodynamics, Yukawa formed the interaction Lagrangian by taking the product of the neutron field, the proton field, and the electron field. At the end of his talk, he concluded a failure of this theory, though not for the reasons a modern reader would expect. The violations of conservation laws resulting from the lack of Lorentz invariance of the Lagrangian did not bother him, since Bohr expected them. What stopped Yukawa was the infinite range he found for the resulting forces, owing to a miscalculation of the Green function of the Dirac operator. (Cf. Darrigol [1988b], pp. 275–279)

When in 1934 Yukawa became aware of Fermi's theory, he immediately tried to replace the electron exchange of his earlier attempt with the exchange of an electron-neutrino pair. Although he agreed with his Western colleagues that the resulting forces were too weak, he disagreed with them on the way to solve this difficulty. Instead of playing with the form of the coupling between the neutron-proton field and the electron-neutrino field, he more daringly decided to replace the exchanged electron-neutrino pairs with the quanta of a new field. Guided by analogy with quantum electrodynamics, he assumed that in a first (non-relativistic) approximation this field was the time-component of a 4-vector, which he called U in analogy with the electric potential V. He found the field created by a point-like nucleon at rest to be proportional to $\exp(-rm_U c/h)$, where r is the distance from the nucleon and m_U the mass of a quantum of the U field. This implies the range $h/m_U c$ for nuclear forces. From the mass defect of the deuterium and from scattering data, Yukawa judged that the mass should be about two hundred times the mass of the electron. (Yukawa [1935]. Cf. Brown and Rechenberg [1996], Chapter 5; Darrigol [1988b], pp. 279–288)

In order to explain β-decay, Yukawa further introduced a coupling between the electron-neutrino field and the U-field. In this view, the basic process responsible for the decay of a neutron is

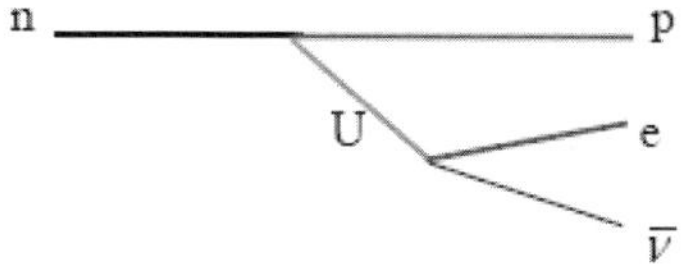

Yukawa investigated the phenomenological consequences of his theory in much greater detail than Stueckelberg. For instance, he showed that the sign choice for the source term in the U-field equation gave the correct spin for the normal state of the deuterium nucleus, whereas Stueckelberg failed to note that his tentative scalar theory gave the opposite value. In general, Stueckelberg tended to favor formal generality over phenomenological adequacy.

These differences and Yukawa's clear priority explain why Stueckelberg is usually not regarded as a co-discoverer of the mesic theory of nuclear forces. Yukawa himself had trouble attracting the attention of Western experts. The first mention of it in a non-Japanese publication appeared in June 1937 in a note by Robert Oppenheimer and Robert Serber on the new cosmic ray particle in the *Physical Review*. It was derogatory. The journal *Nature* rejected a letter by Yukawa of January 1937 in which he identified the new particle with his own U-quantum. However, within a year or so several theorists joined Yukawa's and Stueckelberg's efforts to modify and extend mesic theory. The convoluted history of this theory, with the long-lasting confusion between muon and meson, was only beginning.[12]

5.8. Stueckelberg's B-Field

In 1938, Stueckelberg published a lengthy, three-part memoir on meson theory in the *Helvetica Physica Acta.* The first part emphasized retardation of the mesic field created by the nucleons, and gave a first hint at a vector-meson theory. The second and third part fully developed the latter theory, with the benefit of covariant perturbation theory and of Møller's older method of retarded potentials. Stueckelberg gave a central role to his law of the conservation of the "heavy charge," and sought the most general theory that conserved electric and heavy currents. As in his earlier paper, he explored the full spectrum of possibilities, including the heterodox (Wentzel) case in which the mediating field carries heavy charge, rather than detailing the phenomenological consequences of the most plausible option. (Stueckelberg [**39**, **40**]. Cf. Brown and Rechenberg [1996], p. 211).

One interesting aspect of Stueckelberg's vector-meson theory was the clever way in which he solved an outstanding difficulty in the quantization of massive vector fields.

As was already well-known, the gauge invariance of the electromagnetic Lagrangian $-\frac{1}{4}F^{\mu\nu}F_{\mu\nu}$ leads to difficulties in the canonical quantization procedure. In his foundational memoir of 1929, Fermi circumvented this difficulty by adopting the Lagrangian

$$L = -\frac{1}{2}\partial_\mu A_\nu \partial^\mu A^\nu$$

which leads to the free-field equation

$$\partial^2 A_\mu = 0\,.$$

Maxwell's equations $\partial_\mu F^{\mu\nu} = 0$ follow if the gauge is a posteriori set to $\partial \cdot A = 0$. Quantization then succeeds through the commutation rule

$$\left[A_\mu(x), A_\nu(y)\right] = i g_{\mu\nu}\,\Delta\,(x-y)\,.$$

The quantum counterpart of the gauge condition is the weaker condition[13]

$$\partial \cdot A \mid \psi\rangle = 0$$

for every permitted $\mid \psi\rangle$.

---

[12] Cf. Darrigol [1988b], pp. 288–291. On the vicissitudes of early mesic theory, cf. Monaldi [2005]; Brown and Rechenberg [1996], parts C and D.

[13] Cf. Schweber [1994], pp. 72–75; Darrigol [1982], pp. 95–98, 110–116. As was much later understood this condition is still too strong. Only the annihilating part of $\partial \cdot A$ must be required to vanish on the permitted state-vectors.

The massive vector-field imitation of this procedure leads to the Lagrangian

$$L = -\frac{1}{2}\partial_\mu A_\nu \partial^\mu A^\nu + \frac{1}{2}m^2 A^\mu A_\mu ,$$

and to the field equation

$$\partial^2 A_\mu + m^2 A_\mu = 0 ,$$

which agrees with Proca's field equation

$$\partial_\mu F^{\mu\nu} + m^2 A^\nu = 0$$

if $\partial \cdot A = 0$.

As Stueckelberg realized, this simple procedure leads to various troubles. At the classical level, the condition $\partial \cdot A = 0$ is no longer sufficient to make the energy positive. At the quantum level, the commutation rule

$$\left[A_\mu(x), A_\nu(y)\right] = i g_{\mu\nu} D_m(x-y)$$

leads to

$$\left[\partial \cdot A(x), \partial \cdot A(y)\right] = i\partial^2 D_m(x-y) = -im^2 D_m(x-y)$$

which is incompatible with the (weaker) gauge condition.

Stueckelberg's trick was to replace this condition with the new condition

$$(\partial \cdot A + mB) \mid \psi\rangle = 0$$

where B is an additional scalar field quantized through

$$\left[B(x), B(y)\right] = i D_m(x-y) .$$

The commutation rules for A_μ and B then imply

$$\left[\partial \cdot A(x) + mB(x), \partial \cdot A(y) + mB(y)\right] = i(\partial^2 + m^2) D_m(x-y) = 0$$

which means that the new condition is self-consistent. The complete Lagrangian

$$L = -\frac{1}{2}\partial_\mu A_\nu \partial^\mu A^\nu + \frac{1}{2}m^2 A^\mu A_\mu + \frac{1}{2}\partial_\mu B \partial^\mu B - \frac{1}{2}m^2 B^2 ,$$

leads to the Proca equations

$$\partial_\mu F^{\mu\nu} + m^2 \Omega^\nu = 0$$

for the more physical fields[14]

$$\Omega_\mu = A_\mu - \frac{1}{m}\partial_\mu B \quad \text{and} \quad F_{\mu\nu} = \partial_\mu \Omega_\nu - \partial_\nu \Omega_\mu \,.$$

Again, Pauli was first to understand the value of Stueckelberg's procedure. In a letter to Wentzel of April 1939, he remarked:

> This state of affairs [problem with self-commutation of additional condition] is thoroughly explained (and even in an almost clear manner) in a paper by Stueckelberg, who has constantly given multiple formulations of the meson theory.

The "almost clear" is a hint at the difficulty contemporary theorists had decrypting Stueckelberg's difficult papers. In 1939, Pauli further noted that the Ω_μ field and the action $\int L d^4x$ were invariant under the transformation

$$A_\mu \to A_\mu + \partial_\mu \xi \,, \quad B \to B + m\xi \quad \text{if} \quad (\partial^2 + m^2)\xi = 0 \,.$$

In recent years, there has been a resurgence of interest in Stueckelberg's procedure owing to similar difficulties in modern gauge theories. The invariance noted by Pauli is related to the BRST invariance of some of these theories.[15]

5.9. Canonical Transformations

A last interesting aspect of Stueckelberg's contribution to meson theory is the use he made of canonical transformations to reach more convenient forms of the Hamiltonian. Around 1930, Pauli, Oppenheimer, and Fermi already used canonical transformations to eliminate longitudinal and time-like components of the electromagnetic field and thus get the Coulomb interaction directly in the Hamiltonian. In 1938, Stueckelberg (in parallel with Christian Møller and Léon Rosenfeld) tried something similar for the meson field. He realized that the spin and isospin dependence of the interaction Lagrangian made the elimination of unphysical degrees of freedom much more difficult in the mesic case, and led to additional terms of higher order besides the static Hamiltonian, both at classical and quantum level. In 1940 with his student Jean Patry, he calculated the static terms of higher order, and found them troublesome and divergent in the vector-field case.[16]

---

[14] Stueckelberg [**39**, **40**]. Cf. the in-depth study (Ruegg and Ruiz-Altaba [2004]); and the contribution of Henri Ruegg and Marti Ruiz-Altaba to this volume.

[15] Pauli to Wentzel, 24 Apr 1939, in von Meyenn [1985], vol. 2, pp. 635–636. Cf. Ruegg and Ruiz-Altaba [2004].

[16] Stueckelberg [42, 43, 44], Møller and Rosenfeld [1939]; Stueckelberg and Jean Patry [49]. On earlier uses of canonical transformation in QED, cf. Darrigol [1982], pp. 113–114, 119–124, 198–205.

Despite the negative character of this result, Pauli found the memoir "interesting from a methodological point of view and clearly written," as he wrote to Homi Bhabba in March 1941. As Pauli well knew, canonical transformations that "dress" bare particles with a cloud of virtual photons or mesons were becoming a powerful convenience of applied quantum field theory. It had been used by Felix Bloch and Arnold Nordsieck in 1937 for a first solution of the infrared paradox of quantum electrodynamics, by Pauli himself and Markus Fierz in an improved solution of the same paradox in 1938, and more recently by Gregor Wentzel and by Sin-itiro Tomonaga in the strong-coupling approach to mesic theory. It had a future in Tomonaga's and Schwinger's version of post-war, renormalized quantum electrodynamics.[17]

5.10. A Dream

In July 1939, Stueckelberg published a note in *Nature* in which he proposed a classical point-electron model with a cohesive Yukawa field. Owing to the mutual compensation of electromagnetic and mesic contributions, the self-energy of this electron has a finite value, namely: $e^2/2\lambda$, where e is the charge of the electron, and λ the Yukawa length. In a generalization to particles endowed with spin and isospin, Stueckelberg conjectured that in this case the self-energy was only finite for discrete values of the mass. A determination of the mass of elementary particles seemed within arm's reach.

Plausibly, this exciting speculation was the subject of the telegram that Stueckelberg sent to a few leading physicists around that time. (Stueckelberg [48]; also [42])

On 18 July, Pauli wrote to Dirac: "Stueckelberg [has] a serious nervous breakdown.... His telegram to you and other physicists was at least premature, he had only some vague hopes and not yet anything definitely proved." Three months later, Pauli reported to Kemmer: "Stueckelberg was very ill this summer, but he is much better now. His old illness, you know, seems to be periodical, too bad!" Stueckelberg's manic-depression lasted to the end of his life. Although it repeatedly disrupted his family and work life, it does not seem to have affected his scientific and social ability in the periods of normality. The incident reported by Pauli was exceptional. Stueckelberg usually avoided unwarranted speculation in his publications. The originality of his style should not be confused with fancifulness. Even the dubious letter to *Nature* of July 1939 had solid content: its compensation mechanism later inspired the field mixtures of Shoichi Sakata's group in Japan.[18]

---

[17] Pauli to Bhabba, 20 Mar 1941, in von Meyenn [1985], vol. 3, pp. 83–86.

[18] Pauli to Dirac, 18 Jul 1939, and Pauli to Kemmer, 24 Nov 1939, in von Meyenn [1985], vol. 2. On the Japanese cohesion field, cf. Darrigol [1988], p. 22.

5.11. Conclusions

While developing his grand "field theory of matter" of the late 1930s, Stueckelberg forged some of the basic tools that permitted contemporary and later successes of quantum field theory. He provided manifestly covariant formulations and perturbative expansions; he expressed these expansions in terms of Green functions and propagators; he promoted the view that all interactions should be based on the virtual exchange of mediating particles; he introduced conserved quantum numbers and currents; he applied canonical transformation to the dressing of bare particles; he invented the B-field trick for quantizing massive vector fields; and he introduced the idea of field-mixtures for compensating UV divergencies.

Some of these tools, such as the mediating-field concept and canonical transformations, were independently introduced and used by other theorists around the same time. Some others, such as the manifestly covariant perturbation theory and the recourse to propagators only became popular after Tomonaga, Schwinger, and Richard Feynman used them in their contributions to renormalized quantum electrodynamics. For a long time, Stueckelberg was the only physicist to benefit from them. Possible reasons for this relative isolation have already been mentioned: the difficulty of his writing style, the unfortunate timing of some of his innovations, the excessive generality of his theories, and his lack of taste for working out special cases of phenomenological interest. An additional reason may have been his lack of close partnership with other eminent theorists, except for some of his later disciples.

Whether or not this isolation is judged unfair, it prevented Stueckelberg from inflecting the course of theoretical physics as much as his genuine creativity could have done. His unitary field theory of 1936–39 nonetheless anticipated several features of modern grand-unified theories, and contained Yukawa's theory as a particular case. The tools he developed in this earlier period permitted his and his collaborators' post-war progress on renormalized quantum electrodynamics, including important results on regularization, causal Green functions, and S-matrix theory. Had these results been expressed in a form as seductive as Feynman's, Schwinger, and Tomonaga's, and Freeman Dyson's contemporary writings, Stueckelberg would be more often recognized as a central figure of this extraordinary episode of the history of physics.[19]

---

19 On Stueckelberg's war-time and post-war contributions to QED, cf. Schweber [1994], pp. 576–582; Henri Ruegg's and Marti Ruiz-Altaba's contribution to this volume; Darrigol [1982], pp. 308–312.

Acknowledgment

I thank Jan Lacki for his invitation to the 2005 Stueckelberg's commemoration and for providing some of the necessary sources.

References

Brown, L., The idea of the neutrino, *Physics Today*, vol. 31 (1978), pp. 23–28.

Brown, L. and Rechenberg, H., *The Origin of the Concept of Nuclear Forces*. Bristol: Institute of Physics Publishing, 1996.

Carson, C., The peculiar notion of exchange forces. II: From nuclear forces to QED, 1929–1950, *Studies in History and Philosophy of Modern Physics*, vol. 27 (1996), pp. 99–131.

Darrigol, O., *Les débuts de la théorie quantique des champs*, thèse de 3ème cycle, Université de Paris 1, 1982.

Darrigol, O., La genèse du concept de champ quantique, *Annales de physique*, vol. 9 (1984), pp. 433–501.

Darrigol, O., The origin of quantized matter waves, *Historical Studies in the Physical Sciences*, vol. 16 (1986), pp. 198–253.

Darrigol, O., Elements of a scientific biography of Tomonaga Sin-itiro, *Historical Studies in the Physical Sciences*, vol. 35 (1988), pp. 1–29.

Darrigol, O., The quantum electrodynamical analogy in early nuclear theory or the roots of Yukawa's theory, *Revue d'histoire des sciences*, vol. 41 (1988), pp. 225–297.

Dirac, P., Relativistic quantum mechanics, *Proceedings of the Royal Society of London*, vol. 136 (1932), pp. 453–464;

Dirac, P. A. M., Fock, V., and Podolsky, B., On quantum elecrodynamics, *Physikalische Zeitschrift der Sowjetunion*, vol. 2 (1932), pp. 468–479.

Enz, Ch., Obituary of E. C. G. Stueckelberg, *Physics Today*, vol. 39 (1986), pp. 119–121.

Fermi, E., Tentativo di una teoria dei raggi β, *Nuovo Cimento*, vol. 2 (1934), pp. 1–19.

Fermi, E., Versuch einer Theorie der β-Strahlen, *Zeitschrift für Physik*, vol. 88 (1934), pp. 161–171.

Lacki, J., Ruegg, H. and Telegdi, V., The road to Stueckelberg's covariant perturbation theory as illustrated by successive treatments of Compton scattering, *Studies in History and Philosophy of Modern Physics*, vol. 30 (1999), pp. 457–518.

Heisenberg, W., Die Grenzen der Anwendbarkeit der bisherigen Quantentheorie, *Zeitschrift für Physik*, vol. 110 (1938), pp. 251–260.

Kemmer, N., Zur Theorie der Neutron-Proton Wechselwirkung, *Helvetica Physica Acta*, vol. 10 (1936), pp. 47–67.

Meyenn, K. von (ed), *Wolfgang Pauli, Wissenschaflicher Briefwechsel*, vols. 2–3. New York: Springer, 1985, 1993.

Monaldi, D., Life of μ: The observations of the spontaneous decay of mesotrons and its consequences, *Annals of Science*, vol. 62 (2005), pp. 419–456.

Møller, Ch. and Rosenfeld, L., Theory of mesons and nuclear forces, *Nature*, vol. 143 (1939), pp. 241–242.

Rosenfeld, L., Über eine mögliche Fassung des Diracschen Programms zur Quantenelektrodynamik und deren formalen Zusammenhang mit der Heisenberg–Paulischen Theorie, *Zeitschrift für Physik*, vol. 76 (1932), pp. 729–734.

Ruegg, H. and Ruiz-Altaba, M., The Stueckelberg field, *International Journal of Modern Physics A*, vol. 19 (2004), pp. 3265–3347.

Schweber, S. S., *QED and the Men Who Made It: Dyson, Feynman, Schwinger, and Tomonaga*. Princeton: Princeton Univ. Press, 1994.

Stuewer, R., The nuclear electron hypothesis, in Shea, W. (ed.), *Otto Hahn and the Rise of Nuclear Physics*. Dordrecht: D. Reidel, 1983, pp. 19–67.

Yukawa, H., On the interaction of elementary particles. I, Physico-Mathematical Society of Japan, *Proceedings*, vol. 17 (1935), pp. 48–57.

Wenger, R., *Ernst C. G. Stückelberg von Breidenbach: Etude biographique*. Bibliothèque Section de Physique, Faculté des Sciences de l'Université de Genève, (unpublished MS, 1986).

CHAPTER 6

Stueckelberg 1937–1942: The B-Field and Antiparticles as Time-Reversed Particles

Henri Ruegg and Marti Ruiz-Altaba

We review some of the scientific work done by E. C. G. Stueckelberg just before and just after the outbreak of World War II. In 1938, he introduced what is now known as the Stueckelberg mechanism for giving mass to a gauge field while preserving gauge invariance. In 1941, Stueckelberg introduced the concept of antiparticles as time-reversed particles.

6.1. Introduction

We shall review a subset of the papers published by E. C.G Stueckelberg between 1937 and 1942, which we have classified in two groups:

- Papers about the recent discovery of a new particle – called a heavy electron at the time – leading to the Stueckelberg mechanism [**38**, **39**, **40**]. These papers have been widely cited since 1938 into the 21st century; perhaps, they are more often cited than read.
- Papers on the quantum theory of particle-antiparticle pairs – deep papers submitted in October 1941 which are very rarely quoted [**53**, **54**, **56**].

The 1938 papers on electromagnetic and nuclear interactions were motivated by the discovery of a new particle in cosmic rays which, Stueckelberg argued, could be precisely the particle proposed by Yukawa mediating nuclear reactions. Requiring strict consistency and manifest Lorentz covariance of the quantum field theory for a massive vector field led him to introduce a new quantum field, nowadays called the Stueckelberg field. This extra field is the price to pay for a consistent massive gauge theory with unbroken gauge symmetry. As we shall sketch below, the Stueckelberg mechanism does not extend to a non-Abelian gauge theory, which can only become massive by breaking the gauge symmetry through the Higgs mechanism. A complete review of the Stueckelberg mechanism, and its hypothetical application to the standard model, has appeared elsewhere (Ruegg and Ruiz-Altaba [2004]).

The papers written in 1941 consider a charged massive point particle in external electromagnetic and gravitational fields. Using a slight generalization of Einstein's relativistic mechanics, Stueckelberg shows how to describe the creation and annihilation of a particle-antiparticle pair in a general coordinate invariant way. In Stueckelberg's generalization, an electron's world-line can bend backwards in time – such a time-reversed electron is then reinterpreted as an anti-electron. Very carefully, Stueckelberg proceeds then to (second) quantize this picture. Remarkably, his detailed computations have been applied directly to study the evaporation of black holes (Damour [1986]).

In his construction of QED, Feynman followed the same idea of the positron being an electron going backwards in time (Feynman [1949]), as suggested to him by Wheeler (Feynman [1948]). The two sets of papers, one leading to the Stueckelberg field and the other to the quantum field theory of pair production/annihilation, share an underlying constant passion for generality and covariance. On every occasion, the most general formulation of invariance laws is used.

6.2. Heavy Electrons

In 1937, several experimental groups discovered a new hard component in cosmic rays. Assuming electrodynamics, they announced the existence of a charged particle between one and two-hundred times heavier than the electron (Pais [1986]).

Oppenheimer and Serber on June 1 (Oppenheimer and Serber [1937]), and Stueckelberg on June 6 [**38**], sent letters to *Physical Review* suggesting that the new particle found in cosmic rays could be the meson proposed by Yukawa to mediate nuclear interactions.

Stueckelberg's letter starts as follows:

> Different observers believe they have found evidence for the existence of charged particles whose mass amounts probably to about

> fifty times the electron mass. Furthermore these particles seem to behave according to the Bethe–Heitler theory.
>
> The writer wishes to call attention to an explanation of the nuclear forces, given as early as 1934 by Yukawa, which predicts particles of this sort.
>
> Independently of Yukawa, the writer arrived at the same conclusion. ...

He then plunged into the construction of a five-dimensional vector field on a space-time of the form $\mathcal{M} \times S^1$ (i.e. on a four-dimensional space times a one-dimensional circle), obtaining the electromagnetic potential and a *massive* spin zero bosonic field. By carefully enforcing the conservation of various quantum numbers (which look very much like baryon number and the third component of isospin) on the five-vector's couplings to electrons, neutrinos, protons and neutrons, he found that the massive scalar appearing in nuclear reactions must have positive charge and disintegrate into a (positive) electron and an antineutrino (see the contribution of Olivier Darrigol in this volume).

Let Stueckelberg's own words summarize his conclusions:

> It seems highly probable that Street and Stevenson, and Neddermeyer and Anderson have actually discovered a *new elementary particle*, which has been predicted by theory.
>
> This particle is unstable and can only be of secondary origin, its mass being greater than the sum of the masses of electron plus neutrino.

Stueckelberg was indeed the first to point out that this *new* particle had to be unstable, unlike the neutrino.[1]

6.3. The Stueckelberg Mechanism and the Stueckelberg Field

Stueckelberg's goal was to construct a quantum field theory of nuclear interactions mediated by massive particles. He first considered the interaction of a scalar quantum field with the nucleon fields, but he got the wrong sign for the static potential (repulsive instead of attractive). He then tried with a vector

[1] In 1947, a collaboration of experimentalists and theoreticians found that the "1937 particle" was indeed a heavy electron, now called a muon. It had spin $\frac{1}{2}$ and no strong interactions with the nucleons. Another particle, now called a pion, was discovered in cosmic rays. It is a pseudo-scalar (spin zero and negative parity), heavier than the muon, and it interacts strongly with the proton and the neutron. Hence, it was considered a good candidate for the Yukawa particle (Pais [1986]). Nowadays, the theory of strong interactions is QCD, in which the forces between quarks are mediated by non-Abelian massless vector particles. This shows that ideas in physics often evolve around the same theme.

field. His strategy was to formulate a quantum field theory of massive vector fields as similar as possible to QED. For a more detailed exposition, see Ruegg and Ruiz-Altaba [2004].

The first theory of massive vector fields had appeared for a totally different reason a few months earlier ([Proca, 1936]). Proca wrote down the following equation of motion for the free field $V^\mu(x)$:

$$\partial^\mu\left[\partial_\mu V_\nu(x) - \partial_\nu V_\mu(x)\right] + m^2 V_\nu(x) = 0\,. \tag{6.1}$$

If the massive vector field $V^\mu(x)$ is on-shell, it satisfies automatically the Lorentz condition

$$\partial^\mu V_\mu(x) = 0\,. \tag{6.2}$$

The interaction between the massive vector field and the nucleon fields leads to quadratic divergences at high energy, due to the derivatives of the commutator

$$\left[V_\mu(x), V_\nu^\dagger(y)\right] = -i\left(g_{\mu\nu} + \frac{1}{m^2}\partial_\mu\partial_\nu\right)\Delta_m(x-y) \tag{6.3}$$

where $\Delta_m(x)$ is the Jordan–Pauli function, obeying

$$(\partial^2 + m^2)\Delta_m(x) = 0\,. \tag{6.4}$$

Instead of Proca's equation of motion, Stueckelberg wrote simply

$$(\partial^2 + m^2)A_\mu(x) = 0 \tag{6.5}$$

which derives from the Lagrangian density

$$\mathcal{L} = -\partial_\mu A_\nu^\dagger \partial^\mu A^\nu + m^2 A_\mu^\dagger A^\mu \tag{6.6}$$

which can be quantized canonically with the non-vanishing commutation relations

$$\left[A_\mu(x), A_\nu^\dagger(y)\right] = -i g_{\mu\nu}\Delta_m(x-y)\,. \tag{6.7}$$

This commutator guarantees a much softer behavior at high energies. It has two fatal drawbacks, nevertheless. For one, A^μ is not guaranteed to satisfy the Lorentz condition when on-shell. Secondly, the Hamiltonian density is not positive-definite.

Stueckelberg knew that QED has quite similar problems, and that to solve them one needs to impose a covariant gauge condition on the space of all states to remove states with non-positive definite metric and thus to guarantee that the Hamiltonian density be positive-definite. So the problem reduces really to finding a nice covariant generalization of the Lorentz gauge condition to the massive case.

Stueckelberg solved both evils by introducing an additional scalar field $B(x)$, the "Stueckelberg field". In sharp contrast to most other named fields introduced in physics since then, the Stueckelberg field has a positive metric in Hilbert space.

Stueckelberg's Lagrangian for the free massive vector field $A^\mu(x)$ accompanied by its Stueckelberg field $B(x)$ is then [**39**, **40**]

$$\mathcal{L}_{\text{Stueck}} = -\partial_\mu A^\dagger_\nu \partial^\mu A^\nu + m^2 A^\dagger_\mu A^\mu + \partial_\mu B^\dagger \partial^\mu B - m^2 B^\dagger B \tag{6.8}$$

and the quantization calls for the non-vanishing commutator (6.7) supplemented by

$$\left[B(x), B^\dagger(y)\right] = i\Delta_m(x-y)\,. \tag{6.9}$$

The Hamiltonian density is positive-definite provided that the following subsidiary condition is imposed on the states which make physical sense:

$$\left(\partial_\mu A^\mu(x) + mB(x)\right)|\text{physical}\rangle = 0\,. \tag{6.10}$$

This recovers the standard and awkward Proca Lagrangian (Proca [1936]) with the identification

$$V_\mu(x) = A_\mu(x) - \frac{1}{m}\partial_\mu B(x)\,. \tag{6.11}$$

In 1941, Pauli showed that (Pauli [1941]) $\mathcal{L}_{Stueck}$ is gauge invariant, under the local transformations

$$\begin{aligned} \delta A_\mu(x) &= \partial_\mu \Lambda(x)\,, \\ \delta B(x) &= m\Lambda(x)\,, \end{aligned} \tag{6.12}$$

where the local field Λ satisfies the Klein–Gordon equation of motion

$$(\partial_\mu \partial^\mu + m^2)\Lambda = 0\,. \tag{6.13}$$

In 1988, Delbourgo, Twisk and Thompson showed that the Stueckelberg Lagrangian for *real* fields $A_\mu(x)$ and $B(x)$ is BRST-invariant (Delbourgo et al. [1988]). The advantage of the BRST formulation is that the parameter of the Abelian transformation is not constrained.

Physically, the Stueckelberg field B plays the same role as the Goldstone boson in the Abelian Higgs model.

6.4. Influence of Stueckelberg's 1938 Papers

We sketch the main ideas in the major works on the Stueckelberg mechanism published between 1941 and 2004. Broadly, they fall into one of three categories:

- Renormalizability of massive vector theories. Alternatives to the Higgs mechanism.
- Hidden symmetry.
- Massive photon in $SU(2) \times U(1)$ electroweak theory (Ruegg and Ruiz-Altaba [2004]).

Renormalization of Massive Vector Theories

At the end of the 1940s, the power-counting renormalizability of QED (zero-mass vector theory) was established. Therefore, it was natural that one started working on the renormalizability of massive vector theories. The Stueckelberg decomposition of the Proca field (6.11)

$$V_\mu(x) = A_\mu(x) - \frac{1}{m}\partial_\mu B(x) \tag{6.14}$$

was very convenient. Indeed, the worst divergences at high energy are identified by the terms containing $\partial_\mu B(x)$. One could eliminate these terms by a unitary transformation using equation (6.12). This allowed Matthews [1949] to show that the Stueckelberg model for *real* fields leads to the same divergences as QED.

Umezawa, Kamefuchi and Salam (Salam [1962], Umezawa and Kamefuchi [1961]) generalized the Stueckelberg model to non-Abelian gauge transformations:

$$\begin{aligned} V_\mu(x) &= T_i V^i_\mu(x)\,, \\ [T_i, T_j] &= i f_{ij}{}^k T_k\,, \\ V'_\mu(x) &= U^{-1}(x) V_\mu(x) U(x) + \frac{i}{g} U^{-1}(x)\partial_\mu U(x)\,, \end{aligned} \tag{6.15}$$

Salam used the Ansatz due to Stueckelberg $V_\mu = A_\mu - \frac{1}{m}\partial_\mu B$ in order to localize the divergences. He found that *only the Abelian gauge invariance leads to renormalizable massive vector theories.*

Slavnov provided a more rigorous treatment of the generalized Stueckelberg model using Faddeev–Popov ghosts (Slavnov [1972]).

Kunimasa and Goto proposed a Stueckelberg alternative to the non-Abelian Higgs model (Kunimasa and Goto [1967]).

Curci and Ferrari proposed a non-Stueckelberg, non-Higgs model (Curci and Ferrari [1976]).

Delbourgo, Twisk and Thompson used modern methods to show that (Delbourgo et al. [1988]):

– the Kunimasa–Goto model is unitary but not renormalizable;

– the Curci–Ferrara model is renormalizable but not unitary.

Hurth [1997] agrees with Delbourgo *et al.*

Many other papers dealing with the Stueckelberg model have been written. For a review, see Ruegg and Ruiz-Altaba [2004].

Hidden Symmetry

The idea of inventing new fields in order to uncover or make manifest hidden symmetries, pioneered by Stueckelberg, has been applied in many contexts. These three are beautiful examples:

– The supersymmetric extension of the Stueckelberg formalism (Delbourgo [1975]; Ruegg and Ruiz-Altaba [2004], p. 3325).

– The ten-dimensional superstring fulfills a Stueckelberg symmetry (Bergshoeff and Kallosh [1990]; Ruegg and Ruiz-Altaba [2004], p. 3311).

– String field theory calls for an infinity of Stueckelberg string fields (Ramond [1986], Ruegg and Ruiz-Altaba [2004], p. 3309).

Electroweak Theory with a Massive Photon

The unified theory of weak and electromagnetic interactions is invariant under the gauge group $SU(2) \times U(1)$. In this theory the photon is massless. This is also the case for the matter fields which nevertheless get a mass through the Higgs mechanism upon spontaneous symmetry breakdown.

We gave the photon a mass using a Stueckelberg field, retaining the $SU(2) \times U(1)$ invariance (Ruegg and Ruiz-Altaba [2004]). Our motivation is double. It provides an infrared cut-off independent of the high-energy divergences.[2] Secondly, a non-zero photon mass, if real, would give rise to new physics.

The first step is to give a mass to the $U(1)$ field. After diagonalizing the propagator, both the photon and the Z^0 get a mass, slightly different from the usual one. The photon mass is constrained by the experimental (observational) limit (Luo et al. [2003])

$$m_\gamma < 1.2 \cdot 10^{-17}\,\mathrm{eV}. \tag{6.16}$$

Due to the Stueckelberg mechanism, this procedure preserves indeed the full $SU(2) \times U(1)$ gauge invariance.

2 Private communication by R. Stora and T. Hurth.

The Weinberg angle would change a little bit, and tiny photon neutrino couplings would be induced.[3]

The original papers on the Stueckelberg mechanism are the last which he wrote in German. The ubiquitousness of the Stueckelberg mechanism in removing $U(1)$'s from the low-energy spectrum of string models constitutes a lasting, if unuttered, tribute to a great scientist whose work was largely ignored.

6.5. e^+e^- Pair Creation and Annihilation

Stueckelberg submitted three papers in October 1941 in which he considers a particle coupled to external electromagnetic and gravitational fields. In four-dimensional space-time, the particle follows a one-dimensional world-line. In Einstein's theory, causality imposes that a point mass's world-line can intersect a fixed-time hyperplane only once: for any given t, the particle has only one spatial position. Stueckelberg generalizes the classical theory with an elegant trick. He takes advantage of general coordinate invariance on the world-line to parametrize it not with the "natural" particle's proper time τ, but with an (almost) arbitrary local parameter λ. Without spoiling any invariance, Stueckelberg finds situations in which a particle's world-line crosses twice the hyperplane t =constant.

What is the difference for a local observer between a standard electron and an electron "moving backwards in time"? Carefully, Stueckelberg argues that the sign difference should not be placed on the mass, but rather on the electric charge. Thus an electron with mass m and charge e moving backwards in time appears as a particle moving forward in time with the same mass m but opposite charge $-e$: the anti-electron!

After presenting this revolutionary idea in the classical theory, along with two beautiful diagrams, Stueckelberg turns to the quantum theory, where particles are replaced by wave-packets.

Stueckelberg gave both a classical and a quantum description of pair creation and annihilation in three papers he wrote in October 1941 [**53**, **54**, **56**]. Stueckelberg's three papers on e^+e^- pairs are concise and clear.

Consider Figure 6.1, which appears in [**54**]. In Stueckelberg's notation, the world-line of a point particle is given by

$$x^\mu = q^\mu(\lambda) \quad \mu = 1, 2, 3, 4$$

with λ a real parameter. The space-time metric is $g_{\mu\nu}(q)$. The particle of coordinates $q^\mu(\lambda)$ is subject to Einstein gravity and Maxwell electrodynamics,

3 It is an interesting thesis problem for a student to apply this model to the tantalizing domains of cosmology.

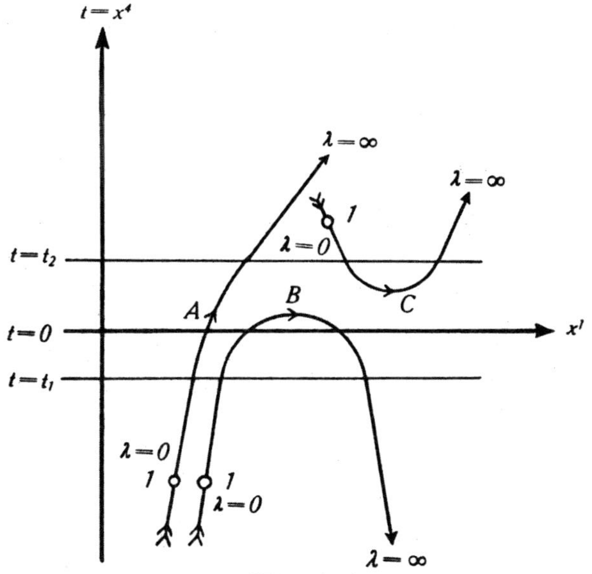

Fig. 1.

Figure 6.1: Reproduction of Figure 1 from [**54**], page 590.

whereby

$$\frac{d^2 q^\mu}{d\lambda^2} = -\Gamma^\mu_{\alpha\beta}\frac{dq^\alpha}{d\lambda}\cdot\frac{dq^\beta}{d\lambda} + eF^{\mu\nu}g_{\alpha\beta}(q)\frac{dq^\beta}{d\lambda} + K^\mu\,.$$

The proper time s defined by

$$ds = \sqrt{g_{\mu\nu}dq^\mu\,dq^\nu}$$

is related to the continuous parameter λ by a factor $\pm m$:

$$ds = \pm m\,d\lambda\,.$$

In the absence of external currents K^μ,

- the parameter m is a constant on the world-line A (the mass of the particle);
- if we set $d\lambda = -m^{-1}\,ds$ instead of $d\lambda = m^{-1}\,ds$, then the equations of motion are unchanged provided we accompany the transformation by a change of sign of the charge $e \to -e$

This means that (in the absence of external fields) there are two kinds of particles, with mass and charge given by

$$(m, e) \quad \text{and} \quad (m, -e) .$$

If external fields are allowed (in the figure, they are non-zero only between times t_1 and t_2), then world-lines like B and C are allowed. Stueckelberg's interpretation of these world-lines is that an electron going backwards in time is a positron (and vice versa). The world-line B thus represents e^+e^- annihilation, whereas the world-line C represents e^+e^- creation.

The passage from ds to $d\lambda$ avoids the use of the square root. This is analogous to the passage in string theory (Green, Schwarz and Witten [1986], p. 22) from the Nambu–Goto to the Polyakov actions. Indeed, the Nambu–Goto action for a relativistic string

$$S = \int d^2\sigma \sqrt{-\partial_\alpha X^\mu \partial^\alpha X_\mu} \tag{6.17}$$

is non-linear in the space-time coordinate fields $X^\mu(\sigma)$. One can introduce a set of algebraic auxiliary two-dimensional fields $g_{\alpha\beta}(\sigma)$, which are the generalization of Stueckelberg's $\tau(\lambda)$ field, such that the action is quadratic in X^μ:

$$S = \int d^2\sigma \sqrt{-\det(g)} g_{\alpha\beta} \partial^\alpha X^\mu \partial^\beta X_\mu . \tag{6.18}$$

This is the celebrated Polyakov action, which is simply the two-dimensional generalization of Stueckelberg's action for a point particle.

6.6. Wave Mechanics à la Stueckelberg

Inasmuch as possible, we keep Stueckelberg's original notation.

It is assumed that $g_{\mu\nu}$ is a constant (no gravitation). Also, let there be no external K_μ force. The electric field E has $E_2 = E_3 = 0$ at all times, whereas $E_1 \neq 0$ only for an infinitesimal amount of time $0 < t < \delta t$.

The electromagnetic potential is chosen as $A_2 = A_3 = A_4 = 0$ and

$$A_1 = \left\{ \begin{array}{ll} 0 & \text{for } t > 0 \\ \frac{\hbar\gamma}{c} & \text{for } t < 0 \end{array} \right\} . \tag{6.19}$$

The interpretation of Figure 6.2 is similar to wave optics: ψ_1 is an incoming electron wave packet. During the interval δt an electric field E acts on the electron, which is then either reflected (ψ_A) or refracted (ψ_B). The wave-function ψ_A corresponds to an electron with the opposite charge (pair creation).

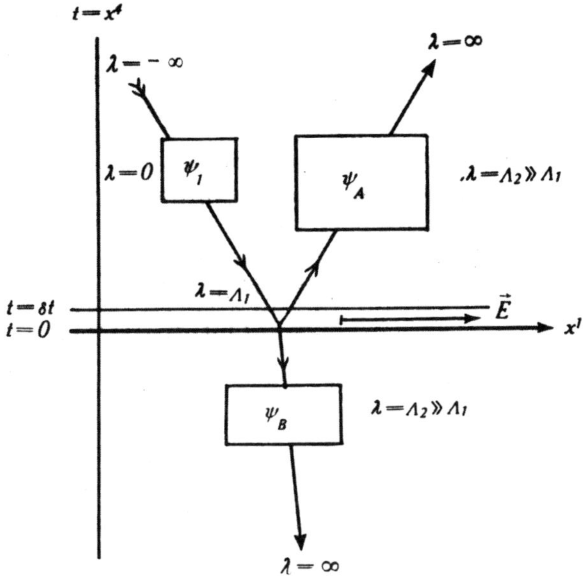

Fig. 2.

Figure 6.2: Reproduction of Figure 2 from [**54**], page 593.

The wave packet

$$\psi(q, \lambda) = \frac{1}{(2\pi)^3} \int d^3k \int d\omega\ u_{k,\omega}(q, \lambda)\varphi(k, \omega) \tag{6.20}$$

satisfies the Klein–Gordon equation

$$i\hbar \frac{\partial}{\partial \lambda} \psi(q, \lambda) = \frac{1}{2} \left(-i\hbar \frac{\partial}{\partial q^\mu} - eA_\mu(q) \right)^2 \psi(q, \lambda) \,. \tag{6.21}$$

The solution to this equation can be written as

$$\psi(q, \lambda) = \psi_1 + \psi_A + \psi_B \,. \tag{6.22}$$

The solution to the Schrödinger Klein–Gordon Stueckelberg equation is (setting $\tau = q^4$):

$$u_{k,\omega}(q, \lambda) = 1\, e^{i(k \cdot q - \omega\tau - \beta\lambda)} + A\, e^{i(k \cdot q + \omega\tau - \beta\lambda)} \,,$$
$$u_{k,\nu}(q, \lambda) = B\, e^{i(k \cdot q - \nu\tau - \beta\lambda)} \,,$$

where the frequencies are

$$\omega = \pm \left(k^2 + \frac{2}{\hbar}\beta\right)^{1/2},$$

$$\nu = \pm \left[(k^1 - \omega)^2 + (k^2)^2 + (k^3)^2 + \frac{2}{\hbar}\beta^2\right]^{1/2}$$

and the (relative) amplitudes are

$$A(k, \omega) = \frac{\omega - \nu}{\omega + \nu},$$

$$B(k, \omega) = \frac{2\omega}{\omega + \nu},$$

with

$$\frac{\omega}{\nu} > 0 . \tag{6.23}$$

Stueckelberg now computes the intensities, to be interpreted as probabilities:

$$W_A = \int d^4q \, |\psi_A|^2 = \overline{A^2} \sim A(k_0, \omega_0)^2 ,$$

$$W_B = \int d^4q \, |\psi_B|^2 = \overline{B^2 \frac{d\omega}{d\nu}} \sim \frac{\omega_0}{\nu_0} B(k_0, \omega_0)^2$$

which satisfy

$$W_A + W_B = 1 ,$$

$$W_A > 0; \quad W_B > 0 .$$

The mean values of the electric charges are

$$\overline{e_1} + \overline{e_A} = -e + eW_A = -eW_B = \overline{e_B} . \tag{6.24}$$

This is interpreted as follows: at $t = 0$ there is a probability W_A that a pair e^-e^+ is created and a probability W_B that e^- existed at $t < 0$ and was accelerated by the electric field at $t = 0$.

To understand the annihilation, one must change t into $-t$. As an application let us observe that a nice interpretation of the Hawking radiation is that if a pair is created on the horizon of a black hole, one particle falls inside while the other escapes outside. Damour showed (Damour [1986]) that a convenient way of calculating the probability for pair creation near a horizon is given by Stueckelberg's reflection coefficient W_A [44].

Feynman cites Stueckelberg [**54**] only in his first paper (Feynman [1949]), where he writes in the abstract:

> The problem of the behavior of positrons and electrons in given external potentials, neglecting their mutual interaction, is analyzed by replacing the theory of holes by a reinterpretation of the solutions of the Dirac equation. [...] In these solutions, the "negative energy states" appear in a form which may be pictured (as by Stueckelberg) in space-time as waves travelling away from the external potential backwards in time.

In an earlier paper dealing with classical electrodynamics (Feynman [1948]) he states:

> This idea that positrons might be electrons with the proper time reversed was suggested to me by Professor J. A. Wheeler in 1941.

To our knowledge, this suggestion of Wheeler was never published. Now, we do not think that the value of Feynman's fabulous contribution to physics is in any way diminished by the historical fact that Stueckelberg had already come up with the classical and quantum interpretation of an antiparticle as a particle going backwards in time. With hindsight, one can only lament that Stueckelberg's papers, published during the war in French in a journal of neutral Switzerland, did not receive the attention they should have merited.

References

Bergshoeff, E. and Kallosh, R., BRST quantization of the Green–Schwarz superstring, *Nucl. Phys.*, vol. B333 (1990) pp. 605–634.

Curci, G. and Ferrari, R., On a class of Lagrangian models for massive and massless Yang–Mills fields, *Nuovo Cim.*, vol. 32A (1976), pp. 151–168,

Damour, T., Strong field effects in general relativity, *Helv. Phys. Acta*, vol. 59 (1986), pp. 292–302.

Delbourgo, R., A supersymmetric Stueckelberg formalism, *J. Phys.*, vol. G8 (1975), pp. 800–804.

Delbourgo, R., Twisk, S. and Thompson, G., Massive Yang–Mills theory: Renormalizability versus unitarity, *Int. J. Mod. Phys.*, vol. A3 (1988), pp. 435–449.

Feynman, R. P., Relativistic cut-off for classical electrodynamics, *Phys. Rev.*, vol. 74 (1948), pp. 939–946.

Feynman, R. P., Theory of positrons, *Phys. Rev.*, vol. 76 (1949), pp. 749–759.

Green, M. B., Schwarz, J. H. and Witten, E., *Superstring Theory*, Vol. 1. Cambridge: Cambridge Univ. Press, 1986.

Hurth, T., Higgs-free massive nonabelian gauge theories, *Helv. Phys. Acta*, vol. 70 (1997), pp. 406–416.

Kunimasa, T. and Goto, T., Generalization of the Stueckelberg formalism to the massive Yang–Mills field, *Prog. Theor. Phys.*, vol. 37 (1967), pp. 452–464.

Luo, J., Tu, L.-Ch., Hu, Z.-K. and Luan, E.-J., New experimental limit on the photon rest mass with a rotating torsion balance, *Phys. Rev. Lett.*, vol. 90 (2003), pp. 081801.1–081801.4

Matthews, P. T., The S-matrix for meson–nucleon interactions, *Phys. Rev.*, vol. 76 (1949), pp. 1254–1255.

Oppenheimer, J. R. and Serber, R., Note on the nature of cosmic ray particles, *Phys. Rev.*, vol. 51 (1937), p. 1113

Pais, A., *Inward Bound. Of Matter and Forces in the Physical World*. Oxford: Clarendon Press, 1986.

Pauli, W., Relativistic field theories of elementary particles, *Rev. Mod. Phys.*, vol. 13 (1941), pp. 203–232.

Proca, A., Sur la théorie ondulatoire des électrons positifs et négatifs, *J. Phys. Radium*, vol. 7 (1936), pp. 347–353

Ramond, P., A pedestrian approach to covariant string theory, *Prog. Theor. Phys. Suppl.*, vol. 86 (1986), pp. 126–134

Ruegg, H. and Ruiz-Altaba, M., The Stueckelberg field, *Int. J. Mod. Phys. A*, vol. 19 (2004), pp. 3265–3347.

Salam, A. Renormalizability of gauge theories, *Phys. Rev.*, vol. 127 (1962), pp. 331–334.

Slavnov, A. A., Massive gauge fields, *Teor. Mat. Fiz.*, vol. 10 (1972), pp. 305–328, english translation in *Theor. Math. Phys.*, vol. 10 (1972), pp. 201–217.

Umezawa, H and Kamefuchi, S, Equivalence theorems and renormalization problem in vector field theory (theYang–Mills field with non-vanishing masses), *Nucl. Phys.*, vol. 23 (1961), pp. 399–429.

CHAPTER 7

Stueckelberg and the S-Matrix Theory

Gérard Wanders

This article describes Stueckelberg's original road to a causal and renormalized S-matrix, from his taking up Heisenberg's idea of the S-matrix in 1944 to the final version of his formalism in 1950. For clarity's sake, the account does not strictly follow Stueckelberg's arguments in all points. This is corrected in an Appendix where three examples of his original work are presented and anlalyzed in detail.

7.1. Introduction

It was generally agreed in the early 1940s that elementary particles were quanta of relativistic quantized fields and that the theory of these particles had to be a relativistic quantum field theory. This theory encountered fundamental difficulties with the description of interacting particles. The main problem came from the occurrence of diverging integrals which no one knew how to get rid off. A radically new approach appeared in 1943 with Heisenberg's idea of the S-matrix (Heisenberg [1943a,b]). He started from the observation that, apart from the spectrum of their bound states, the properties of elementary particles were exclusively displayed through their behaviour in collision processes. These processes are characterized by transition probabilities which Heisenberg postulated to be the only observables of the subatomic world. The task of the theoretician was thus reduced to the prediction of these probabilities. Heisenberg collected the quantum probability amplitudes of all possible collisions in a unitary operator he called the S-matrix. He was looking for algorithms which would allow a direct computation of the elements of the S-matrix without getting lost in the maze of relativistic quantum field theory.

After many efforts, perturbative versions of this programme were achieved in the 1950s, in particular by R. P. Feynman with his famous formalism for quantum electrodynamics (Feynman [1949a,b]). Stueckelberg was also in search of new methods and Heisenberg's ideas appealed to him at once. He worked out his own approach and got the same final results as Feynman. This paper presents an outline of his undertaking.

It may first be worthwhile to mention a particular Feynman–Stueckelberg connexion. Feynman derived his electron propagator in 1949 (Feynman [1949b]) by using the idea that positrons can be considered as electrons moving backwards in time. This idea was already proposed by Stueckelberg in 1941 [**54**, **56**], years before the emergence of the S-matrix (see the contribution of Henri Ruegg and Marti Ruiz-Altaba to this volume).

7.2. Causality and the S-Matrix

The S-matrix is a unitary operator S which connects state vectors Ψ_{in} and Ψ_{out} representing collections of particles before and after their collisions:[1]

$$\Psi_{out} = S\Psi_{in}\,. \tag{7.1}$$

According to Heisenberg, the S-matrix had to be conceived as an entirely new framework which would enable a totally new approach. It should eliminate all the difficulties encountered in quantum field theory at that time.

The S-matrix being unitary, Heisenberg wrote:

$$S = \exp(i\eta) \tag{7.2}$$

where η is a hermitian Lorentz invariant operator. This raised the problem of the possible requirements this η had to fulfil (besides relativistic invariance and hermiticity) in order to provide a physically acceptable S-matrix. In spite of the complex nature of the true phase operator η, Heisenberg attempted to get some insight into the problem by constructing S-matrices produced by simple Ansatzes for η (Heisenberg [1943b]). Stueckelberg soon joined in this game. The note he published in *Nature* in January 1944 [**67**] deserves a close look. We can easily get its main point by rephrasing its argument as follows.

[1] Equation (7.1) is written in the interaction representation which Stueckelberg was one of the first to use. The field operators behave as free fields and the interactions produce an evolution of the state vectors.

Consider two types of massive spin zero neutral particles which are the quanta of two scalar fields φ and u and use the following Ansatz for η:[2]

$$\eta = \varepsilon \int dx^4 \varphi(x)\big[u(x)\big]^2 . \tag{7.3}$$

The product of field operators in the integrand is Wick-ordered: all creators are on the left of all annihilators. The factor ε measures the strength of the interaction, which is meant to be weak.

Integrals such as in equation (7.3) are first extended to a finite domain of space-time. The limit of infinite space-time is taken after the computation of the integrals. Stueckelberg investigated this procedure much later in great detail.[3]

The first-order term of the S-matrix is $(1+i\eta)$. In second order it contains the term

$$S_2 = \epsilon^2 \int dx^4 \int dy^4 D^1_\varphi(x-y)\big[u(x)u(y)\big]^2 \tag{7.4}$$

where D^1_φ is a known invariant real symmetrical solution of the homogeneous Klein–Gordon equation (m_φ = mass of the φ-particle, in suitable units):

$$\left[\left(\frac{\partial}{\partial x^0}\right)^2 - \sum_{k=1}^{3}\left(\frac{\partial}{\partial x^k}\right)^2 + m_\varphi^2\right] D^1_\varphi(x-y) = 0 . \tag{7.5}$$

This function describes correlations between the space-time points x and y produced by the field φ. We may ask if this S_2 is physically acceptable. One of the processes it describes is the creation of a pair of outgoing u-quanta at the space-time point x correlated by the function $D^1_\varphi(x-y)$ with the annihilation of a pair of incoming u-quanta at y. The probability amplitude of this process gets contributions from the integral in equation (7.4) not only if x is in the future of y but also if x is in its past. There is a non-vanishing probability for the outgoing pair to be produced before the occurrence of its cause, the annihilation of the incoming pair. Obviously, this contradicts causality, which requires that a cause always occurs before its effects. Stueckelberg would thus conclude that the form (7.3) of η has to be rejected, being incompatible with causality.

If we consider the energy-momentum aspects of the annihilation-creation process, we see that the annihilation of the incoming pair releases positive

---

[2] Notation: $x = (x^0, x^1, x^2, x^3)$, (x^1, x^2, x^3) = cartesian space coordinates, $x^0 = ct$, t = time, c = velocity of light, $(dx)^4 = dx^0 dx^1 dx^2 dx^3$.

[3] The infinite space-time limit is delicate. Boundary effects can lead to divergences which have to be avoided. Stueckelberg resolved these difficulties in [**84**] and with T. A. Green in [**85**]. This has been acknowledged by W. Pauli (von Meyenn, part II [1996], p. 925 and N. N. Bogoliubov and D. V. Shirkov (Bogoliubov and Shirkov [1959]).

energy which is absorbed in the creation of the outgoing pair.[4] The correlation function controls the propagation of this positive energy: it propagates into the future if the process is causal and into the past if it is acausal.

If we put aside the phase operator η and identify the integrand in the right-hand side of equation (7.3) with an interaction hamiltonian density, we can compute the resulting S-matrix by solving perturbatively a Schrödinger equation. It produces a new second-order term S_2 which results from the previous one by replacing $D^1_\varphi(x-y)$ in equation (7.4) by a solution $D^c_\varphi(x-y)$ of a singular inhomogneneous version of equation (7.5):

$$\left[\left(\frac{\partial}{\partial x^0}\right)^2 - \sum_{k=1}^{3}\left(\frac{\partial}{\partial x^k}\right)^2 + m_\varphi^2\right] D^c_\varphi(x-y) = -\delta^4(x-y)\,, \qquad (7.6)$$

where $\delta^4(x-y)$ is the four-dimensional delta distribution.[5]

This new function has the fundamental property of transferring positive energy from y to x only if x is in the future of y. If x is in the past of y, the transferred energy is negative. As a result, the new S_2 is fully consistent with causality as it should be, being produced by a Schroedinger equation. The function D^c_φ is in fact Stueckelberg's causal propagator.

Stueckelberg was quite aware, as early as 1944, that causality cannot be ignored in the construction of the S-matrix. The features of the space-time evolution in collision processes must be taken into account. This contradicted strongly Heisenberg's view that the S-matrix had to be characterized exclusively by means of observables. In fact Heisenberg considered space-time evolution at the subatomic level as fundamentally unobservable.

Although it was obvious that macroscopic causality, involving macroscopic time intervals, was required, it was not absolutely certain that microscopic causality was also at work. Dirac [1938] developed a classical electron theory where premonitions are produced by the interaction with the radiation field. Analogous phenomena were suspected at quantum level. Heisenberg accepted microscopic acausalities and hoped that a fundamental length would show up in the S-matrix and would prevent the occurrence of divergences. Note that the above acausalities are microscopic ones: the premature appearance of the outgoing pair can only occur slightly before the disappearance of the incoming one. It would be easy to modify η in equation (7.3) in such a way that macroscopic acausalities would also show up. The situation was still

---

4 When dealing with energy-momentum conservation in localized processes, one has to be careful and avoid statements which contradict the uncertainty relations. This point has been discussed by Fierz in Fierz [1950].

5 $\delta^4(x)$ = four-dimensional delta distribution with $\int (dx)^4 \delta^{(4)}(x) f(x) = f(0)$ for an arbitrary test function f.

unclear to Stueckelberg in 1946, when he attended an *International Conference on Fundamental Particles and Low Temperature*, held at the Cavendish Laboratory. He concluded his report in the following way [79]:

> The condition of unitarity and relativistic invariance is by no means sufficient to determine S. The theory is too general and it contains numerous effects contradicting macroscopical causality. We therefore conclude that additional conditions must be looked for limiting the choice of S or η. They must be of a form so as to admit acausality (= premonition) only for microscopical time intervals ($\approx \tau_0$).

While he knew that causality constrains the phase operator η, Stueckelberg thought that other conditions, such as correspondence principles, should be applied. This led him to massive constructions like "functional mechanics" or "asymptotic mechanics" [71, 72, 73]. From a short paper he delivered at a meeting of the Swiss Physical Society in 1946 [**78**], we may guess that he came very close to the discovery that microcausality alone, combined with relativistic invariance and unitarity, determines the S-matrix in perturbation theory. It turns out that the causal propagator D^c is the only correlation function ensuring consistency with causality. It is not mentioned in this paper [**78**], but it is explicitly used in an unpublished manuscript (1947) concerned with various aspects of macrocausality. The propagator D^c is explicitly mentioned for the first time in 1948, in an announcement of the results of D. Rivier's thesis (Rivier [1949]), in a *Letter to Physical Review* [**81**].[6]

Pauli and Fierz are among the few people who were aware of Stueckelberg's concern with causality. The following episode may be found in Pauli's correspondence (von Meyenn, part I [1996], pp. 101, 103). In 1950, Heisenberg presented a new Ansatz for the S-matrix (Heisenberg [1950]). Pauli was very sceptical about it and suspected it violates Stueckelberg's causality condition. He asked Fierz to check this and it came out that macroscopic causality was indeed violated. Heisenberg's Ansatz could thus be rejected.

---

6 While D. Rivier completed his thesis (Rivier [1949]), Pauli was working with Villars on the elimination of divergences by regularization (Pauli and Villars [1949]). Pauli took interest in Rivier's work and found that it needed some clarifications. This led to correspondence between them in summer 1948 ([von Meyenn, 1993, pp. 543, 548, 551, 567], pp. 543, 548, 551, 567.)

7.3. Perturbative Construction of the S-Matrix

In 1949, Stueckelberg and D. Rivier presented their final programme for the perturbative construction of the S-matrix [**82**]:

$$S = 1 + \sum_{n=1}^{\infty} S_n \,. \tag{7.7}$$

This programme is an iterative procedure, the first steps are:

1. Select an anti-hermitian S_1 specifying the interaction.
2. The S_n have to satisfy the perturbative unitarity conditions:

$$\begin{aligned} S_1 + S_1^\dagger &= 0\,, \quad S_2 + S_2^\dagger = -S_1 S_1^\dagger\,, \\ S_3 + S_3^\dagger &= -(S_1 S_2^\dagger + S_2 S_1^\dagger)\,, \\ &\dots \end{aligned} \tag{7.8}$$

   The second one determines the hermitian part of S_2 ($S^\dagger$ = adjoint of S). This hermitian part is not causal.

3. Add an antihermitian causal correction to the hermitian part of S_2 such that the resulting S_2 allows only causal processes.

The causal correction is constructed in such a way that S_2 contains exclusively causal propagators D^c as correlation functions. This last point is due to the fact that D^c is the only invariant solution of equation (7.6) which has the right frequency spectrum.

The use of the unitarity conditions (7.8) and the construction of antihermitian causal corrections determine successively all terms of the expansion (7.7). The invention of causal corrections is one of the major breakthroughs in Stueckelberg's work. We can say, without exaggeration, that [**82**] is an extraordinary paper. Within a few pages and without detailed explanations, it provides the complete recipe for the covariant construction of the causal S-matrix produced by an elementary interaction.

The probability amplitude of a collision process is given by sums of matrix elements of the approximations S_n. As a result of the construction of the S_n via unitarity and causal corrections, these matrix elements are multiple space-time integrals involving networks of causal propagators D^c, interaction vertices and wave packets of the incoming and outgoing particles. The integrands can be represented by means of the famous Feynman diagrams (Feynman [1949a]). Feynman's propagator D_F and Stueckelberg's causal propagator D^c are identical.

7.4. Renormalization and Renormalization Group

Once the formal construction of the S_n has been achieved, divergences are still present in the causal corrections, the antihermitian parts of the S_n. Apparently, Heisenberg's expectation did not come true. Stueckelberg and Rivier devised a method to eliminate these divergences within their S-matrix programme in Rivier [1949]. An improved technique was presented in 1949 at the *Basel–Como Konferenz über Kernphysik und Quantenelektrodynamik* in Basel, it is described in [**83**].

The most simple divergences are due to the occurrence of squares like $[D^c(x-y)]^2$ which are singular at $x = y$. From a mathematical point of view, the propagator D^c is a distribution and, in general, a product of distributions is not uniquely defined. If D^c is the propagator of a scalar field, the correctly defined $[D^c(x-y)]^2$ contains an arbitrary additive term and has the form

$$\left[D^c(x-y)\right]^2 = F(x-y) + c \cdot \delta^{(4)}(x-y)\,, \tag{7.9}$$

where $F(x-y)$ is a known distribution, $\delta^{(4)}(x-y)$ is again the four-dimensional delta function and the coefficient c is real, finite and arbitrary. With all singular products well defined, the S_n are given by sums of convergent integrals. They are finite, but contain arbitrary factors, like c in equation (7.9). These factors are related to the strengths of coinciding events, ($x = y$ in (7.9)). Clearly, causality has nothing to say about such events. This means that causality, combined with unitarity, does not determine uniquely the S-matrix generated by an elementary interaction.

This proper way of eliminating the divergences was worked out by Stueckelberg with T. A. Green [**85**] and with A. Petermann [**86**]. The general outcome is that the S_n are uniquely determined by unitarity and causality, up to finite linear combinations of delta functions and derivatives of delta functions. The constants in these combinations have to be determined by imposing normalization conditions to ensure, for instance, that the electron has its observed mass and charge. With suitable mathematical treatment and use of microscopic causality, divergences can be traded for finite arbitrary constants! Although Stueckelberg's renormalization scheme is mathematically clean and elegant it did not get much attention and is not mentioned in the textbooks on quantum field theory.

At this point it may be worthwhile to come back to Heisenberg's objectives when he devised his S-matrix. He aimed for a formalism liberated from Schroedinger-type evolution equations and he thought that divergences could be avoided in this way. If we look at Stueckelberg's formalism, we see that evolution equations have indeed been replaced by the specification of correlation functions. Whereas differential evolution equations would determine the state of the evolution everywhere in space and time, Stueckelberg's causality condition does not concern coinciding space-time events. As a result, the S-

matrix can be constructed in such a way that divergences never actually show up. Heisenberg's wishes have finally been realized in an unexpected way, without fundamental length!

One year after the Basel Conference, Stueckelberg and T. A. Green produced examples of the construction of causal corrections in second-, third- and fourth-order quantum electrodynamics [**85**]. To my knowledge, there is no explicit proof of the equivalence to all orders of Feynman's and Stueckelberg's formalisms.

Between 1951 and 1953, Stueckelberg worked with A. Petermann on renormalization, i.e. the determination of the arbitrary constants in the S_n. This procedure is based on prescriptions involving normalization conditions which fix the values of these constants. The prescriptions are not unique, but the final result should be independent of their choice. This led Stueckelberg and Petermann to the discovery of the famous renormalization group [**86**, **87**, 91]. This topic will not be discussed here.

7.5. Acausality of Non-Local Interactions

One of the last contributions of Stueckelberg to the S-matrix theory concerns non-local interactions. They are obtained by replacing local interactions as in equation (7.3) by non-local ones including real Lorentz invariant form factors (Kristensen and Möller [1952]). The main advantage is that divergences do not appear at all. Microscopic acausalities are unavoidable, but a suitable choice of the form factor could prevent macroscopic ones. This can be achieved in second-order approximation, but Stueckelberg showed with G. Wanders that macroscopic acausalities are unavoidable in higher orders [92], Wanders [1956]. Consequently, interactions have to be local, at least in perturbation theory.

7.6. Concluding Remarks

The main limitation of Stueckelberg's approach is that it is perturbative by nature. While it is appropriate in quantum electrodynamics, it does not apply to strong interactions. Feynman's and Stueckelberg's formalisms produce nevertheless paradigms which are helpful even when the perturbative treatment is not valid. For instance, the concept of a virtual particle is a product of the perturbation expansion, but we think in terms of virtual particles even if this expansion does not work.

Efficient non-perturbative approaches started to appear in quantum field theory around 1955, with significant contributions from N. N. Bogoliubov and his collaborators (Bogoliubov and Shirkov [1959]) and from H. Lehmann, K. Symanzik and W. Zimmermann (Lehmann, Symanzik and Zimmermann

[1955]). Practical outcomes include analyticity properties of scattering amplitudes leading to the existence of dispersion relations. In a way, these properties are consequences of causality. These developments were not followed by Stueckelberg, who lost active interest in theoretical particle physics.

Stueckelberg took part in the adventure of the S-matrix theory as an outsider, in his own distinctive way. Unfortunately, we know almost nothing about the course of investigations which led him from the first recognition, in 1944, of the role of causality [**67**] to the 1949 causal S-matrix [**82, 83**]. Apart from an unpublished manuscript of 1947, the only available documents are his published articles. The relevant ones are very short and do not indicate how the results were obtained. The two papers [**82**, **83**] which describe the whole construction of the S-matrix are short eight-page and four-page notes (in French) published in *Helvetica Physica Acta.* This brevity was an achievement in a sense, but it did not help to disseminate Stueckelberg's results. In general, Stueckelberg's style, language and choice of journals did not favour the propagation of his ideas. He was quite isolated and his work progressed in parallel with those of celebrities like R. P. Feynman, J. Schwinger and S. Tomonaga. These circumstances explain why Stueckelberg's contributions are so little known and may seem marginal. It is, in fact, unquestionable that some of his findings have a deep and lasting importance, such as his showing that collision processes cannot be understood if their causal time evolution is ignored.

7.7. Appendix: Three S-Matrix Papers

I discuss here in a more detailed way the three articles [**67**, **82** and [**83**].

7.7.1. An Unambiguous Method of Avoiding Divergence Difficulties in Quantum Theory [67]

The paper's initial purpose is a connexion between the usual hamiltonian formalism and Heisenberg's S-matrix theory. Unexpectedly, causality shows up for the first time in Stueckelberg's work.

Two scalar fields φ and u interact via the hamiltonian H given in equation (3). The state vector $\Psi(t)$ is written in the interaction representation by means of an operator $\alpha(t)$ according to equation (5). The phase η of the S-matrix [$S = \exp(-i\eta)$] is proportional to the $t \to \infty$ limit of $\alpha(t)$.

Equations (8) provide the first terms of the perturbative expansion of $\alpha(t)$. The explicit form of the integral in the second equation (8) is

$$\int_{-T}^{t} dx^0 \int (dx)^3 \int_{-T}^{t} dy^0 \int (dy)^3 \big(u(x)\big)^2 D_\varphi^{\text{ret}}(x-y)\big(u(y)\big)^2 ,$$

where $x = (x, x^0)$ and D_φ^{ret} is the retarded correlation function of the field φ.

The product of field operators in H being well-ordered (all creators on the left of all annihilators) the same is true for the factors $(u)^2$ and $(u\varphi)$ in equations (8). Their integrands are therefore partially well-ordered. Inspired by Heisenberg (Heisenberg [1943a]), Stueckelberg redefines drastically the operator $\alpha(t)$ by replacing all products of field operators with totally ordered ones. This eliminates all divergences at one stroke and explains the title of the paper. It goes without saying that this wild proposal was unsuccessful. In our actual context the interest of the paper comes from the fact that its amazing treatment of divergences leads to a first explicit mention of the causality concept.

In the last part of the paper, Stueckelberg examines modified S-matrices obtained by altering the phase η. Each term in the expansion of $\alpha(t)$ defines its own unitary S-matrix. If one keeps only $(\epsilon^2\alpha^{(2)}R)$ the ε^2-term of the resulting S-matrix is given by expression (1). Stueckelberg recognizes that this expression implies acausalities, an outgoing pair can be created at y before the annihilation of an ingoing pair at x.

The last four sentences in the second section of page 144 are quite enigmatic and possibly afflicted by misprints. In spite of that they establish that Stueckelberg was aware, already in 1944, that causality plays a role in the construction of the S-matrix, although it was not clear at the time whether acausalities may occur at short distances in nuclear forces.

The paper closes with statements on the results of line width calculations.

7.7.2. Causalité et Structure de la Matrice S [82]

The first sections display three representations of the S-matrix describing an evolution between two times τ' and τ''. After a reminder of the classical dynamics of charged particles, Section 1 deals with the quantum transition amplitudes relating particle detections at times τ' and τ'', localized at points $\zeta'$and ζ''. Equation (1.4) gives the general form of the contribution to these amplitudes produced by the action of a local interaction Γ in infinitesimal space-time domains $dx^{(k)}$ ($dx^{(k)}$ stands for $dx^0_{(k)}dx^1_{(k)}dx^2_{(k)}dx^3_{(k)}$). The positive frequency correlation functions D^+ propagate the incoming and outgoing particles and a network of causal propagators D^c connects the localized interactions. The form (1.5) of D^c is dictated by the causality requirement that $D^c(x/y)$ has to be proportional to $D^+(x/y)$ if $x^0 > y^0$ (D^c has to propagate positive energy into the future).

The infinitesimal dS in equation (1.4) is the basic germ of the S-matrix in the Stueckelberg–Rivier approach. The originality of this approach lies in the fact that its germ is postulated without any reference to a hamiltonian formalism and its differential equations. Notice that the quantum version of Figure 1 could be the sketch of a Feynman diagram.

Section 2 deals with particles localized in wave packets. The general term of the perturbative expansion of their transition amplitude is displayed in equation (2.9). The S-matrix is presented in Section 3 as a unitary operator acting on a free particle Hilbert space (equation (3.4)). It involves products of operators a annihilating incoming particles and their conjugates $a^\dagger$ creating outgoing ones.

The first three sections characterize the structure of the perturbative expansion of the S-matrix, without specifying the way it is effectively constructed. This is done in Section 4, the decisive section of the paper. The Stueckelberg–Rivier integral method is outlined after some statements about the differential hamiltonian formalism. Following Heisenberg (Heisenberg [1943a]), the hermitian phase operator α of the S-matrix is introduced in equation (4.3). Its perturbative expansion is based on an elementary local interaction that is specified by the first order term $\alpha_{(1)}$. The square of this term produces a second order contribution to the S-matrix (equation (4.5) which has not the causal form required in equation (1.4). Causality is restored by a suitable choice of the second-order term $\alpha_{(2)}$ of the phase operator. It plays the role of a causal correction of $\alpha_{(1)}$.

In its turn, the third-order contribution of $(\varepsilon\alpha_{(1)} + \varepsilon^2\alpha_{(2)})$ produces also acausalities. They are removed by a third-order causal correction $\alpha_{(3)}$. Iterating this procedure the higher order's $\alpha_{(n)}(n \geq 4)$ are determined successively as causal corrections and the resulting perturbative expansion of the S-matrix has the causal form required by equation (1.4).

It is claimed without explanation in the last paragraph of Section 4 that the Stueckelberg–Rivier approach avoids the divergent integrals of the hamiltonian treatment and replaces them by arbitrary finite constants.

Section 4 provides a schematic recipe for the perturbative construction of a causal S-matrix. It does not go into any details. In particular, nothing is said on the explicit determination of the causal corrections and on the normalization of the arbitrary constants. This is partly done in D. Rivier's thesis (Rivier [1949]) and in [**83**, **85**].

7.7.3. A Propos des Divergences en Théorie des Champs Quantifiés [83]

In September 1949, Stueckelberg and Rivier attended a Conference in Basel (*Basel–Como Konferenz über Kernphysik und Quantenelektrodynamik*) where they presented a new method for the elimination of divergences in S-matrix calculations. Their paper appeared in the Proceedings of the Conference, in *Helvetica Physica Acta*.

The account of the method is restricted to a scalar second-order one-loop self-energy kernel $\Delta(x/y)$.[7] Relativistic invariance requires that Δ is a covariant function of $(x - y)$. It has to be integrable because it is used in integrals, as in equation (3). Second-order perturbation theory tells us that Δ is proportional to the square $(D^c)^2$ of a causal propagator D^c. Whereas $D^c(x - y)$ is a distribution with an integrable singularity at $x = y$, its square is not integrable and produces divergences. Consequently Δ has to be redefined in a suitable way. To achieve this, Stueckelberg and Rivier use a function defined as the product of Δ with a monomial θ in the components of $(x - y)$, as given in equation (2). This monomial is chosen in such a way that $\Delta\theta$ is integrable. Once this has been done, an integrable Δ is constructed by inverting skilfully the multiplication by θ. This is best performed in terms of Fourier transforms.

The Fourier transform of Δ is a function of an external momentum k formally defined by means of a (divergent) integral over an internal momentum p:

$$\Delta(k) = \int dp D^c(p) D^c(k - p)$$

where $D^c(k)$ stands for the Fourier transform of the causal propagator. Multiplication by the monomial θ corresponds in Fourier space to a multiple derivative $\theta(\partial/\partial k)$ with respect to the components of k. $\theta\Delta(k)$ is defined as

$$\theta\Delta(k) = \int dp D^c(p)\theta\left(\frac{\partial}{\partial k}\right) D^c(k - p)\,.$$

Remember that the propagator $D^c(k)$ behaves as $1/k^2$ for large k^2. The order of the derivatives in θ has to ensure a decrease in p of $\theta(\partial/\partial k)D^c(k-p)$ ensuring the convergence of the right-hand side integral. This leads to a well-defined Fourier transform of $\theta\Delta$. The transform of Δ is obtained by means of the inverse of $\theta(\partial/\partial k)$, i.e. a multiple integral on the components of k. This integral being indefinite, its result contains, besides a well-behaved function $\Delta(k)$, an arbitrary polynomial $P(k)$ annihilated by $\theta(\partial/\partial k)$.

Clearly, the whole procedure has to preserve covariance and in the scalar case $P(k)$ is a polynomial in k^2. The order of the derivatives in $\theta(\partial/\partial k)$ is chosen in such a way that the degree of $P(k)$ is minimal. The final result in Fourier space has the form of equation (10).

In x-space the arbitrariness gives rise to a combination of local terms, equation (11), whose coefficients are fixed by normalization conditions.

---

7 Notation: x, y = four-vectors specifying space-time points, components x^α, $\alpha = 0, 1, 2, 3$, dx stands for $dx^0 dx^1 dx^2 dx^3$.

k, p = energy-momentum four-vectors, k^2 = abreviation for $(k, k) = (k^1)^2 + (k^2)^2 + (k^3)^2 - (k^0)^2$. The kernels D^c, D^1 and D^s are displayed in [Rivier, 1949].

In their article, Stueckelberg and Rivier do not discuss the square $(D^c)^2$ but its non-integrable component Δ^s produced by the second-order causal complement defined in [**82**]. Their treatment is explicitly covariant at each step: the expression (6a) of $\Delta^s(k)$ involves only k^2 and p^2 and θ is a derivative with respect to k^2.

The paper closes with the statement of results obtained in second- and third-order quantum electrodynamics. An expression for the electron anomalous magnetic moment established by J. Schwinger is reproduced.

The Stueckelberg–Rivier method has been applied to quantum electrodynamics up to fourth-order approximation by Stueckelberg and Green [**85**]. It was formulated by Stueckelberg and Petermann with the full machinery of the theory of distributions in [91].

The methods used around 1950 for the elimination of divergences were based on regularization techniques, (see for instance Pauli and Villars [1949]). In contrast, Stueckelberg and Rivier reach their goal directly, without the need of any regularization. This is quite remarkable.

References

Bogoliubov, N. N. and Shirkov, D. V., *Introduction to the Theory of Quantized Fields*. New York: Interscience Publishers, 1959.

Dirac, P. A. M., Classical theory of radiating electrons, *Proc. Roy. Soc.*, vol. 157 (1938), pp. 148–169.

Feynman, R. P., Space-time approach to quantum electrodynamics, *Phys. Rev.*, vol. 76 (1949), pp. 769–789.

Feynman, R. P., The theory of positrons, *Phys. Rev.*, vol. 76 (1949), pp. 749–759.

Fierz, M., Über die Bedeutung der Funktion D in der Quantentheorie der Wellenfelder, *Helv. Phys. Acta*, vol. 23 (1950), pp. 731–739.

Heisenberg, W., Die "beobachtbaren Grössen" in der Theorie der Elementarteilchen, *ZS. f. Phys.*, vol. 120 (1943), pp. 513–538.

Heisenberg, W., Die beobachtbaren Grössen in der Theorie der Elementarteilchen. II, *ZS. f. Phys.*, vol. 120 (1943), pp. 673–702.

Heisenberg, W., Zur Quantentheorie der Elementarteilchen, *Z. Naturforschung*, vol. 5a (1950), pp. 251–259.

Kristensen, P. and Möller, C., On a convergent meson theory. I, *Kgl. Danske Vid. Selsk.*, vol. 27 (1952), p. 51.

Lehmann, H., Symanzik, K. and Zimmermann, W., Zur Formulierung quantisierter Feldtheorien, *Nuovo Cimento*, vol. 1 (1955), pp. 205–224.

Meyenn, K. von (ed), *Wolfgang Pauli, Scientific Correspondence with Bohr, Einstein, Heisenberg, a.o.*, Volume III: 1940–1949. Berlin: Springer, 1993.

Meyenn, K. von, *Wolfgang Pauli, Scientific Correspondence with Bohr, Einstein, Heisenberg, a.o.*, Volume IV, Part I: 1950–1952. Berlin: Springer, 1996.

Meyenn, K. von, *Wolfgang Pauli, Scientific Correspondence with Bohr, Einstein, Heisenberg, a.o.*, Volume IV, Part II: 1953–1954. Berlin: Springer, 1996.

Pauli, W. and Villars, F., On the invariant regularization in relativistic quantum theory, *Rev. Mod. Phys.*, vol. 21 (1949), pp. 434–444.

Rivier, D., Une méthode d'élimination des infinités en théorie des champs quantifiés. Applications au moment magnétique du neutron, *Helv. Phys. Acta*, vol. 22 (1949), pp. 265–318.

Wanders, G., Kausale Formulierung der S-Matrixtheorie, Fortschritte der Physik, vol 4 (1956), pp. 611–629.

CHAPTER 8

Relativistic Thermodynamics

Werner Israel

8.1. Introduction

It is an honour to be participating in these centenary celebrations for E. C.G Stueckelberg, one of the authentic geniuses of 20th century physics, whose visionary contributions to quantum electrodynamics, S-matrix theory and the renormalization group are still insufficiently appreciated.[1]

In the early 1960s, perhaps motivated in part by an abiding fascination with the mystery of the arrow of time, Stueckelberg turned his attention to a purely classical topic: relativistic thermodynamics. Like his pioneering work on quantum field theory, this bore all of Stueckelbergs hallmarks: a wholly new and original approach, a broad line of attack with general principles like covariance placed firmly on the foreground, and, of course, an unconventional and inimitable notation. For easier comparison with current work and to facilitate access to the underlying ideas, I shall take the liberty of transcribing his notation into that now commonly in use.

[1] Talk given at the Symposium "E. C. G. Stueckelberg (1905–1984), Symposium for the Centenary of his Birth", Geneva University, 2nd–3rd December 2005.

To put Stueckelberg's work in context, I shall begin with a brief sketch of the chequered development of relativistic thermodynamics from its early, pre-Minkowskian beginnings to the 1970s. This will clear the way for a concise description of the substantially different Stueckelberg–Wanders formulation [**89**]. Finally, I will sketch how their approach and methods were partially rediscovered "unbekannterweise" in the last decade or two, and incorporated into mathematically sophisticated extensions of the conventional theory.

8.2. Historical Remarks

The advent of special relativity forced reassessment of several branches of physics in light of the new requirement of Lorentz invariance.

Within a year of Einstein's 1905 paper, Planck and his young student, Kurt von Mosengeil were examining how thermodynamical quantities like temperature are affected by Lorentz transformations.[2] In themselves, such studies could be hardly more than formal exercises, since they were confined to thermal equilibrium situations, for which a privileged frame, the rest frame of the system, always exists. It was a further misfortune that they fell into the three-year gap before Minkowski reformulated Einstein's theory of spacetime, and they were not recast in four-dimensional language until many years later (van Dantzig [1939]) (Mosengeil died in 1906 and Planck never returned to the subject).

Consequently, it was in their half-developed, non-covariant form that the Mosengeil–Planck results were reported in Pauli's famous 1921 *Enzyklopädie* article (Pauli [1921]). Decades later, this was to lead to a long drawn-out debate over the question whether a moving body appears cool (see e.g., several contributions in Stuart, Gal-Or and Brainard [1970]). Since this question has an operational meaning only insofar as it concerns the thermal *interaction* of two bodies in relative motion, and none of the participants explicitly considered it an interaction problem, much of this debate was a futile exercise in formalism.

The matter was definitively settled in 1968 by van Kampen [1968] who derived a covariant expression of the second law for a system of arbitrarily moving and weakly interacting bodies, each close to its own internal equilibrium. The total change of entropy in an infinitesimal interaction is given by

$$0 \leq \delta S = -\sum_A (\alpha_A \delta N_A + \beta_{A\lambda} \delta p_A^\lambda) \,, \qquad (8.1)$$

[2] See Pauli [1921]: This work includes references to all of the early work on the subject.

where δp^λ_A and δN_A are the 4-momentum and number of particles acquired by component A in the interaction; $\alpha_A = \mu_A/T_A$, $\beta^\lambda_A = \mu^\lambda_A/T_A$ and μ^λ_A, T_A, are its 4-velocity, rest-frame temperature and its chemical potential (or injection energy per particle, including rest-mass energy).

Since (8.1) is a vectorial relation, involving transfer of momentum as well as energy, it cannot be expressed purely in terms of the relative *speed* of the bodies and the temperature measured in their rest-frames. Whether a moving body appears cool depends on the circumstances of the experiment.

Returning to the early years, in 1911 Jüttner obtained the relativistic form of Maxwell's distribution function for a classical gas, and extended this to quantum gases in 1928 (Jüttner [1928]). The foundations of a relativistic kinetic theory were laid down in the 1930s by Synge, Walker and Lichnerowicz (Lichnerowicz and Marrot [1940], Synge [1934], Walker [1936]).

Meanwhile, equilibrium thermodynamics had been studied in a general-relativistic context by Ehrenfest and Tolman and by Oskar Klein (Klein [1949], Tolman and Ehrenfest [1930]). They determined how equilibrium temperature and chemical potential vary with depth in a gravitational field.

Nonequilibrium relativistic thermodynamics was first treated in a groundbreaking 1940 paper by Carl Eckart [1940] for the case of a simple fluid. His work was extended a decade later by Sybren de Groot and his Amsterdam group to include mixtures, chemical reactions and polarized fluids (Kluitenberg and de Groot [1955]).

A formulation superficially different from Eckart's was given independently in 1959 by Landau and Lifshitz in the *Fluid Mechanics* volume of their famous *Course of Theoretical Physics*. This difference in formal appearance stems from different choices for the definition of hydrodynamical 4-velocity in terms of mean particle flux in the former, mean energy flux in the latter case. However, it can be shown that, to the first order in the deviations from equilibrium, the physics is independent of this choice.

In the 1959 survey of the thermodynamics of irreversible processes that Meixner and Reik contributed to the *Handbuch der Physik*, the relativistic work of Eckart and the Amsterdam group is duly referenced and outlined. Had it included also a reference to the papers of Stueckelberg and his collaborators which had begun to appear in *Helvetica Physica Acta* in 1953 (see [**89**]), it might have helped to save them from complete obscurity. Regrettably, they were overlooked.

8.3. Covariant Thermodynamics: Conventional Formulation

Let us proceed now to a brief description of relativistic thermodynamics as conventionally presented nowadays (see e.g., articles in Anile and Choquet-Bruhat [1989]).

For an N-component fluid one associates with an arbitrary non-equilibrium state a set of so-called "primary variables": N 4-currents J^{μ}_{A} describing the flows of electric charge, baryon-number, and other conserved charges; a symmetric conserved stress-energy $T^{\lambda\mu}$ and an entropy flux S^{μ} with non-negative divergence, expressing positivity of entropy production:

$$\nabla_{\mu} J^{\mu}_{A} = \nabla_{\mu} T^{\lambda\mu} = 0, \nabla_{\mu} S^{\mu} \geqslant 0 \,. \tag{8.2}$$

There is no mention of a 4-velocity or rest-frame. These are secondary objects which could be defined in a variety of ways, for example in terms of the fluxes of energy of entropy or the various kinds of charge.

However, all of these definitions would agree in assigning a unique 4-velocity u^{μ} to *equilibrium* states (indicated by subscript 0)

They form an $(N+4)$ dimensional subspace E of the infinite-dimensional space of all states, which can be parametrized by an inverse-temperature 4-vector $\beta^{\mu} = u^{\mu}/T$ and N "thermal potentials" $\alpha_A = \mu_A/T$. The equilibrium equation of state specifies the functional form of the fugacity 4-vector $\Phi^{\lambda}(\alpha, \beta^{\mu})$, which generates all equilibrium quantities via the covariant Gibbs–Duhem relations

$$d\Phi^{\lambda} = \sum_{A} J^{\mu}_{A(0)} d\alpha_A + T^{\lambda\mu}_{(0)} d\beta^{\mu} \,, \tag{8.3}$$

$$S^{\lambda}_{(0)} = \Phi^{\lambda} - \sum_{A} \alpha_A \delta J^{\lambda}_{A(0)} - T^{\lambda\mu}_{(0)} \beta^{\mu} \,, \tag{8.4}$$

which imply

$$\delta S^{\lambda} = -\sum_{A} \alpha_A \delta J^{\lambda}_{A} - \beta_{\mu} \delta T^{\lambda\mu} \,. \tag{8.5}$$

The equilibrium fluid pressure is given by

$$P\beta^{\lambda} = \Phi^{\lambda} \,. \tag{8.6}$$

It follows from these relations that all equilibrium quantities have a perfect-fluid form:

$$J^{\lambda}_{A} = n_A u^{\lambda} \,, \quad S^{\lambda}_{(0)} = s u^{\lambda} \,, \quad T^{\lambda\mu}_{(0)} = \rho u^{\lambda} u^{\mu} + P \Delta^{\lambda\mu} \,, \tag{8.7}$$

where $\Delta^{\lambda\mu} = g^{\lambda\mu} + u^{\lambda} u^{\mu}$ projects vectors onto the 3-space orthogonal to u^{λ}.

Thermal equilibrium is characterized by zero entropy production,

$$\nabla_{\mu} S^{\mu}_{(0)} = 0 \,. \tag{8.8}$$

Together with the conservation laws (8.2) this leads to

$$\partial_{\mu} \alpha = 0 \,, \quad \nabla_{\mu} \beta_{\lambda} + \nabla_{\lambda} \beta_{\mu} = 0 \,, \tag{8.9}$$

which imply that the gravitational field is stationary (i.e. the geometry is invariant under translations along the timelike direction β^λ), that the fluid motion is rigid, and validity of the Ehrenfest–Tolman–Klein relations (Tolman and Ehrenfest [1930])

$$T\sqrt{-g_{00}} = \text{constant}\,, \quad \mu\sqrt{-g_{00}} = \text{constant} \tag{8.10}$$

in the fluid rest-frame, where the spatial components of u^μ vanish.

In the rest-frame, (8.3, 8.4 and 8.5) reduce to the Gibbs–Duhem relations in their familiar form:

$$\begin{aligned} d(\beta P) &= \sum_A n_A d\alpha_A - \rho d\beta\,, & s &= \beta(\rho + P) - \sum_A \alpha_A n_A\,, \\ ds &= -\sum_A \alpha_A dn_A + \beta d\rho\,, & \beta &= T^{-1}\,. \end{aligned} \tag{8.11}$$

What is the point of dressing them up in the pretentiously formal disguise (8.3)–(8.5)? There are several reasons.

First, the covariant equations (8.3)–(8.5) actually contain a bit more than the usual relations: they also describe the effects of a change of 4-velocity. Thus, the covariant formulation incorporates the fluid velocity as an additional thermodynamical variable.

Secondly, there are situations where there is no alternative to the covariant formulation because a rest-frame does not exist – for example, Hawking radiation at a black hole horizon, where the rest-frame is moving outward (with respect to free-falling inertial observers) at the speed of light; or a pencil of radiation issuing from a pinhole in a wall of a hot cavity.

Thirdly, since the covariant equations are expressed directly in terms of the *conserved* quantities J^λ_A, $T^{\lambda\mu}$, they can be manipulated that much more easily and transparently.

Finally – and most importantly – as we shall see in a moment, equations (8.4)–(8.6), *remain valid* to first order for small deviations from equilibrium. And (8.4) then *automatically includes* the standard linear relation between entropy flux, heat flux and diffusion. (Several of these advantages of a fully covariant formulation were already anticipated by Stueckelberg and Wanders in their first (1953) publication [**89**]).

As just indicated, only one assumption is needed to extend this formalism off equilibrium. One postulates that the covariant Gibbs relation (8.5) holds, not just for transitions between equilibrium states, i.e., displacements tangent to E, but for *arbitrary* infinitesimal displacements $\delta J^\lambda_A, \delta T^{\lambda\mu}$ from an equilibrium state $(\alpha_A, \beta^\lambda_A)$.

Then, if the displacement $(\delta J^\lambda_A, \delta T^{\lambda\mu}, \ldots)$ from a point of E is small but finite, addition of (8.4) and (8.5) gives

$$S^\lambda = P(\alpha_A, \beta)\beta^\lambda - \sum \alpha_A J^\lambda_A - T^{\lambda\mu}\beta_\mu - Q^\lambda \tag{8.12}$$

where the added term Q^λ is of second order in the deviations δJ^λ_A, $\delta T^{\lambda\mu}$ and generally will also depend (quadratically) on other, "auxiliary" variables that vanish in equilibrium. Equation (8.12) gives the entropy of an off-equilibrium situation described (partially) by the primary variables ($T^{\lambda\mu}$, J^λ_A, S^λ). The parameters (α_A, β^λ) that enter (8.12) refer to an *arbitrarily chosen* neighbouring equilibrium state.[3]

It is now a simple matter to derive the linear transport laws. I will just outline the procedure (see e.g., articles in Anile and Choquet-Bruhat [1989]). Taking the divergence of (8.12) and recalling (8.2) and (8.3), we obtain

$$0 \leq \nabla_\lambda S^\lambda = -\sum_A J^\lambda_{A(1)} \partial_\lambda \alpha_A - T^{\lambda\mu}_{(1)} \nabla_\lambda \beta_\mu - \nabla_\lambda Q^\lambda \,, \tag{8.13}$$

where

$$J^\lambda_{A(1)} = J^\lambda_A - J^\lambda_{A(0)} \,, \quad T^{\lambda\mu}_{(1)} = T^{\lambda\mu} - T^{\lambda\mu}_{(0)} \tag{8.14}$$

are deviations from equilibrium, representing diffusion, heat flux and viscous stresses.

The requirement that the right-hand side of (8.13) reduce to a positive-definite quadratic form in the deviations then leads – ignoring for a moment the second-order term Q^λ – to relativistic forms of Fick's and Fourier's laws and the Navier–Stokes law, relating the diffusive and heat fluxes and the viscous stresses linearly to spatial grandients of the thermal potentials, temperature and velocity respectively.

The early relativistic theories of Eckart and Landau–Lifshitz did in fact assume $Q^\lambda = 0$. That leads, just as in the non-relativistic case, to parabolic equations and infinite propagation speeds – particularly embarrassing in a relativistic context. Beginning in the mid-1970s, attempts were made (see e.g., articles in Anile and Choquet-Bruhat [1989]) to ameliorate this situation by retaining Q^λ, assuming for example that it is quadratic in the deviations (8.14). This results in propagation equations of the telegraph form, hence finite propagation speeds, at the cost of introducing additional phenomenological coefficients and greater complexity.

[3] It is useful to relate this local equilibrium state to the actual state by requiring that they have the same particle and energy densities, i.e., that the scalar parameters α_A, β satisfy the $N+1$ fitting conditions

$$J^\lambda_{(1)A} u_\lambda = T^{\lambda\mu}_{(1)} u_\lambda u_\mu = 0 \,.$$

That still leaves the 4-velocity u^λ arbitrary to first order. This is a characteristic feature of the relativistic theory with no classical analogue: $E = mc^2$ means that a change of rest-frame produces a flux (largely rest-mass) energy. The effect is a unification of Fourier's and Ficks laws in the relativistic theory.

8.4. Stueckelberg–Wanders 1953

With this background it becomes quite easy to give a concise sketch of the Stueckelberg–Wanders theory [**89**], which – please bear in mind – preceded the covariant formulation I have just outlined by more than 20 years.

Apart from the little-known work of van Dantzig [1939], covariant treatments of thermodynamics were far from being the norm in 1953. The presumptive existence of a privileged rest-frame seemed to negate any advantages of covariance, and the practice was to break up the stress-energy from the start into spatial and temporal pieces easily interpreted in Newtonian terms.

In 1952, the *Physical Review* published a paper by Boris Leaf, an instructor at Kansas State College, that was very much in this (3+1) tradition. It is remarkable solely in being the only external reference cited by Stueckelberg and Wanders. Since the two papers could not be more different in approach and style, it seems fair to speculate that it may have served as a goal and trigger, a sample of the pedestrian approach to the subject. In this setting the appearance of Stueckelberg–Wanders in the following year must have come as something of a shock. (I recall coming across one of the later papers in the series in the early 1960s and finding it a challenge).

Stueckelberg and Wanders set out from the same basis as in the last section. For an N-component fluid they postulate existence of a set of primary variables J^μ_A, $T^{\lambda\mu}$, S^μ satisfying

$$\nabla_\mu J^\mu_A = \nabla_\mu T^{\lambda\mu} = 0\,, \quad \nabla_\mu S^\mu = \sigma(x) \geq 0\,. \tag{8.15}$$

But from here their procedure was rather different. They introduced two postulates:

(a) An arbitrary state of the fluid is completely specified by just $N + 4$ variables – the same number as for an equilibrium state.

(b) These variables, say f_i $(i = 1, \dots, N + 4)$ can be chosen so that the primary variables depend at most linearly on their gradients $\nabla_\mu f_i$ (on the other hand, $\sigma(x)$ in (8.15) is strongly constrained by the requirement to be a positive-definite quadratic form in $\nabla_\mu f_i$.

Equations (8.15) constitute a set of $N + 5$ conditions on the $N + 4$ variables f_i. Thus, they must be linked by an identity of the form (linear by postulate (b))

$$\sum_A \alpha_A \nabla_\mu J^\mu_A + \beta_\lambda \nabla_\mu T^{\lambda\mu} \equiv \sigma(x) - \nabla_\mu S^\mu \tag{8.16}$$

with some set of $N + 4$ coefficients α_A, β_λ.

Now proceed as follows: choose α_A, β_λ as the state variables f_i. Write the most general tensorial expression for

$$J_A^\mu = J_{A(0)}^\mu + J_{A(1)}^\mu, \quad T^{\lambda\mu} = T_{(0)}^{\lambda\mu} + T_{(1)}^{\lambda\mu} \tag{8.17}$$

with the subscript-1 terms linear in the gradients $\nabla_\mu f^i$ and subscript-0 terms dependent on f_i only. Substitute into the left-hand side of the identity (8.16) and segregate the terms into pure divergences (these are equated to $-\nabla_\mu S^\mu$) and those quadratic in $\nabla_\mu f_i$ (they correspond to $\sigma(x)$). Any residual terms must have zero coefficients. The results are:

The terms linear in $\nabla_\mu f_i$ (arising from $J_{A(0)}^\mu, T_{(0)}^{\lambda\mu}$) identify the coefficients in (8.16) to be

$$\alpha_A = \mu_A / T, \quad \beta^\lambda = u^\lambda / T, \tag{8.18}$$

as already anticipated by the notation.

The quadratic terms lead to the linear transport laws of Fick, Fourier and Navier–Stokes. The results are equivalent (to first order in deviations from equilibrium) to those obtained by Eckart and by Landau–Lifshitz, but the formulation is more general in that the fluid 4-velocity u^α is not tied a priori to a flux of particles or energy, but is left open to first order.

8.5. Sound Speed, Causality and Stability

Stueckelberg returned to the subject nine years later with a dedicated treatment of fluids in thermal equilibrium [105]. Here the Tolman–Klein laws (8.10) were rediscovered, but he also found something new and unexpected.

He considered a fluid at rest or in steady rotation in a stationary gravitational field, so that the spacetime geometry is independent of an appropriate time co-ordinate t. The condition that the entropy of the system

$$S = -\int S^\mu d\Sigma_\mu ,$$

be a maximum subject to the constraints of fixed 4-momentum and particle numbers,

$$P^\lambda = -\int T^{\lambda\mu} d\Sigma_\mu , \quad N_A = -\int J_A^\mu d\Sigma_\mu , \tag{8.19}$$

requires maximization of

$$\Psi = S + \sum_A \alpha_A N_A + \beta_\lambda P^\lambda ,$$

where α_A, β_λ are Lagrange multipliers.[4]

The integrals are conveniently taken over a 3-space of constant t, so that

$$d\Sigma_\mu = n_\mu d\Sigma = \frac{1}{\sqrt{(-g^{(00)})}}\delta^0_\mu d\Sigma .$$

The first variation, $\delta^{(1)}\Psi = 0$ yields

$$\delta S^\mu + \sum_A \alpha_A \delta J^\mu_A + \beta_\lambda \delta T^{\lambda\mu} = 0 ,$$

which is the covariant Gibbs relation (8.4) and establishes the physical meaning of the Lagrange multipliers α_A, β_λ.

Stueckelberg went on to consider the second variation, $\delta^{(2)}\Psi \leq 0$. I quote the most interesting part of his result, the effect of variations of the fluid velocity u^i $(i, j, \ldots = 1, 2, 3)$. A somewhat lengthy calculation gives

$$\delta^{(2)}\Psi = -\int d\Sigma T^{-1}\{(\rho + P)(g^{ij} + \gamma u^i u^j) - a\gamma u^i u^j\}\delta u^i \delta u^j .$$

Here, $\gamma = (1 - v^2/c^2)^{-\frac{1}{2}}$ is the relativistic boost factor, and (for a one-component fluid)

$$a = n(\partial P/\partial n)_{s/n} .$$

Using the thermodynamical identities

$$dP = s dT + n d\mu , \quad T = \frac{\partial\rho(n,s)}{\partial s} , \quad \mu = \frac{\partial\rho(n,s)}{\partial n} ,$$

it is straightforward to show that the speed of sound c_s can be expressed in terms of a by

$$c_s^2 = \left(\frac{\partial P}{\partial \rho}\right)_{s/n} = \frac{a}{(\rho + P)} .$$

Stueckelberg concludes that the stability condition $\delta^{(2)}\Psi \leq 0$ requires the sound speed to be subluminal, $c_s^2 \leq 1$. He adds,

> To our knowledge, this result is new. Pauli [1921] criticizes the procedure of Herglotz and Lamla (who impose, from the condition 'maximum signal velocity = 1', an upper limit on the elastic modulus a) with the words '...the principle of relativity cannot make any statements on the magnitude of the cohesive forces'. He expects that, at this upper limit for a, '...the phenomenological equations become incorrect'. We were therefore rather surprised that phenomenological thermodynamics leads, for stability reasons, to this upper limit for a.

4 As written in (8.19) for brevity, the first integral needs more careful definition in a curved space.

I must admit that my own first reaction to this conclusion was skeptical surprise. The state variables and transport coefficients employed in phenomenological thermodynamics are not limited by any micromechanics, causal or otherwise. As already noted, the speed of thermal and viscous disturbances is actually infinite in the theories of Eckart, Landau–Lifshitz, Stueckelberg and, indeed, in any "first-order" theory in which entropy flux depends at most linearly on heat flux and other deviations from equilibrium (i.e., $Q^\lambda = 0$ in (8.12)). It seemed a priori unlikely that such theories would impose a speed limit on sound waves.

But, on futher thought and with the benefit of hindsight, it seems entirely reasonable on general grounds to expect a close link between faster-than-light signals and instability.

The crucial point is that, in a Lorentz-invariant theory, it is not possible to distinguish between superluminal acausality (faster-than-light signal propagation) and chronological acausality (response preceding cause). If signals can travel faster than light then, at least for some observers, effect will precede cause. The further link of chronological acausality to instability has long been known. A familiar example is Dirac's third-order equation of motion for the classical electron, where pre-acceleration effects go together with runaway solutions. If the source of an effect can be delayed, it should be possible for a system to borrow energy from its ground state, and this implies instability.

It was, indeed, found in 1985 by Hiscock and Lindblom [1985] that all first-order theories are subject to catastrophic instabilities because they admit superluminal thermal and viscous waves. They can be straightforwardly extended, however, to theories that are stable by including the second-order terms Q^λ in (8.12).[5]

8.6. Stueckelberg–Wanders Rediscovered: Phenomenological Theories of "Divergence Tyype"

Although the Stueckelberg–Wanders papers have lain forgotten for over fifty years, the underlying idea was rediscovered in 1986 (Geroch and Lindblom [1990], Liu, Müller and Ruggeri [1986]) and applied in the context of a somewhat broader class of theories known as "theories of divergence type". I shall conclude with a brief description of these developments.

Formally, these phenomenological theories are modelled on the Grad approximation in kinetic theory (see e.g., articles in Anile and Choquet-Bruhat

---

[5] The linear phenomenological laws are inferred from the bilinear expression (8.13) for the entropy production. Thus, the extra second-order terms $\nabla_\lambda Q^\lambda$ in (8.13) lead to changes to the phenomenological laws at the linear order.

[1989]). For a simple gas of particles with 4-momentum $p^\mu = mv^\mu$, the relativistic Boltzmann equation for the distribution function $N(x^\alpha, p^\beta)$ is

$$v^\mu \partial_\mu N(x, p) = C[N], \tag{8.20}$$

where $C[N]$ represents the collision integral. Since collisions conserve particle number and 4-momentum, the two lowest moments of (8.20) simply express the conservation of

$$J^\mu = \int N v^\mu d\omega \quad \text{and} \quad T^{\lambda\mu} = \int N p^\lambda v^\mu d\omega ,$$

where the integrals are taken over the mass shell. The third moment is

$$\nabla_\mu \int N p^\alpha p^\beta v^\mu d\omega = \int C[N] p^\alpha p^\beta d\omega . \tag{8.21}$$

The Grad approximation works with just the first three moments of (8.20), cut off at (8.21).[6] The number of unknown functions of position is correspondingly reduced by assuming a special form for the distribution function $N = N_0(1 + f)$, where

$$N_0(x, p) = \alpha(x) e^{\beta_\mu(x) p^\mu}$$

is a local equilibrium distribution, and $f(x, p)$, representing small deviations, is restricted to be quadratic in momentum.

The phenomenological theories we are considering here follow this scheme very closely. For a simple fluid they postulate a set of variables $J^\mu, T^{\lambda\mu}, S^\mu, \sigma, A^{\alpha\beta\mu}$ and $I^{\alpha\beta}$ (symmetric and trace-free in α, β) satisfying the identities

$$\nabla_\mu J^\mu = 0 , \quad \nabla_\mu T^{\lambda\mu} = 0 , \quad \nabla_\mu A^{\alpha\beta\mu} = I^{\alpha\beta} , \tag{8.22}$$

$$\nabla_\mu S^\mu = \sigma . \tag{8.23}$$

They further postulate that $A^{\alpha\beta\mu}$, $I^{\alpha\beta}$, S^μ and σ are algebraic functions of J^μ and $T^{\lambda\mu}$.

Then, for (8.23) to be a consequence of (8.22), there must be a linear relation between their left-hand sides:

$$\alpha \nabla_\mu J^\mu + \beta_\lambda \nabla_\mu T^{\lambda\mu} + \gamma_{\alpha\beta} \nabla_\mu A^{\alpha\beta\mu} + \nabla_\mu S^\mu \equiv 0 \tag{8.24}$$

(this should be compared with the Stueckelberg–Wanders identity (8.16)).

---

[6] Note that the trace of (8.21) just reproduces $\nabla_\mu J^\mu = 0$.

ksNow define the fugacity 4-vector

$$\Phi^\mu = S^\mu + \alpha J^\mu + \beta_\lambda T^{\lambda\mu} + \gamma_{\alpha\beta} A^{\alpha\beta\mu} \,. \tag{8.25}$$

The divergence of (8.25) combined with (8.24) leads to

$$\begin{aligned} 0 \equiv & \left(\frac{\partial \Phi^\mu}{\partial \alpha} - J^\mu \right) \nabla_\mu \alpha + \left(\frac{\partial \Phi^\mu}{\partial \beta^\lambda} - T^{\lambda\mu} \right) \nabla_\mu \beta_\lambda \\ & + \left(\frac{\partial \Phi^\mu}{\partial \gamma_{\alpha\beta}} - A^{\alpha\beta\mu} \right) \nabla_\mu \gamma_{\alpha\beta} \,, \end{aligned}$$

an identity for arbitrary local values of the gradients (taking α, β_λ and $\gamma_{\alpha\beta}$ as new dynamical variables).

The symmetry of $T^{\lambda\mu}$ implies that Φ^μ and thence all thermodynamical variables can be derived from a single scalar potential $\Phi(\alpha, \beta, \gamma)$:

$$\begin{aligned} \Phi^\mu &= \frac{\partial \Phi}{\partial \beta_\mu} \,, & J^\mu &= \frac{\partial^2 \Phi}{\partial \alpha \partial \beta_\mu} \,, \\ T^{\lambda\mu} &= \frac{\partial^2 \Phi}{\partial \beta_\lambda \partial \beta_\mu} \,, & A^{\alpha\beta\mu} &= \frac{\partial^2 \Phi}{\partial \gamma_{\alpha\beta} \partial \beta_\mu} \,. \end{aligned}$$

From a mathematical point of view, this is an extremely elegant and economical formalism. For nongaseous materials, of course, the physical meaning and status of the new variables $\gamma_{\alpha\beta}$ and $A^{\alpha\beta\mu}$ is not entirely clear. I am not sure what Stueckelberg would think of these developments. But he would be pleased, I think, to know that some of the ideas and techniques he introduced into the field fifty years ago are now actively in use, even if their original authorship is still unrecognized.

Acknowledgment: I should like to express appreciation to Dr Jan Lacki for organizing this stimulating Symposium in honour of E. C.G Stueckelberg, and for his warm hospitality in Geneva.

References

Anile, A. and Choquet-Bruhat, Y. (eds), *Relativistic Fluid Dynamics*, Springer Lecture Notes in Mathematics # 1385. Berlin: Springer, 1989.

Eckart, C., The thermodynamics of irreversible processes. III. Relativistic theory of the simple fluid, *Phys. Rev.*, vol. 58 (1940), pp. 919–924.

Van Dantzig, D., On the phenomenological thermodynamics of moving matter, *Physica*, vol. 6 (1939), pp. 673–704.

Geroch, R. and Lindblom, L., Dissipative relativistic fluid theories of divergence type, *Phys. Rev. D*, vol. 41 (1990), pp. 1855–1861.

Hiscock, W. A. and Lindblom, L., Generic instabilities in first-order dissipative relativistic fluid theories, *Phys. Rev. D*, vol. 31 (1985), pp. 725–733.

Jüttner, F., Die relativistische Quantentheorie des idealen Gases, *Z. Physik*, vol. 47 (1928), pp. 542–546.

Kampen, N. G. van, Relativistic thermodynamics of moving systems, *Phys. Rev.*, vol. 173 (1968), pp. 295–301.

Klein, O., On the thermodynamic equilibrium of fluids in gravitational fields, *Rev. Mod. Phys.*, vol. 21 (1949), pp. 531–533.

Kluitenberg, G. A. and de Groot, J. R., Relativistic thermodynamics of irreversible processes, *Physica*, vol. 21 (1955), p. 148–168, and earlier references cited there.

Lichnerowicz, A. and Marrot, R., Propriétés statistiques des ensembles de particules en relativité restreinte, *Comptes Rendus Acad. Sci. Paris*, vol. 210 (1940), pp. 759–761.

Liu, I.-S., Müller, I. and Ruggeri, T., Relativistic thermodynamics of gases, *Annals of Physics*, vol. 169 (1986), pp. 191–219.

Pauli, W., Relativitätstheorie, in *Enzyklopädie der Mathematischen Wissenschaften*, vol. 5, Part 2. Leipzig, B. G. Teubner, 1921. Translated as *Theory of Relativity*. New York: Pergamon Press, 1958.

Stuart, E. B., Gal-Or B. and Brainard, A. J. (eds), *A Critical Review of Thermodynamics*. Baltimore: Mono Book Corp., 1970.

Synge, J. L., *Trans. Roy. Soc. Canada*, vol. 28 (3) (1934), p. 127.

Tolman, R. C. and Ehrenfest, P., Temperature equilibrium in a static gravitational field, *Phys. Rev.*, vol. 36 (1930), pp. 1791–1798.

Walker, A. G., The Boltzmann equations in general relativity, *Proc. Edinb. Math. Soc.*, vol. 4 (1936), pp. 238–253.

III

Stueckelberg's Selected Papers

Theorie der unelastischen Stösse zwischen Atomen [19]

Helvetica Physica Acta vol. 5 (1932), pp. 369–422

Theorie der unelastischen Stösse zwischen Atomen

von **E. C. G. Stueckelberg** in Basel.

(28. XI. 32.)

Kurze Inhaltsangabe.

§ 1. **Einleitung:** Dieser § enthält die Problemstellung und die aus der Arbeit hervorgehenden Resultate in kurzer Zusammenfassung.

§ 2. **Die Formulierung des Problems. (Zweiatomiges Molekül.)**

§ 3. **Grenzbedingungen und Wirkungsquerschnitt:** Dieser § enthält die nötigen Definitionen zur Errechnung der Wirkungsquerschnitte.

§ 4. **Die Formulierung des Problems nach der Born-Dirac'schen Theorie:** Diese Formulierung unterscheidet sich von § 2 in für die Arbeit wichtigen Punkten.

§ 5. **Beispiel einer strengen Lösung.**

§ 6. **Die angenäherten Lösungen:** Hier wird das Störungsverfahren kurz diskutiert.

§ 7. **Vergleich der London'schen mit der Born-Dirac'schen Näherung:** Dieser Abschnitt vergleicht die Ansätze des § 2 und § 4.

§ 8. **Das Anschlussverfahren der Näherungsfunktionen in den Umkehrpunkten:** Hier werden die für die folgenden § verwendeten Näherungsfunktionen aufgestellt und das W. K. B.[1]) Verfahren von einem neuen Gesichtspunkte aus erläutert.

§ 9. **Das Anschlussverfahren der Näherungsfunktionen in der Überschneidungsgegend:** (Fall II): Der § diskutiert den Fall sich schneidender Potentialkurven.

§ 10. **Das Anschlussverfahren für den Fall I (ψ auf der reellen Axe überall grösser als null):** Hier wird der Fall sich nicht schneidender Potentialkurven behandelt.

§ 11. **Abschätzung der Matrixelemente $|\eta_{1J}|^2$ nach dem Störungsverfahren:** Für einige Fälle wird hier das Verfahren des § 6 angewendet.

§ 12. **Bestimmung des Wirkungsquerschnittes in Fall I (kein Überschneiden der Kurven).**

§ 13. **Der Wirkungsquerschnitt im Resonanzfall.**

§ 14. **Bestimmung des Wirkungsquerschnittes im Fall II (Überschneidungsfall).**

§ 15. **Die Ionisation von Edelgasen durch Alkaliionen:** Die Resultate des § 14 werden auf Ionenstossversuche angewendet.

§ 16. **Zusammenfassung.**

§ 1. Einleitung.

Das Problem des unelastischen Stosses zwischen zwei Atomen reduziert sich, wenn es erlaubt ist nur zwei Elektronenzustände jedes Atoms zu berücksichtigen, immer auf unendlich viele Systeme von je zwei gekoppelten linearen Differentialgleichungen zweiter Ordnung mit einer unabhängigen Variabeln r, welche den Abstand zwischen den beiden Systemschwerpunkten darstellt.

Ist die für die Übergänge verantwortlich zu machende Wechselwirkungsenergie zwischen den beiden Elektronenzuständen im Bereiche reeller Geschwindigkeiten klein gegen die vom Abstand abhängige Termdifferenz, so kann man mit Franck von potentiellen Energiekurven der beiden Elektronenzustände sprechen. Die Begriffe „sich schneiden" und „sich nicht schneiden" haben dann im folgenden den Sinn, dass die Kurven für *fehlende Wechselwirkungsenergie* sich schneiden oder nicht schneiden. Die Tatsache, dass Übergänge auftreten bedeutet ja schon, dass das Bild der potentiellen Energiekurven nicht mehr streng richtig sein kann.

Das Auftreten einer Wechselwirkung bewirkt immer, dass die beiden Kurven sich im Überschneidungsgebiet abstossen. Die Wechselwirkung zwischen zwei Termen kann in zwei Teile zerlegt werden: in einen von der Rotationsenergie unabhängigen und in einen von der Rotationsenergie (Quantenzahl J) abhängigen. Die Franck'schen Kurven zeichnet man im ganzen für $J = 0$. Haben die Elektronenterme verschiedene Symmetrieeigenschaften oder verschiedene Quantenzahlen Λ (resp. Ω) (= Elektronendrehimpulsquantenzahl um die Kernverbindungsachse), so verschwindet der zweite Teil[1]). Ist das nicht der Fall, so bewirkt dieser zweite Teil, dass die Kurven auch für $J = 0$ sich nicht schneiden. Die Kurven kommen sich also nur nahe. In Figur 1 ist der Fall des „Überschneidens" und „Nahekommens" gezeichnet. Man kann also in beiden Fällen annehmen, dass die Kurven sich schneiden, je nachdem man die Äste A_1B_1 und A_0B_0 zuordnet, da im Überschneidungsgebiet die Kurven ihre Bedeutung verlieren.

Wählt man das Bild der Überschneidung, und ist die durch die Wechselwirkung bedingte Übergangswahrscheinlichkeit klein gegen 1, so kann man das Franck'sche Bild in erster Näherung auch im Überschneidungsgebiet gelten lassen und die Übergangswahrscheinlichkeiten nach der Störungstheorie ermitteln. Ist dies jedoch nicht mehr der Fall, so kann man insofern von „schneiden" sprechen, als man links und rechts die Wellenfunktionen der Elektronenkonfiguration in Beziehung bringen kann. Die

[1]) J. v. Neumann und E. Wigner, Phys. Zeitschr. **30**, 467 (1929).

„Schnittgegend" hat dann, für die asymptotische Darstellung der Wellenfunktion ähnliche Eigenschaften wie die Umkehrpunkte der klassisch-mechanischen Bewegung. Im § 9 wird ein dem WENTZEL, KRAMERS, BRILLOUIN'schen[2]) analoges Anschlussverfahren beschrieben, welches gestattet die Übergangswahrscheinlichkeiten für beliebig grosse Wechselwirkungsenergien zu ermitteln, wenn die Schnittgegend mit reeller Geschwindigkeit durchfahren wird. Betrachten wir einen Wahrscheinlichkeitsstrom der Stärke 1,

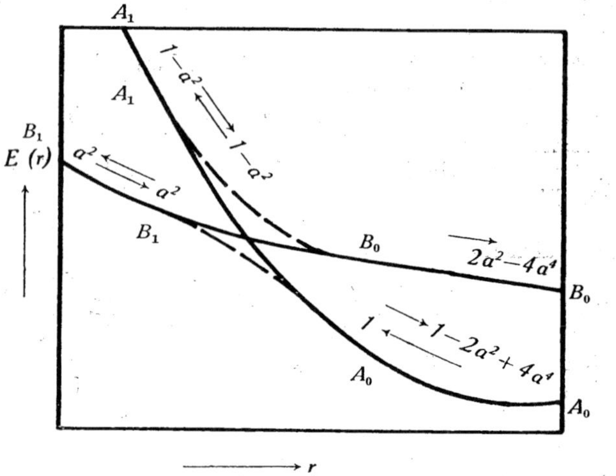

Fig. 1.

Sich „schneidende" Energiekurven für den Stoss zweier Atomsysteme. A_0 und A_1 haben dieselben charakteristischen Elektronenkonfigurationen. Die Pfeile stellen die Ströme dar, welche auf den einzelnen Kurvenästen fliessen, wenn der Strom 1 auf dem Ast A_0 ankommt, und wenn jeder Strom sich im Schnittpunkt so verteilt, dass der Teil $1-a^2$ auf dem Ast gleicher Konfiguration weiterläuft, während der Teil a^2 auf dem andern Aste weiterfliesst.

der „auf" der einen potentiellen Energiekurve (mit A_0 bezeichnet) ankommt, so teilt er sich im „Schnittpunkt" in zwei Ströme: den Strom $1-a^2$ auf der Kurve A_1 und den Strom a^2 auf der Kurve B_1. Die Ströme fliessen im ankommenden Sinne weiter bis an ihre Umkehrpunkte. Dort werden sie reflektiert und beide kreuzen den Umkehrpunkt ein zweites Mal. Jeder der beiden Ströme verteilt sich dort wieder so, dass der Teil $(1-a^2)$ seiner Intensität auf der Kurve gleicher Elektronenkonfiguration weiter-

[2]) G. WENTZEL, Zeitschr. f. Phys. **38**, 518 (1926); A. H. KRAMERS, Zeitschr. f. Phys. **39**, 828 (1926); L. BRILLOUIN, C. R. Juli (1926).

fliesst, während der Teil a^2 auf der andern Kurve fortläuft. Der auslaufende Strom $4 \cdot |\eta_1|^2$ auf der Kurve B_1 beträgt also:

$$4 \cdot |\eta_1|^2 = (1-a^2)a^2 + a^2(1-a^2) = 2a^2 \cdot (1-a^2)$$

Nennt man a^2 Übergangswahrscheinlichkeit, so nähert diese sich für grosse Wechselwirkungsenergien dem Wert 1. Bezeichnet man sie aber mit $4|\eta_1|^2$ wie dies im folgenden geschehen soll, so erreicht sie für $a^2 = \frac{1}{2}$ ihr Maximum mit dem Werte $4 \cdot |\eta_1|^2 = \frac{1}{2}$, und wird für $a^2 = 1$ zu Null.

Schneiden sich die Kurven nicht und kann ihr Abstand angenähert durch eine konstante Termdifferenz dargestellt werden, so liegt die „Schnittgegend" dort, wo die Wechselwirkungsenergie der Termdifferenz gleich wird. Befindet sich diese Gegend bei imaginären Geschwindigkeiten, so ist die Störungsmethode gerechtfertigt. Liegt sie aber bei reellen Geschwindigkeiten so erhält man einen ähnlichen Fall wie wenn sich die Kurven im reellen Geschwindigkeitsbereich schneiden (§ 10).

Das Verfahren wird zur Bestimmung von Wirkungsquerschnitten von Stössen zweiter Art verwendet. Man findet, dass der Fall, wo keine Überschneidung im oben definierten Sinne stattfindet, in Grössenordnung und Resonanzcharakter die experimentellen Ergebnisse besser darstellt, als der Fall der Überschneidung. Das plötzliche Einsetzen der Ionisation durch schnelle Alkaliionen, wie es von BEECK, MOUZON und NORDMEYER[3]) gefunden wurde, lässt sich sehr gut durch die Überschneidung der Potentialkurven erklären[4]).

Die Durchrechnung des Problems ist bis jetzt von verschiedenen Gesichtspunkten aus erfolgt. Eine erste, halbklassische Rechnung von KALLMANN und LONDON[5]) hatte den Nachteil, dass nur die Elektronenbewegung wellenmechanisch erfasst wurde. Die zum Vorgang wichtige kinetische Energie der Kerne konnte daher nicht in Erscheinung treten. MORSE und STUECKELBERG[6])

[3]) O. BEECK und J. C. MOUZON, Ann. d. Phys. **11**, 737, 858 (1931) siehe auch O. BEECK, Ann. d. Phys. **6**, 1001 (1930); C. J. BRASEFIELD, Phys. Rev. **42**, 11 (1932); M. NORDMEYER, erscheint demnächst in den Ann. d. Phys., der Verfasser ist Herrn Dr. NORDMEYER für die Mitteilung seiner Ergebnisse zu Dank verpflichtet. J. C. MOUZON, Phys. Rev. **41**, 605 (1932).

[4]) W. WEIZEL und O. BEECK, Zeitschr. f. Phys. **76**, 250 (1932).

[5]) H. KALLMANN und F. LONDON, Zeitschr. f. Phys. Chem. **2** (B), 220 (1929).

[6]) P. M. MORSE und E. C. G. STUECKELBERG, Ann. d. Phys. **9**, 579 (1931). Für eine zusammenfassende Darstellung der elastischen und inelastischen Stösse sei auf die Arbeit von P. M. MORSE Rev. Mod. Phys. **4**, 577 (1932) verwiesen, s. auch O. K. RICE, Phys. Rev. **38**, 1943 (1931).

wendeten das Störungsverfahren von BORN und DIRAC unter Berücksichtigung der endlichen Ausdehnung der Atome an.

Die Theorie des unelastischen Stosses zwischen sich langsam bewegenden schweren Massenteilchen muss sich aber, wie LONDON[7]) später gezeigt hat, wesentlich von der Theorie des unelastischen Elektronenstosses unterscheiden.

Wendet man nämlich formal dasjenige Störungsverfahren an, in welchem die gesamte elektrische Wechselwirkung der beiden Atome als Störungsfunktion angesetzt wird so erhält man eine Grösse, welche den pro Zeiteinheit eintretenden Zuwachs der Endsubstanz angibt. Diese Grösse enthält, in LONDON's Ausdrucksweise, zwei Vorgänge:

1. den Anteil des „adiabatischen" Vorgangs, welcher sich nach dem Stosse wieder vollständig in die Anfangssubstanz zurückverwandelt und daher nur zur elastischen Reflexion beiträgt.
2. den Anteil des unelastischen Stosses.

Beim Elektronenstoss wird der Anteil 1 verschwindend klein, während beim Atomstoss unter Umständen der Teil 1 wesentlich grösser als der Teil 2 wird. Eine Methode zur Trennung von 1 und 2 wurde von ihm angegeben. Sie besteht darin (siehe § 7 dieser Arbeit und § 5 bei LONDON), dass er von einer den adiabatischen Vorgang bereits enthaltenden Näherung ausgeht. Auf diese Näherungen wendet er dann das übliche Störungsverfahren an. Bei der Auswertung der hierbei auftretenden Matrixelemente machte (LONDON's § 7) er aber wieder eine Vernachlässigung, die in gewissen Fällen nicht erlaubt ist (siehe § 11 dieser Arbeit).

LONDON[7]) und auch die Arbeit von MORSE und dem Verfasser[6]) nehmen den „allgemeinen" Fall an, dass sich die potentiellen Energiekurven, auf welchen sich die elastische Bewegung abspielt, *nicht schneiden.* Dem entgegen behandelt LANDAU[8]) den Fall, dass die Kurven *sich schneiden.* Er glaubt folgern zu dürfen, dass dieser letzte Fall bedeutend grössere Wirkungsquerschnitte ergibt als der erste „allgemeine" Fall.

Die vorliegende Arbeit behandelt nun beide Fälle nach einer neuen Methode, welche nicht auf einem Störungsverfahren beruht, sondern als eine Lösung des Problems im Sinne des WENTZEL, KRAMERS, BRILLOUIN'schen[2]) Verfahrens bezeichnet werden kann. Zum Vergleich wird in jedem Fall auch die Auswertung des entsprechenden Matrixelementes nach den Verfahren[6]) [7]) und [8]) im § 11 diskutiert.

---

[7]) F. LONDON. Zeitschr. f. Phys. **74**, 143 (1932).

[8]) L. LANDAU, Sow. Phys. **1**, 89 (1932), s. auch O. K. RICE, Phys. Rev. **37**, 1187 (1931).

§ 2. Die Formulierungen des Problems (zweiatomiges Molekül).

Die Behandlung des zweiatomigen Moleküls gestaltet sich am einfachsten im Koordinatsystem der Eulerschen Winkel. Auf das unelastische Stossproblem können wir, formal wenigstens, diese Betrachtungen übertragen.

KRONIG[9]) zeigt, dass in erster Näherung das Problem in diesen Koordinaten folgendermassen separierbar ist:

$$\psi_{\sigma J \Lambda M T} = \frac{1}{r}\, v_{\sigma J \Lambda T}(r) \cdot \Theta_{J M \Lambda}(\vartheta) \cdot e^{i M \varphi} \cdot \frac{\sin}{\cos} \Lambda \varphi \cdot \Phi_{\sigma \Lambda}$$
$$\cdot \exp\left[-\frac{2\pi i}{h}\left(E_{\Lambda \sigma \infty} + T\right) t\right].$$

Hier bezeichnet r den Abstand der beiden Kerne, ϑ den Winkel zwischen der Verbindungsaxe der Kerne (ζ-Axe) und einer im Raum festen Richtung (z-Axe). ψ resp. φ die Winkel von der Knotenlinie (Schnitt der xy mit der $\xi\eta$-Ebene) nach der positiven x- resp. ξ-Axe. ξ bedeutet in den Formeln auch die Gesamtheit der Elektronenkoordinaten $\xi_i \eta_i \zeta_i$.

σ bezeichnet die Quantenzahlen des Elektronenzustandes, Λ ihre Drehimpulsquantenzahl um die ζ-Axe (Λ-Type doubling!), J die Gesamtimpulsquantenzahl des Systems und M ihre Projektion auf die z-Axe. $E_{\sigma\infty}$ bedeutet die Energie der Elektronen (Termwert) und T die kinetische Energie der Kerne, beides für $r = \infty$. Dann genügen die verschiedenen Funktionen den Gleichungen:

$$\left\{\Delta_{\xi_i \eta_i \zeta_i \ldots \varphi} + \frac{8\pi^2 m}{h^2}\left[E_{\sigma\Lambda}(r) - V(\xi_i\, \eta_i\, \zeta_i \ldots, r)\right]\right\} \Phi(\xi_i \ldots, r) \cdot \frac{\sin}{\cos} \Lambda \varphi$$
$$= 0\,. \qquad (1)$$

(Dies ist die Lösung bei festen Kernen im Abstand r (r als Parameter).)

$$\left\{\frac{1}{\sin\vartheta}\frac{\partial}{\partial\vartheta}\left(\sin\vartheta\,\frac{\partial}{\partial\vartheta}\right) - \frac{(M - \Lambda\cos\vartheta)^2}{\sin^2\vartheta}\right.$$
$$\left. + \left(J(J+1) - \Lambda^2\right)\right\} \Theta_{J M \Lambda}(\vartheta) = 0 \qquad (2)$$

$$\left\{\frac{\delta^2}{\delta r^2} + \frac{8\pi^2 M}{h^2}\left[T - E_{\sigma\Lambda}(r) - \frac{J(J+1) - \Lambda^2}{r^2} \cdot \frac{h^2}{8\pi^2 M}\right]\right\} v_{\sigma J \Lambda T}(r) = 0^{*}). \qquad (3)$$

---

[9]) R. DE L. KRONIG, Zeitschr. f. Phys. **46**, 814 (1928) und **50**, 247 (1928).

*) In Formel (3) ist der Kürze halber $E_{\sigma\Lambda\infty} = 0$ gesetzt.

Die Störungen (fehlende Glieder) sind dann gegeben durch die Operatoren ($\frac{\partial}{\partial r}$ wirke auf r in $\Phi(\xi_i \ldots, r)$ und $\frac{\delta}{\delta r}$ auf r in $v(r)$):

$$\mathrm{I} + \mathrm{II} = \frac{h^2}{8\pi^2 M}\left\{\frac{\partial}{\partial r^2} + 2\frac{\partial}{\partial r} \cdot \frac{\delta}{\delta r}\right\}$$

$$+ \frac{h^2}{8\pi^2 M} \cdot \frac{1}{r^2}\left\{\cos\varphi \cdot \frac{\partial}{\partial \varphi} \cdot \frac{\partial}{\partial \vartheta} + \ldots\right\} \Xi(\xi\,\eta\,\zeta), \tag{4}$$

wo Ξ Operatoren der Art ξ / ζ und $\eta\partial / \partial\xi$ enthält.

(4) zerfällt daher in zwei Teile I und II. Idealisiert man wieder das Problem auf *nur zwei Elektronenzustände* $\sigma = 0$ und $\sigma = 1$ mit Λ_0 und Λ_1, so hat man folgende Fälle zu unterscheiden:

Fall 1: $\Lambda_0 = \Lambda_1$:

Man erhält, nach linksseitiger Multiplikation mit

$$\overline{\Phi}_0 \cdot e^{-iM_0\psi} \cdot {\sin \atop \cos} \Lambda_0 \varphi \cdot \Theta_{J_0 M_0 \Lambda_0}$$

und Integration über die Elektronenkoordinaten ξ und über ψ, φ und ϑ, (wenn $'$ die Ableitung nach r bezeichnet, und wenn man J_0^2 statt $J_0(J_0+1) - \Lambda_0^2$ schreibt was ja für grosse J_0 immer angängig ist):

$$v''_{0JT_0} + \frac{8\pi^2 M}{h^2}\left[T_0 - (E_{0\Lambda}(r) - E_{0\Lambda\infty}) - \frac{h^2}{8\pi^2 M} \cdot \frac{J_0^2}{r^2}\right] v_{0JT_0}$$

$$+ \int d\xi \,(\overline{\Phi}_0 \Phi''_0 v_{0J_0T_0} + 2\overline{\Phi}_0 \Phi'_0 v'_{0J_0T_0})$$

$$= -\int d\xi\,(\overline{\Phi}_0 \Phi''_1 v_{1J_0T_1} + 2\overline{\Phi}_0 \Phi'_1 v'_{1J_0T_1}) \tag{5}$$

und eine entsprechende Gleichung (wenn 0 und 1 vertauscht werden) für $v_{1J_0T_1}$. Es ist $T_0 + E_{0\Lambda\infty} = T_1 + E_{1\Lambda\infty}$. Dies sind die *zwei gekoppelten Gleichungen,* welche London[7]) seiner Betrachtung zugrunde legt. Die Kopplung durch die unter II aufgeführten höheren Glieder ist Null oder doch im allgemeinen von kleinerer Grössenordnung, da $\cos\varphi \frac{\partial}{\partial\varphi}$ keine diagonalen Matrixelemente in Λ hat.

Fall II: $\Lambda_0 \neq \Lambda_1$:

Dies ist der Fall, welchen Landau[8]) behandelt. Die Kopplung geschieht nur durch das zweite Störungsglied II. Bei $J \gg 1$ kann das erste Glied von II als allein wesentlich betrachtet werden. (Näheres darüber siehe bei Kronig[9]).

Man kann nämlich, für grosse J, setzen:

$$\left| \int \Theta_{J\Lambda_0} \frac{\partial}{\partial \vartheta} \Theta_{J\Lambda_1} d(\cos\vartheta) \right| \sim J\,.$$

Der Faktor $\cos\varphi \frac{\partial}{\partial\varphi}$ bewirkt:

1. dass $\Lambda_1 = \Lambda_0 \pm 1$ sein muss, und
2. Kopplungen nur zwischen $\sin \Lambda_0 \varphi$ und $\cos \Lambda_1 \varphi$ auftreten.

Ferner kann die Kopplung nur zwischen Funktionen mit gleichem M (da kein $\frac{\cos}{\sin}\,\psi$ auftritt) und gleichem J (Erhaltung des Gesamtdrehimpulses) eintreten.

Das Kopplungsglied ist

$$\frac{h^2}{8\pi^2 M} \cdot \frac{1}{r^2} \cdot \int d\xi\, \overline{\Phi}_0\, \Xi(\xi)\, \Phi_1 \cdot \int \frac{\cos}{\sin} \Lambda_0 \varphi \cdot \cos\varphi \cdot \frac{\sin}{\cos} \Lambda_1 \varphi \cdot d\varphi$$

$$\cdot \int \Theta_{J\Lambda_0} \frac{\partial}{\partial\vartheta} \Theta_{J\Lambda_1} d(\cos\vartheta) \simeq \frac{h^2}{8\pi^2 M} \cdot \Lambda(r) \cdot J \cdot \frac{1}{r^2} = W^{\mathrm{II}}_{01}(r) \qquad (7)$$

wo $\Lambda(r)$ noch Funktion von r ist. $\Lambda(r)$ kann für kleine Distanzen den Wert der Quantenzahl Λ erreichen.

Das Matrixelement W^{II}_{01} muss selbstverständlich hermitisch sein und, da wir die Φ reell voraussetzen können, ist das Matrixelement der reellen physikalischen Grösse

$$\cos\varphi \cdot \frac{h}{2\pi i} \frac{\partial}{\partial\varphi} \cdot \frac{h}{2\pi i} \frac{\partial}{\partial\vartheta} = \cos\varphi \cdot p_\varphi \cdot p_\vartheta$$

ebenfalls reell. Wir haben also wieder $W^{\mathrm{II}}_{01} = W^{\mathrm{II}}_{10}$, und ebenfalls nur *zwei gekoppelte Gleichungen:*

$$v''_{0J_0T_0} + \frac{8\pi^2 M}{h^2} \left[T_0 - (E_{0\Lambda_0}(r) - E_{0\Lambda_0\infty}) - \frac{h^2}{8\pi^2 M} \cdot \frac{J_0^2}{r^2} \right] v_{0J_0T_0}$$

$$= W^{\mathrm{II}}_{10}(r) \cdot \frac{8\pi^2 M}{h^2} v_{1J_0T_1} \qquad (5\mathrm{b})$$

(und entsprechend für $v_{1J_0T_1}$, wenn 1 und 0 vertauscht werden).

Betrachtungen von Neumann und Wigner[1]) ergeben (wie Landau[8]) zeigte) folgendes:

Ist $\Lambda_0 = \Lambda_1$, so überschneiden sich Terme nur bei verschiedenem Gesamtspinn, dann sind aber die Wechselwirkungsenergien W^{I}_{01} gering. Die Wechselwirkungsenergien W^{II}_{01} sind im allgemeinen für kleine J kleiner als die W^{I}_{01}, aber, weil $\Lambda_0 \neq \Lambda_1$, so ist ein Überschneiden der potentiellen Energiekurven möglich. Landau[8]) schliesst nun in einer sehr geschickten Weise, dass in Fall I, wo kein Überschneiden für reelle Geschwindigkeiten eintritt, die Übergangswahrscheinlichkeiten mit wachsendem Elektronentermabstand exponentiell verschwinden, während sie im Fall II beim Überschneiden eine andere Form haben.

In §§ 9 und 10 werden diese beiden Fälle nach dem neuen Verfahren ausgewertet. Dabei wird sich zeigen, dass die Landau'sche Behauptung des exponentiellen Verlaufes von Fall I zwar richtig ist, dass aber die Wirkungsquerschnitte von Fall I *bei guter Resonanz* doch wesentlich grösser als die von Fall II trotz Überschneidung, werden können (§ 12).

§ 3. Grenzbedingungen und Wirkungsquerschnitt.

Wir wollen annehmen, dass Λ klein sei gegen J, dann heissen die Lösungen von (2) : (auf 1 normiert)

$$\Theta_{JM,\,\Lambda=0} = \sqrt{\frac{(2\,J+1)\,(J-M)\,!}{4\,\pi\,(J+M)\,!}} \cdot P_J^M(\cos\vartheta)\,.$$

Die Wellenfunktion der Kernbewegung im Zustande 0 soll eine ebene einfallende und eine ausgestrahlte (elastisch reflektierte) Kugelwelle darstellen. Denken wir uns die v_J welche sich ja im ∞ wie Bessel'sche Funktionen verhalten müssen, wie diese normiert d. h.:

$$\lim_{r=0} v_{0J} = r^{\frac{1}{2}} \left[J_{J+\frac{1}{2}}\left(p_0\, r + \mu_{0J}\cdot\frac{\pi}{2}\right) + \eta_{0J}\, H^{(1)}_{J+\frac{1}{2}}\left(p_0\, r + \mu_{0J}\cdot\frac{\pi}{2}\right)\right];$$

$$p_0^2 = \frac{8\,\pi^2\,M}{h^2}\,T_0 \tag{8a}$$

so stellt

$$F_0 = \sum_J A_J \cdot \Theta_{J00}\cdot\frac{1}{r}\cdot v_{0J}(r) \tag{9}$$

mit

$$A_J = \left[\left(J+\frac{1}{2}\right)\frac{(2\,\pi)^3\,M}{h\,p_0}\right]^{\frac{1}{2}} \cdot e^{i\frac{\pi}{2}(J+\mu_{0J})}$$

im Unendlichen eine so beschaffene Welle der Stromdichte 1 cm^{-2} sec^{-1} dar. μ_{0J} ist eine Phasenkonstante, welche vom Potentialfeld $E_0(r) - E_{0\infty}$ abhängt und die *elastische Streuung* bedingt (Dispersion).

Bei fehlender Kopplung muss v_{0J} überall eine reelle Funktion sein damit kein von Null verschiedener Strom fliesst, d. h. $\eta_{0J} = 0$.

Für v_{1J} lautet die Grenzbedingung im Unendlichen (nur ausgestrahlte Kugelwelle):

$$\lim_{r=\infty} v_{1J} = \eta_{1J} \cdot r^{\frac{1}{2}} \cdot H^{(1)}_{J+\frac{1}{2}}\left(p_1 r + \mu_{1J} \cdot \frac{\pi}{2}\right); \; p_1^2 = \frac{8\pi^2 M}{h^2} \cdot T. \quad (8b)$$

Die Funktion

$$F_1 = \sum_J A_J \Theta_{J00} \frac{1}{r} v_{1J} \xrightarrow[r=\infty]{} \sum_J A_J \Theta_{J00} \cdot \eta_{1J} \cdot \frac{1}{r^{\frac{1}{2}}} H^{(1)}_{J+\frac{1}{2}}$$

stellt dann die auslaufende Kugelwelle dar. Der durch die unendlich weite Kugelfläche fliessende Strom von F_1 ist, wegen der Normalisierung von F_0 (Stromdichte 1 cm^{-2} sec^{-1}) gleich dem Wirkungsquerschnitt ($d\omega$ bedeutet das Differential des räumlichen Winkels):

$$q = \frac{h}{8\pi M} \cdot \int \operatorname{imag}(\bar{F}_1 \nabla F_1)\, d\omega = \frac{8\pi}{p_0^2} \sum_J (J + \tfrac{1}{2}) \cdot |\eta_{1J}|^2 . \quad (10)$$

Es handelt sich also darum die gekoppelten Gleichungen (5), (5a) oder (5b) mit den Grenzbedingungen (8a) und (8b) zu lösen, und so die $|\eta_{1J}|^2$ zu bestimmen. Bei fehlender Kopplung stellt v_0 eine stehende Welle dar ($J_{J+\frac{1}{2}}$), welche als eine Überlagerung zweier fortschreitender ($H^{(1)}_{J+\frac{1}{2}}$ und $H^{(2)}_{J+\frac{1}{2}}$) gedacht werden kann. $|\eta_{1J}|^2$ ist dann identisch Null. Sobald die Kopplung einsetzt, werden $|\eta_{0J}|^2$ und $|\eta_{1J}|^2$ von 0 verschieden, und wir haben fortschreitende Wellen in beiden Fällen. Da $J_n = \frac{1}{2}(H^{(1)}_n + H^{(2)}_n)$ ist, so wird der Maximalwert von $|\eta_{1J}| = \frac{1}{2}$; $|\eta_{1J}|^2 = \frac{1}{4}$ bedeutet also „Übergangswahrscheinlichkeit = 1" (siehe § 1).

§ 4. Die Formulierung des Problems nach der Born-Dirac'schen Stosstheorie.

Einschaltungsweise legen wir hier nicht das zweiatomige Molekül unseren Betrachtungen zugrunde, sondern gehen (mit Rücksicht auf eine spätere Anwendung) vom Grenzfall der getrennten Atome aus. Wir setzen wieder in nullter Näherung die Wellenfunktion gleich einem Produkt aus den Wellenfunktionen der Elektronenbewegung und derjenigen der Kernbewegung. Diese beiden Funktionen, mit $\chi(\xi)$ und $\frac{1}{r} \cdot u(r)$ bezeichnet, haben aber ganz andere Bedeutungen als die mit $\Phi(\xi, r)$ und $\frac{1}{r} v(r)$ bezeichneten des § 2.

$$\Psi_{\sigma J \Lambda M T} = \frac{1}{r} u_{\sigma J \Lambda M T}(r) \cdot \Theta_{J M \Lambda}(\vartheta)\, e^{i M \varphi} \cdot {\sin \atop \cos} \Lambda\varphi \cdot \chi_{\sigma\Lambda}(\xi)$$
$$\cdot \exp\left[-\frac{2\pi i}{h}(E_{\sigma\Lambda\infty} + T)\, t\right]$$

wo χ die Gleichung

$$\left\{\Delta_{\xi_i \eta_i \zeta_i \ldots \varphi} + \frac{8\pi^2 m}{h^2}\left[E_{\sigma\Lambda\infty} - V(\xi_i \eta_i \zeta_i \ldots, r = \infty)\right]\right\}$$
$$\chi_{\sigma\Lambda}(\xi_i \eta_i \zeta_i \ldots) {\sin \atop \cos} \Lambda\varphi = 0 \qquad (1a)$$

befriedigt. Die dem ersten Störungsglied I in (4) entsprechende Wechselwirkungsenergie heisst jetzt:

$$W^{\mathrm{I}}(\xi_i \eta_i \zeta_i, r) = V(\xi_i \eta_i \zeta_i, r) - V(\xi_i \eta_i \zeta_i, r = \infty). \qquad (4a)$$

An Stelle des Operators haben wir jetzt eine gewöhnliche Funktion. Es treten natürlich auch noch dem Gliede II in (4) analoge Operatoren auf, welche $\frac{\partial}{\partial\varphi}$ und $\frac{\partial}{\partial\vartheta}$ enthalten. Die χ sind natürlich die Eigenfunktionen der getrennten Atome*).

An Stelle von (5) steht jetzt im Falle I

$$u''_{0 J_0 T_0} + \frac{8\pi^2 M}{h^2}\left(T_0 - W^{\mathrm{I}}_{00}(r)\right) u_{0 J T_0} = \frac{8\pi^2 M}{h^2} W^{\mathrm{I}}_{01}(r)\, u_{1 J T_1} \qquad (5a)$$

und die entsprechende Gleichung für $u_{1 J T_1}$, wobei

$$W^{\mathrm{I}}_{ik} = \int d\xi \cdot \overline{\chi}_i(\xi) \cdot W^{\mathrm{I}}(\xi) \cdot \chi_k(\xi) + \frac{h^2}{8\pi^2 M} \delta_{ik} \frac{J_i^2}{r^2};\ \delta_{ik} = {0\ i \neq k \atop 1\ i = k}.$$

Das Resultat ist auch hier *zwei gekoppelte Gleichungen,* nur haben sie den Vorteil gegenüber (5), dass die Kopplung statisch und nicht dynamisch erscheint. Die gegenseitigen Vorteile dieser zwei Ausgangsgleichungen werden im § 7 diskutiert werden. Die Grenzbedingungen für die u sind selbstverständlich dieselben wie die für die v (§ 3, Gl. (8a) und (8b).

---

*) Streng genommen[7]) die Eigenfunktionen bei fehlender Wechselwirkung. Diese Definition unterscheidet sich eventuell gegen diejenige im Text. Die Darstellung (4a) hat dann wie (1a), rein formale Bedeutung.

§ 5. Beispiel einer strengen Lösung*).

Da nach Voraussetzung überhaupt nur zwei Elektronenzustände eine Rolle spielen sollen, so müssen die strengen Lösungen der Probleme im § 2 (5) und im § 4 (5a) dasselbe Resultat liefern. Wir wollen in diesen Paragraphen vorübergehend einige weitere einfache Annahmen machen, ohne ihre Beziehungen zur Wirklichkeit zu diskutieren. Wir gehen von Gleichung (5a) aus und nehmen an, dass

$$W_{00}(r) = W_{11}(r) = \frac{\beta^2}{r^2} \cdot \frac{h^2}{8\pi^2 M}$$

sei, ferner dass

$$W_{01} = \text{const} = \frac{h^2}{8\pi^2 M} \cdot \alpha$$

sei für $r < r_0$ und gleich Null für $r > r_0$. Setzt man noch

$$\tfrac{1}{2}(p_0^2 - p_1^2) = \psi > 0.$$

$$\tfrac{1}{2}(p_0^2 + p_1^2) = p^2$$

$$p^2 \pm \sqrt{\psi^2 + \alpha^2} = \nu_{\substack{0\\1}}^2$$

$$f_0 = \frac{\alpha}{p_0^2 - \nu_1^2}\,; \quad f_1 = \frac{\alpha}{p_1^2 - \nu_0^2}\,; \quad f^2 = f_0 f_1$$

dann lauten (5a), wenn $J^2 = J\,(J+1) - \Lambda^2 + \beta^2$

$$\left.\begin{aligned} u_0'' + \left(p_0^2 - \frac{J^2}{r^2}\right) u_0 &= \alpha u_1 \\ u_1'' + \left(p_1^2 - \frac{J^2}{r^2}\right) u_1 &= \alpha u_0 \end{aligned}\right\} \quad \alpha = 0 \text{ für } r > r_0.$$

Die Lösungen, welche im Nullpunkt endlich sind lauten:

$$\text{für: } r < r_0 \quad u_0 = r^{\frac{1}{2}} \cdot [c_0 \cdot J_J(\nu_0 r) + c_1 \cdot f_0 \cdot J_J(\nu_1 r)]$$

$$u_1 = r^{\frac{1}{2}} \cdot [c_0 \cdot f_1 \cdot J_J(\nu_0 r) + c_1 \cdot J_J(\nu_1 r)]$$

und wegen (8a) und (8b).

$$\text{für: } r > r_0 \quad u_0 = r^{\frac{1}{2}} \cdot [J_J(p_0 r) + \eta_{0J} \cdot H_J^{(1)}(p_1 r)]$$

$$u_1 = r^{\frac{1}{2}} \cdot [\eta_{1J} \cdot H_J^{(1)}(p_1 r)].$$

Gleich setzen von beiden Werten u_0 (resp. u_1) in Grösse und Neigung

*) Dieses Beispiel findet sich ausführlich bei: P. M. MORSE, Rev. Mod. Phys. **4**, 632 (1932), dem der Verfasser das Beispiel zur Verfügung stellte.

im Punkte $r = r_0$, gibt 4 lineare unhomogene Gleichungen zur Bestimmung der 4 Konstanten c_0, c_1, η_0, η_1. Die Lösung für η_{1J} schreibt sich in der Form einer Determinante aus Bessel'schen Funktionen und deren Ableitungen. Der Zähler ist:

$$f_1 p_0 \cdot [H_J(p_0 r_0) \cdot J'_J(p_0 r_0) - H'_J(p_0 r_0) \cdot J_J(p_0 r_0)]$$
$$\cdot [J_J(\nu_0 r_0) \cdot \nu_1 J'_J(\nu_1 r_0) - \nu_0 J'_J(\nu_0 r_0) \cdot J_J(\nu_1 r_0)]$$
$$= \frac{2 i f_1}{\pi r_0^2} (\nu_0^2 - \nu_1^2) \int_0^{r_0} r\,dr \cdot J_J(\nu_0 r) \cdot J_J(\nu_1 r).$$

Der Nenner hat für $\alpha^2 \ll \psi^2$ $(|f^2| \ll 1)$ den Wert:

$$-\frac{4}{\pi^2 r_0^2}$$

und für $\alpha^2 \gg \psi^2$ $(f^2 = -1)$ den Wert:

$$-\frac{8}{\pi^2 r_0^2}$$

(solange als $\nu_1 r_0$ reell und grösser als J ist).

Man erhält daher, in beiden Fällen wegen der Definition von ν_i und f_i:

$$|\eta_{1J}| = \frac{\pi}{2} \cdot \int_0^\infty r\,dr \cdot \alpha \cdot J_J(\nu_0 r) \cdot J_J(\nu_1 r). \tag{12}$$

Unter α ist hier eine Funktion verstanden, welche konstant $= \alpha$ ist für $0 < r < r_0$ und $= 0$ für $r > r_0$. Für imaginäres ν_1, wo $\nu_1 r_0 > iJ$ ist wird

$$|\eta_{1J}|^2 = \frac{\sin x}{\sqrt{\frac{2\alpha}{p_0 p_1}} \cdot \cos\left(x + \frac{\pi}{4}\right) - i \sqrt{\frac{p_0 + p_1}{p_0 p_1}} \cdot \cos x};$$
$$x = \nu_1 r - \frac{2J - 1}{4}.$$

§ 6. Die angenäherten Lösungen (Störungsrechnung).

Wir können die Gleichungen (5), (5a), (5b) alle in der Form schreiben:

$$\begin{aligned} y''_0 + K_0(r)\, y_0 &= L(r)\, y_1 \\ y''_1 + K_1(r)\, y_1 &= L(r)\, y_0 \end{aligned} \tag{13}$$

wo L in (5) ein linearer Operator ist. Für *kleine Störungen* $L(r)$ (schwache Kopplung), setzen wir in erster Näherung $y_1 = 0$ und wählen für y_0 die im Nullpunkte endliche Lösung, welche wir, nach Wentzel, Kramers, Brillouin[2]) für positive K_0 wie

$$y_0 = y_0^{\mathrm{I}} \simeq \sqrt{\frac{2}{\pi}}\,\frac{1}{K_0^{\frac{1}{4}}}\cos\left(\int_{K_0=0}^{r} K_0^{\frac{1}{2}}\,dr - \frac{\pi}{4}\right)$$

normieren können. (Im Falle $W_{00} \infty 1/r^2 : y_0 = r^{\frac{1}{2}} \cdot J_J(p_0 r)$). Die homogene Gleichung für y_1 hat ebenfalls eine im Nullpunkt endliche Lösung y_1^{I} und eine unendliche Lösung y_1^{II}. Wir wählen die beiden so, dass

$$y_1^{\mathrm{I}} + y_1^{\mathrm{II}} \simeq \sqrt{\frac{2}{\pi}} \cdot \frac{1}{K_1^{\frac{1}{4}}} \cdot \exp\left[i\left(\int_{K_1=0}^{r} K_1^{\frac{1}{2}}\,dr - \frac{\pi}{4}\right)\right]$$

normiert ist. Wegen (13) gilt streng[10]):

$$y_1^{\mathrm{I}} \cdot y_1^{\mathrm{II}\prime} - y_1^{\mathrm{I}\prime} \cdot y_1^{\mathrm{II}} = \frac{2}{\pi}.$$

Die Lösung der unhomogenen Gleichung heisst[10]):

$$y_1 = \left[c_1 + \frac{\pi}{2}\int_{\infty}^{r} d\varrho \cdot y_0^{\mathrm{I}}(\varrho) \cdot L(\varrho) \cdot y_1^{\mathrm{II}}(\varrho)\right] \cdot y_1^{\mathrm{I}}(r)$$

$$+ \left[c_2 - \frac{\pi}{2}\int_{\infty}^{r} d\varrho \cdot y_0^{\mathrm{I}}(\varrho) \cdot L(\varrho) \cdot y_1^{\mathrm{I}}(\varrho)\right] \cdot y_1^{\mathrm{II}}(r).$$

Die Grenzbedingungen im Unendlichen geben $c_2 = i c_1 = i\eta_{1J}$. Für $r = 0$ müssen wir verlangen, dass

$$\lim_{r=0}\left[\int_{\infty}^{r} d\varrho \cdot y_0^{\mathrm{I}} \cdot L \cdot y_1^{\mathrm{II}}\right] \cdot y_1^{\mathrm{I}}(r) = 0,$$

was eine gewisse Beschränkung auf die Störungsfunktion $L(r)$ auferlegt. Da aber die Lösungen y_0^{I} für $r = 0$ stark verschwinden (Abstossung der Kerne), so wird dieser Bedingung im ganzen genügt werden. Ferner muss der Faktor von $y_1^{\mathrm{II}}(r)$ für $r = 0$ zu null werden. Wegen (8b) gibt das:

$$-i c_2 = \eta_{1J} = +i\frac{\pi}{2}\int_0^{\infty} dr \cdot y_1^{\mathrm{I}} \cdot L \cdot y_0^{\mathrm{I}}. \tag{14}$$

[10]) Franck und R. v. Mises, Dif. Gleich. d. Phys. p. 299 und p. 300 (1930).

Auf den *Fall des vorhergehenden Paragraphen* angewandt, gibt (14):

$$\eta_{1J} = \frac{\pi}{2}\int\limits_0^\infty \alpha \cdot r\,dr : J_J(p_0 r) \cdot J_J(p_1 r). \qquad (15)$$

Der Unterschied zwischen (12) und (15) besteht darin, dass der Näherungsausdruck andere Frequenzen p_i enthält, an Stelle von ν_i im exakten Ausdruck (12). Da das Integral im wesentlichen von der Resonanzschärfe $\frac{1}{2}(p_0^2 - p_1^2) = \psi$ abhängt, so entspricht der Fehler in (15) der Verwendung von ψ an Stelle von $\sqrt{\psi^2 + \alpha^2}$. Wir wollen das nun noch allgemeiner beweisen.

§ 7. Vergleich der London'schen mit der Born-Dirac'schen Näherung.

Dazu wendet man das im vorangehenden Paragraphen beschriebene Verfahren auf die Gleichung (5) (London[7])) und (5a) (Morse und Stueckelberg[6])) an. Die Lösungen von (1a) seien als bekannt vorausgesetzt. Haben wir nur zwei Zustände in Betracht zu ziehen, so können die beiden (1) genügenden *adiabatischen molekülaren Wellenfunktionen* $\Phi_\sigma(r, \xi)$ durch die beiden Funktionen $\chi_\sigma(\xi)$ (1a) exakt ausgedrückt werden ($\sigma = 0{,}1$). Wir machen mit London[7]) den, vorerst willkürlichen Ansatz.

$$\begin{aligned} \Phi_0 &= \chi_0 \cos g + \chi_1 \sin g \\ \Phi_1 &= -\chi_0 \sin g + \chi_1 \cos g \end{aligned} \qquad (16)$$

Die χ_σ hängen definitionsgemäss nur von ξ ab. (Dabei stellen wir uns vor, dass die ξ_i eines zum Atom I gehörenden Elektrons vom Schwerpunkt des Atoms I aus gezählt werden usw. Näheres darüber siehe bei London[7]) in seinem § 2 und bei Morse, Stueckelberg[6]) in ihrem § 3). Die vorerst noch unbestimmte Funktion g hänge nur von r ab. Setzen wir die χ_σ als normiert und orthogonal voraus, so sieht man leicht, dass auch die beiden Φ_σ für jeden beliebigen reellen Wert von $g(r)$ normal und orthogonal sind. Das gilt selbstverständlich nur solange als man annehmen darf, dass nur zwei Elektronenzustände existieren.

Der Hamilton'sche Operator der Wellengleichung der Elektronen für getrennte Atome lautet ((1a) in § 4):

$$H(\xi_1 \ldots \zeta_n, \varphi) = -\frac{h^2}{8\pi^2 m} \Delta_{\xi_1 \ldots \zeta_n, \varphi} + V(\xi_1 \ldots \zeta_n, r = \infty).$$

Der entsprechende Operator der Molekülgleichung (1) im § 2 ist dann

$$H(\xi_1 \ldots \zeta_n, \varphi) + W(\xi_1 \ldots \zeta_n, r),$$

wobei W durch Gleichung (4a) im § 4 plus die dort erwähnten von $\frac{\partial}{\partial \varphi}$ und $\frac{\partial}{\partial \vartheta}$ abhängigen Operatoren definiert ist. Die Φ_σ müssen der folgenden Gleichung genügen. (Die χ_σ und daher die Φ_σ seien reell.):

$$\int \Phi_i (H + W) \Phi_k \, d\xi_1 \ldots d\zeta_n = \delta_{ik} E_i(r)$$

$$\int \chi_i H \chi_k \, d\xi_1 \ldots d\zeta_n = \delta_{ik} E_i(\infty) = \delta_{ik} E_{i\infty}.$$

Die Indizes i und k stehen an Stelle der Indices $\sigma_0 = 0$ und $\sigma_1 = 1$ in §§ 2 und 4. Der hier belanglose Index Λ ist weggelassen. Ausser $E_i(r)$ und $E_{i\infty}$ tritt noch das (für den Fall I (W^{I}_{01}) bereits im § 4 definierte) Integral (Gleichung 5a)) auf*):

$$W_{ik}(r) = \int \chi_i \, W(\xi_1 \ldots \zeta_n, r) \, \chi_k \, d\xi_1 \ldots d\zeta_n.$$

Aus der Gleichung für $i \neq k$ d. h. $\delta_{ik} = 0$ erhält man

$$\operatorname{tang} 2g(r) = \frac{2 W_{01}(r)}{[E_{0\infty} + W_{00}(r)] - [E_{1\infty} + W_{11}(r)]}$$

und aus den beiden Gleichungen für $i = k = 0$ resp. $= 1$, d. h. $\delta_{ik} = 1$ bestimmen sich die Termwerte zu:

$$E_{\substack{0\\1}}(r) = \frac{[E_{0\infty} + W_{00}] + [E_{1\infty} + W_{11}]}{2}$$
$$\pm \sqrt{\frac{\{[E_{0\infty} + W_{00}] - [E_{1\infty} + W_{11}]\}^2}{4} + W_{01}^2}.$$

Die beiden Termwerte (Franck'sche Kurven) können sich also in dieser Betrachtungsart bei von Null verschiedenen W_{01} für reelle Abstände r nicht überschneiden, sondern sich nur nahe kommen (siehe [1])). Wie in § 5 erleichtern folgende Abkürzungen die Schreibweise (W_{01} steht für W^{I}_{01} oder W^{II}_{01}):

$$\frac{8\pi^2 M}{h^2} \cdot \frac{[E_{0\infty} + W_{00}] - [E_{1\infty} + W_{11}]}{2} - \psi(r); \quad \psi(\infty) \geqslant 0;$$

*) In Gl. (5a) enthalten die Diagonalelemente W_{ii} noch das von der Rotation der Kerne herrührende Zusatzglied $h^2 J^2 / 8\pi^2 M r^2$. In der Schreibweise dieses Paragraphen sind diese Glieder *nicht* enthalten, sondern sie werden erst in (17) explicite aufgeführt.

$$\frac{8\,\pi^2 M}{h^2} W_{01} = \alpha(r);$$

$$\frac{8\,\pi^2 M}{h^2} T_i = p_i^2;\; p^2 = \tfrac{1}{2}(p_1^2 + p_0^2);$$

$$g(r) = \tfrac{1}{2}\operatorname{tang}^{-1}\frac{\alpha}{\psi}$$

$$\nu^2_{\substack{0\\1}} = p^2 - \frac{8\,\pi^2 M}{2\,h^2}\left[W_{00}(r) + W_{11}(r)\right] - \frac{J^2}{p^2} \pm \sqrt{\psi^2(r) + \alpha^2(r)};$$

$$\mu^2_{\substack{0\\1}} = p^2_{\substack{0\\1}} - \frac{8\,\pi^2 M}{h^2} W_{\substack{00\\11}}(r) - \frac{J^2}{r^2} = \varphi_{\substack{0\\1}}.$$

Wie bereits erwähnt, sind die Φ_i orthogonal und normal, wenn die χ_i es waren. Es ist weiter noch:

$$\int d\,\xi\,\Phi_{\substack{0\\1}}\,\Phi'_{\substack{0\\1}} = 0;\quad \int d\,\xi\,\Phi_{\substack{0\\1}}\,\Phi'_{\substack{1\\0}} = \mp g'$$

$$\int d\,\xi\,\Phi_{\substack{0\\1}}\,\Phi''_{\substack{0\\1}} = -g'^2;\quad \int d\,\xi\,\Phi_{\substack{0\\1}}\,\Phi''_{\substack{1\\0}} = \mp g''$$

(5) wird, wenn man die Glieder mit g'^2 gegen ν_i^2 vernachlassigt:

$$\begin{aligned} v_0'' + \nu_0^2\,v_0 &= 2\,g'\,v_1' + g''\,v_1 \\ v_1'' + \nu_1^2\,v_1 &= -2\,g'\,v_0' - g''\,v_0 \end{aligned} \tag{18}$$

entsprechend wird (5a):

$$\begin{aligned} u_0'' + \mu_0^2\,u_0 &= \alpha\,u_1 \\ u_1'' + \mu_1^2\,u_1 &= \alpha\,u_0 \end{aligned} \tag{18a}$$

Die *Gleichungen* (18) und (18a) sind *äquivalent.* Die *angenäherte Bestimmung von* $|\eta_{1J}|^2$ (14) auf (18) und (18a) angewandt kann natürlich trotzdem zu zwei *verschiedenen Resultaten* führen.

Aus (18) folgt durch partielle Integration, da g und alle vorkommenden Ableitungen von g für $r = \infty$ zu Null werden, und v_i und alle vorkommenden Ableitungen von v_i für $r = 0$ Null sind*):

$$\int_0^\infty v_0\,v_1\,g''\,d\,r = \int_0^\infty g\,d\,r\,[v_0''\,v_1 + 2\,v_0'\,v_1' + v_0\,v_1'']$$

$$2\int_0^\infty v_0'\,v_1\,g'\,d\,r = \int_0^\infty g\,d\,r\,[-2\,v_0''\,v_1 - 2\,v_0'\,v_1']$$

*) Unter v_i und u_i sind hier, der Anwendung von § 6, Gl. (14) entsprechend, die Lösungen der homogenen Gleichungen verstanden, welche im Nullpunkt endlich sind, d. h. y_i^{I}.

Wegen (14):

$$|\eta_{1J}| = \frac{\pi}{2}\int_0^\infty \tfrac{1}{2} \cdot \mathrm{tg}^{-1}\frac{\alpha}{\psi} \cdot (v_0 v_1'' - v_0'' v_1) \cdot d\,r$$

oder wegen (18), und weil $\frac{1}{2}(\nu_0^2 - \nu_1^2) = \sqrt{\psi^2 + \alpha^2}$:

$$|\eta_{1J}| = \frac{\pi}{2}\int_0^\infty \sqrt{\psi^2 + \alpha^2}\,\mathrm{tg}^{-1}\frac{\alpha}{\psi} \cdot v_0 v_1 \cdot d\,r. \tag{19}$$

Entsprechend erhält man aus (18a)

$$|\eta_{1J}| = \frac{\pi}{2}\int_0^\infty \alpha \cdot u_0 u_1 d\,r. \tag{19a}$$

In (19) ist aber: $\sqrt{\psi^2 + \alpha^2}\ \mathrm{tg}^{-1}\frac{\alpha}{\psi} \simeq \alpha$ für $\alpha^2 < \psi^2$ und $\cong \frac{\pi}{2}\alpha$ für $\alpha^2 \gg \psi^2$ oder $\sqrt{\psi^2 + \alpha^2}\ \mathrm{tg}^{-1}\frac{\alpha}{\psi} = \delta \cdot \alpha$, wo δ ein Faktor der Grössenordnung 1 ist ($1 < \delta < 1.57$), eine genügende Annäherung. Der Unterschied zwischen (19) und (19a) ist also von derselben Art, wie derjenige zwischen der exakten Lösung (12) und der angenäherten (15). Wie London[7]) zeigte, stellt (19a) die Summe von adiabatischer und nicht adiabatischer Zustandsänderung ($0 \rightarrow 1$) dar, während das *wegen schlechterer Resonanz kleinere* (19) allein die, von uns gesuchten, nicht adiabatischen Übergänge $0 \rightarrow 1$ enthält. *Ist also im Bereiche, wo u_{1J} von Null verschieden ist $\alpha \ll \psi$, so gibt die dem (Born-Dirac'schen) Näherungsverfahren von* Morse *und dem Verfasser*[6]) *zugrunde gelegte Gleichung* (19a) *schon die richtige Antwort.*

Da q in eine Summe über J zerfällt, so darf für alle Glieder J, wo

$$|\alpha_{\max}(r)| = |\alpha(r_{\min})| = \left|\alpha\left(\frac{J}{p_0}\right)\right| \ll |\psi|,$$

das Verfahren von Morse und Stueckelberg[6]) angewandt werden*).

*) In Gl. (17) dieses Paragraphen steht W_{01} für W_{01}^{I} und W_{01}^{II}. Wird dieser Paragraph als logische Entwicklung der Paragraphen 2 und 4 gelesen, so darf das in (17) und weiter oben vorkommende W_{01} nur W_{01}^{I} bedeuten. Für das Folgende wollen wir eine etwas andere Übereinkunft treffen:

Die Parallelität zwischen u_i mit den Frequenzen μ_i und mit v_i mit den Frequenzen ν_i, wollen wir so verallgemeinern, dass die μ_i^2 jeweils ein Mass für die Termgrössen ohne die für die Übergänge verantwortliche Störung $\alpha = 8\pi^2 M W_{01}^{\mathrm{I}}/h^2$ oder $8\pi^2 M W_{01}^{\mathrm{II}}/h^2$ bedeuten, während die ν_i^2 die Termgrössen mit eingeschalteter Störung darstellen.

Wir werden daher im folgenden von der *exakten adiabatischen Lösung* (Gl. (5) für Fall I und Gl. (5b) für Fall II) ausgehen, soweit sich das als möglich erweist. Beim Fall II wird das immer möglich sein, während beim Fall I eine Schwierigkeit auftreten wird, die darin besteht, dass die nicht-adiabatische Wechselwirkung, welche die Übergänge induziert, nicht lediglich eine Funktion sondern ein Operator ist. Dieser Schwierigkeit können wir nur entgehen, wenn wir eine London'sche Näherung einschalten, die in der Tat (siehe (18)) die dynamische in eine statische Wechselwirkung überführt.

§ 8. Das Anschlussverfahren der Näherungsfunktionen in den Umkehrpunkten.

Wenn man versucht das Problem der gekoppelten Gleichungen streng zu lösen, so ist es wie in § 5 methodisch gleichgültig, ob man (5), (5a) oder (5b) betrachtet, sofern nur die Kopplung durch eine Funktion, nicht einen Operator dargestellt wird. Beide Gleichungspaare sind von Typus:

$$u_0'' + \varphi_0 u_0 = \alpha u_1$$
$$u_1'' + \varphi_1 u_1 = \alpha u_0,$$

wo α im Fall I die Funktion $8\pi^2 M W^{\mathrm{I}}_{01}/h^2$ und im Fall II die Funktion $8\pi^2 M W^{\mathrm{II}}_{01}/h^2$ darstellt. Im Fall II legen wir unseren Rechnungen also die exakten adiabatischen Näherungsfunktionen zugrunde. Im Fall I können wir das nicht tun, weil die folgende Auflösungsmethode des Gleichungspaares nicht auf Operatorenkopplung anwendbar ist. Wir gehen daher von den Gleichungen (18a) aus, in welchen die Operatorenkopplung durch eine Funktionenkopplung ersetzt ist. Wie im § 7 gezeigt wurde, bedeutet das, dass die exakten adiabatischen Näherungsfunktionen durch die London'schen Näherungsfunktionen ersetzt sind.

Durch Elimination erhalten wir in beiden Fällen:

$$u_0^{\mathrm{IV}} - 2\frac{\alpha'}{\alpha}\cdot u_0^{\mathrm{III}} + \left[\varphi_0 + \varphi_1 - \frac{\alpha''}{\alpha} - 2\frac{\alpha'^2}{\alpha^2}\right]\cdot u_0''$$
$$+ \left[2\varphi_0' - 2\frac{\alpha'\varphi_0}{\alpha}\right]\cdot u_0' + \left[\varphi_0\varphi_1 - \alpha^2 - 2\alpha'\left(\frac{\varphi_0}{\alpha}\right)' - \alpha''\frac{\varphi_0}{\alpha}\right]\cdot u_0 = 0 \quad (20)$$

und

$$u_1 = \frac{1}{\alpha}\cdot[u_0'' + \varphi_0 u_0]. \quad (21)$$

Wir erinnern uns, dass φ_i und α Grössen bedeuten, welche mit $\frac{1}{h^2}$ multipliziert sind. Wir schreiben deshalb statt (20), (20a).

$$u_0^{\mathrm{IV}} + D \cdot u_0^{\mathrm{III}} + \frac{A}{h^2} \cdot [1 + h^2 \cdot A_2 + \ldots] \cdot u_0'' + \frac{B}{h^2} \cdot u'$$
$$+ \frac{C}{h^4} \cdot [1 + h^2 \cdot C_2 + \ldots] \cdot u_0 = 0, \qquad (20a)$$

wo

$$A \cdot \frac{1}{h^2} = \varphi_0 + \varphi_1 = 2\varphi \qquad A_2 = \ldots$$
$$B \cdot \frac{1}{h^2} = 2 \cdot \left[\varphi_0' - \frac{\varphi_0}{\alpha}\,\alpha'\right]$$
$$C \cdot \frac{1}{h^4} = \varphi_0\,\varphi_1 - \alpha^2 \qquad C_2 = \ldots$$
$$D = -2\,\frac{\alpha'}{\alpha} \qquad \text{ist.}$$

Entsprechend dem W. K. B.[2]) Verfahren, setzt man auch hier

$$u_0 = e^{\frac{1}{h} \cdot (S_0 + h\,S_1 + h^2\,S_2 + \ldots)}$$

und findet für die erste Näherung:

$$S_0'^4 + A\,S_0'^2 + C = 0; \quad \frac{S_0'^2}{h^2} = i^2\,\nu^2 = i^2\,[\varphi \pm \sqrt{\psi^2 + \alpha^2}];$$

$$S_1' = -\tfrac{1}{2}\,\frac{\nu'}{\nu} - \tfrac{1}{2}\,\frac{(\nu^2 - \varphi_0)' - \frac{\alpha'}{\alpha}\,(\nu^2 - \varphi_0)}{\nu^2 - \varphi}$$

usw.*). Es treten also die gleichen Frequenzen ν_0 und ν_1 auf, wie im § 5 Gl. (17). In zweiter Näherung lautet daher die Lösung von (24) oder von (18a) nach einigem Umformen, wenn $t = \frac{\psi}{\alpha}$:

$$u_0 \simeq \Bigg[\Big(c_+ \cdot e^{i\int \nu_0\,dr} + c_- \cdot e^{-i\int \nu_0\,dr}\Big) \cdot \exp \tfrac{1}{2}\left(-\int \frac{d\,\nu_0}{\nu_0} + \int \frac{d\,t}{\sqrt{1 + t^2}}\right)$$
$$+ \Big(d_+ \cdot e^{i\int \nu_1\,dr} + d_- \cdot e^{-i\int \nu_1\,dr}\Big) \cdot \exp \tfrac{1}{2}\left(-\int \frac{d\,\nu_1}{\nu_1} - \int \frac{d\,t}{\sqrt{1 + t^2}}\right)\Bigg] \cdot e^{-\frac{1}{2}\int \frac{t\,dt}{1 + t^2}} \qquad (26)$$

*) Wenn man die Glieder D und B/h^2 schon in nullter Näherung berücksichtigt, so erhält man vier voneinander verschiedene komplexe Grundfrequenzen, wo nicht $\nu_{0+} = -\nu_{0-}$ ist. Das bedeutet für das Folgende, dass die Umkehrpunkte $\nu_{0+} = \nu_{0-}$ im Komplexen liegen. Die übrige Behandlung bleibt sich aber gleich.

resp., wenn $z = \frac{\alpha}{\psi}$:

$$u_0 \simeq \Big[(c_+ \cdot e^{i\int \nu_0 dr} + c_- \cdot e^{-i\int \nu_0 dr}) \cdot \exp \tfrac{1}{2}\Big(-\int \frac{d\nu_0}{\nu_0} - \int \frac{dz}{z\sqrt{1+z^2}}\Big)$$

$$+ (d_+ \cdot e^{i\int \nu_1 dr} + d_- \cdot e^{-i\int \nu_1 dr}) \cdot \exp \tfrac{1}{2}\Big(-\int \frac{d\nu_1}{\nu_1}$$

$$+ \int \frac{dz}{z\sqrt{1+z^2}}\Big)\Big] \cdot e^{\frac{1}{2}\int \frac{dz}{z(1+z^2)}}$$

Die Annäherung ist überall gut, ausser in der Nähe der Punkte $\nu_0 = 0$, $\nu_1 = 0$, $\nu_0 \simeq \nu_1$, da dort die von ν_i oder t resp. z abhängigen Amplituden rasch variieren. Beim Durchgang resp. Umfahren von diesen Punkten treten sprungweise Änderungen in den Konstanten c_+, c_-, d_+, d_- auf. Für eine Gleichung zweiten Grades (Punkte $\nu_i = 0$) wurde das Problem von KRAMERS[2]) für den Durchgang auf der reellen Axe, und von ZWAAN[11]) für die Umfahrung auf der komplexen r-Ebene durchgeführt. Wir wollen hier ein dem Zwaan'schen ähnliches Verfahren anwenden, welches sich auch auf die Punkte $\nu_0 = \nu_1$ ausdehnen lässt. Dazu muss angenommen werden, dass die Funktionen φ_0, φ_1 und α im betrachteten kleinen Gebiet analytisch sind. Dann muss auch u_0 analytisch sein. Beim Umfahren jedes Punktes $\nu_0 = 0$, $\nu_1 = 0$, $\nu_0 = \nu_1$ muss darum u_0 auf seinen Anfangswert zurückkehren. Die Näherungsausdrücke tun das aber nicht. Darum müssen die Konstanten sich sprungweise an bestimmten Stellen ändern. Da sich ν_0 mit $-\nu_0$ beim Umfahren von $\nu_0 = 0$, und $\nu_0 - \nu_1$ mit $-(\nu_0 - \nu_1)$, d. h. ν_0 mit ν_1 beim Umfahren von $\nu_0 = \nu_1$ vertauscht, so können diese Sprünge nur proportional den vorhandenen Konstanten $c_+ \, d_+ \, c_- \, d_-$ sein (siehe STOKES[12])).

Beim Umfahren von $\nu_0 = 0$ betrachtet man

$$\nu_0^{-\frac{1}{2}} \cdot \left[c_+ \cdot e^{i\int_0^x \nu_0 dx} + c_- \cdot e^{-i\int_0^x \nu_0 dx}\right].$$

Da ν_0 die Wurzel aus einer im Punkte $\nu_0 = 0$, $x = 0$ verschwindenden analytischen Funktion ist, so muss, wenn c_+', c_-' die Konstanten nach der Umfahrung bedeuten:

$$c_+' = c_- \cdot e^{i\frac{\pi}{2}}; \quad c_-' = c_+ \cdot e^{i\frac{\pi}{2}}.$$

---

11) A. ZWAAN, Diss., Utrecht 1929.

12) STOKES, Collected papers, siehe auch G. N. WATSON, Theory of Bessel Functions, Cambridge 1922.

Es stellt sich jetzt die Frage, wo die Konstanten springen können. STOKES[11][12]) zeigt, dass die verwendeten Näherungen, welche den asymptotischen Darstellungen entsprechen, in den Richtungen vom Punkte $x = 0$ aus, wo das Verhältnis der beiden Funktionen

$$\exp\left(\pm i \int_0^x \nu_0 \, d\, x\right)$$

ein Extremum wird, der Fehler in der grösseren Funktion grösser ist, als der absolute Betrag der Kleinern. Die Konstante der Kleinern darf sich also dort sprungweise ändern. *Dies gilt nur, wenn die betrachteten Unstetigkeitslinien, ohne von andern Punkten kommende Unstetigkeitslinien zu durchkreuzen, sich zu solchen Entfernungen verfolgen lassen, wo*

$$\exp\left(\pm \int_0^x \nu_0 \, d\, x\right)$$

beliebig gross resp. klein wird. Die Gleichung der Unstetigkeitslinien von c_+ ist*):

$$\arg\left(+ i \int_0^x \nu_0 \, d\, x\right) = \pi + (\text{ganze Zahl} \times 2\,\pi).$$

Bezeichnet $\lambda(x)$ das Argument von $\nu_0\,(x)$ und φ' das Argument von $d\,x$, so führt (wenn $x = \varrho \cdot e^{i\varphi}$ ist),

$$\lambda\,(\varrho\,,\ \varphi) + \varphi' = \pi + (\text{ganze Zahl} \times 2\,\pi)$$

auf die Differentialgleichung:

$$\operatorname{tang}^{-1}\left(\frac{\varrho\, d\,\varphi}{d\,\varrho}\right) + \varphi + \lambda\,(\varrho,\, \varphi) - \pi = (\text{ganze Zahl} \times 2\,\pi) + 0.$$

Man sieht leicht, dass die Extremalrichtungen den Punkt $x = 0$ unter den Winkeln

$$\varphi = \frac{\pi}{3} \text{ und } \frac{5\,\pi}{3}$$

für Sprünge von c_+, und unter $\varphi = \pi$ für Sprünge von c_- verlassen (siehe Fig. 2). Wir nehmen jetzt an, dass beim Überschreiten dieser Richtungen c_+ sich proportional c_- ändert und umgekehrt.

---

¹) Ist a eine komplexe Zahl $a = |a| \cdot e^{i\,\psi}$, so bedeutet

$$\arg\ a - (\log\ a - \log|a|)/i - \varphi.$$

In diesem und in folgenden § sind die Buchstaben φ, ψ und ϑ als Argumente von komplexen Zahlen gebraucht Eine Verwechslung mit den Grössen φ und ψ, definiert in Gleichung (17) § 7, welche Termwerte resp. -differenzen darstellen, ist nicht zu befürchten; ebensowenig eine mit den Eulerschen Winkeln der § 2 u. § 4.

(Würde man die Sprünge noch proportional sich selbst und proportional den Konstanten d_+ und d_- annehmen, so erhalten nämlich diese weiteren Proportionalitätskonstanten den Wert Null.)

Wir haben daher: Anfangswert: c_+; c_-

nach Überschreiten von $\varphi = \pi/3$:

$$c_+ + \alpha\, c_-\,;\; c_-{}^{*)}$$

nach Überschreiten von $\varphi = \pi$:

$$c_+ + \alpha\, c_-\,;\; \beta c_+ + (1 + \alpha\beta)\, c_-$$

nach Überschreiten von $\varphi = 5\,\pi/3$:

$$c_+ (1 + \beta\gamma) + c_- (\alpha + \gamma + \alpha\beta\gamma);\; \beta c_+ + (1 + \alpha\beta)\, c_-$$

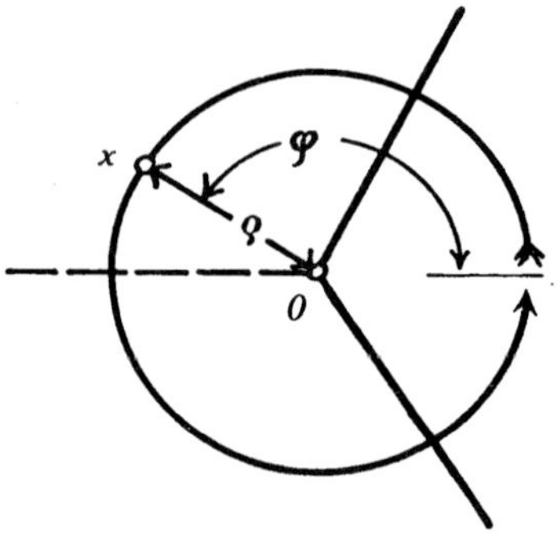

Fig. 2.
x-Ebene in der Umgebung von $x = 0$. Die Unstetigkeitslinien verlassen den Punkt $x = 0$ unter den Winkeln: $\pi/3$, π, $5\,\pi/3$.

Die letzten Werte müssen gleich $c_- \cdot e^{i\frac{\pi}{2}}$ resp. $c_+ \cdot e^{i\frac{\pi}{2}}$ sein. Die Konstanten α, β, γ sind also bis auf den belanglosen Faktor $e^{2\pi i}$ bestimmt zu:

$$\alpha = \beta = \gamma = e^{i\frac{\pi}{2}}\,.$$

Verlangt man jetzt, dass auf der negativen reellen x-Axe, wo ν_0 imaginär ist, nur der negative reelle Exponent auftritt ($c_+ = 0$, $c_- = c$), so lautet die Lösung auf der positiven x-Axe, nach Überschreiten von $\varphi = 5\,\pi/3$:

$$\nu_0^{-\frac{1}{2}} \cdot c \left[e^{i\int\limits_0^x \nu_0\, dx - i\frac{\pi}{2}} + e^{-i\int\limits_0^x \nu_0\, dx + i\frac{\pi}{2}} \right].$$

*) Die Grösse $\alpha = 8\,\pi^2\, M\, W_{01}/h^2$ darf nicht mit den hier verwendeten Zeichen α, β, γ für die Sprungkonstanten verwechselt werden.

Auf der reellen Axe tritt noch die Nullstelle von ν_1 auf. Dieselbe behandelt sich in gleicher Weise. Sähe man von den Nullstellen von $\psi^2 + \alpha^2$, die im Komplexen liegen, ab, so wäre u_0 im Unendlichen gleich dem obigen Ausdruck, und u_1 gleich einem analogen. Die Grenzbedingungen (8a) und (8b) lassen sich also nur für $\eta_0 = \eta_1 = 0$ (d. h. beide Ströme $= 0$) erfüllen. *Das Stossphänomen muss also durch die Punkte* $\nu_0 = \nu_1$, *die immer im Komplexen, aber für kleine Störung nahe reellen Axe liegen, erklärt werden.*

Da sich die Behandlung des Falles II (LANDAU[8])) übersichtlicher gestaltet, so betrachten wir diesen im folgenden Paragraphen zuerst.

§ 9. Das Anschlussverfahren der Näherungsfunktionen in der Überschneidungsgegend (Fall II).

In einem kleinen Bereich um die Überschneidungsstelle $\psi = 0$, $r = r_0$ dürfen wir $\alpha =$ konstant und $\psi = \psi' \cdot (r - r_0)$ setzen. Dann sieht man, dass die Punkte $\nu_0 - \nu_1 = 0$ symmetrisch um und nahe bei der reellen r-Axe liegen. Die Unstetigkeitslinien liegen, wie weiter unten gezeigt wird so, dass nur eine von je drei Linien die reelle Axe schneidet und die beiden Punkte verbindet. Die vier andern laufen in Paaren in die positiv und negativ imaginäre Halbebene. In Fig. 3 ist dieser Fall II dargestellt. Zum Anschlussverfahren müssen wir das kleine Gebiet etwa in der gezeichneten Art umfahren. Die Ausführungen dieses § haben aber auch, wie sich zeigen wird, noch in vielen Fällen Geltung wo α nicht klein ist. (Also auch wenn sich die Kurven nach dem in der Einleitung und im § 7 skizzierten Verfahren von NEUMANN und WIGNER[1]) nicht schneiden.) Die Bedingung ist nur, dass α und ψ sich in dem jetzt benötigten Gebiet durch eine analytische Funktion darstellen lassen. Um diesen Fall gleich einzuschliessen nehmen wir $t = \frac{\psi}{\alpha}$ als unsere unabhängige Variable. Ist das Gebiet dann klein, so bedeutet das nur eine Masstabänderung, da dann $t = \frac{\psi'}{\alpha} \cdot (r - r_0) = \text{konstant} \cdot (r - r_0)$. Die untere Grenze der Integrale in den Exponenten von (22) kann durch Abspaltung eines in die Konstanten eingehenden Faktors beliebig gewählt werden. Wir wählen z. B. den in Fig. 4 gezeichneten Punkt B. Dann zerlegt man noch

$$\int_B^t \nu_0 \, dr - \int_B^t \frac{\nu_0 + \nu_1}{2} \, dr \qquad \int_B^t \frac{\nu_0 - \nu_1}{2} \, dr$$

und entsprechend auch $\int_B^t \nu_1 \, dr$.

Das erste Integral hat keine Extremalrichtungen, welche von der Überschneidungsstelle ausgehen und kehrt nach der Umfahrung *jedes einzelnen* der beiden Punkte auf seinen Anfangswert zurück. Das zweite aber hat Extremalrichtungen und sein Vorzeichen ändert sich nach der Umfahrung eines der Punkte $t = \pm i$.

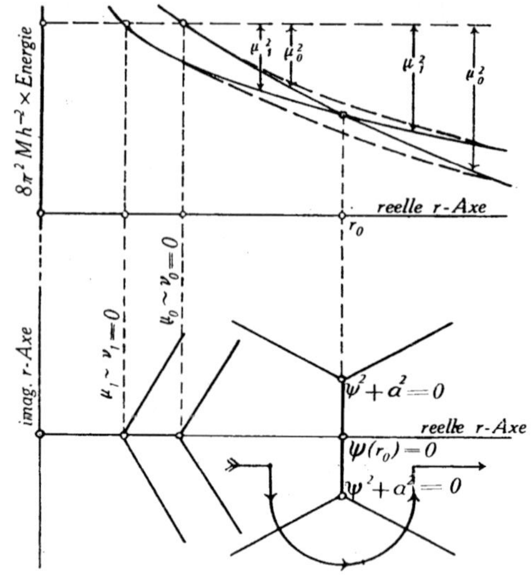

Fig. 3.

Die *obere Hälfte* der Figur stellt das bei Vernachlässigen der Kopplung α auftretende Überschneiden der potentiellen Energiekurven μ_0^2 und μ_1^2 dar. Die gestrichelten Kurven stellen ν_0^2 und ν_1^2 dar. Sie überschneiden sich nicht, sondern kommen sich nur nahe. Da $\nu_0 \sim \mu_0$ und $\nu_1 \sim \mu_1$ werden soll, wenn wir weit vom Schnittpunkt der μ_i entfernt sind, so verlieren die ν_i ihre Identität am Schnittpunkte.
Die *untere Hälfte* stellt die dazugehörige komplexe r-Ebene für kleine Störungen α dar. Beschreibt man die gezeichnete Umfahrung, so ändert sich das Vorzeichen von $\sqrt{\psi^2 + \alpha^2}$. Das auf der linken Seite kleinere ν_0 wird jetzt auf der rechten Seite grösser als ν_1. Die in der oberen Figurenhälfte verlangte Zuordnung ergibt sich also zwanglos. Die Unstetigkeitslinien der Umkehr- und der Überschneidungspunkte sind (starke Linien) eingezeichnet.

Wir haben jetzt vier Glieder mit den Exponenten

$$\pm i \int^{t} \frac{\nu_0 - \nu_1}{2} \, dr = \pm i \int^{t} T(t) \cdot \sqrt{1 + t^2} \cdot dt ,$$

$$\text{wo } T(t) = \frac{\alpha}{t' (\nu_0 + \nu_1)} \text{ ist.}$$

Die Grössen α, t', ν_0 und ν_1 sind als Funktionen von t gedacht. Nimmt man T (t) als im zur Umfahrung der Punkte $t = \pm i$ benötigten Gebiet konstant an, so erhält man Fig. 4. Die ausgezogenen Linien stellen die Kurven

$$\arg\left(+i \int_{\pm i}^{t} \sqrt{1+t^2} \cdot dt\right) = \pi + m \cdot 2\pi\,,$$

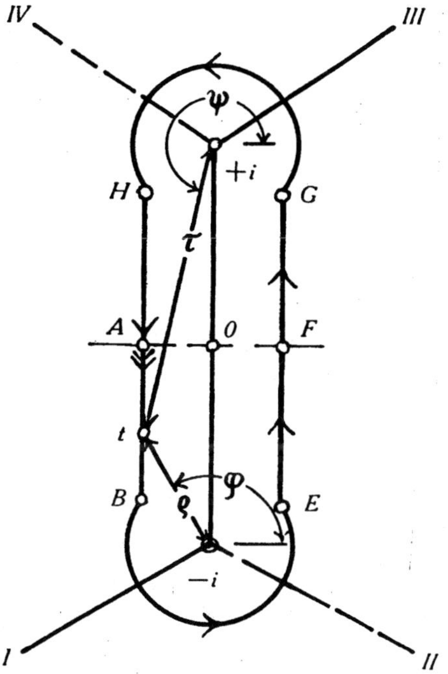

Fig. 4.

t-(ψ/α)-Ebene in der Nähe von $t = 0$ für ψ' und α *konstant*. Die Unstetigkeitslinien verlassen die Punkte $\pm i$ unter Winkeln, die um je $2\pi/3$ voneinander entfernt sind.

und die Gestrichelten die Kurven

$$\arg\left(-i \int_{\pm i}^{t} \sqrt{1+t^2}\, dt\right) = \pi + m \cdot 2\pi$$

dar. Dabei ist festgesetzt

$$\sqrt{1+t^2} = \sqrt{(t-(-i))\cdot(t-(+i))} = \varrho^{\frac{1}{2}} \cdot e^{i\frac{\varphi}{2}} \cdot \tau^{\frac{1}{2}} \cdot e^{i\frac{\psi}{2}} \,{}^{*)}$$

mit

$$\frac{\pi}{2} < \varphi < \frac{5\pi}{2}$$

und

$$\frac{3\pi}{2} < \psi < \frac{7\pi}{2}$$

(siehe Fig. 4). Die Bestimmung der Sprungkoeffizienten lässt sich streng durchführen, gestaltet sich aber ziemlich kompliziert. Es zeigt sich dabei, dass die Konstante von $\exp\,(+\,i\int \nu_0\,d\,x)$ nur proportional der Konstanten von $\exp\,(+\,i\int \nu_1\,d\,x)$ springen usw. Die Bestimmung gestaltet sich wesentlich einfacher, wenn man dieses Resultat voraussetzt. Zur Vereinfachung der Schreibweise lassen wir die belanglose Funktion

$$\exp\left(i\int \frac{\nu_0+\nu_1}{2}\,d\,r\right)$$

weg. Dann haben wir nur die Kopplung zwischen den beiden Summanden in

$$\left[\nu_0^{-\frac{1}{2}}\cdot C_+\cdot e^{\,i\int_B^t dt\cdot T\cdot\sqrt{1+t^2}+\frac{1}{2}\int_B^t\frac{dt}{\sqrt{1+t^2}}} + \nu_1^{-\frac{1}{2}}\cdot D_+\cdot e^{-i\int_B^t dt\cdot T\cdot\sqrt{1+t^2}-\frac{1}{2}\int_B^t\frac{dt}{\sqrt{1+t^2}}}\right]\cdot\exp\left(-\frac{1}{2}\int_B^t\frac{t\,dt}{1+t^2}\right)$$

wo

$$C_+ = c\cdot e^{\,i\int_0^B \nu_0\,dx - i\frac{\pi}{2}} \quad \text{und} \quad D_+ = d\cdot e^{\,i\int_0^B \nu_1\,dy - i\frac{\pi}{2}}$$

ist, zu betrachten. Die Linie, welche i mit $-\,i$ verbindet, *darf aber nicht zur Umfahrung* benützt werden, da dort

$$\left[i\int_{t=\pm i}^{x}\frac{\nu_0-\nu_1}{2}\,d\,x\right]$$

nicht beliebig gross werden kann. Der in Fig. 4 gezeichnete Weg

*) Siehe Anmerkung auf Seite 390.

A—B—E—F—G—H—A ist erlaubt, wenn die Unstetigkeitslinien der Punkte ν_0 und $\nu_1 = 0$ die gezeichneten Linien I, II, III und IV *nicht* oder erst (bei x_∞) für sehr grosse Werte der Exponenten schneiden. Dann ist der Fehler von der Grössenordnung

$$\exp\left(-\left|\, i\int_{t\pm i}^{x_\infty}(\nu_0-\nu_1)\,d\,x\,\right|\right).$$

Beschreibt die Lösung den Weg von B nach E (in Fig. 4), so springt zuerst C_+ auf $C_+ + aD_+$, und dann D_+ auf $b\,C_+ + (1+ab)\,D_+$. Auf dem Weg von B über F nach G multipliziert sich das erste Glied ausserdem mit

$$M_1 = \exp\left(i\int_B^G T\cdot dt\cdot\sqrt{1+t^2}+\int_B^G\frac{dt}{2\sqrt{1+t^2}}\right)$$

und das zweite Glied mit $N_1 = M_1^{-1}$, so dass die Lösung bei C lautet:

$$\left[\nu_0^{-\frac{1}{2}}\cdot\Gamma_+\cdot e^{i\int_G^t dt\cdot T\cdot\sqrt{1+t^2}+\frac{1}{2}\int_G^t\frac{dt}{\sqrt{1+t^2}}}+\nu^{-\frac{1}{2}}\cdot\Delta_+\cdot e^{-i\ldots}\right]$$

$$\cdot\exp\left(-\tfrac{1}{2}\int_B^t\frac{t\,d\,t}{1+t^2}\right)$$

$$\Gamma_+ = (C_+ + a\,D_+)\cdot M_1;\ \Delta_+ = (b\,C_+ + (1+a\,b)\,D_+)\cdot N_1.$$

Beim Umfahren von $+i$ von G nach H ändert sich Γ_+ in $\Gamma_+ + \alpha\,\Delta_+$ und Δ_+ in $\beta\Gamma_+ + (1+\alpha\beta)\,\Delta_+$*). Auf dem Wege von G nach B über A, treten entsprechende Faktoren M_2 und N_2 hinzu.

Es ist:

$$\log M_1 M_2 = \log N_1 N_2 = \frac{1}{2}\oint\frac{d\,t}{\sqrt{1+t^2}} + i\oint T\sqrt{1+t^2}\cdot dt = i\pi - 2\,\delta;$$

$$\log M_1 N_2 = \log M_2 N_1 = 0$$

$$\frac{1}{2}\oint\frac{t\,d\,t}{1+t^2} = i\,\pi\,.$$

($\oint$ bedeutet das Linienintegral auf dem Weg A—B—E—F—G—H—A)

---

*) Siehe Anmerkung auf Seite 391.

Die positiv reelle Grösse δ bestimmt sich folgendermassen:

Das Integral $i \oint T \cdot \sqrt{1+t^2} \cdot dt$ ist, da T analytisch $= T_0 + t\,T_0' + \ldots$,

$$i \oint T \cdot \left(t + \frac{1}{2} \cdot \frac{1}{t} + \ldots\right) \cdot dt = i \oint T_0 \cdot \frac{1}{2} \cdot \frac{dt}{t} = -\pi \cdot T_0 = -2\,\delta,$$

da $|t| > 1$ auf dem ganzen Weg gewählt werden kann. Nach der Umfahrung gehen ν_0 und ν_1 ebenso $\nu_0 - \nu_1$ (im Gegensatz zum Punkte $\nu_0 = 0$) auf ihre Anfangswerte zurück. Man erhält die beiden Bestimmungsgleichungen:

$$-[C_+ (\alpha\, b - e^{-2\,\delta}) + D_+ (\alpha\,(1 + a\, b) - a\, e^{-2\,\delta})] = C_+$$
$$-[C_+ (\beta - b\,(1 + \alpha\,\beta)\, e^{+2\,\delta})$$
$$+ D_+ (\beta\, a - (1 + a\, b)\,(1 + \alpha\,\beta)\, e^{+2\,\delta})] = D_+ \,.$$

Dies gibt *vier Gleichungen* zur Bestimmung der *vier Konstanten.* Man findet (bis auf den Faktor $e^{2\pi i}$):

$$\alpha = \beta = a = b = e^{i\frac{\pi}{2}} \cdot \sqrt{1 - e^{-2\,\delta}}.$$

(Hätte man das strenge Problem durchgeführt, so hätten wir im ganzen $4 \times 4 = 16$ Konstanten zu bestimmen, von denen die 8 Null sind, welche ν_0 mit $-\nu_1$ und umgekehrt koppeln.)

Die Konstanten, welche C_- und D_- koppeln, haben denselben Wert wie α, β, a, b.

Ist $e^{-2\,\delta} \ll 1$, so heissen wir die Punkte weit voneinander getrennt oder *isoliert.* Der isolierte Punkt hat also 3 Unstetigkeitslinien, und die Sprungkoeffizienten sind $e^{i\frac{\pi}{2}}$. Das deckt sich mit dem für $\nu_0 = 0$ (Fig. 2) gefundenen Werte. *Fallen die Punkte zusammen* ($T_0 = \frac{\delta}{2\,\pi} = 0$), so sind die Sprungkoeffizienten alle $= 0$. ($T_0 = 0$ bedeutet aber $\alpha = 0$, d. h. keine Wechselwirkung.) Wir kennen die Lösung links von $\psi = 0$, d. h. auf der negativen reellen t-Axe; sie ist gegeben durch Gleichung (22), wenn $c_\pm = c \cdot e^{\mp i \frac{\pi}{4}}$ und $d_\pm = d \cdot e^{\mp i \frac{\pi}{4}}$ gesetzt wird. Beschreibt man jetzt den Weg A—B—E—F, und berücksichtigt, dass aus Symmetriegründen:

$$i \int_A^F \frac{\nu_0 - \nu_1}{2}\, dr = \frac{i}{2} \oint T(r) \cdot \sqrt{1+t^2} \cdot dt = -\delta$$

ist, so lautet die Lösung auf der reellen Axe rechts von $\psi = 0$ (siehe Fig. 4), d. h. auf der positiven reellen t-Axe:

$$u_0 = \frac{1}{\nu_0^{\frac{1}{2}}} \cdot \left[\frac{\psi + \sqrt{\psi^2+\alpha^2}}{\sqrt{\psi^2+\alpha^2}}\right]^{\frac{1}{2}} \cdot \left[(c \cdot e^{-\delta} + d \cdot \sqrt{1-e^{-2\delta}} \cdot e^{-i\tau})\, e^{i \cdot \!\!\!\!\!\!\!\!\int\limits_0^x \nu_0\, dx - i\frac{\pi}{4}} \right.$$

$$\left. + (c \cdot e^{-\delta} + d \cdot \sqrt{1-e^{-2\delta}} \cdot e^{+i\tau})\, e^{-i \cdot \!\!\!\!\!\!\!\!\int\limits_0^x \nu_0\, dx + i\frac{\pi}{0}} \right]$$

$$- \frac{1}{\nu_1^{\frac{1}{2}}} \left[\frac{\alpha^2}{\psi\sqrt{\psi^2+\alpha^2}+\psi^2+\alpha^2}\right] \cdot \left[(-c \cdot \sqrt{1-e^{-2\delta}} \cdot e^{i\tau} + d \cdot e^{-\delta})\, e^{i \cdot \!\!\!\!\!\!\!\!\int\limits_0^y \nu_1\, dy - i\frac{\pi}{4}} \right.$$

$$\left. + (-c \cdot \sqrt{1-e^{-2\delta}} \cdot e^{-i\tau} + d \cdot e^{-\delta})\, e^{-i \cdot \!\!\!\!\!\!\!\!\int\limits_0^y \nu_1\, dy + i\frac{\pi}{4}} \right]; \qquad (24)$$

darin bedeutet

$$\displaystyle\;-\!\!\!\!\!\!\int\limits_{\nu_{0,1}=0}^{r} \nu_{0,1}\, dr = \int\limits_{\nu_{0,1}=0}^{A} \nu_{0,1}\, dr + \int\limits_A^r \frac{\nu_0+\nu_1}{2}\, dr \pm \int\limits_A^r \frac{\nu_0-\nu_1}{2}\, dr\,.$$

Die Schreibweise $-\!\!\!\!\int$ deutet an, dass der Integrand ν_0 resp. ν_1 sich an der Stelle A—F sprungweise von

$$\sqrt{\varphi \mp |\sqrt{\psi^2+\alpha^2}|} \quad \text{in} \quad \sqrt{\varphi \pm |\sqrt{\psi^2+\alpha^2}|}$$

verändert. Diese Definition der Identität der ν_i vor und nach der Überschneidung bedeutet, dass für negative t resp. $(r - r_0)$, $\nu_0 < \nu_1$ und für positive t resp. $(r - r_0)$, $\nu_0 > \nu_1$ ist, d. h. wir folgen der Bezeichnungsweise, welche in nullter Näherung eine reelle Überschneidung darstellt. (Wählte man die andere Möglichkeit: ν_0 überall $> \nu_1$, so hätte man einen kontinuierlichen Integranden, aber in nullter Näherung nur ein Nahekommen der Kurven, siehe [2]) und Fig. 3). Ferner ist τ die reelle Grösse:

$$\tau = \int\limits_{\nu_0=0}^{A} \nu_0\, dr - \int\limits_{\nu_1=0}^{A} \nu_1\, dr \quad (\tau < 0,\ \text{da } \nu_0 < \nu_1 \text{ für negative reelle } t)\,.$$

Für grosse Werte von $\frac{\psi}{\alpha}$ wird der Faktor des ersten (ν_0)-Gliedes zu $\sqrt{2}$ und der des zweiten (ν_1) zu $\left(\frac{\alpha}{\psi\sqrt{2}}\right)$. Die Grössen ν_0

und ν_1 sind dort schon praktisch konstant. Dann gilt auch in (21):

$$e^{\pm i\int^r \nu_i dr} \cdot \frac{1}{\alpha} \cdot \left(\frac{d^2}{dr^2} + \varphi_0\right) e^{\pm i\int^r \nu_i dr} = \frac{\varphi_0 - \nu_i^2}{\alpha} = \begin{matrix} -\frac{1}{2}\frac{\alpha}{\psi} \\ +2\frac{\psi}{\alpha} \end{matrix} \text{ für } \begin{matrix} i=0 \\ i=1 \end{matrix}$$

Geht man jetzt zur Grenze $\frac{\psi}{\alpha} = \infty$ über, was ja bei fehlender Resonanz $(\alpha(r = \infty) = 0,\ \psi(r = \infty) > 0)$ immer möglich ist, so hat man:

$$u_0 = \sqrt{\frac{2}{\nu_0}} \cdot \left[(c \cdot e^{-\delta} + d \cdot \sqrt{1 - e^{-2\delta}} \cdot e^{-i\tau}) \cdot e^{i\int_0^x \nu_0 dx - i\frac{\pi}{4}}\right.$$
$$\left.(c \cdot e^{-\delta} + d \cdot \sqrt{1 - e^{-2\delta}} \cdot e^{+i\tau}) \cdot e^{-i\int_0^x \nu_0 dx + i\frac{\pi}{4}}\right] \qquad (25)$$

und wegen (21)

$$u_1 = -\sqrt{\frac{2}{\nu_1}} \cdot \left[(-c \cdot \sqrt{1 - e^{-2\delta}} \cdot e^{+i\tau} + d \cdot e^{-\delta})\, e^{i\int_0^y \nu_1 dy - i\frac{\pi}{4}}\right.$$
$$\left.(-c \cdot \sqrt{1 - e^{-2\delta}} \cdot e^{-i\tau} + d \cdot e^{-\delta})\, e^{-i\int_0^y \nu_1 dy + i\frac{\pi}{4}}\right]. \qquad (26)$$

Die Grenzbedingungen (8a) und (8b) lauten, wenn statt der Bessel'schen Funktionen die W.K.B.[2])-Näherungen verwendet werden (die im ∞ natürlich asymptotisch in Bessel- und Hankelfunktionen übergehen), und wenn

$$\left.\begin{aligned} \lim_{r=\infty}\left(\int_{\nu_0=0}^r \nu_0\, dr - \int_{\mu_0=0}^r \mu_0\, dr\right) &= \beta \\ \lim_{r=\infty}\left(-\int_{\nu_1=0}^r \nu_1\, dr + \int_{\mu_1=0}^r \mu_1\, dr\right) &= \gamma \end{aligned}\right| \qquad (27)$$

gesetzt wird:

$$\lim_{r=\infty} u_0 = \sqrt{\frac{2}{\pi p_0}} \cdot \left[(\eta_0 + \tfrac{1}{2}) \cdot e^{i\beta} \cdot e^{i\int_0^x \nu_0 dx - i\frac{\pi}{4}}\right.$$
$$\left. + \tfrac{1}{2} e^{-i\beta} \cdot e^{-i\int_0^x \nu_0 dx + i\frac{\pi}{4}}\right] \qquad (28a$$

$$\lim_{r=\infty} u_1 = \sqrt{\frac{2}{\pi p_1}} \cdot \left[\eta_1 \cdot e^{-i\gamma} \cdot e^{i\int_0^y \nu_1 dy - i\frac{\pi}{4}}\right]. \qquad (28b)$$

Es gilt ferner für $r = 0$: $\lim \nu_0 = \lim \mu_0 = p_0$ und $\lim \nu_1 = \lim \nu_1 = p_1$. Setzt man (29) resp. (30) gleich (32a) resp. (32b) in Grösse und Ableitung nach r (oder x resp. y), so erhält man *vier* Gleichungen zur Bestimmung der *vier* Unbekannten c, d, η_0, η_1. Sie bestimmen sich zu

$$\eta_0 = i \cdot e^{-i(2\beta+\gamma)} \cdot a^2 \cdot \sin\tau - i \cdot e^{-i\beta} \cdot \sin\beta$$

$$\eta_1 = -e^{i(\beta-\gamma)} \cdot a \cdot e^{-\delta} \cdot \sin\tau, \qquad \text{wo } a = i \cdot \sqrt{1 - e^{-2\delta}}.$$

(Die Grössen a^2, β, γ, τ sind reell, a^2 bedeutet $(e^{-2\delta} - 1)$). Der Strom der Funktion u_0 ist proportional*)

$$\frac{2}{\pi} \cdot u_0' \, \overline{u}_0 = \text{Realteil von } (\eta_0 + \eta_0 \overline{\eta}_0) = -(1 - e^{-2\delta}) \cdot e^{-2\delta} \cdot \sin^2\tau,$$

und der Strom der Funktion u_1 proportional

$$\frac{2}{\pi} \cdot u_1' \, \overline{u}_1 = \eta_1 \overline{\eta}_1 = (1 - e^{-2\delta}) \cdot e^{-2\delta} \cdot \sin^2\tau.$$

Mittelt man über τ, so erhält man für $|\eta_1|^2$:

$$\left\{\begin{aligned} &|\eta_1|^2 = \tfrac{1}{2} \cdot e^{-2\delta} \cdot \left(1 - e^{-2\delta}\right) \\ &2\,\delta = \pi \cdot T_0 = \frac{\alpha^2}{\psi'\,\nu} \cdot \frac{\pi}{2} \\ &\nu = \frac{\nu_0 + \nu_1}{2}. \end{aligned}\right. \tag{33}$$

Die Funktionswerte α^2, ψ' und ν sind an der Stelle $\psi = 0$ verstanden. Für Werte von $2\,\delta \ll 1$ erhält man

$$|\eta_1|^2 = \frac{\pi}{4} \cdot \frac{\alpha^2}{\psi'\,\nu} = \tfrac{1}{2} \cdot 2\,\delta \tag{29a}$$

in Übereinstimmung mit der Landau'schen Formel[8]). Landau integriert aber in seiner ersten Arbeit[8]) später über die $|\eta_{1J}|^2$ in solcher Weise, dass er eine beträchtliche Anzahl von $|\eta_{1J}|^2$ erhält, wo $|\eta_{1J}|^2 > 1/4$ ist, da er bis zu J_m geht, wo an der Überschneidungsstelle

$$\nu_{J_m} = \sqrt{p^2 - \frac{J_m^2}{r_0^2}} = 0.$$

Die vorliegende Methode gilt allerdings auch nicht mehr streng, wenn $\nu \simeq 0$ wird, weil dann die Unstetigkeitslinien von $\nu_0 = 0$ sich mit denen von $t = \pm i$ überkreuzen. Immerhin ist (29) länger richtig als (29a), da $|\eta_{1J}|^2$ immer $< 1/4$ ist. (Das Maximum von

*) Die Grössen β und γ sind selbstverständlich hier und im Folgenden diejenigen aus Gleichung (27) und nicht etwa die Springkonstanten.

$e^{-2\delta}(1 - e^{-2\delta})$ liegt bei $e^{-2\delta} = \frac{1}{2}$ und beträgt $^1/_4$.) Wird $\delta \gg 1$, so eignet sich die Darstellung des Nahekommens der Kurven besser als die Darstellung des Überschneidens. $|\eta_{1J}|^2$ wird dann gleich $\frac{1}{2} \cdot e^{-2\delta}$. Diesen andern Grenzfall erhält nun LANDAU in seiner neuen Arbeit[13]) auch, jedoch auf eine andere, wie mir scheint weniger übersichtliche Weise.

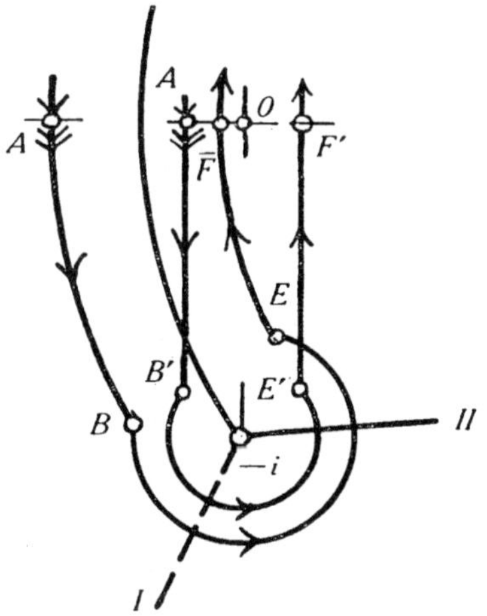

Fig. 4a.
t-(ψ/α)-Ebene in der Nähe von $t = 0$ für *nicht konstante* ψ' und α.

Die Voraussetzung $T(t) =$ konstant, welche wir zur Zeichnung der Figur 4 annahmen, ist übrigens nicht nötig. Man muss nur $T(t)$ im benötigten Gebiet als überall von 0 verschiedene analytische Funktionen von t approximieren können. Für

$$\int^t t\,dt/(1 + t^2) \quad \text{und} \quad \int^t dt/\sqrt{1 + t^2}$$

wählt man A' oder F', statt A und F als untere Grenze (siehe Fig. 4a) und beschreibt den Weg $A'-B'\ldots$ usw., während für

$$\int^t dt \cdot T(t) \cdot \sqrt{1 + t^2}$$

der alte Weg $A-B\ldots$, welcher jetzt deformiert ist, genommen wird.

---

[13]) L. LANDAU. Sow. Phys. **2**, 46 (1932). Zu ähnlichem Resultat für grosse α^2 kommt auch C. ZENER (Proc. Roy. Soc. **137 A**, **696** (1932)) durch eine ganz verschiedene Betrachtungsweise.

§ 10. Das Anschlussverfahren der Näherungsfunktion für den Fall I (ψ auf der reellen Axe überall > 0).

Im Fall I trete im ganzen reellen Geschwindigkeitsbereich auf der reellen Axe keine Überschneidung auf. Es wird sich dann zeigen, dass die Gegend, wo $W^I_{01} = \frac{1}{2}(E_0 - E_1)$ d. h. wo $|\alpha| = \psi$ wird, ähnliche Eigenschaften hat, wie die Überschneidungsstelle im vorhergehenden Paragraphen. Das ist verständlich, denn dort beginnen die potentiellen Energiekurven für $\alpha = 0$ (entsprechend den Grössen μ_1^2 und μ_2^2 in Gl. (17) im § 7) sich wesentlich von denjenigen mit $\alpha \simeq \psi$ (entsprechend den Grössen ν_0^2 und ν_1^2 in Gl. (17) § 7) zu unterscheiden. Die letztgenannten Kurven (ν_0^2 und ν_1^2) haben nämlich dort bereits den nahezu doppelten Termabstand $2\sqrt{\psi^2 + \alpha^2} \simeq 2\sqrt{2\psi^2}$ als die ungestörten Kurven. Für kleinere Abstände, wo $|\alpha| > \psi$ wird, wächst der Termabstand wie 2α und für grössere Abstände der Atome ($\alpha \to 0$) wird er konstant $= 2\psi$. Die potentiellen Energiekurven verlieren also auch in dieser Gegend in beiden Fassungen (ν_i oder μ_i Identifizierung) ihren Sinn. Das Bild ähnelt etwa dem des vorigen Paragraphen, wenn man α und damit δ sehr gross werden lässt, so dass die Neumann-Wigner'sche Darstellung des Nahekommens vorteilhafter als die der Überschneidung wird. Das bedeutet die am Schluss von § 9 in Klammern erwähnte Identifizierung der Kurven. Das Schlussresultat dieses Paragraphen wird auch dem des vorhergehenden sehr ähnlich (Gl. (30)), wenn man annimmt, dass α und ψ ihre Rollen vertauscht haben. Dies zeigt sich schon darin, dass wir die Betrachtung nicht in der doch in roher Weise r ähnlichen t-Ebene durchführen können, sondern von vorne herein die abstrakte zu t reziproke Grösse z heranziehen müssen. Wir verwenden die Form (23) der Lösung des § 8. Der Punkt $z = \frac{\alpha}{\psi} = 0$ entspricht dann dem Ort $r = \infty$. Die Punkte $z = \pm i$ behandelt man gleich wie oben und bestimmt die Unstetigkeitslinien:

$$\arg\left(\pm i \int_{\pm i}^{z} Z(z) \cdot \sqrt{1 + z^2} \cdot dz\right) = \pi\,; \quad Z = \frac{\psi}{2\,\nu\, z'}\,; \quad \nu = \frac{\nu_0 + \nu_1}{2}.$$

Der Fall I lässt sich aber nicht so allgemein behandeln wie Fall II. Die Wechselwirkungsenergien $W^I_{01} - \frac{h^2}{8\pi^2 M} \cdot \alpha(r)$ müssen genauer betrachtet werden. Die Änderung der Elektronenkonfiguration für $r = \infty$ von $\Phi_0 = \chi_0$ nach $\Phi_1 = \chi_1$ gibt uns das Kriterium für den stattgehabten Stoss. χ_0 bedeutet aber: Atom I im Zustand m, Atom II im Zustand n. χ_1 in gleicher Weise: Atom I im

Zustand m' und II in n'. Die Wechselwirkungsenergie W_{01}^{I} lässt sich dann in vielen Fällen für grosse Distanzen nach Potenzen von $1/r$ entwickeln. Für beiderseits neutrale Systeme ist das erste Glied

$$\frac{d_{\mathrm{I}} \cdot d_{\mathrm{II}}}{r^3},$$

das zweite Glied

$$\frac{d_{\mathrm{I}} \cdot q_{\mathrm{II}} + q_{\mathrm{I}} \cdot d_{\mathrm{II}}}{r^4}$$

usw., worin d_{I}, q_{I} etc. das mit dem Übergang $m\,m'$ verbundene Dipolmoment bzw. Quadrupolmoment des Atoms I bedeutet und entsprechend für Atom II. Für beiderseits optisch erlaubte Übergänge wird das Glied mit $1/r^3$ für grosse Distanzen ausschlaggebend sein usw. Dieses Resultat erhält man durch klassische Überlegungen. (Näheres über die wellenmechanische Begründung findet sich in der ersten Arbeit von MORSE und dem Verfasser[6]) im § 3.) Wir beschränken uns im folgenden auf Funktionen α, welche für grosse r wie $\varepsilon \cdot r^{-(\lambda+1)}$ verlaufen, wo $\lambda \geqslant 2$ ist.

Dann ist, wenn

$$\nu = \frac{\nu_0 + \nu_1}{2}$$

und ψ im Bereiche $|z| < 2$ konstant sind:

$$\alpha' = -(\lambda + 1) \cdot \frac{1}{r}\,\alpha;\quad r = \left(\frac{\varepsilon}{\alpha}\right)^{\frac{1}{\lambda+1}}$$

$$\text{oder } z' = -\frac{\lambda + 1}{r} \cdot z = -(\lambda + 1) \cdot \frac{\psi^{\frac{1}{\lambda+1}} \cdot z^{1+\frac{1}{\lambda+1}}}{\varepsilon^{\frac{1}{\lambda+1}}}.$$

Wir setzen $n = 1 + \frac{1}{\lambda+1}$ und haben:

$$Z(z) = -\frac{\psi^{1-\frac{1}{\lambda+1}}}{(\lambda+1) \cdot \varepsilon^{\frac{1}{\lambda+1}}} \cdot \frac{1}{z^n};\quad 1 < n < 1.5.$$

Unter der z-Ebene müssen wir, da n eine nicht ganze Zahl ist, eine Riemann'sche Fläche verstehen, welche der negativen reellen

Axe entlang aufgeschnitten ist. Die Richtungen der Unstetigkeitslinien bestimmen sich, wenn $z = |z| \cdot e^{i\vartheta}$, $z - (+i) = \varrho \cdot e^{i\varphi}$,

$$z - (-i) = \tau \cdot e^{i\psi} \quad \text{und} \quad dz = |dz| \cdot e^{i\varphi'} \text{ ist,*)}$$

aus:

$$\arg\left(\pm i \int_{z=+i}^{r} \frac{\nu_0 - \nu_1}{2} dr\right)$$

$$= \arg\left[\mp i \int_{+i}^{z} \frac{dz}{z^n} \cdot (z - (+i))^{\frac{1}{2}} \cdot (z - (-i))^{\frac{1}{2}}\right] = \pi + m \cdot 2\pi$$

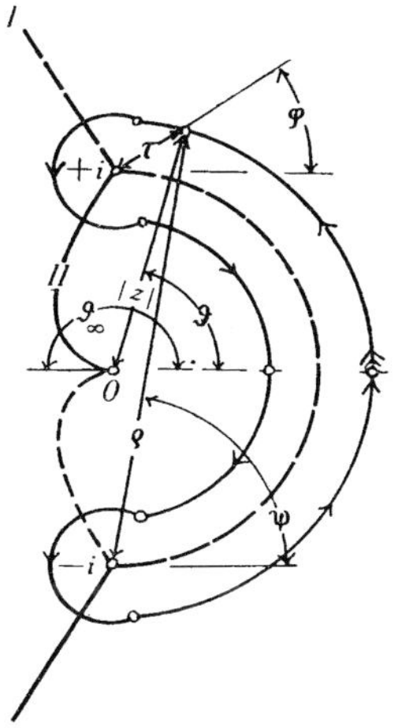

Fig. 5.

z-(α/ψ)-Ebene in der Nähe von $z = 0$ für ψ konstant und $\alpha = \varepsilon/r^2$. Die z-Ebene ist eine Riemann'sche Fläche, welche der negativen reellen Axe entlang aufgeschnitten ist.

oder, wenn $\lambda_\mp$ das Argument von $\pm i \int_{+i}^{r} (\nu_0 - \nu_1)\, dr/2$ bedeutet:

$$\lambda_\mp + \varphi' = \pi + m \cdot 2\pi,$$

wo $\lambda_\mp = \frac{1}{2} \cdot (\varphi + \psi - 2n\vartheta \pm \pi)$ ist.**)

Wir setzen fest $0 < \varphi, \psi < 2\pi$, und auf dem nullten Blatt der Riemann'schen Fläche $-\pi < \vartheta < \pi$. Für die Ausgangsrichtungen

*) Siehe Anmerkung auf S. 390.

**) λ_+ und λ_- ist nicht mit dem Exponenten λ ohne Index in $\alpha = \varepsilon \cdot r^{-(\lambda+1)}$ zu verwechseln.

ist $\varphi' = \varphi \equiv \varphi_0$ zu wählen. Das gibt für φ_0, da $\vartheta_0 = \psi_0 = \frac{\pi}{2}$ ist: für D_+ und C_- Sprünge (λ_-):

$$\varphi_0 = \frac{1 + 8\,m + 2\,n}{6}\,\pi\,.$$

Das Bild für $\alpha = \frac{\varepsilon}{r^2}$, $\lambda = 1$ und $n = 1.5$ ist in Fig. 5 gegeben: Bezeichnet man φ_0, für D_+ Sprünge und $m = 1$, mit

$$\varphi_0 = 2\,\pi - \sigma, \text{ und } \sigma = \frac{2\,n - 3}{6}\,\pi > 0$$

so ist $\sigma = 0$ für $n = 1.5$. Der Halbkreis $|z| = 1$ von $-\frac{\pi}{2} < \vartheta < \frac{\pi}{2}$ genügt der Gleichung, da dort $\varphi + \psi = \frac{3\,\pi}{2} + 2\,\psi$, und $\psi = \frac{\vartheta}{2} + \frac{\pi}{4}$. Daher $\varphi + \psi = 2\,\pi + \vartheta$ und $\lambda_- = \frac{3\,\pi}{2} + \vartheta$. Der Tangentenwinkel ist aber in jedem Punkte $\varphi' = \frac{3\,\pi}{2} - \vartheta$, so dass $\lambda_- + \varphi' = \pi + 2\,\pi$ ist. Die Kurve $\lambda_- + \varphi' = \pi$, wo C_- und D_+ springen (I) geht nach $z = \infty$ mit der asymptotischen Richtung $\varphi'_\infty = \pi$. Die Kurve $\lambda_+ + \varphi' = \pi$, wo C_+ und D_- springen, (II) nähert sich $z = 0$ von der Richtung $\vartheta_0 = \pi$ her. Wir können also I sowohl wie II an einer Stelle überschreiten, wo

$$\pm\, i \int\limits_{z=i}^{r} \frac{\nu_0 - \nu_1}{2}\, d\,r$$

über alle Grenzen wächst, wenn die Punkte ν_1 und $\nu_0 = 0$ weit entfernt sind. Die Verhältnisse um den Punkt $-i$ sind spiegelbildlich*) wie sich leicht nachweisen lässt. Zur Bestimmung der Sprungkonstanten wählen wir wie im vorhergehenden Fall den Weg A—B—E—F—G—H—A**).

Da uns die Lösung für $\lim r = \infty$, d. h. für $\lim z = +0$ interessiert, so wählen wir, nach Bestimmung der Sprungkonstanten den Weg A—B—E—F zum Anschlussverfahren.

Ist $\alpha = \frac{\varepsilon}{r^{\lambda+1}}$ und $\lambda > 1$, $1 < n < 1{,}5$, so ändert sich das Bild nur wenig. (Fig. 6). Der Halbkreis wird zu einer Kurve K_w, die vom Punkte i unter dem Winkel $2\,\pi - \sigma$ $(\sigma = (2\,n - 3)\,\pi/6 > 0)$ verlässt. Wir zeichnen den Kreis K_a für $\sigma = 0$ aus Fig. 5 in Fig. 6 ein, und einen Kreis K_i, welcher in den Punkten $z = \pm\, i$ die gleichen Tangenrichtungen wie K_w $(2\,\pi - \sigma$ resp. $3\,\pi - \sigma)$ hat. Be-

*) Spiegelbildlich in dem Sinne, dass die Kurven spiegelbildlich verlaufen, aber I und II ihre Bedeutnng vertauscht haben, wenn man den gezeichneten (Fig. 5) Weg einschlägt.

**) Der Übersichtlichkeit halber sind die Buchstaben A—B...—F in Fig. 5 *nicht* eingezeichnet, sondern nur die *Punkte* und der *Weg* durch einen geschlossenen gefiederten Pfeil (wie in Fig. 4).

zeichnet man mit φ_i' und φ_a' die Tangenrichtungen der Kreise K_i und K_a als Funktion von $\vartheta\left(\frac{\pi}{2} > \vartheta > -\frac{\pi}{2}\right)$, und bezeichnet man mit φ' die Richtung, wie sie sich aus $\lambda_- + \varphi' = 3\pi$ ergibt, so ist:

$$\varphi' - \varphi_i' = 3\sigma\left[\frac{1}{2\sigma}\sin^{-1}(\sin\sigma\cdot\sin\vartheta) - \frac{\vartheta}{\pi}\right]$$

$$\cong 3\sigma\left[\frac{\sin\vartheta}{2} - \frac{\vartheta}{\pi}\right] > 0$$

und

$$\varphi' - \varphi_a' = -3\sigma\left[\frac{\vartheta}{\pi}\right] < 0\,.$$

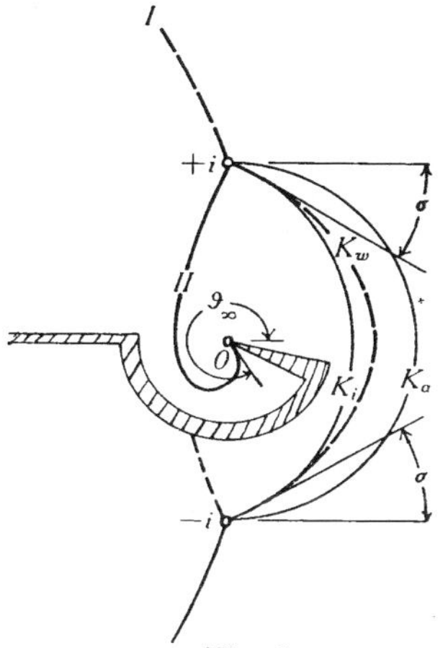

Fig. 6.

z-(α/ψ)-Ebene in der Nähe von $z = 0$ für $\alpha = \varepsilon/r^{\lambda+1}$, $(\lambda > 1)$. Die z-Ebene ist eine der negativen reellen Axe entlang aufgeschnittene Riemann'sche Fläche. In der Figur ist das nullte Blatt $(-\pi < \vartheta < +\pi)$ und ein Stück des plus ersten Blattes $(+\pi < \vartheta < 3\pi)$ gezeichnet. Die schraffierte Fläche stellt den Aufschnitt dar.

Die wahre Kurve K_w, auf welcher $\varphi' - \varphi_w' \overset{id}{=} 0$, muss also zwischen K_i und K_a liegen, und zwar, da der Absolutwert des ersten Ausdruckes überall kleiner ist als der des zweiten, nahe bei K_i. Die Kurve I läuft nach wie vor ins Unendliche mit einer Asymptotenrichtung $\varphi_\infty = (n - \frac{1}{2})\pi$. Die Kurve II nähert sich spiralförmig auf der Riemann'schen Fläche dem Punkte $z = 0$ mit der Asymptotenrichtung

$$\vartheta_\infty = \frac{\pi}{2}\cdot\frac{1}{n-1} = (\lambda+1)\cdot\frac{\pi}{2}\,.$$

Die Ausführung der Sprungkonstantenbestimmung und des Anschlusses geht genau gleich wie im Fall II. Man erhält für $|\eta_1|^2$ nach der Phasenmittelung wieder:

$$|\eta_1|^2 = \tfrac{1}{2} \cdot e^{-2\delta} (1 - e^{-2\delta})$$

$$\delta = \left| i \int_{-i}^{+i} Z(z) \cdot \sqrt{1+z^2} \cdot dz \right| = \frac{\psi^2}{2\nu\alpha'} \cdot M_\lambda$$

wo unter α' der Absolutwert der Ableitung von α an der Stelle $\alpha^2 = \psi^2$ verstanden ist, und wo

$$M_\lambda = \left| i \int_{-i}^{+i} \frac{1}{z^{1+\frac{1}{\lambda+1}}} \cdot \sqrt{1+z^2} \cdot dz \right|$$

eine Zahl der Grössenordnung 1 ist. (Für $\lambda = 1$ beträgt der Wert 3,41.) Wir haben also:

$$|\eta_1|^2 = \frac{1}{2} \cdot \left(e^{-M_\lambda \frac{\psi^2}{\nu\alpha'}} - e^{-2M_\lambda \frac{\psi^2}{\nu\alpha'}} \right). \qquad (30)$$

Diese Formel gilt nur, wenn im Bereiche, wo $|\alpha| \simeq |\psi|$ wird, ν reell und nahezu konstant ist d. h. wenn $\nu_0 = 0$ weit entfernt ist. Mit Annäherung gilt (34) überall wo $|\alpha_{\max}| > |\psi|$.

Wie in der Einleitung zu diesem Paragraphen bemerkt, entspricht diese Gleichung (30) abgesehen vom Zahlenfaktor M_λ genau Gleichung (29), wenn man die Bedeutungen von α und ψ vertauscht.

§ 11. Abschätzung der Matrixelemente $|\eta_{1J}|^2$ nach dem Störungsverfahren.

Ist die *Kopplung sehr schwach*, so müssen die durch das Anschlussverfahren ermittelten Übergangswahrscheinlichkeiten $|\eta_{1J}|^2$ ((29) im § 9 und (30) im § 10) in die nach der Störungstheorie ((14) § 6) ermittelten Werte übergehen.

Im Fall I können wir diesen Übergang allerdings *nicht* verfolgen, da die im § 10 gegebene Ableitung nur für den Fall gilt, wo $|\alpha_{\max}| > |\psi|$. In diesem Fall ist aber die Kopplung nicht mehr schwach.

Ist $|\alpha|$ im ganzen reellen Geschwindigkeitsbereich $> |\psi|$, so gilt auch, wenigstens für grosse J; näherungsweise

$$v_i = u_i \simeq \sqrt{r} \cdot J_J(p_i r).$$

Beschränkt man sich wieder auf $\alpha = \varepsilon \cdot r^{-(\lambda+1)}$ und ist $p_1 < p_0$, so gilt (14):

$$\eta_{1J} = \frac{\pi\,\varepsilon}{2} \int\limits_0^\infty \frac{J_J(p_0 r) \cdot J_J(p_1 r) \cdot d r}{r^\lambda} = \frac{\pi\,\varepsilon}{2} \frac{1}{2^\lambda \cdot \Gamma\left(\frac{\lambda+1}{2}\right)}$$

$$\cdot \frac{\Gamma\left(J - \frac{\lambda-1}{2}\right)}{\Gamma(J+1)} \cdot \left(\frac{p_1}{p_0}\right)^J \cdot F\left(-\frac{\lambda-1}{2},\ J - \frac{\lambda-1}{2},\ J+1,\ \left(\frac{p_1}{p_0}\right)^2\right).$$

Für die Γ-Funktionen verwenden wir (grosse J) Stirlings Formel, und setzen ferner[6]):

$$F \simeq \left[1 - \frac{J}{J + \left(\frac{\Gamma(\lambda)}{\Gamma\left(\frac{\lambda+1}{2}\right)}\right)^{\frac{2}{\lambda-1}}} \cdot \left(\frac{p_1}{p_0}\right)^2\right]^{\frac{\lambda-1}{2}}, \quad c_1 = \frac{\pi^2 \cdot \Gamma(\lambda)^2}{4^{\lambda+1} \cdot \Gamma\left(\frac{\lambda+1}{2}\right)^4}$$

$$c_2 = \left[\frac{\Gamma\left(\frac{\lambda+1}{2}\right)}{\Gamma(\lambda)}\right]^{\frac{2}{\lambda-1}}$$

dann wird:

$$|\,\eta_{1J}|^2 = c_1 \cdot \varepsilon^2 \cdot p_0^{2\lambda-2} \cdot \frac{1}{J^{2\lambda}}\, e^{-\frac{\psi}{p_0^2} J} \cdot \left[1 + c_2 \cdot \frac{\psi}{p_0^2} J\right]^{\lambda-1}. \quad (31)$$

Setzt man nun $\alpha_{\max} = \varepsilon \cdot p_0^{\lambda+1} / J^{\lambda+1}$

$$\text{und} \left|\alpha'_{\max}\right| = \left|\frac{\lambda+1}{r_{\min}} \cdot \alpha_{\max}\right| = \left|\frac{(\lambda+1)\,p_0}{J} \cdot \alpha_{\max}\right|$$

so wird:

$$\frac{\psi}{p_0^2} J = \frac{\psi\,\alpha_m}{p_0\,\alpha'_m} (\lambda+1)$$

bunter α_m und α'_m sind immer die Absolutwerte verstanden, m (edeutet max.)

Man hat also mit guter Annäherung für $|\alpha_{\max}| < |\,\psi\,|$:

$$|\,\eta_{1J}|^2 = c_1 \cdot \left[\frac{\alpha_m^2}{p_0\,\alpha'_m} (\lambda+1)\right]^2 \cdot e^{-\frac{\psi\,\alpha_m}{p_0\,\alpha'_m}(\lambda+1)}$$

$$\cdot \left[1 + c_2 \cdot (\lambda+1) \frac{\psi\,\alpha_m}{p_0\,\alpha'_m}\right]^{\lambda-1}. \quad (31a)$$

Je kleiner ψ/α', umso grösser wird $|\,\eta\,|^2$, wie es ja auch London[7]) verlangte.

Wir bezeichnen jetzt für diese Teilwelle J, wo

$$\alpha_m = |\alpha(r_{\min})| = \psi \text{ wird,}$$
$$r_{\min} = p_1^{-1} \cdot J \text{ mit } r_0 \text{ und das } J \text{ mit } J_m.$$

Zum Werte $J = J_m$ gehört also, der im § 3 vorgenommenen Zerlegung in sphärische Harmonische entsprechend, diejenige Teilwelle, welche einer Annäherung der beiden Atomzentren bis zu dem Punkte entspricht, wo die Störungsenergie gleich der Resonanzunschärfe wird.

Für die $|\eta_{1J}|^2$, wo $J > J_m$ ist lautet in (31a) die charakteristische Exponentialfunktion:

$$\exp\left[-(\lambda+1)\frac{\psi\,\alpha_m}{p_0\,\alpha_m'}\right],$$

während sie für $J < J_m$, wo $\alpha_m > \psi$ ist, nach (30) in § 10 lautet:

$$\exp\left[-M_\lambda\,\frac{\psi\,\psi}{p_0\,\alpha_m'}\right].$$

Es sieht also so aus, als ob abgesehen von den Zahlenfaktoren M_λ und $\lambda + 1$, sich das beim Verkleinern von J wachsende α_m im Zähler nur bis zum Werte $\alpha_m = \psi$ vergrössert ($J = J_m$) und dann konstant bleibt, während das α_m' im Nenner weiter wächst. Diese Erscheinung kann auch auf eine andere, allerdings sehr wenig strenge Weise, plausibel gemacht werden:

Idealisiert man α, indem man setzt $\alpha = \alpha' \cdot (r_0 - r) = \alpha' \cdot x$ und $\alpha = 0$ für $r > r_0$, so hat man in der Gegend $r < r_0$:

$$v_0\,v_1 = \frac{2}{\pi}\frac{1}{\sqrt{\nu_0\,\nu_1}}\cdot\tfrac{1}{2}\left[\cos\left\{\int_0^x(\nu_0-\nu_1)\,d\,x\right\}\cdot\cos\tau\right.$$
$$\left.+\sin\left\{\int_0^x(\nu_0-\nu_1)\,d\,x\right\}\cdot\sin\tau\right] \tag{32}$$

plus einem schnell oszillierenden Teil mit

$$\begin{matrix}\cos\\ \sin\end{matrix}\left\{\int_0^x(\nu_0+\nu_1)\,d\,x\right\}.$$

τ ist eine (siehe § 10) hier belanglose Phasenkonstante.

Für $\sqrt{\psi^2+\alpha^2}\ \mathrm{tg}^{-1}\frac{\alpha}{\psi}$ setzen wir (siehe § 7) $\alpha \simeq \alpha' \cdot x$.

Die Gleichsetzung von:

$$\int_0^x (\nu_0 - \nu_1)\, d\,x = \frac{\alpha'}{\nu} \int_0^x \sqrt{\frac{\psi^2}{\alpha'^2} + x^2} \cdot d\,x$$

$$= \frac{\alpha'}{2\,\nu} \cdot x \cdot \sqrt{\frac{\psi^2}{\alpha'^2} + x^2} \equiv \frac{\psi^2}{4\,\nu\,\alpha'} \cdot t$$

ist erlaubt bei Vernachlässigung logarithmischer Glieder.

Dann wird

$$\alpha' \cdot x\, d\,x = \frac{\psi^2}{2\,\alpha'} \cdot \frac{t\,d\,t}{\sqrt{1+t^2}},$$

und auf Grund von (19) im § 7:

$$|\,\eta_{1J}\,|^2 = \frac{\psi^2}{4\,\nu\,\alpha'} \cdot \int_0^\infty \frac{t\,d\,t}{\sqrt{1+t^2}} \begin{matrix}\cos\\ \sin\end{matrix} \left[\frac{\psi^2}{4\,\nu\,\alpha'} \cdot t\right] \simeq e^{-\frac{\psi^2}{4\,\nu\,\alpha'} + \log \cdot \mathrm{Gl.}} \tag{33}$$

Grössenordnungsweise tritt also der Exponent $\frac{\psi^2}{4\,\nu\,\alpha'}$ auf, doch wissen wir hier nicht an welcher Stelle (bei nicht konstantem α') α' zu nehmen ist.

Endlich sei, der Vollständigkeit halber, noch das Landau'sche[8]) Matrixelement erwähnt. Man erhält es aus (19a) und (21), wenn man berücksichtigt, dass für $x = 0 : \psi = \psi' \cdot x$ ist wo in (32) u_i statt v_i und μ_i statt ν_i zu setzen ist, (die Phasenkonstante ist allerdings wegen des Unterschiedes zwischen ν_i und μ_i von derjenigen des § 9 verschieden).

$$\int_0^x (\mu_0 - \mu_1)\, d\,x = \frac{\psi'}{2\,\mu} \cdot x^2$$

und dass

$$\int_0^\infty \begin{matrix}\cos\\ \sin\end{matrix} \left[\frac{\psi'}{2\,\mu}\, x^2\right] \cdot d\,x = \left[\frac{2\,\pi\,\mu}{\psi'}\right]^{\frac{1}{2}}.$$

Dann ist nach Mitteilung über die Phase:

$$|\,\eta_{1J}\,|^2 = \frac{\pi}{4} \frac{\alpha^2\,(r_0)}{\mu_J\,(r_0) \cdot \psi'\,(r_0)}; \quad \psi\,(r_0) = 0 \tag{34}$$

Diese Auswertung von Landau[8]) ist natürlich nur erlaubt, wenn $|\,\eta_{1J}\,|^2 \ll 1/4$. Im „Resonanzfall" d. h. sehr kleines $\psi'\,(r_0)$ und für grosse α^2 resp. kleine μ_J wird daher sein Resultat ungültig. Die Formel (34) deckt sich (da ja für kleine α und grosse ψ' $\mu_i = \nu_i$ ist) mit (29a).

§ 12. Bestimmung des Wirkungsquerschnittes im Fall I (keine Überschneidung der Kurven).

Im Fall I zerfällt der Wirkungsquerschnitt in zwei Teile. Wir ersetzen die Summation $\sum_J J$ durch $\int J\,d\,J$. Für $J \ll J_m$ (wenn $J_m/p_1 \simeq r_0$ und wenn $\alpha(r_0) = \psi$ ist) gilt sicher die Formel (30). Für $J \simeq J_m$ haben wir keinen gültigen Ausdruck, da für $J \simeq J_m$ und $J > J_m$ sich die Unstetigkeitslinien von $\nu_1 = 0$ und von $z = \pm i$ im Endlichen überschneiden. Der Ausdruck liegt aber auf alle Fälle in der gleichen Grössenordnung. Für $J \gg J_m$ ist aber sicher die Born'sche[6]) (oder London'sche[7]), da sie zusammenfallen) Näherung (31) oder (31a) gut. Verwenden wir (30) und (31a) bis zu $J = J_m$, so machen wir keinen grossen Fehler.

Wir setzen daher (35)

$$q = q_< + q_>, \text{ wo } \lesseqgtr : J \lesseqgtr J_m \text{ bedeutet.} \tag{35}$$

Für den Teil $q_<$ wollen wir nur den Fall „guter" Resonanz betrachten, d. h. wir setzen an der Stelle r_0

$$r_0 = J_m / p_m; \ (p_m^2 - p_0{}^2 - 8\,M\pi^2 \cdot W_{00}(r_0) / h^2)$$

$$\nu_0{}^2 \simeq \nu_1{}^2 \simeq \left[\frac{\nu_0 + \nu_1}{2}\right]^2 \simeq \nu^2 \simeq p_m^2 - \frac{J^2}{r_0{}^2},$$

so dass $\nu^2 (J = J_m) = 0$ an der Stelle $r = r_0$.

Führen wir die Grössen

$$z = \frac{1}{\sqrt{1 - \frac{J^2}{p_m^2\, r_0{}^2}}}, \quad k = 3\,\frac{\psi^2}{p_0\,\alpha}$$

ein*) und definieren die Funktion

$$f_1 = 2\left[\int_1^\infty \frac{d\,z}{z^3}\left(e^{-kz} - e^{-2kz}\right)\right]$$

so haben wir:

$$q_< = \pi\, r_0{}^2 \cdot 2\left[\frac{p_m}{p_0}\right]^2 \cdot f_1\left(3 \cdot \frac{\psi^2}{p_m\,\alpha'}\right). \tag{36}$$

Die Funktion $f_1(k)$ steigt für $k < 1$ wie $2\,k$ an, erreicht bei $k = 0{,}45$ ein flaches Maximum vom Werte 0,22 und nähert sich für grosse Werte von k der Funktion $2 \cdot e^{-k}/k$.

---

*) Der Faktor M_λ des § 10, welcher für $\lambda = 2$ den Wert 3,41 hat, ist der Einfachheit halber im folgenden überall $= 3$ gesetzt.

Für den Teil $q_>$ setzen wir $z = \frac{J}{J_m}$ und erhalten aus (31):

$$q_> = \pi r_0^2 \cdot 8 c_1 \left[\frac{\psi}{p_0^2} J_m\right]^2 \cdot \sum_{k=0}^{k=\lambda-1} \binom{\lambda-1}{k} \cdot c_2^k \cdot \left[\frac{\psi}{p_0^2} \cdot J_m\right]^k \cdot \int_1^\infty \frac{dz}{z^{2\lambda-1-k}} \cdot e^{-\frac{\psi}{p_0^2} J_m \cdot z}.$$

Man sieht, dass für sehr schlechte Resonanz

$$\frac{\psi^2}{p_0^2 \alpha'} \gg 1,$$

$q_>$ überwiegt, da in $q_>$ die Exponentialfunktion mit der grossen Zahl

$$\left(\frac{\psi}{p_0^2} J_m\right)^2$$

multipliziert ist, während in $q_<$ die grosse Zahl

$$k = \left(\frac{\psi}{p_0^2} J_m\right) = 3 \frac{\psi^2}{p_0 \alpha'}$$

im Nenner auftritt*).

Eine allgemeine Diskussion würde sehr weitläufig. Wir wollen daher den konkreten Fall $\alpha = \varepsilon / r^3$, $\lambda = 2$ betrachten.

Die Funktionen f_3 und f_4 seien definiert als

$$f_3(k) = 2 \int_1^\infty \frac{dz}{z^3} \cdot e^{-k \cdot z} = e^{-k} - k \cdot \{e^{-k} - k[-Ei(-k)]\};$$

$$[-Ei(-k)] = \int_k^\infty \frac{e^{-u}}{u} du$$

und

$$f_4(k) = \int_1^\infty \frac{dz}{z^2} \cdot e^{-kz} = e^{-k} - k[-Ei(-k)].$$

*) $q_>$ ist nichts anderes als der von MORSE und dem Verfasser[6]) berechnete Querschnitt, wenn das dortige $\beta = J_m$ und $\frac{W}{E} = \frac{2\psi}{p_0^2}$ gesetzt wird. Auf den Kurven der Figuren $D_\beta\left(\frac{W}{E}\right)$ im § 5 der alten Arbeit, bedeutet das, dass man bei zunehmender Resonanz sich auf Kurven mit grösserem β begeben muss. Allerdings zeigt die Rechnung, dass $J_m = \beta$ leicht die Grössenordnung 100 übersteigt. Die „Resonanzkurven" werden daher flacher. Dazu tritt aber noch der mit schärferer Resonanz wachsende Wert von $q_<$ hinzu, welcher diesen Effekt aufhebt und schliesslich sogar überwiegt.

Dann gilt:

$$f_1(k) = f_3(k) - f_3(2k)$$

$$\tfrac{1}{2}\lim_{k=\infty} f_1(k) = \tfrac{1}{2}\lim_{k=\infty} f_3(k) = \lim_{k=\infty} f_4(k) = \frac{1}{k}\cdot e^{-k}$$

$$f_3(0) = f_4(0) = 1;\ \lim_{k=0} f_1(k) = 2\,k.$$

Es ist, da

$$\lambda + 1 = 3:\ c_1 = 1/4;\ c_2 = \pi/4\quad \text{(siehe (31))}.$$

$$q_> = \pi\, r_0^2 \cdot 2 \cdot \left(3\,\frac{\psi^2}{p_0\,\alpha'}\right)^2 \cdot \left[\tfrac{1}{2}\cdot f_3\left(3\,\frac{\psi^2}{p_0\,\alpha'}\right) + \frac{\pi}{4}\left(3\,\frac{\psi^2}{p_0\,\alpha'}\right)\cdot f_4\left(3\,\frac{\psi^2}{p_0\,\alpha'}\right)\right] \tag{37}$$

oder aus (36), wenn $p_m \sim p_0$, und aus (37):

$$q = \pi\, r_0^2 \cdot \left[2\, f_1(k) + k^2 \cdot f_3(k) + \frac{\pi}{2}\, k^3 \cdot f_4(k)\right] \tag{38}$$

$$k = \frac{\psi}{p_0^2}\, J_m = \frac{\psi^{\frac{2}{3}}\cdot \varepsilon^{\frac{1}{3}}}{p_0},\quad \text{da}\ r_0 = \left(\frac{\varepsilon}{\psi}\right)^{\frac{1}{3}}\ \text{ist.}$$

In Fig. 7 sind der Klammerausdruck (ausgezogen) und der erste Summand (gestrichelt) gezeichnet. Dieser erste Summand, welcher für gute Resonanz überwiegt, stellt den Anteil $q_<$ dar, während die Differenz zwischen der ausgezogenen und der gestrichelten Kurve den Anteil $q_>$ bedeutet. Für kleine k hat man:

$$q = \pi\, r_0^2 \cdot 4\, k \tag{39}$$

und für grosse k:

$$q = \pi\, r_0^2 \cdot \frac{\pi}{2}\, k^2 \cdot e^{-k} \tag{40}$$

Ist $\lambda \neq 2$, so gilt (39) immer noch, während in (40) sich der numerische Faktor $\frac{\pi}{2}$ in $8\, c_1 c_2^{\lambda-1}$ verwandelt, und statt k^2, k^λ steht. k bedeutet dann in (39): $M_\lambda \psi^2/p_m \alpha'$, und in (40): $(\lambda+1)\ \psi^2/p_m \alpha'$.

Die Formeln (29) und (30) wurden aus den Lösungen der Differentialgleichung (18a) oder (20) und den Grenzbedingungen (28a) und 28b) erhalten. Dabei wurde vorausgesetzt, dass $r = +$ reell ∞ einem Punkte $t = +$ reell ∞ für Fall II, und $z = +$ reell 0 im Fall I entspricht, der weit von der Unstetigkeitslinie, welche

$+i$ mit $-i$ verbindet, und weit von z (oder t) $= \pm i$ liegt. Weit heisst hier, dass

$$\left| \, i \int_{z,\,t = \pm i}^{r} (\nu_0 - \nu_1)\, d\, r \, \right|$$

eine grosse Zahl ist. Das ist immer der Fall, wenn $\nu_0 \neq \nu_1$ für $r = \infty$ ist, d. h. wenn $\psi_\infty \neq 0$ ist.

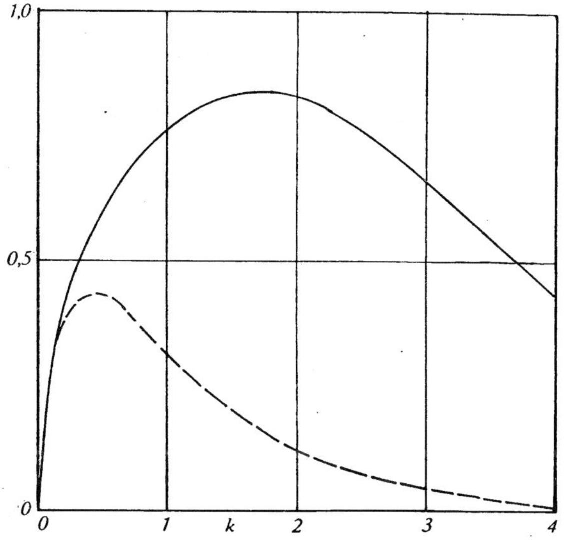

Fig. 7.
Die in Gl. (36) verwendete Funktion $2\, f_1\,(k)$ gestrichelt (------) und die Summe in Gl. (38) $[2\, f_1\,(k) + k^2 \cdot f_3\,(k) + \pi k^3 \cdot f_4(k)/2]$ ausgezogen (———) als Funktion von k gezeichnet. Die Summe verläuft für sehr kleine k wie $4\, k$ und für sehr grosse k wie $\pi\, k^2 \cdot \exp\,(-k)/2$.

Dann gilt nämlich:

$$\lim_{r = R = \infty} \left| \, i \int_{z,\,t = \pm i}^{r} (\nu_0 - \nu_1)\, d\, r \, \right| > \left| \int_0^\infty \frac{\psi_\infty}{p_0}\, d\, r \, \right| > \frac{\psi_\infty}{p_0} \cdot R$$

R bedeutet den maximalen Abstand, d. h. unsere Betrachtungen beziehen sich alle auf $\lim R = \infty$.

Formel (39) gilt daher für beliebig kleine ψ_∞ und damit kleine k solange als

$$\lim_{R = \infty} \frac{\psi_R \cdot R}{p_0} = \infty\,.$$

Wir bezeichnen diesen Fall für beliebig kleine ψ_∞ als *Grenzfall* im Gegensatz zum im § 13 behandelten *Resonanzfall.*

Im *Grenzfall* gilt also

$$\lim q = \frac{12\pi \cdot \psi_\infty^{1-\frac{3}{1+\lambda}} \cdot \varepsilon^{\frac{3}{1+\lambda}}}{p_0 \cdot (\lambda+1)} .^{*)}$$

Für $\lambda + 1 < 3$ erhalten wir beim Verschärfen der Resonanz beliebig grosse Werte von q.

Für $\lambda + 1 = 3$ (den betrachteten Fall) strebt q einem festen Grenzwert zu. Der Grenzquerschnitt von $\lambda + 1 = 3$ ist von gleicher Grössenordnung wie die Maximalquerschnitte von $\lambda + 1 > 3$ (siehe unten für $\lambda + 1 > 3$).

Für $\lambda + 1 > 3$ durchläuft q beim Verkleinern von ψ_∞ ein Maximum. Dort hat die (38) analoge Klammer $[2 f_1 + \ldots]$ die Grössenordnung 1. Das Maximum beträgt also etwa

$$q_{\max} = \pi r_{00}^2 \times 1 .$$

Dieses Resultat gilt für alle Wechselwirkungen, wo $\lambda + 1 > 3$. Die Grössen $r_{00}(k_0)$ und k_0 sind bestimmt durch

$$\frac{d}{dk}\left\{\left[2 \cdot f_1(k) + k^2 \cdot \sum_l c_l k \cdot f_{3+l}\left(\frac{3}{\lambda+1}\right)\right] \cdot r_0(k)\right\} = 0$$

k_0 hat ebenfalls die Grössenordnung 1, so dass r_{00} sich grössenordnungsweise durch:

$$\alpha'(r_{00}) = \frac{\psi^2}{p_m} \text{ bestimmt.}$$

Der Grenzquerschnitt wird hier $= 0$.

Allgemein gilt:

Die Geschwindigkeitsabhängigkeit wird nur durch den Klammerausdruck (Fig. 7, für $\lambda = 2$) gegeben, da r_0 dann konstant, und k proportional $\frac{1}{p_0}$ ist. Der W. Q. erreicht also ein Maximum von πr_0^2 für $p_0 = \psi^2/\alpha'(r_0)$, und fällt dann wie $1/p_0$ zu null ab.

Die Grössenordnung der Wirkungsquerschnitte erreicht die der gaskinetischen Querschnitte und übersteigt sie sogar.

Fig. 8 (Kurve a) gibt eine „Resonanzkurve" für den Fall:

$$T_0 = 1 \text{ Volt}, \quad W_{01}^{\mathrm{I}} = \frac{(e\, a_0)^2}{r^3},$$

$M = 10\, M_H$. (e = Elektronenladung, a_0 Bohr'scher Kreisbahnradius, M_H Masse von Atomgewicht 1).

*) An Stelle von 12 steht streng genommen $4\, M_\lambda$.

§ 13. Der Wirkungsquerschnitt im Resonanzfall.

Die Ableitung der Grössen $|\eta_{1J}|$ erfolgte in den §§ 9 und 10 immer unter der Annahme, dass

$$\lim_{r=\infty} \frac{\psi(r)\cdot r}{p_0} = \infty$$

ist. Für $\psi_\infty = 0$ erhalten wir dementsprechend eine Reihe ver-

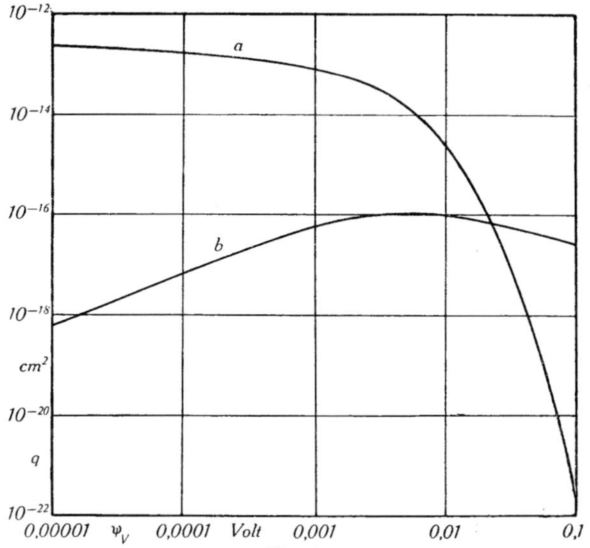

Fig. 8.

Wirkungsquerschnitte für Stösse zweiter Art zwischen Atomen in logarithmischer Skala: q in cm² gegen ψ_V in Volt. Kurve a ist berechnet für sich nicht überschneidende Potentialkurven (Fall I) mit der Wechselwirkung $W^{I} = (ea)^2/r^3$. Kurve b für sich schneidende Potentialkurven mit $\psi'_V = \psi_V 10^{-8}$ Volt cm^{-1} und mit der Wechselwirkungsenergie $W^{II} = h^2 \Lambda\, J/8\,\pi^2\, Mr^2$ ($\Lambda = 1$). Die Relativgeschwindigkeit beträgt in beiden Fällen ein Elektron-Volt und die reduzierte Masse 10 mal die Masse vom Atomgewicht eins.

schiedener Fälle je nach dem Grenzwert, dem $\psi \cdot r/p_0$ zustrebt. Im vorhergehenden Paragraphen ist der *Grenzfall* $\psi \cdot R = \infty$ behandelt.

London[7]) betrachtet noch den Resonanzfall identischer Systeme: ψ identisch $= 0$. Dann zerfallen die gekoppelten Gleichungen (18a) in zwei ungekoppelte, wenn man die Funktionen

$u_0 + u_1$ und $u_0 - u_1$ einführt. Sein Resultat ist, wenn man die Phasenkonstanten β und γ (Gl. 27 § 9) benützt (da im Endlichen keine Überschneidung auftreten soll, ist dort $\text{-}\!\!\!\int = \int$):

$$| \eta_{1J}|^2 = \frac{1}{4} \cdot \sin^2 (\beta + \gamma). \tag{41}$$

Diese Gleichung lässt sich auch allgemeiner beweisen. Wir verlangen nicht dass ψ identisch verschwinde, sondern nur dass

$$\lim_{r=\infty} \frac{\psi}{\alpha} = \lim_{r=\infty} t = 0$$

sei. Wir nehmen z. B. unsere Gleichung (22) als Lösung von (20), wo wir wissen, dass $c_\pm = c \cdot \exp(\mp i \frac{\pi}{4})$ und $d_\pm = d \exp(\mp i \frac{\pi}{4})$ ist. Die untere Grenze der ν_i-Integrale sei $\nu_i = 0$ und die der t-Integrale $t = 0$. Für kleine $t = \frac{\psi}{\alpha}$ wird dann:

$$\lim_{r=\infty} u_0 = \nu_0^{-\frac{1}{2}} \cdot c \cdot \left(1 + \tfrac{1}{2} \frac{\psi}{\alpha} \ldots\right) \cos\left(\int^r \nu_0 \, d r - \frac{\pi}{4}\right)$$
$$+ \nu_1^{-\frac{1}{2}} \cdot d \cdot \left(1 - \tfrac{1}{2} \frac{\psi}{\alpha} \ldots\right) \cos\left(\int^r \nu_1 \, d r - \frac{\pi}{4}\right).$$

Aus der Gleichung (21) erhält man, da für

$$t = 0: \lim \frac{\varphi_0 - \nu_0^2}{\alpha} = -1, \quad \lim \frac{\varphi_0 - \nu_1^2}{\alpha} = +1 \text{ ist:}$$

$$\lim_{r=\infty} u_1 = -\nu_0^{-\frac{1}{2}} \cdot c \cdot (1 - \ldots) \cos\left(\int^r \nu_0 \, d r - \frac{\pi}{4}\right)$$
$$+ \nu_1^{-\frac{1}{2}} \cdot d \cdot (1 + \ldots) \cos\left(\int^r \nu_1 \, d r - \frac{\pi}{4}\right).$$

Beobachtet man jetzt, dass $\nu_0 = \nu_1$ wird für $r = \infty$, und setzt

$$\tau = \int_{\nu_0 = 0}^{t=0} \nu_0 \, d r - \int_{\nu_1 = 0}^{t=0} \nu_1 \, d r ,$$

so findet man durch Befriedigen der Grenzbedingungen (28a, b):

$$| \eta_{1J}|^2 = \frac{1}{4} \sin^2 \tau .$$

Wenn ψ identisch 0 wird, so wird μ_0 identisch gleich μ_1 und nach Gl. (27) $\tau = \beta + \gamma$.

Im speziellen Fall $\psi \overset{id}{=} 0$ und $\alpha = \varepsilon/r^2$ erhält man für

$$|\eta_{1J}|^2 = \frac{1}{4} \sin^2 \left(\sqrt{J^2 + \varepsilon^2} - \sqrt{J^2 - \varepsilon^2}\right)\pi .$$

Für grosse J wird dies $= \frac{\pi^2}{4} \frac{\varepsilon}{J^2}$ (wie es auch die Matrizenauswertung gibt) so dass

$$q \infty \sum_J |\eta_{1J}|^2 \cdot J$$

nicht konvergiert. Im Falle $\lambda \geqq 1$ konvergieren also die Wirkungsquerschnitte weder im *Grenzfall* (§ 12) noch im *Resonanzfall.* Dass der *Resonanzfall* $\psi R = 0$ sich ganz anders verhält als der *Grenzfall* $\psi R = \infty$ ist auch aus London's Arbeit ersichtlich: Seine Funktion ω (bei uns in § 7 mit g bezeichnet) macht beim Übergang von $\psi R = \infty$ nach $\psi R = 0$ einen Phasensprung.

Ausser zwischen identischen Systemen, wird unseres Erachtens, nur der Grenzfall für *kleine* ψ_∞ von Bedeutung sein, d. h. der im § 12 behandelte Fall

$$\lim_{\substack{R = \infty \\ \psi_R = 0}} \psi_R R = \infty .$$

§ 14. Bestimmung des Wirkungsquerschnittes im Fall II (Überschneidungsfall).

Hier machen wir wie Landau[8]) die Annahme, dass, wenn der Schnittpunkt bei imaginären v liegt, die Übergangswahrscheinlichkeiten $|\eta_{1J}|^2$ verschwinden im Vergleich zu denen bei reellen. Für sehr kleine reelle Schnittpunktgeschwindigkeiten

$$v_J = \sqrt{p_m^2 - \frac{J^2}{r_0^2}} \simeq 0 ,$$

d. h. für $J \simeq p_m r_0 = J_m$ würden die Landau'schen $|\eta_{1J}|^2 \gg 1/4$. Unsere Methode (§ 9) gilt zwar für ganz kleine v auch nicht mehr, aber sie gilt doch solange als

$$\frac{\alpha^2}{\psi' p_m} < \int_{v_0 = 0}^{r} v_0 \, d r < p_m \cdot (r_0 - r_{v_0 = 0}) .$$

Jedoch geht unser $|\eta_{1J}|^2$ *nie* über $1/4$ hinaus. Es ist also auf alle Fälle länger richtig als Landaus Formel. Führt man die Funktion

$$f_2(k) = \int_0^1 d z \cdot \left(e^{-k \frac{z}{1-z}} - e^{-2k \frac{z}{1-z}}\right)$$

ein, und setzt

$$z = \frac{J^2}{p_m^2 r_0^2} = \frac{J^2}{J_m^2} ,$$

so ist der W. Q., wegen (7), (10), (17) und (29):

$$q = \pi r_0^2 \cdot 2 \left(\frac{p_m}{p_0}\right)^2 \cdot f_2\left(\pi \frac{A^2 p_m}{r_0^2 \psi'}\right); \quad q \cong 0 \text{ für } p_m^2 < 0. \quad (42)$$

Die Funktion f_2 ist für $k \ll 1$ gleich $4\,k/3$, erreicht für $k = 0{,}8$ ein breites Maximum im Werte von 0,164, und geht für grosse k wie $1/2\,k$. Für kleine k geht (42) in den Landau'schen Ausdruck (Landau's Formel (27)) über. *Der maximale W. Q. im Falle II hat also die Grössenordnung von πr_0^2, wo r_0 den Radius der Überschneidungsstelle bedeutet.*

Wollen wir eine „Resonanzkurve" zeichnen um die Grössenverhältnisse der Fälle I und II vergleichen zu können, so setzen wir

$$r_0 = 10^{-8}\,\text{cm}, \quad M = 10\,M_H.$$

Für ψ' setzen wir $\psi_\infty \cdot 10^8$, da ja ψ_∞ meistens schon bei $r = 2$ bis 3 Atommessern (10^{-8} cm) erreicht sein dürfte. Dann kann man q aus Formel (42) mit den q aus (38) vergleichen. Dieser Vergleich ist in Fig. 6 durchgeführt.

Man sieht sofort, dass der Überschneidungsfall (42) *nicht* den Charakter eines „Resonanzvorganges" hat, da der W. Q. für wachsende Resonanzschärfe nur wie $1/\psi$ abfällt. Die experimentellen Ergebnisse (eine Zusammenstellung findet sich bei KALLMANN und LONDON[5]), neuere Messungen liegen vor von ZEMANSKY[14])) zeigen aber deutlich Resonanzcharakter.

Was die Grössenordnung anbetrifft, so zeigt Fig. 8, dass, entgegen Landau's Auffassung, die W. Q. in der Nähe der Resonanz ohne Schnittpunkt meist grösser sind, als diejenigen mit Schnittpunkt. Die ersteren übersteigen sogar die gaskinetischen W. Q.*)

§ 15. Die Ionisation von Edelgasen durch Alkalionen.

Der Fall der Überschneidung hat aber noch eine andere Bedeutung. WEIZEL und BEECK[4]) erklären nämlich die Versuche von BEECK und MOUZON[3]) durch Überschneidung von Potentialkurven. Die experimentellen Ergebnisse sind kurz folgende:

Die Ionisation eines Edelgases durch Stoss positiver Ionen steigt bei einer Ionengeschwindigkeit, deren kinetische Energie

[14]) M. W. ZEMANSKY, Phys. Rev. **36**, 933 (1930).

*) Der W. Q. und Resonanzcharakter lässt sich im Fall II dem Fall I annähern, wenn man r_0 mit kleiner werdendem ψ_∞ entsprechend wachsen lässt.

erheblich über der Ionisationsenergie liegt, plötzlich stark an. Unterhalb dieser kinetischen Geschwindigkeit ist die beobachtete Ionisation von kleinerer Grössenordnung. Die gewöhnliche Ionisation durch Stoss schwerer Teilchen kann, wenigstens qualitativ, durch das Verfahren von MORSE und STUECKELBERG[6]) (§ 4, Gleichung (36) und Fig. 3 der betreffenden Arbeit) dargestellt werden. (In dem dortigen Bild ist die Figur $\beta = 10$ zu wählen.) Von einem plötzlichen Einsatz kann dort nicht die Rede sein. Es muss sich aber hier um einen Stoss erster Art handeln, da die auftretenden langsamen Elektronen beobachtet werden.

WEIZEL und BEECK[4]) erklären nun den Vorgang durch ein Überschneiden der Potentialkurven von $K^+ + A$ und $K^+ + A^+$. Allerdings erwähnten WEIZEL und BEECK nicht, dass es sich dabei um ein Überschneiden der Potentialkurve $K^+ + A$ und der, den verschiedenen reellen Geschwindigkeiten v des abgetrennten Elektrons im Unendlichen entsprechenden Schar von Potentialkurven $K^+ + A^+$ handelt. Für jede einzelne dieser unendlich nahe beieinander liegenden Kurven hat man aber einen andern Schnittpunkt und ein anderes Matrixelement.

Dieses Problem im Komplexen durchzuführen scheint hoffnungslos. Für die Näherungsmethode in (42) ($f_2(k) = \frac{4}{3} k$) hat man aber, wenn $\frac{d Q_v}{d v}$ die Anzahl von Elektronenzuständen per dv ist:

$$q = \int d\, q_v = \frac{8\,\pi^2}{3} \cdot \int_0^\infty \frac{p_m^3}{p_0^2} \cdot \frac{\Lambda_v^2}{\psi'_V} \cdot \frac{d\, Q_v}{d\, v} \cdot d\, v\,.$$

Wir wissen aber, dass $\Lambda_v^2 \frac{dQ}{dv}$ mit steigenden v schnell abnimmt (siehe MORSE und STUECKELBERG[3]) p. 597), und für $v = 0$ seinen Maximalwert hat[15]). Wir ersetzen daher das Integral $\Lambda_v^2 \cdot \frac{dQ}{dv} \cdot dv$ durch eine mittlere Grösse Λ^2. Das heisst mit anderen Worten wir ersetzen die unendliche Schar von Potentialkurven durch eine mittlere Kurve, welche sehr nahe bei der Kurve, für $v = 0$ liegt.

Ohne auf die Diskussion, ob die Kurven sich schneiden einzugehen (darüber siehe (14)), wollen wir einmal annehmen, dass die Überschneidung eintritt. Dann haben wir, wenn V_0 und r_{0A}, in Volt resp. Å gemessen, die Koordinaten der Überschneidungs-

[15]) E. C. G. STUECKELBERG und P. M. MORSE, Phys. Rev. **36**, 16 (1930).

stelle*) sind und wenn ψ'_V in Volt per Å gemessen ist und V die Voltgeschwindigkeit der positiven Ionen bedeutet, statt (42),

$$q = 3{,}78 \times 10^{-17} \cdot \frac{\Lambda^2}{\psi'_V \cdot \sqrt{m}} \cdot \frac{(V - V_0)^{\frac{3}{2}}}{V} \cdot \frac{3}{4\,k} \cdot f_2(k) \tag{43}$$

$$k = 4{,}48 \times 10^{-2} \cdot \frac{\Lambda^2}{\psi'_V \cdot \sqrt{m}} \cdot \frac{1}{r^2_{0\,A}} \cdot (V - V_0)^{\frac{1}{2}}$$

$$m = M/10\, M_H \,.$$

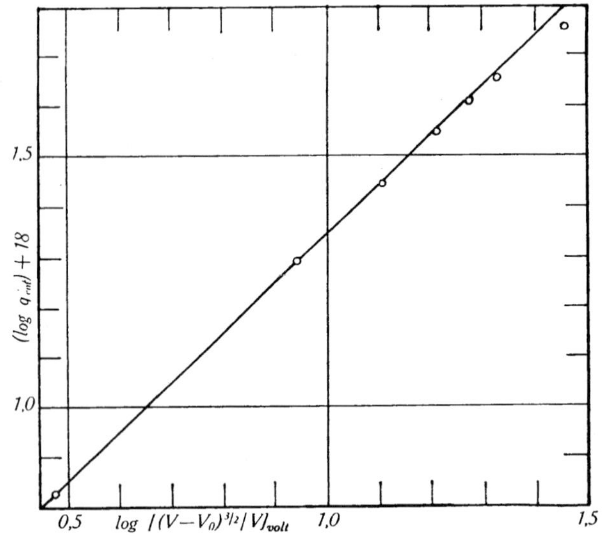

Fig. 9.

Wirkungsquerschnitt q für Ionisation von Argon durch Kalium-Ionen als Funktion von $(V - V_0)^{3/2}/V$ in cm² und $\sqrt{\text{Volt}}$ logarithmisch aufgetragen. ($\psi'_V/\Lambda^2 = 23{,}8$ Volt/cm²). Die gerade Linie liegt unter 45°. Die eingetragenen Punkte sind die Messpunkte von Nordmeyer.

Nordmeyer[3]) hat ebenfalls neuere Messungen an $K^+ + A \rightarrow K^+ + A^+$ angestellt, die die Resultate von Beeck und Mouzon[3]) im Ganzen bestätigen, aber auf höhere Geschwindigkeiten ausdehnen (bis 1000 Volt). Seine Messpunkte sind in Fig. 9 mit Gleichung (43) logarithmisch verglichen. Für V_0 wurde ein solcher Wert gewählt, dass die ersten Punkte auf einer Geraden unter

*) Im Diagramm der Potentiellen Energiekurven (Fig. 1). V_0 ist dann $= E(r_0) - E(\infty)$ auf der Kurve A_0, gemessen in Volt.

45^0 liegen (kleine k: $3\, f_2(k)/4k = 1$). Da $M \simeq 40 \times 40/80 = 20$ ist, ist $m = 2$. Daraus bestimmt sich

$$V_0 = 55 \text{ Volt}; \quad \frac{\Lambda^2}{\psi_V'} = 0{,}42 \text{ Å Volt}^{-1}.$$

Die höheren Messpunkte geben die Abweichung von $3\, f_2/4\, k$ gegen 1 an. Für 1000 Volt entspricht diese Abweichung einem k-Werte von der Grössenordnung 0,02. Setzt man $\psi_V' = 16$ Volt/1 Å, so bestimmt sich Λ^2 zu 0,67. Die grössenordnungsweise Bestimmung von k ergibt r_0 zu 1,5 Å.

Die Weizel-Beeck'sche Erklärung scheint also auch quantitativ richtige Ergebnisse zu liefern. Es wäre interessant die Versuche zu höheren Geschwindigkeiten auszudehnen, um das bei $k = 1$ liegende Maximum zu bestimmen. Wenn die Abschätzung von r_0 richtig ist so liegt dasselbe allerdings erst bei ca. 500000 Volt, sein Wert beträgt aber nur ungefähr das Doppelte von dem bei 1000 Volt.

§ 16. Zusammenfassung.

Ein dem W. K. B.[2]) Verfahren entsprechende Anschlussmethode der Näherungsfunktionen zweier gekoppelter Differentialgleichungen wird ausgearbeitet.

Dieses Verfahren lässt sich auf Stösse zweiter Art zwischen Atomen anwenden. Es sind zwei Fälle zu unterscheiden. Im Fall I (§ 10) schneiden sich die für die elastische Bewegung verantwortlich zu machenden potentiellen Energiekurven in nullter Näherung nicht. In Fall II (§ 9) schneiden sie sich bei reellen Geschwindigkeiten. In gewissen Fällen darf das Anschlussverfahren durch das übliche Störungsverfahren ersetzt werden (§ 6 und § 11).

Zur Bestimmung der Wirkungsquerschnitte für Stösse zweiter Art angewendet zeigt sich, dass im Ganzen Fall I für diese Erscheinung verantwortlich zu machen ist (§ 14). Für Stösse zwischen positiven Ionen und Atomen hingegen dürfte der Fall II von Bedeutung sein (15).

Eine weitere Anwendungsmöglichkeit des beschriebenen Verfahrens bieten die Prädissoziation und eventuell die Vorgänge beim Zusammenstoss von α-Teilchen und Atomkernen.

Basel, Physikalische Anstalt der Universität.

Relativistisch invariante Störungstheorie des Diracschen Elektrons. I. Teil: Streustrahlung und Bremsstrahlung [20]

Annalen der Physik, 5. Folge, vol. 21 (1934), pp. 367–389.
Copyright Wiley-VCH Verlag GmbH & Co. KGaA.

Relativistisch invariante Störungstheorie des Diracschen Elektrons

I. Teil: Streustrahlung und Bremsstrahlung

Von E. C. G. Stueckelberg

(Mit 2 Figuren)

Einleitung und Zusammenfassung

Die übliche Störungstheorie der Quantenmechanik entwickelt die Lösung des gestörten Problems nach Eigenfunktionen eines ungestörten Systems und bestimmt die Abhängigkeit der *variabeln Entwicklungskoeffizienten* von der Zeit durch ein Näherungsverfahren.

Der elastische Stoß eines Elektrons an einem Kern (elektromagnetisches Feld) wird dann für hohe Geschwindigkeiten durch die erste Näherung, und seine Ausstrahlung in einem elektromagnetischen Felde durch die zweite Näherung gegeben, wenn als nullte Näherung das freie Elektron betrachtet wird. Ist das störende Feld z. B. dasjenige einer Lichtwelle, so erhält man die Klein-Nishina-Formel (K.-N.-F.). Ist es das Feld eines Kernes, so erhält man die Formel für die Bremsstrahlung.

Für viele Zwecke ist es nun vorteilhaft die durch Störungsverfahren gewonnenen Ausdrücke für Bewegung und Ausstrahlung eines Elektrons invariant zu schreiben. Dies wird erreicht durch eine Fourierentwicklung der variablen Koeffizienten nach der Zeit. Die Lösung schreibt sich dann als vierdimensionale Fourierreihe mit *konstanten Entwicklungskoeffizienten*, die durch Näherungsverfahren bestimmt werden. Da Zeit und Raum in gleicher Weise in die Rechnung eingehen, sind die Resultate relativistisch invariant und man umgeht z. B. damit das Problem der Lorentztransformation in der Berechnung der K.-N.-F. für bewegte Elektronen aus derjenigen für ruhende (§ 5)[1]).

Die Bremsstrahlung eines Elektrons im beliebigen elektromagnetischen Felde kann durch eine Eich- und Lorentztransformation aus der verallgemeinerten K.-N.-F. hergeleitet werden (§ 3), wenn in deren Ableitung die Lichteigenschaften (Lichtgeschwindigkeit und Transversalität) der Primärwellen nicht benützt werden. Das Resultat stimmt mit demjenigen von Bethe und Heitler und von Sauter[2]) überein. Für große Geschwindigkeiten des Elektrons kann die Nichttransversalität und Unterlichtgeschwindigkeit vernachlässigt werden[3]) (§ 6 und § 7). Die Anwendung der Methode auf das Problem der Paarerzeugung soll in einem *zweiten Teile* erfolgen.

§ 1. Die Wellengleichung

Die Wellengleichung schreiben wir in der Form*):

$$\left[\frac{1}{i}\left(\gamma, \frac{\partial}{\partial x}\right) + C + V(x)\right] \psi(x) = 0. \tag{1,1}$$

Hier und im folgenden bedeuten kleine Buchstaben $a, b \ldots$ Weltvektoren [z. B. $x = (x_1, x_2, x_3, x_4)$; $x_4 = i c t$] mit imaginärem Zeitanteil. (a, b) ist ihr skalares Produkt. $\vec{a} = (a_1, a_2, a_3)$ stellt den räumlichen Anteil von a in einem bestimmten Lorentzsystem dar. $(\vec{a}, \vec{b})$ ist das räumliche skalare Produkt. Die reelle Größe a_0 ist der durch die imaginäre Einheit i dividierte Zeitanteil von a. γ ist der aus den hermiteischen Diracschen Operatoren gebildete Weltvektor:

$$\gamma = (\beta \vec{\alpha}; i \beta).$$

Seine Komponenten genügen den Relationen:

$$\left\{\begin{aligned} \gamma_\mu \gamma_\nu + \gamma_\nu \gamma_\mu &= -2\delta_{\mu\nu}, \\ \gamma_\mu^+ &= -\gamma_\mu. \end{aligned}\right. \tag{1,2}$$

Ein $^+$ bedeutet „hermiteisch konjugiert“.

V ist die Wechselwirkungsenergie von Elektron und Feld (dividiert durch $\hbar c$), wenn die Feldenergie eliminiert wird[4]) und $C = \frac{\mu c}{\hbar}$ ist die reziproke Comptonwellenlänge. (Es bedeutet $\hbar$ die Plancksche Konstante durch 2π, μ die Elektronenmasse und c die Lichtgeschwindigkeit.) Beschreiben wir in der üblichen Weise nur den transversalen Feldanteil durch die Quantenelektrodynamik, so treten dort die Operatoren Γ_k und Γ_k^+ auf. Denkt man sich ψ nach Eigenfunktionen des Strahlungsfeldes $u(N^j)$ entwickelt, wo

$$N^j = (N_1^j \ldots N_k^j \ldots)$$

die Gesamtheit der Photonenzahlen in der durch k numerierten

*) Vgl. z. B. bei Pauli, Handb. d. Physik XXIV Formel (II) S. 220. Die γ_μ der vorliegenden Rechnungen sind die γ_μ von Pauli mit i multipliziert.

Lichtwellen darstellt, und der Index j alle Verteilungsmöglichkeiten durchläuft, so ist:

$$\psi = \sum_j \varphi^j(x)\, u(N^j). \tag{1,3}$$

Die Operatoren sind definiert durch:

$$\Gamma_k \cdot u(N_1 \ldots, N_k, \ldots) = \sqrt{N_k} \cdot u(N_1 \ldots, N_k + 1, \ldots),$$
$$\Gamma_k^+ \cdot u(N_1 \ldots N_k \ldots) = \sqrt{N_k + 1} \cdot u(N_1 \ldots, N_k - 1, \ldots).$$

Die Fourierzerlegung des transversalen Anteils von V lautet dann

$$V^t + V^{t+} = \sum_{\vec{k}} T_k(\sigma^k, \gamma)\left[\Gamma_k e^{i(k,x)} + \Gamma_k^+ e^{-i(k,x)}\right]. \tag{1,4}$$

Die Summation ist nur über ein dreidimensionales Gebiet zu erstrecken, da immer

$$(k, k) = 0\,; \quad k_0 > 0 \tag{1,5}$$

gilt, und über zwei gegenseitig und zu k senkrechte Polarisationsrichtungen σ^k. Es ist also

$$(k, \sigma^k) = 0. \tag{1,6}$$

Wählen wir

$$(\sigma^k, \sigma^k) = 1$$

und denken uns die Fourierzerlegung in einem bestimmten Lorentzsystem durchgeführt, so wird, wenn G dort einen räumlichen Periodizitätsbereich und e die elektrische Elementarladung bedeutet:

$$T_k^2 = \frac{2\pi e^2}{G k_0 c \hbar}. \tag{1,7}$$

Der von Ladungen herrührende Teil kann durch

$$V^L = \sum_{\vec{k}} M_k(\sigma^k, \gamma)\, e^{i(k,x)} \tag{1,8}$$

dargestellt werden.

Wird die Summation in demjenigen Lorentzsystem ausgeführt, wo dieser Anteil ein kugelsymmetrisches Feld mit einem skalaren Potential

$$\frac{Ze}{r} f(r)$$

darstellt, so ist $\vec{\sigma}^k = 0$, $\sigma_4^k = i$ und

$$\left\{\begin{aligned} M_k &= \frac{4\pi Z e^2}{G\,|\vec{k}|^2 \hbar c}\, g(|\vec{k}|), \\ g(|\vec{k}|) &= \lim_{a=0} \int_0^\infty e^{-ar} f(r)\, |\vec{k}| \sin |\vec{k}|\, r\, dr. \end{aligned}\right. \tag{1,9}$$

Für ein Coulombfeld wird $f = g = 1$.

§ 2. **Die Störungstheorie**

Wir fassen die Fourierentwicklungen der Felder V^t, V^{t+} und V^L als V_k, die Operatoren Γ_k, Γ^+_{-k} und 1 als P_k zusammen, und definieren die Matrixelemente als:

$$(2,1) \qquad P_k\, u(N^j) = \sum_i u(N^i)\, P_{kij}$$

(z. B. ist im Falle von V^t: $P_k = \Gamma_k$; $P_{kij} = \sqrt{N^j_k}$, wenn die Verteilung N^i sich von N^j durch die Vermehrung der Lichtquantenzahl um 1 in k unterscheidet. In allen anderen Fällen ist $P_{kij} = 0$; $V_k = T_k$).

Die Wellenfunktion ψ, als vierdimensionale Fourierreihe geschrieben, lautet:

$$(2,2) \qquad \psi(x, N) = \sum_j \int d l^4\, e^{i(l,x)} A^j(l)\, u(N^j) = \sum_j \varphi^j(x)\, u(N^j)$$

$\int d l^4$ bedeutet das vierfache, über l_1, l_2, l_3 und l_4 erstreckte Integral. $\int d \vec{l}^{\,3}$ ist das Symbol für die Integration über l_1, l_2 und l_3. Definieren wir jetzt die Operatoren:

$$(2,3) \qquad \begin{cases} H(l) = \quad (l,\gamma) + C\,, \\ K(l) = -(l,\gamma) + C\,, \\ K(l) H(l) = H(l) K(l) = R(l) = (l,\, l) + C^2 \end{cases}$$

und setzen (2,2) in die Wellengleichung (1,1), so gilt:

$$(2,4) \qquad \begin{cases} \sum_j \int d l^4\, e^{i(l,x)}\, u(N^j) \Big[H(l)\, A^j(l) \\ \qquad + \sum_{k,i} V_k P_{kji} (\sigma^k,\, \gamma) A^i(l-k) \Big] = 0\,. \end{cases}$$

Als nullte Näherung wählen wir die ungestörten Eigenfunktionen ($V = 0$), deren Entwicklungskoeffizienten für jedes j

$$(2,5) \qquad \int d l^4\, e^{i(l,x)} H(l)\, A^j(l) = 0$$

genügen müssen. Die $A^j(l)$ sind Spinoren. Die (2,5) erfüllende Bedingung

$$H(l)\, A^j(l) = 0$$

entspricht vier homogenen Gleichungen mit vier Unbekannten. Damit ihre Determinante verschwindet, muß das in (2,3) definierte

$$(2,6) \qquad R(l) = 0$$

sein. Im reellen l Raume liegen diese Punkte auf einem zweischaligen Hyperrotationshyperboloid, entsprechend den möglichen Zuständen positiver und negativer Energie des Diracelektrons. Die dreidimensionale Mannigfaltigkeit der (2,6) erfüllenden Vektoren der einen Schale sei mit $\bar{l}$ bezeichnet. Dann genügt:

$$A^j(l) = \frac{1}{\pi} \frac{B^j(l)}{R(l)}, \tag{2,7}$$

welches auf dem Hyperboloid singulär ist, der Bedingung (2,5), wenn $B^j(l)$ eine für reelle $\vec{l}$ und l_0 singularitätenfreie stetige Funktion ist, die auf der einen Hyperboloidschale $l_4 = \bar{l}_4(\vec{l})$ der Bedingung

$$H(\bar{l})\, B^j(\bar{l}) = 0 \tag{2,8}$$

genügt und in großer Entfernung davon genügend stark verschwindet*). Dann kann die Integration über dl_4 durch Ausweichen in die komplexe l_4-Ebene ausgeführt werden**):

$$\int d l_4\, e^{i l_4 x_4} A^j(l) = \frac{B^j(\bar{l})}{\bar{l}_0} e^{i \bar{l}_4 x_4},$$

(2,7) erfüllt also (2,5). Die räumliche Fourierentwicklung lautet

$$\varphi^j = \int d\vec{l}^3\, e^{i(\vec{l}, x)} \frac{B^j(\bar{l})}{\bar{l}_0}. \tag{2,9}$$

Die Teilchenzahl ist:

$$n^j = \int \varrho^j d\vec{x}^3 = \frac{1}{i}\int \varphi^{j+} \gamma_4 \varphi^j d\vec{x}^3 = \frac{(2\pi)^3}{i}\int d\vec{l}^3 \frac{B^j(\bar{l})^+}{\bar{l}_0} \gamma_4 \frac{B^j(\bar{l})}{\bar{l}_0} \tag{2,10}$$

und ist (immer in nullter Näherung) selbstverständlich eine Invariante.

Für unsere Probleme wählen wir insbesondere eine solche Lösung nullter Näherung, für welche die $A^j(l)$ für alle Verteilungen j Null sind, außer für eine bestimmte Lichtquantenverteilung, welche durch den Index $j = 0$ gekennzeichnet sei. (2,4) wird dann in nullter Näherung (2,5) mit $j = 0$.

Größen erster Näherung sind dann das erste Glied von (2,4) summiert über $j \neq 0$, und im zweiten Glied diejenigen Summanden, für welche $i = 0$ ist. Die $A^j(j \neq 0)$ werden dann in erster Näherung durch die bekannten A^0 bestimmt, wenn

---

*) $\bar{l}$ bedeutet den Vierervektor mit den Komponenten l_1, l_2, l_3 und $\bar{l}_4 = \bar{l}_4(\vec{l}) = i\bar{l}_0(\vec{l}) = \pm i \sqrt{C^2 + (\vec{l}, \vec{l})}$. Je nach der Wahl des Vorzeichens durchläuft $\bar{l}$ die Gesamtheit von Punkten (d. h. von möglichen Zuständen des Dirac Elektrons) positiver oder negativer Energie.

**) Bei der Integration für positive Zeiten muß in der komplexen l_0-Ebene der Punkt $l_0 = \bar{l}_0$ im positiven Sinne umfahren werden, wenn $\bar{l}_0$ positiv ist, und im negativen Sinne, wenn $\bar{l}_0$ negativ ist. Bei der Umfahrung im umgekehrten Sinne gibt die Integration den Wert Null.

die Klammer von (2,4) gleich Null gesetzt wird. Werden die Operatoren aus (2,3) verwendet, so ist in erster Näherung:

$$(2{,}11)\quad A^j(l) = -\frac{1}{R(l)}\cdot K(l)\sum_k{}' V_k\, P_{kj0}(\sigma^k,\gamma)\, A^0(l-k)\,.$$

Schreiben wir dies in der Form (2,7), so hat das $B^j(l)$ erster Näherung jetzt eine Singularität auf dem um ein festes k verschobenen Hyperboloid $R(l-k)=0$. Die Koeffizienten der dreidimensionalen Fourierentwicklung φ^j werden daher zeitabhängig. Die Bedeutung der Formel (2,11) gibt Fig. 1:

Der reelle $\vec{l}$, l_0-Raum ist im Schnitt l_3, l_0 gezeichnet. Wir betrachten nur die von der $k$ten Welle des Feldes hervorgerufene Störung. Die Gegend, in welcher die $A^0(l)$ wesentlich von Null verschieden sind, liegt wegen $R(l)$ in (2,7) auf dem Hyperboloid. Stellt die nullte Näherung ein Wellenpaket mit mittlerem

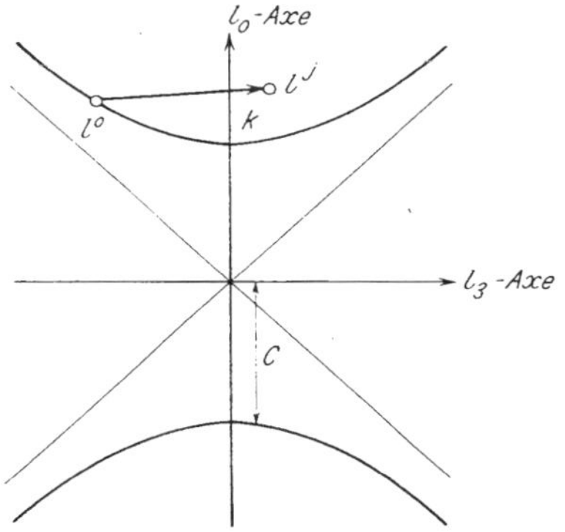

$l_0\, l_3$-Schnitt durch den (reellen) Energieimpulsraum. Die durch die k-te Fourierkomponente hervorgerufene Störung bildet in erster Näherung das Gebiet l^0 auf das Gebiet $l^j = l^0 + k$ ab. Nur wenn l^j auf der Hyperbel $l_0^2 - l_3^2 = C^2$ liegt, tritt eine Störung erster Näherung auf

Fig. 1

Impuls $\vec{l^0}$ und mittlerer Energie l_0^0 dar, so ist A^0 nur in der Gegend l^0 von Null verschieden. Dann sagt (2,11), daß $A^j(l)$ nur in der Gegend $l = l^j = l^0 + k$ von Null verschiedene Werte annehmen kann. Wegen $R(l)$ in (2,11) ist aber dieser Wert unendlich viel größer, wenn l^j auch auf dem Hyperboloid liegt, als wenn dies nicht der Fall ist. Ist das Störungsfeld im gezeichneten Raum–Zeitsystem ein zeitlich konstantes Feld, so liegt dieses, durch die Gesamtheit aller k aus l^0 erzeugte, Gebiet l^j, wegen $k_4 = 0$, auf dem von der, zu $\vec{l}$ parallelen, Hyperebene durch l^0 mit dem Hyperhyperboloid erzeugten Hyperkreis. Dem entspricht die elastische Streuung des Elektrons ($l_4^j = l_4^0$). Sind die

k-Lichtvektoren, die wegen $(k, k) = 0$ parallel zu den Mantellinien des Asymptotenhyperkegels des Hyperhyperboloids sind, so liegt keine der Gegenden $l^0 + k$ auf dem Hyperhyperboloid. Eine Wechselwirkung zwischen Licht und Elektron tritt also in erster Näherung nicht auf.

Die zweite Näherung erhalten wir aus der ersten in gleicher Weise, wie die erste aus der zweiten entwickelt wurde:

$$A^j(l) = \frac{1}{R(l)} \sum_{pk}' V_p V_k P_{pji} P_{kio} \, \Omega(l^0) \, A^0(l^0), \tag{2,12}$$

$$\Omega(l^0) = K(l) \left\{ \frac{(\sigma^p, \gamma) \, K(l-p) \, (\sigma^k, \gamma)}{R(l-p)} + \frac{(\sigma^k, \gamma) \, K(l-k) \, (\sigma^p, \gamma)}{R(l-k)} \right\} \tag{2,13}$$

mit $l^0 = l - k - p$.

Die Fig. 2 stellt die Beziehungen der Formel (2,12) dar.

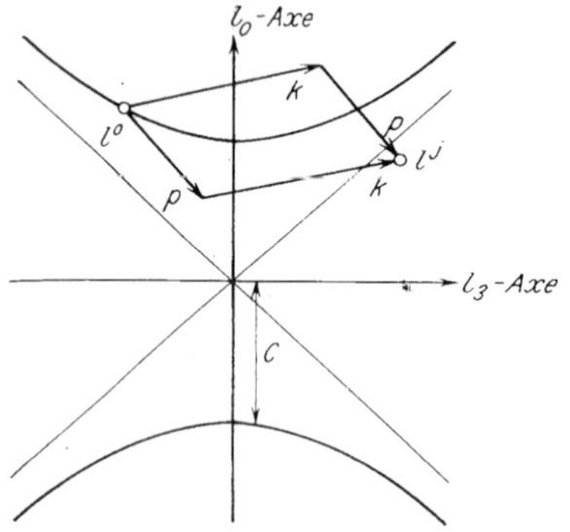

Wie Fig. 1. Die durch die k-te und p-te Fourierkomponente hervorgerufene Störung bildet das Gebiet l^0 auf das Gebiet

$$l^j = l^0 + k + p$$

ab. Nur wenn l^j auch auf der Hyperbel liegt, tritt eine Störung zweiter Näherung auf. Die Zwischenzustände $l^{i_1} = l^0 + k$ und $l^{i_2} = l + p$ liegen dann *nicht* auf der Hyperbel. (Die Symbole l^i der Zwischenzustände sind in die Figur nicht eingetragen)

Fig. 2

Auch hier ist $A^j(l)$ auf dem Hyperhyperboloid unendlich viel größer als in anderen Gegenden. Die zwei Wege in der Fig. 2 entsprechen der Summe von zwei Gliedern im Operator Ω. Das Herausziehen der Matrizen der P-Operatoren aus der Summe ist erlaubt, weil immer $P_{pji_1} P_{ki_1 0} = P_{kji_2} P_{pi_2 0}$ ist. Die Summation über i kann in (2,12) weggelassen werden, da für jedes k nur *ein* Matrixelement P_{kji} existiert.

Die zeitabhängigen Koeffizienten der dreidimensionalen Fourierzerlegung werden aus der Beziehung

$$\frac{B^j(\bar{l}; x_4)}{\bar{l}_0} = \int d\, l_4 \, e^{i(l_4 - \bar{l}_4) x_4} A^j(l) \tag{2,14}$$

durch Verlegen des Integrationsweges ins Komplexe erhalten*). [Die Beziehung (2,14) auf A^0 angewandt, gibt trivialerweise $B^0(\bar{l}; x_4) = B^0(\bar{l})$.]

Aus (2,12) folgt:

$$(2,15)\quad \frac{B^j(\bar{l}; x_4)}{l_0} = \sum_{kp} V_k V_p P_{kji} P_{pio} \frac{1 - e^{i s_4 x_4}}{s_4} \frac{\Omega(\bar{l}^0)}{l_4} \frac{B^0(\bar{l}^0)}{l_0^0}$$

mit

$$(2,16)\quad s_4 = l_4 - (l_4^0 + k_4 + p_4).$$

Die zeitabhängige Teilchenzahl im Zustande j wird

$$(2,17)\quad n^j(x_4) = \frac{1}{i}\int d\vec{x}^3 \varphi^{j+} \gamma_4 \varphi^j = \frac{(2\pi)^3}{i}\int d\vec{l}^3 \frac{B^j(\bar{l}; x_4)}{l_0} \gamma_4 \frac{B^j(\bar{l}; x_4)}{l_0}.$$

§ 3. Die Lichtstreuung am freien Teilchen (verallgemeinerte K.-N.-F.)

Als Verteilungsfunktion nullter Näherung wird

$$N^0 = (0\,0\,0 \ldots 0,\ N_k,\ 0\ldots)$$

gewählt und die zu

$$N^{j_m} = (0\,0\,0 \ldots 0,\ 1_m,\ 0\ldots 0,\ N_k - 1,\ 0\ldots)$$

gehörige Teilchenzahl bestimmt. Den Zwischenzuständen entsprechen

$$N^i(0\,0\,0 . 0,\ N_k - 1, 0 \ldots) \text{ und } (0\,0\,0 \ldots 0, 1_m,\ 0 \ldots 0,\ N_k,\ 0 \ldots).$$

In (2,15) werden k und $-p = m$ Lichtvektoren. Dies ist die einzige mögliche Störung, wenn Übergänge zu Zuständen auf der anderen Hyperhyperboloidschale nicht berücksichtigt werden. [Zustände negativer Energie werden von solchen positiver Energie aus erreicht, wenn $-k$ und $-p$ Lichtvektoren sind. Dem Vorgang entspricht in der Diracschen Theorie des positiven Elektrons die Rekombinationsstrahlung (Aussendung von zwei Lichtquanten) von positiven und negativen Elektronen, wenn ein unbesetzter Zustand negativer Energie als positives Elektron betrachtet wird.] Die Matrizen P_{kji} P_{pio} werden gleich $\sqrt{N_k}$.

An Stelle von p schreiben wir $-m$, so daß k und m Lichtvektoren sind: $(m, m) = (k, k) = 0$; m_0 und $k_0 > 0$. Dem „Anfangszustand" entspricht dann das Vorhandensein von N_k Quanten in der $k$ten Lichtwelle, welche eine Stromdichte in Richtung $\vec{k}$ von $N_k c/G$-Lichtquanten der Frequenz $\nu_k = k_0 c$ erzeugen. Dem „Endzustande" entspricht das Vorhandensein von N_k 1 Quanten in k und einem Quant in m. Die Summation über k und p bzw. m in (2,15) fällt daher weg. Fragen wir nach der

*) Vgl. zweite Anmerkung auf S. 371.

Teilchenzahl, welche dem Auftreten eines Lichtquants bestimmter Polarisation σ^m im Raumwinkel $d\,\omega_m$ entspricht, so ist $n^{j_m}(x_4)$ aus (2,17) über alle in $d\,\omega_m$ liegenden m zu summieren. Dieser Summation entspricht die Integration über $(2\pi)^{-3}\,G\,m_0^2\,dm_0\,d\omega_m$. Ersetzt man noch zuerst dm_0 durch ds_4 und dann $d\vec{l}^{\,3}$ durch $d\vec{l^0}^{\,3}$, wobei die Funktionaldeterminanten

$$\frac{dm_0}{ds_4} = \frac{\bar{l}_4^0\, m_0}{(\bar{l}^0,\, m)} \tag{3,1}$$

und (aus $\vec{l} = \vec{l^0} + \vec{k} - \vec{m}$ bei festem $\vec{k}$ und $\vec{m}_0$)

$$\frac{d\vec{l}^{\,3}}{d\vec{l^0}^{\,3}} = \frac{\bar{l}_4}{\bar{l}_4^0}\,\frac{(\bar{l}^0,\, m)}{(\bar{l},\, m)} \tag{3,2}$$

auftreten und integriert, so erhält man wegen der Integration über den Resonanznenner $|s_4|^2$ ein lineares Anwachsen der Teilchenzahl mit der Zeit x_4. Die Zunahme dieser Größe in der Zeiteinheit mal der Energie $h\,c\,m_0$ des einzelnen Quantes ist aber der in den Raumwinkel $d\,\omega_m$ gestreute Energiestrom J^m der Frequenz $\nu_m = m_0\,c$. Durch Einsetzen der Größen T_k und T_m und der Energiestromdichte

$$S^k = \frac{N_k\, h\, c^2\, k_0}{G} \tag{3,3}$$

erhält man die K.-N.-F. in der üblichen Form.

$$J^m\,d\,\omega_m = d\,\omega_m \cdot S^k n^0 \cdot \frac{e^4}{2\,\mu^2\,c^4} \cdot \frac{m_0^3\,\overline{C^2(\gamma_4\Omega^+\,\Omega)}}{2\,k_0^2\,l_0\,|(\bar{l},\,m)|}\,. \tag{3,4}$$

Der überstrichene Operator stellt dessen Erwartungswert im ungestörten Zustande dar [vgl. unten Formel (3,5) und § 4]. (3,4) folgt direkt aus (2,15) und der dazu adjungierten Gleichung für $B^{j+}(\bar{l};\,x_4)$, (2,16) und (2,17) unter Verwendung von (3,1), (3,2) und den Definitionen (1,7) und (3,3). n^0 ist die Teilchenzahl im Anfangszustande gemäß (2,10).

In der zu (2,15) adjungierten Gleichung wurde die Definition $A(l)^+$ bzw. $B(l)^+$ verwendet. Die Diracgleichung in der Form (1,1) bestimmt eine vierkomponentige Wellenfunktion (Spinor) ψ, deren adjungierte ψ^+ durch

$$\psi^+ = \psi^*\,\gamma_4$$

definiert ist. ψ^* ist eine vierkomponentige Funktion, deren Komponenten konjugiert komplex zu denen von ψ sind. Dann gilt natürlich auch für den Spinor B

$$B^+ = B^*\,\gamma_4; \quad B^* = -\,B^+\,\gamma_4\,.$$

In der zu (2,15) adjungierten Gleichung tritt $(\Omega B^0)^+$ auf, wo der Operator Ω (2,13) eine vierdimensionale quadratische Matrix bedeutet. Es ist dann also*):

$$(\Omega B^0)^+ = (\Omega B^0)^* \gamma_4 = B^{0*} \Omega^+ \gamma_4 = - B^{0+} \gamma_4 \Omega^+ \gamma_4 .$$

Im Ausdruck für die Teilchenzahl erscheint dann der Erwartungswert des Operators $\gamma_4 \Omega^+ \Omega$ in nullter Näherung:

$$(3,5)\quad \begin{cases} \overline{\gamma_4 \Omega^+ \Omega} = \dfrac{1}{n^0} \displaystyle\int \overrightarrow{d x}^3 \varphi^{0+} \gamma_4 \Omega^+ \Omega \varphi^0 \\ \qquad = \dfrac{(2\pi)^3}{n^0} \displaystyle\int \overrightarrow{d \bar{l}}^{0\,3} \dfrac{B^0(\bar{l}^0)^+}{\bar{l}_0^0} \gamma_4 \Omega^+(\bar{l}^0)\, \Omega(\bar{l}^0) \dfrac{B^0(\bar{l}^0)}{\bar{l}_0^0} . \end{cases}$$

Dieser Erwartungswert, welcher in allen Problemen auftritt, ist im folgenden Paragraphen ausgerechnet.

Es sei noch bemerkt, daß die vorkommenden Zwischenzustände *nicht* mit denjenigen bei Waller[5]) übereinstimmen. Für die Rückwärtsstreuung z. B. sind die Beiträge, welche die Zwischenzustände i_1 und i_2 allein geben würden, ihrem Betrage nach beinahe gleich. Ihre Phase ist aber beinahe um π verschieden. Ihre Superposition ergibt darum einen Betrag, welcher von kleinerer Größenordnung ist als die Beträge der einzelnen Beiträge.

Aus später ersichtlichen Gründen wollen wir eine zweite Ableitung der K.-N.-F. geben. Diese mehr korrespondenzmäßige Ableitung beschreibt die Primärwelle k auf klassische Weise durch ein elektromagnetisches Feld der Art (1,8)

$$(3,6)\qquad W = M_k (\sigma^k, \gamma)\, 2 \cos (k, x) .$$

Als Anfangszustand wählen wir $N^0 = (0, 0, \ldots 0)$, d. h. es sind keine Lichtquanten vorhanden. Unter Einwirkung der Störung W werden dann Lichtquanten emittiert.

Die Rechnung geht ganz analog der oben angeführten quantenelektrodynamischen Behandlung. An Stelle von $\sqrt{N_k}\, T_k$ steht überall M_k. Die Störung W ist die Wechselwirkung zwischen dem Elektron und einer elektromagnetischen Welle mit dem Viererektorpotential:

$$(3,7)\qquad a^k = \frac{2 \hbar c}{e} M_k \sigma^k \cos (k, x) .$$

Berechnen wir jetzt die Anzahl Lichtquanten, welche unter Einwirkung dieses Potentials in den Raumwinkel $d\omega_m$ gestreut werden, so erhalten wir statt (3,5)

$$(3,8)\qquad \frac{J^m d\omega_m}{\hbar \nu_m} = m_0^2\, d\omega_m\, n^0 M_k^2 \frac{e^2}{8\pi\hbar} \frac{\overline{\gamma_4 \Omega^+ \Omega}}{\bar{l}_0 \,|(\bar{l}, m)|} .$$

*) Das Zeichen $^+$ bedeutet bei einem Operator „hermiteisch konjugiert“ und bei einem Spinor „adjungiert“.

Wir wollen diese Form der K.-N.-F., in welcher rechts die *Amplitude der Primärwelle* und nicht die *Energiestromdichte* steht, die *verallgemeinerte K.-N.-F.* nennen. Der zeitliche Mittelwert der Energiestromdichte der Primärwelle beträgt:

$$(3,9)\qquad S^k = \frac{\hbar^2\, c^3\, M_k^2}{2\pi\, e^2}\left\{ k_0 \left[\vec{\sigma}^k, [\vec{k}, \vec{\sigma}^k]\right] - \sigma_0^k \left[\vec{k}, [\vec{k}, \vec{\sigma}^k]\right]\right\}$$

$[\vec{a}, \vec{b}]$ bedeutet das Vektorprodukt von $\vec{a}$ und $\vec{b}$.

Ist k eine Lichtwelle $\left((k, k) = (\sigma^k, k) = 0\right)$ und wird $(\sigma^k, \sigma^k) = 1$ gesetzt, so gilt

$$(3,10)\qquad M_k^2 = \frac{2\pi\, e^2\, S^k}{\hbar^2\, c^3\, k_0^2} = S^k\, \frac{e^2}{2\mu^2\, c^4}\, \frac{4\pi\, C^2}{c\, k_0^2}.$$

Wird dieser Ausdruck in die verallgemeinerte K.-N.-F. eingesetzt, so folgt wieder (3,4). Man erkennt, daß (3,8) eine allgemeinere Form der K.-N.-F. ist, welche erst durch die für Licht gültige Beziehung (3,10) zwischen S^k und M_k in die gewöhnliche Form der K.-N.-F. (3,4) übergeht.

§ 4. Der Erwartungswert von $\gamma_4\, \Omega^+\, \Omega$

Aus der Vertauschungsrelation (1,2) folgen sofort die Beziehungen

$$(4,1)\qquad \begin{cases} (a, \gamma)\,(b, \gamma) + (b, \gamma)\,(a, \gamma) = -\,2\,(a, b), \\ K(l)\,(a, \gamma) = (a, \gamma)\,H(l) + 2\,(a, l), \\ (a, \gamma)\,K(l) = H(l)\,(a, \gamma) + 2\,(a, l) \end{cases}$$

und aus der Beziehung $\gamma_\mu^+ = -\,\gamma_\mu$ in (1,2)

$$(4,2)\qquad \begin{cases} K(l)\,\gamma_4 = \gamma_4\, K(l)^+, \\ (a, \gamma)\,\gamma_4 = \gamma_4\,(a, \gamma)^+. \end{cases}$$

Ferner ist immer, wegen (2,8)

$$(4,3)\qquad H(\bar{l}^0)\, B^0(\bar{l}^0) = B^0(\bar{l}^0)^+\, H(\bar{l}^0) = 0\,.$$

Bildet man den zu (2,13) hermitetisch konjugierten Operator Ω^+, so hat dieser am Ende $K(\bar{l})^+$ stehen. Nun ist aber wegen (1,2) und weil l_4 imaginär ist

$$K(l)^+ = H(l) - 2\, l_4\, \gamma_4\,.$$

Schreibt man jetzt $\gamma_4\, \Omega^+\, \Omega$ und berücksichtigt die letzte Gl. (2,3) und daß $R(\bar{l}) = 0$ ist, so kann man unter Verwendung von (4,2) das in der „Mitte" stehende γ_4 an den Anfang ziehen, und erhält:

$$
(4,4)\quad \begin{cases} \frac{1}{2}\gamma_4\, \Omega^+\, \Omega = \\ \qquad \bar{l}_4 \left\{ \dfrac{(\sigma^m, \gamma)\, K\,(\bar{l}^0 - m)\,(\sigma^k, \gamma)}{R\,(\bar{l}^0 - m)} + \dfrac{(\sigma^k, \gamma)\, K\,(\bar{l} + m)\,(\sigma^m, \gamma)}{R\,(\bar{l} + m)} \right\} \\ \quad K\,(\bar{l}) \left\{ \dfrac{(\sigma^k, \gamma)\, K\,(\bar{l}^0 - m)\,(\sigma^m, \gamma)}{R\,(\bar{l}^0 - m)} + \dfrac{(\sigma^m, \gamma)\, K\,(\bar{l} + m)\,(\sigma^k, \gamma)}{R\,(\bar{l} + m)} \right\}. \end{cases}
$$

(4,4) wird mit Hilfe der Relationen (4,1) umgeformt. Dabei können solche Glieder, bei welchen $H\,(\bar{l}^0)$ als erster oder letzter Faktor steht, wegen (4,3) weggelassen werden. Ferner ist $R\,(\bar{l}) = 0$ zu berücksichtigen. Zur Vereinfachung werde noch angenommen, daß gelte:

$$
(4,5)\qquad (m, m) = (m, \sigma^m) = 0\,; \quad (\sigma^m, \sigma^m) = 1\,.
$$

Für die Primärwelle k wollen wir diese Einschränkung *nicht* machen. Durch die Umformung kann erreicht werden, daß der Operator schließlich die γ_μ nur noch in erster Potenz enthält.

Hat das Wellenpaket nullter Ordnung große Linearausdehnungen und daher einen scharfen definierten Impuls $\vec{l}^0$ und eine scharf definierte Energie l_0^0, so wird der Erwartungswert von γ_μ ($\mu = 1, 2, 3$)

$$
\overline{\gamma_\mu} = \frac{l_\mu^0}{l_0^0} = \frac{v_\mu}{c}\,; \qquad \overline{\gamma_4} = 1\,,
$$

wo $\vec{v}$ die Geschwindigkeit des Wellenpaketes im gewählten Lorentzsystem ist. Man erhält so den Ausdruck:

$$
(4,6)\quad \begin{cases} \dfrac{l_4^0}{2\,l_4}\;\overline{\gamma_4\, \Omega^+\, \Omega} \\ = \dfrac{(\sigma^m, l^0)^2}{(m, l^0)^2} \left\{ 2\,(\sigma^k, l)^2 + \frac{1}{2}(\sigma^k, \sigma^k)\,(k, k) - 2\,(\sigma^k, l)\,(\sigma^k, k) \right\} \\ + \dfrac{(\sigma^m, l)^2}{(m, l)^2} \left\{ 2\,(\sigma^k, l^0)^2 + \frac{1}{2}(\sigma^k, \sigma^k)\,(k, k) + 2\,(\sigma^k, l^0)\,(\sigma^k, k) \right\} \\ + 2\,\dfrac{(\sigma^m, l^0)\,(\sigma^m, l)}{(m, l^0)\,(m, l)} \left\{ 2\,(\sigma^k, l^0)\,(\sigma^k, l) + \frac{1}{2}(\sigma^k, \sigma^k)\,(k, k) - (\sigma^k, m)\,(\sigma^k, k) \right\} \\ + 2\,\dfrac{(\sigma^m, l^0)}{(m, l^0)}\,(\sigma^m, \sigma^k)\,(\sigma^k, 2\,l - k) \\ \qquad\qquad - 2\,\dfrac{(\sigma^m, l)}{(m, l)}\,(\sigma^m, \sigma^k)\,(\sigma^k, 2\,l^0 + k) + 2\,(\sigma^k, \sigma^m)^2 \\ + \dfrac{1}{2\,(m, l^0)\,(m, l)} \left\{ (\sigma^k, \sigma^k)\,(m, k)^2 - 2\,(\sigma^k, m)\,(\sigma^k, k)\,(m, k) + (\sigma^k, m)^2\,(k, k) \right\}. \end{cases}
$$

Dabei wurde l statt $\bar{l}$ geschrieben. Folgende Beziehungen wurden gebraucht

$$(4,7)\quad \begin{cases} l^0 + k = l + m\,, \\ R(l+m) = 2(l, m), \\ R(l^0 - m) = -2(l^0, m), \\ (l,\ m) = (l^0, k) + \tfrac{1}{2}(k, k), \\ (l^0, m) = (l, k) - \tfrac{1}{2}(k, k). \end{cases}$$

Auch hier und im folgenden § 5 ist unter l immer $\bar{l}$ verstanden. Die rechte Seite von (4,6) ist selbstverständlich lorentzinvariant, da nur skalare Produkte von Weltvektoren auftreten. Sie muß auch eichinvariant sein, d. h. gegenüber jeder der beiden Substitutionen

$$\sigma^{k0} = \sigma^k + \text{const} \cdot k$$

und

$$\sigma^{m0} = \sigma^m + \text{const} \cdot m$$

invariant sein. Diese letzte Invarianzeigenschaft bietet eine Kontrolle zur Vermeidung von Rechenfehlern in der Ausrechnung von (4,6).

§ 5. **Die Klein-Nishina-Formel für bewegte Elektronen**

Führen wir im Anschluß an Pauli die Abkürzungen

$$D_k = 1 - \frac{1}{c}\left(\frac{\vec{k}}{k_0}, \vec{v}\right)$$

und entsprechend D_m ein, so erhält man die K.-N.-F. für ein mit der Geschwindigkeit $\vec{v}$ bewegtes Elektron, wenn über die Polarisationsrichtungen σ^k gemittelt und über σ^m summiert wird, am einfachsten auf folgende Weise:

Wir eichen σ^k und σ^m so, daß $(\sigma^k, l^0) = (\sigma^m, l^0) = 0$ ist (Transversalidität im Ruhsystem des Elektrons). Ferner ist $(k, k) = (\sigma^k, k) = 0$; $(\sigma^k, \sigma^k) = 1$. Dann verschwinden die ersten fünf Glieder von (4,5). Das letzte Glied wird wegen der ersten und letzten Beziehung (4,7) und weil

$$(l^0, m) = -l_0^0\, m_0\, D_m$$

ist:

$$\frac{\nu_m D_m}{\nu_k D_k} + \frac{\nu_k D_k}{\nu_m D_m} - 2\,.$$

Das zweitletzte Glied wird im gestrichenen Lorentzsystem, wo das Elektron ruht

$$2\frac{(\vec{k'}, \vec{m'})^2}{k_0'^2 m_0'^2} = 2\cos^2 \vartheta'.$$

Dieses läßt sich aber leicht transformieren, da

$$\frac{(\vec{k'}, \vec{m'})}{k_0' m_0'} = \frac{(k, m)(l^0, l^0)}{(k, l^0)(m, l^0)} - 1 = \frac{C^2}{l_0^0}\left(\frac{1}{m_0 D_m} - \frac{1}{k_0 D_k}\right) - 1.$$

Die K.-N.-F. lautet daher

$$(5,1)\left\{\begin{aligned} J^m d\omega_m &= d\omega_m \cdot S^k \cdot n^0 \cdot \frac{e^4}{2m^2c^4}\left(\frac{mc^2}{E^0}\right)^2 \frac{1}{D_k} \cdot \left(\frac{\nu_m}{\nu_k}\right)^3 \\ &\quad \cdot \left\{\frac{\nu_m D_m}{\nu_k D_k} + \frac{\nu_k D_k}{\nu_m D_m} - 2\sin^2\vartheta'\right\} \end{aligned}\right.$$

mit

$$\sin^2\vartheta' = \frac{(mc^2)^2}{\hbar E^0}\left(\frac{1}{\nu_m D_m} - \frac{1}{\nu_k D_k}\right)\left[2 - \frac{(mc^2)^2}{h E^0}\left(\frac{1}{\nu_m D_m} - \frac{1}{\nu_k D_k}\right)\right].$$

E^0 ist die Energie des Elektrons vor dem Streuprozeß. Diese Formel stimmt mit der mittels einer Lorentztransformation erhaltenen Formel (22) und (19) von Pauli[1]) überein. Beim Vergleich ist darauf zu achten, daß die Formel (22) von Pauli wegen seiner Formel (7) mit D_k multipliziert werden muß, um mit unserem Ausdruck verglichen zu werden.

§ 6. Die Ableitung der Bremsformel aus der verallgemeinerten Klein-Nishina-Formel

Die Formel für die Bremsstrahlung im elektrostatischen Feld wurde von Bethe und Heitler und von Sauter[2]) in zweiter Näherung abgeleitet. Durch eine qualitative Überlegung erhielt v. Weizsäcker[3]) eine Näherungsformel, welche für große Anfangsenergie des Elektrons und große Energie der emittierten Lichtquanten im Vergleich zu μc^2 mit der Formel von Bethe und Heitler übereinstimmt. Seine Überlegung kann in unserem Formalismus folgendermaßen dargestellt werden:

Im Raum–Zeit-System, in welchem das Feld statisch erscheint (d. h. wo z. B. der Kern ruht) gelte die Fourierzerlegung des Feldes (1,8) in ruhende Partialwellen. Dieses System werde *Kernsystem* genannt. Das Elektron bewegt sich im Kernsystem mit der Geschwindigkeit $\vec{v}$. Die Spaltung eines Weltvektors nach diesem System werde durch ungestrichene Komponenten gekennzeichnet. k in (1,8) hat also im Kernsystem nur von Null verschiedene räumliche Komponenten.

Im Raum–Zeit-System, wo das Elektron anfangs ruht (kurz mit *Elektronsystem* und durch gestrichene Komponenten von Weltvektoren bezeichnet), bewegen sich dann die Partialwellen mit einer Geschwindigkeit $\vec{v}' = -\vec{v}$. Ist die Anfangsenergie $\gg \mu c^2$, so ist $|\vec{v}'|$ *beinahe* gleich c. Eine Umeichung des Polarisationsvektors σ^k, welcher gemäß § 2 im Kernsystem nur eine von Null verschiedene Zeitkomponente hat, in einen Weltvektor σ^{k0}, der im Elektronsystem rein räumlich erscheint, ergibt für die wichtigsten Partialwellen mit kleinem k *beinahe* transversale Wellen.

v. Weizsäcker wendet einfach auf diese Quasilichtwellen im Elektronsystem die K.-N.-F. für ruhende Elektronen [(5,1) mit $D_m = D_k = 1$] an.

Die Beziehung zwischen Amplitude und Energiestromdichte (3,9) wird nämlich auch *beinahe* die für Lichtwellen (3,10), da $(k, \sigma^k) \cong 0$ und $|\vec{k}|^2 - k_0^2 \cong 0$ wird. Auf der rechten Seite von (3,10) tritt allerdings noch $(\sigma^{k0'}, \sigma^{k0'})^{-1}$ auf. Da aber $\overline{\gamma_4 \Omega^+ \Omega}$ in σ^k bilinear ist, so bedeutet das Auftreten dieses Faktors einfach, daß in $\overline{\gamma_4 \Omega^+ \Omega}$ σ^k als Einheitsvektor zu wählen ist.

Die Bremsstrahlung im Elektronsystem erscheint dann als die inkohärente Überlagerung der Streustrahlung der einzelnen primären Partialwellen k.

Daß hier nur die inkohärente Überlagerung berücksichtigt zu werden braucht, rührt davon her, daß wir über das im Vergleich zum Kernfelde große Wellenpaket mitteln. Bei dieser Mittelung heben sich die kohärenten Effekte durch Interferenz weg. (Der Beweis findet sich in § 7.)

In diesem Paragraphen soll gezeigt werden, wie die genaue Bremsformel dieser Näherung aus der verallgemeinerten K.-N.-F. (3,8) auf gleiche Weise abgeleitet werden kann, wie die für hohe Geschwindigkeiten gültige Näherung aus der gewöhnlichen K.-N.-F. (3,5) durch v. Weizsäcker erhalten wurde.

Dazu bilden wir zuerst die Partialwellen im Elektronsystem. Wird in (1,8) die Summation im Kernsystem über die rein räumlichen k durch eine Integration ersetzt, so ist:

$$V^L = (2\pi)^{-3} \int M_k G(\sigma^k, \gamma) e^{i(k,x)} d\vec{k}^3 . \tag{6,1}$$

Hier ist außer $d\vec{k}^3$ alles invariant, da das Grundgebiet des Kernsystems G herausfällt und M_k nur von (k, k) abhängt.

$d\vec{k}^3$ kann aber unter Verwendung der δ-Funktion invariant geschrieben werden:

$$d\vec{k}^3 = \delta(k_4)\,d k^4 .$$

$d k^4$ ist invariant. Die Lorentztransformation für die vierte Komponente heißt

$$k_4 = \left(1 - \frac{|\vec{v}|^2}{c^2}\right)^{-1/2}\left(-\frac{i}{c}(\vec{v'},\vec{k'}) + k'_4\right).$$

Daraus folgt

$$d\vec{k}^3 = \left(1 - \frac{|\vec{v}|^2}{c^2}\right)^{1/2} d\vec{k'}^3 \tag{6,2}$$

und, wenn anstatt der Integration über $d\vec{k'}^3$, im Periodizitätsbereich G' eine Summation über k' ausgeführt wird,

$$V^L = \sum_{\vec{k'}} M'_k(\sigma^k, \gamma)\, e^{i(k,x)} .$$

Wegen (1,9) und (6,2) ist

$$M'_k = \frac{4\pi Z e^2 g\left(\sqrt{(k,k)}\right)}{G'(k,k)\hbar c}\left(1 - \frac{|\vec{v'}|^2}{c^2}\right)^{1/2}. \tag{6,3}$$

Setzt man diese Amplitude in die verallgemeinerte K.-N.-F. im Elektronsystem ein und summiert inkohärent (bzw. integriert) über alle Partialwellen, so erhält man für die in der langen Zeit t' in den Raumwinkel $d\omega_m$ gestreute Gesamtzahl von Lichtquanten (aller Frequenzen):

$$\left\{\begin{aligned}\sum_{\vec{k'}} \frac{J^{m'} d\omega'_m t'}{\hbar \nu'_m} &= t'\left(1 - \frac{|\vec{v}|^2}{c^2}\right)^{1/2} \cdot m_0'^2\, d\omega'_m \cdot \left(\frac{c\, n^0}{G'\, l_0^{0'}}\right) \cdot \frac{Z^2 e^6}{4\pi^2 \hbar^3 c^3} \\ &\cdot \int \left(1 - \frac{|\vec{v}|^2}{c^2}\right)^{1/2} d\vec{k'}^3 \frac{g^2}{(k,k)^2}\left\{\frac{1}{|(l,m)|}\frac{l_0^{0'}}{l'_0}\overline{\gamma_4\,\Omega^+\,\Omega}\right\}.\end{aligned}\right. \tag{6,4}$$

Wenn (6,3) im Quadrat eingesetzt wird, und wenn die Summation über $\vec{k'}$ durch die Integration ersetzt wird, so bleibt in (6,4) ein G' im Nenner stehen. Dieses hat den Charakter eines Normierungsfaktors, der ein Volumen im Elektronensystem darstellt. Im § 7 wird gezeigt werden, daß dieses Volumen mit dem Volumen $L'_1 L'_2 L'_3$ des Wellenpaketes identisch ist. Als Zeit t' wählen wir die Zeit, welche der Kern braucht, um mit der Geschwindigkeit $\vec{v'}$ durch das Wellenpaket des ruhenden Elektrons hindurchzustreichen. Hat das Wellenpaket im Vergleich zum Kernfeld große Linearausdehnungen, so erfolgt die Emission der Quanten gleichmäßig und nur während dieser Zeit. Der Ausdruck (6,4) ist also eine Invariante.

Im Kernsystem ist t die Zeit, während welcher das mit $\vec{v} = -\vec{v}'$ bewegte und daher verkürzt erscheinende Wellenpaket das Kernfeld umgibt. Diese Zeit ist in gleicher Weise verkürzt wie die Linearausdehnung des Wellenpaketes in der Bewegungsrichtung. Es ist also

$$t = t' \left(1 - \frac{|\vec{v}|^2}{c^2}\right)^{1/2}. \tag{6,5}$$

Ferner gilt (6,2) und

$$m_0^2 \, d\,\omega_m = m_0'^2 \, d\,\omega_m'. \tag{6,6}$$

Die Größe n^0/G' ist, wenn man G' mit dem Volumen des Wellenpaketes identifiziert, die Elektronendichte im Elektronsystem. $c n^0/G'$ ist also die reelle vierte Komponente $d_0^{0\prime}$ des Vierervektors der Teilchendichte d^0, welcher definitionsgemäß dem Vektor der Vierergeschwindigkeit und damit dem Vierervektor l^0 parallel ist. Somit ist auch

$$\frac{|\vec{d^0}|}{|\vec{l^0}|} = \frac{d_0^{0\prime}}{\bar{l}_0^{0\prime}} = \frac{c\, n^0}{G' \, \bar{l}_0^{0\prime}}. \tag{6,7}$$

Die letzte Klammer in (6,4) ist wegen (4,6) invariant.

Werden die gestrichenen Größen in (6,4) vermittels (6,2), (6,5), (6,6) und (6,7) durch ungestrichene Größen ausgedrückt und werden in der letzten Klammer, wegen der Invarianz von (4,6), die Striche weggelassen, so stellt das durch t dividierte (6,4) die Anzahl Lichtquanten dar, welche ein Elektronenstrom der Stromdichte $\vec{d^0}$ in der Zeiteinheit als Bremsstrahlung in den Raumwinkel $d\,\omega_m$ emittiert. Dividieren wir noch $|\vec{d^0}|$, so erhalten wir den Wirkungsquerschnitt.

Um die Bethe-Heitlersche Form zu erhalten, ersetzen wir die Integration über $d\vec{k}^3$ durch die über $d\vec{l}^3$ bei festem l^0. Man findet für die Funktionaldeterminante $\left(\text{aus } \vec{l} = \vec{l^0} + \vec{k} - \vec{m}\right.$ und der Nebenbedingung $\left. m_0 = \bar{l}_0^0 - \bar{l}^0\right)$:

$$\frac{d\vec{k}^3}{d\vec{l}^3} = \frac{(\bar{l}, m)}{\bar{l}_0 \, m_0}.$$

Aus $R(\bar{l}) = 0$ und der Nebenbedingung folgt, wenn $d\,\omega_l$ den Raumwinkel bedeutet, in welchen das Elektron gestreut wird:

$$d\vec{l}^3 = d\omega_l \, |\vec{l}|^2 d|\vec{l}| = |\vec{l}| \, \bar{l}_0 \, d\omega_l \, dm_0.$$

Werden diese Größen in das transformierte (6,4) eingesetzt, so erhalten wir den differentiellen Wirkungsquerschnitt dQ eines kugelsymmetrischen Kraftfeldes gegenüber einem Elektron der Energie $\bar{l}_0^0$, welcher die Emission eines Quants bestimmter Polarisation im Frequenzbereich zwischen m_0 und $m_0 + d\,m_0$ in den Raumwinkel $d\,\omega_m$ und die gleichzeitige Ablenkung des Elektrons in den Raumwinkel $d\,\omega_l$ bewirkt:

$$(6,8)\quad \begin{cases} d\,Q = d\,\omega_m\, d\,\omega_l\, \dfrac{d\,m_0}{m_0}\, \dfrac{|\vec{l}\,|}{|\vec{l^0}|}\, \dfrac{Z^2 e^4}{(2\pi^2)\cdot 137\cdot \hbar^2 c^2}\, \dfrac{g^2(|\vec{k}|)}{|\vec{k}|^4} \\ \qquad \cdot \left\{ m_0^2\, \dfrac{l_4^0}{l_4}\, \overline{\gamma_4\, \Omega^+\, \Omega} \right\}. \end{cases}$$

Hier und in den folgenden Formeln ist $\bar{l}$ wieder durch l ersetzt. Den Klammerausdruck entnimmt man (4,6). Wir wählen die Eichung so, daß im Kernsystem σ^m rein räumlich und σ^k rein zeitlich erscheint. Dann verschwindet das 4., 5. und 6. Glied von (4,6). Summiert man noch über die beiden Polarisationsrichtungen $\vec{\sigma^m}\,(\perp \vec{m})$ und führt den räumlichen Einheitsvektor $\vec{\mu}$ in Richtung von $\vec{m}$ ein, so wird der letzte Faktor von (6,3)

$$(6,9)\quad \begin{cases} m_0^2\, \dfrac{l_4^0}{l_4}\, \overline{\gamma_4\, \Omega^+\, \Omega} = \dfrac{\left|[\vec{\mu}, \vec{l^0}]\right|^2 (4\, l_0^2 - |\vec{k}|^2)}{\left(l_0^0 - (\vec{\mu}, \vec{l^0})\right)^2} \\ \qquad + \dfrac{\left|[\vec{\mu}, \vec{l}\,]\right|^2 (4\, l_0^{0\,2} - |\vec{k}|^2)}{\left(l_0 - (\vec{\mu}, \vec{l}\,)\right)^2} \\ \qquad - \dfrac{2\left([\vec{\mu}, \vec{l^0}], [\vec{\mu}, \vec{l}\,]\right)(4\, l^0 l - |\vec{k}|^2)}{\left(l_0^0 - (\vec{\mu}, \vec{l^0})\right)\left(l_0 - (\vec{\mu}, \vec{l}\,)\right)} \\ \qquad + \dfrac{2\left|[\vec{m}, \vec{l^0} - \vec{l}\,]\right|^2}{\left(l_0^0 - (\vec{\mu}, \vec{l^0})\right)\left(l_0 - (\vec{\mu}, \vec{l}\,)\right)}. \end{cases}$$

(6,8) und (6,9) ist genau die bereits von Bethe und Heitler und von Sauter[2]) gefundene Formel. Die quantitative Verschärfung der Überlegungen von v. Weizsäcker[3]) führt also auf die richtige zweite Näherung des Störungsverfahrens.

Die Integration der Bremsformel sowie ihre Diskussion steht bei Bethe und Heitler[2]).

Die v. Weizsäckersche Näherung erhalten wir, wenn $\sigma^{k\,0}$ so geeicht wird, daß es im Elektronensystem ($l^0 =$ rein zeitlich) rein räumlich erscheint. Ist σ^k der im Kernsystem rein zeitliche Vektor vom Betrage $(\sigma^k, \sigma^k) = -1$, so ist

$$\sigma^{k0} = \sigma^k - \frac{(\sigma^k, l^0)}{(k, l^0)} k. \tag{6,10}$$

Rechnet man im Elektronensystem, wie v. Weizsäcker es tut, so ist für hohe Geschwindigkeiten der Vektor $\vec{\sigma}^{k0\prime}$ beinahe senkrecht zu $\vec{k}'$. Die Bedeutung der Quasitransversalität im Elektronsystem bedeutet im Kernsystem folgende Näherung:

Wird das durch (6,10) definierte σ^{k0} in den eich- und lorentzinvarianten Ausdruck (4,6) unter Vernachlässigung von (k, k) und (σ^{k0}, k) eingesetzt und im Kernsystem ausgewertet, so stimmt es mit dem Ausdruck (6,9) überein, wenn

1. (k, k) gegen $l_0^{0\,2}$ vernachlässigt werden kann, und

2. $$\frac{(\sigma^k, m)^2}{(k, m)^2} = \frac{(\sigma^k, l^0)^2}{(k, l^0)^2}$$

wird. Die erste Bedingung ist für die, wegen $(k, k)^{-2}$ in (6,8) hauptsächlich wirksamen, großen Wellenlängen der Fourierzerlegung des Coulombfeldes für große Energien l_0^0 und m_0 erfüllt. Die zweite Bedingung ist ebenfalls erfüllt. Sie verlangt nämlich, daß im Kernsystem $\vec{m}$ und $\vec{l}^0$ nahezu parallel sind, und daß l^0 beinahe ein Lichtvektor ist. Dies ist tatsächlich bei hohen Energien l_0^0 für die kleinen (k, k) Werte mit großer Annäherung der Fall.

§ 7. Die strenge Ableitung der Bremsformel

In der im vorhergehenden Paragraphen gegebenen Ableitung der Bremsformel aus der verallgemeinerten K.-N.-F. bedürfen zwei Punkte einer Rechtfertigung:

Erstens die Identifizierung des Normierungsfaktors G' mit dem Volumen des Wellenpaketes $L_1' L_2' L_3'$ in den Amplituden der einzelnen Partialwellen

und zweitens die inkohärente Überlagerung ihrer Wirkungen.

Die Richtigkeit der Überlegungen ergibt sich aus folgendem:

In (2,15) setzen wir $V_k = M_k$, $P_k = 1$ und $V_{-p} = T_m$. Als Verteilung N^0 wählen wir $(0, 0, 0 \ldots)$ (keine Lichtquanten vorhanden). Die Endverteilung und die Verteilung in den Zwischenzuständen ist $(0, 0, 0 \ldots 1_m \ldots 0)$. Dadurch wird $P_{-m\,i\,0} = 1$.

Die Summation über $\vec{k}$ ersetzen wir durch eine Integration und setzen ferner wegen $s_4 = \bar{l}_4 - \bar{l}_4^0 + m_4$ und $l_3^0 = \bar{l}_3 - k_3 + m_3$

$$d k_3 = \left(\frac{-\bar{l}_4^0}{\bar{l}_3^0}\right) d s_4 .$$

Wird jetzt die Integration über ds_4 in (2,15) für Zeiten, welche groß im Vergleich zu $t = \frac{L_3}{v_3}$ sind [L_3 ist die Ausdehnung des Wellenpaketes in der Bewegungsrichtung (x_3-Achse)], ausgeführt, so wird

$$(7,1) \quad \frac{B^{jm}(\bar{l}\,\infty)}{\bar{l}_0} = - \frac{T_m \pi}{(2\pi)^3} \iint d k_1 \, d k_2 \frac{\bar{l}_4^0}{l_3^0} \frac{M_k \, G \, \Omega(\bar{l}^0) \, B^0(\bar{l}^0)}{\bar{l}_4 \, \bar{l}_0^0}$$

eine Konstante.

Gemäß (2,7) entspricht dieser Konstante eine zeitunabhängige Teilchenzahl im Endzustande: Das Wellenpaket hat das Kernfeld vollständig passiert.

Im Bereich $d\omega_m$ ist dann diese Teilchenzahl

$$d\omega_m \int n^{jm}(\infty) \frac{G}{(2\pi)^3} m_0^2 \, d m_0 = \sum_m \frac{J^m \, d\omega_m \, t}{h \nu_m} .$$

Da die B^{jm} selbst Integrale über $d k_1 \, d k_2$ sind, so addieren sich die Wirkungen der einzelnen Partialwellen vorerst *kohärent*. Um zur Ausgangsformel (6,4) des vorhergehenden Paragraphen zu gelangen, ersetzen wir $d m_0$ durch $d k_3$ bei festen $\bar{l}$:

$$d m_0 = \frac{m_0 \, l_3^0}{(m, l^0)} d k_3 .$$

Die Integration über $d\vec{l}^3$ in (2,17) können wir wieder, wie früher, bei festgehaltenem k vermittelst (3,2) durch $d\vec{l^0}^3$ ersetzen. Dann wird die Teilchenzahl:

$$(7,2) \quad \left\{ \begin{aligned} \sum \frac{J^m \, d\omega_m \, t}{h \nu_m} &= m_0^2 \, d\omega_m \frac{Z^2 e^6}{4\pi^2 \hbar^8 c^3} \int d\vec{k}^3 \frac{g(|\vec{k}|)}{(k, k)} \\ &\cdot (2\pi)^3 \frac{1}{4\pi^2} \int d q_1 \, d q_2 \int d\vec{l^0}^3 \frac{B^0(\bar{l}^q)^+}{\bar{l}_0^q} \\ &\cdot \frac{g(|\vec{q}|)}{l_3^q \, (q, q)} \frac{\bar{l}_4^q}{|(\bar{l}, m)| \, \bar{l}_4} \gamma_4 \, \Omega^+(\bar{l}^q) \, \Omega(\bar{l}^0) \frac{B^0(l^0)}{\bar{l}_0^0} . \end{aligned} \right.$$

q hat dabei dieselbe Bedeutung wie k in der zu (7,1) adjungierten Gleichung, d. h. $q = (q_1, q_2, k_3, 0)$, wenn $k = (k_1, k_2, k_3, 0)$ ist. Ferner ist $l^q = l - q + m$ analog zu $l^0 = l - k + m$. (7,2) unterscheidet sich von (6,4) durch die *kohärente* Superposition der Wirkung der Partialwellen. Geben wir dem auf n^0 normierten Wellenpaket einen rechteckigen Querschnitt $L_1 L_2$ und legen wir die Längsachse des Paketes durch den Kern, so wird der dreidimensionale Fourierkoeffizient $B^0(\bar{l}^0)/l_0^0$ in seiner Abhängigkeit von l_1^0:

$$(7,3) \qquad \frac{B^0(\bar{l}^0)}{l_0^0} = \frac{\sin\frac{l_1^0 L_1}{2}}{\sqrt{L_1}\cdot l_1^0} F,$$

wo der Spinor F nur von l_2 und l_3 abhängt, und $H(\bar{l})F = 0$ genügt. Das Integral auf der rechten Seite von (3,5) welches in (6,4) auftritt, nennen wir den inkohärenten Erwartungswert. An Stelle des nur von $\bar{l}^0$ abhängigen Operators $\gamma_4\, \Omega(\bar{l}^0)^+\, \Omega(\bar{l}^0)$ in (3,5) wählen wir entsprechend (7,2) einen von zwei Vektoren l^q und l^0 abhängigen Operator $\Xi(l_1^q, l_1^0)$. Die unter Verwendung von (7,3) angeführte Integration über l_1^0 ergibt dann als *inkohärenten Erwartungswert:*

$$(7,4) \qquad \int dl_1^0 \frac{B^0(\bar{l}^0)^+}{l_0^0} \Xi(l_1^0, l_1^0) \frac{B^0(\bar{l}^0)}{l_0^0} = \frac{1}{2}\int dx \frac{\sin^2 x}{x^2} F^+ \,\Xi\left(\frac{2x}{L_1}, \frac{2x}{L_1}\right) F.$$

Der in (7,2) vorkommende kohärente Erwartungswert wird, wenn an Stelle von q_1 als Integrationsvariable $l_1^q = k_1 - q_1$ bei festen k gewählt sind:

$$(7,5) \qquad \begin{cases} \dfrac{1}{2\pi}\displaystyle\iint dl_1^q\, dl_1^0 \frac{B^0(\bar{l}^q)^+}{l_0^q} \Xi(l_1^q, l_1^0) \frac{B^0(\bar{l}^0)}{l_0^0} \\ \quad = \dfrac{1}{2\pi L_1}\displaystyle\iint dx\, dy \frac{\sin x \sin y}{xy} F^+ \,\Xi\left(\frac{2x}{L_1}, \frac{2y}{L_1}\right) F. \end{cases}$$

Da $\int \frac{dx \sin x}{x} = \int \frac{dx \sin^2 x}{x^2} = \pi$ ist, sind im Limes $L_1 = \infty$ *), wo das Wellenpaket unendlichen Querschnitt und einen scharf definierten Impuls ($l_1^0 = 0$) besitzt, der kohärente Erwartungswert (7,5) gleich dem inkohärenten (7,4) dividiert durch L_1. Die gleiche Überlegung wird für die x_2-Richtung durchgeführt. Hat das Wellenpaket in der Bewegungs-(x_3)-Richtung die im Vergleich zur Reichweite des Kernfeldes große Ausdehnung L_3, so werden die Lichtquanten während der langen Zeit $t = \frac{L_3}{v_3} = \frac{L_3 l_0^0}{c\, l_3^0}$ gleichmäßig ausgestrahlt. Die inkohärente Überlagerung kommt also dadurch zustande, daß sowohl in

---

*) Die Anwendung des Dirichletschen Integrals im Limes $L_1 = \infty$ bedeutet: „L_1 groß gegen Reichweite des Kernfeldes".

der Längsrichtung (7,1) wie in den beiden Querrichtungen (7,5) über einen Raum (Wellenpaket) gemittelt wird, der groß im Vergleich zum Wirkungsbereich des Kernfeldes ist.

Eliminiert man das l_3^0 im Nenner von (7,2) durch t und setzt den Erwartungswert der inkohärenten Überlagerung aus (3,5) ein, so erhält man ($l^q = l^0$ und $q = k$ gesetzt) tatsächlich die Form (6,2) mit ungestrichenen (Kernsystem!) Größen, da $\left(\frac{c\,n^0}{L_1 L_2 L_3}\right) = d_0^0$ ist*). $L_1 L_2 L_3$ tritt also im Nenner der strengen Ableitung in diesem Paragraphen an der Stelle auf, wo im vorhergehenden Paragraphen der Normierungsfaktor G (bzw. im Elektronsystem G') auftrat. Hiermit ist die Ableitung des vorhergehenden Paragraphen, insbesondere die Berechtigung der inkohärenten Superposition und der Identifizierung von G' mit $L_1' L_2' L_3'$ gezeigt.

Da in den Rechnungen, insbesondere in Formel (4,6) eine Spezialisierung auf positive Energien nirgends erfolgt ist, lassen sich noch folgende drei Probleme auf Grund der Diracschen Deutung der Zustände negativer Energie berechnen:

1. Die *Erzeugung eines Elektronenpaares aus zwei Lichtquanten.* Dieser Fall ist ganz analog der Ableitung der K.-N.-F. in § 3 zu behandeln. $\bar{l}^0$ stellt dann einen Zustand *negativer Energie* dar und anstatt m ist $p = -m$ ein Lichtvektor.

2. Die *Erzeugung eines Elektronenpaares durch das Zusammenwirken eines Lichtquantes und des Kernfeldes.* Der hierzu inverse Vorgang ist, wie Bethe und Heitler[2]) gezeigt haben, eine „Bremsstrahlung", bei der der Endzustand $\bar{l}$ ein Zustand negativer Energie ist.

3. Die *Erzeugung eines Elektronenpaares durch den Stoß eines schnellen Teilchens am Kernfelde.* Das Feld des stoßenden Teilchens wird nach Fourier zerlegt, und dann die Paarerzeugung durch jede einzelne Partialwelle nach 2. berechnet. Die inkohärente Superposition der Beiträge der einzelnen Wellen gibt das gewünschte Resultat, solange man nur Stöße zuläßt, bei welchen sich der Impuls des stoßenden Teilchens nur um einen im Vergleich zu seinem anfänglichen Impuls kleinen Betrag ändert.

Diese angedeuteten Rechnungen sollen in einem *zweiten Teile* ausgeführt werden.

---

*) Die Rechnungen dieses Paragraphen können natürlich ebensogut im (gestrichenen) Elektronsystem durchgeführt werden. Die Parallele zum vorhergehenden Paragraphen kommt dann besser zum Ausdruck, jedoch wird die Schreibweise etwas komplizierter.

Die Anregung zur Ausarbeitung der vorliegenden invarianten Störungstheorie verdanke ich Herrn Prof. G. Wentzel, dem ich an dieser Stelle meinen Dank aussprechen möchte.

Literatur

1) W. Pauli, Helv. Phys. Acta **6**. S. 279. 1933.
2) H. Bethe u. W. Heitler, Proc. Roy. Soc. [A] **146**. S. 83. 1934; F. Sauter, Ann. d. Phys. [5] **20**. S. 404. 1934.
3) C. F. v. Weizsäcker, Ztschr. f. Phys. **88**. S. 612. 1934.
4) Vgl. z. B. P. A. M. Dirac, V. A. Fock u. B. Podolsky, Sow. Phys. **2**. S. 468. 1932.
5) I. Waller, Ztschr. f. Phys. **61**. S. 837. 1930.

Zürich, Physik. Institut der Universität, 31. August 1934.

(Eingegangen 10. September 1934)

Austauschkräfte zwischen Elementarteilchen und Fermische Theorie des β-Zerfalls als Konsquenzen einer möglichen Feldtheorie der Materie [26]

Helvetica Physica Acta, vol. 9 (1936), pp. 389–404

Austauschkräfte zwischen Elementarteilchen und Fermi'sche Theorie des β-Zerfalls als Konsequenzen einer möglichen Feldtheorie der Materie

von **E. C. G. Stueckelberg.**

(11. V. 36.)

Inhalt: Elektron, Neutrino, Proton und Neutron werden als vier verschiedene Quantenzustände einer einzigen Elementpartikel angesehen. Quantensprünge zwischen diesen Zuständen erklären den β-Zerfall (gemäss der Theorie von Fermi) und geben zur Heisenberg-Majorana'schen Neutron-Proton-Austauschkraft Anlass. Die Festsetzung, dass negatives Elektron und positives Proton „Partikel"-Zustände (im Gegensatz zu „Antipartikel") sind, verbietet Zerstrahlungsprozesse der schweren Teilchen. Die umgekehrte Festsetzung (positives Elektron und positives Proton sind Partikel) führt zu Zerstrahlungsprozessen (siehe Zusammenfassung).

1. Vorbemerkung. Der Wunsch nach einer einheitlichen Feldtheorie wurde von verschiedenen Autoren[1]) ausgesprochen. Ihr Ziel war, das Gravitationsfeld und das elektromagnetische Feld als verschiedene Äusserungen eines einzigen Feldes zu erklären.

Die Quantentheorie Schrödinger's führt eine neue Feldgrösse ein, die ψ-Funktion. Diese erste Fassung genügte aber nicht der Forderung nach relativistischer Covarianz. Auf zwei Arten konnte Covarianz erreicht werden:

1. Durch die Schrödinger-Gordon'sche Wellengleichung. Sie hat, wie die Schrödinger'sche Theorie, eine skalare Feldstärke ψ als abhängige Variable. Wie Pauli und Weisskopf[2]) gezeigt haben, erklärt sie die Entstehung von Partikelpaaren (entgegengesetzter elektrischer Ladung, als „Partikel" und „Antipartikel" zu bezeichnen) durch Einwirkung des elektromagnetischen Feldes auf das Vakuum. Hingegen gibt sie keine Erklärung des Spins und des magnetischen Momentes der Partikel und erlaubt nur die Einstein-Bose-Statistik für die Teilchen.

2. Durch die Dirac'sche Wellengleichung. Dieselbe führt eine vierkomponentige Feldstärke ψ ein. Sie erklärt Spin und magnetisches Moment und erlaubt nur die Fermi-Dirac'sche Statistik für die Teilchen. Andererseits gibt sie zu den bekannten Niveaus negativer Energie Anlass, welche wir im Sinne von Dirac als „aufgefüllt" ansehen. Dirac und Heisenberg[3]) haben

*

gezeigt, dass man den Begriff „alle Niveaus negativer Energie sind besetzt bis auf eine endliche Zahl" mathematisch sinnvoll definieren kann. Die endliche Anzahl unbesetzter Niveaus verhalten sich dann wie Partikel entgegengesetzter Ladung, deren Energie und Impuls die Energieimpulsgrössen der betreffenden Niveaus sind mit umgekehrten Vorzeichen. Diese Löcher wollen wir als „Antipartikel" bezeichnen im Gegensatz zur „Partikel". Die Theorie ist symmetrisch in bezug auf Vertauschung des Begriffes Antipartikel und Partikel. Wir wollen daher im folgenden willkürlich das *positive Elektron als Partikel und das negative als Antipartikel* definieren. Das elektromagnetische Feld kann dann auch aus dem Vakuum (= alle negativen Niveaus aufgefüllt) ein Paar (Partikel und Antipartikel) erzeugen, indem eine Partikel aus einem ausgefüllten Niveau negativer Energie in ein solches positiver Energie springt. Die Analogie in bezug auf Partikelpaare zwischen der SCHRÖDINGER-GORDON'schen und der DIRAC-HEISENBERG'schen Theorie gilt auch in quantitativer Hinsicht, wie PAULI und WEISSKOPF gezeigt haben[2]).

Trotz der Unschönheit der Löchertheorie ist sie, wegen Spin und Statistik, der skalaren Theorie vorzuziehen.

Das Programm der einheitlichen Feldtheorie ist also dahin zu erweitern, dass Gravitationsfeld (g_{ik}-Feld), elektromagnetisches Feld ($\vec{E}\,\vec{B}$ resp. F_{ik}-Feld) und das materielle Feld (ψ_μ-Feld) als Äusserungen ein und derselben mehrkomponentigen Feldstärke zu erklären sind.

Nun transformieren sich (bei Vernachlässigung der Gravitationseffekte) die vier Komponenten des materiellen Feldes einer Partikelsorte ψ_μ bei Lorentztransformationen anders (unäquivalent) als die Komponenten c_i eines Weltvektors[4]). Trotzdem ist es möglich, aus den Spinoren ψ Vektoren c_i mit Hilfe gewisser Konstanten $\alpha_{i\,\mu\,\nu}$ zu bilden:

$$c_i = \psi_\mu^+ \alpha_{i\,\mu\,\nu} \psi_\nu = \psi^+ \alpha_i \psi . \qquad (1)$$

Über doppelt vorkommende lateinische oder griechische Indizes (hier μ und ν) ist jeweils zu summieren. Transformiert sich dann ψ nach der Regel der Spinortransformation, ψ^+ nach derjenigen der adjungierten Spinoren, und bleiben die Zahlen $\alpha_{i\,\mu\,\nu}$ konstant, so transformiert sich c_i wie ein Vierervektor[4]).

Das Programm der einheitlichen Feldtheorie sei nunmehr reduziert auf den Versuch, materielles Feld ψ_μ und elektromagnetisches Feld (z. B. gegeben durch die Potentiale A_i ($A_1, A_2, A_3 =$ Vektorpotential, $A_4 = i\Phi =$ skalares Potential) als Äusserung desselben Feldes zu verstehen. Wir vernachlässigen also alle Gravitationseffekte.

In dieser Richtung liegen zwei Versuche vor: ein erster stammt von DE BROGLIE[5]), welcher die A_i durch das spinorielle ψ-Feld zu erklären sucht. Unter dem Namen „Neutrinotheorie des Lichtes" hat sich diese Theorie zwar sehr weit entwickelt[6]), ist aber wegen verschiedener Schwierigkeiten noch nicht als endgültig zu betrachten.

Einen zweiten Weg schlägt BORN[7]) ein. Er erklärt die Existenz materieller Partikel als Äusserungen des vektoriellen A_i-Feldes. Trotz der grossen Erfolge der Born'schen Theorie scheint mir der de Broglie'sche Weg richtiger; da Formeln von ähnlichen Bau wie (1) gestatten, die Vektoren A_i durch die Spinoren ψ darzustellen, nicht aber ψ durch A_i.

Die BORN'sche Vektor-Theorie gestattet, die Schwierigkeiten der unendlichen Selbstenergie geladener Partikel zu umgehen; aber sie kann aus dem angeführten Grunde den halbzahligen Spin der Partikel nicht erklären. Die Spinor-Theorie dagegen erklärt Spin und Statistik der Partikel und Lichtquanten. Sie gibt die Bewegungsgleichungen (*Dirac-Gleichungen* für ψ) und, so hofft man wenigstens, die MAXWELL-Gleichung für A_i. Sie enthält aber zurzeit die bekannten Selbstenergieschwierigkeiten sowie den Schönheitsfehler der Darstellung von Antipartikeln durch Löcher.

Die folgenden Ausführungen sollen zeigen, was aus einer solchen unitären Spinortheorie für Folgerungen zu ziehen sind, wenn man, entsprechend der Existenz von vier Elementarpartikeln (Elektron, Neutrino, Proton, Neutron), eine $4 \times 4 = 16$ komponentige Spinorfeldstärke ψ annimmt.

2. Bezeichnungen. Lateinische Indizes i ($= 1, 2, 3, 4$) bezeichnen Vektorkomponenten im MINKOWSKI'schen Raum. x bedeutet $x_1, x_2, x_3, x_4 = ict$. $\vec{x}$ bedeutet in einem bestimmten Raum-Zeit-System die Ortskoordinaten. $dx^4 = dx_1\, dx_2\, dx_3\, dx_4$; $d\vec{x}^3 = dx_1\, dx_2\, dx_3$. Griechische Indizes μ, ν ($= 1, 2 \ldots 16$) bedeuten Spinorkomponenten. An Stelle von $\psi^+_\mu\, \Gamma_{i\,\mu\,\nu}\, \psi_\nu$ ist jeweils $\psi^+ \Gamma_i \psi$ geschrieben. e ist die Elementarladung des positiven Elektrons m, μm, μ'm und μ''m sind die Massen von Elektron, Neutrino, Proton und Neutron. $2\pi h$ ist die PLANCK'sche Konstante und c die Lichtgeschwindigkeit. Für die ψ-Funktion wird oft von der gespaltenen Schreibweise

$$\psi = \begin{pmatrix} \varphi \\ \chi \\ u \\ v \end{pmatrix} \tag{2.1}$$

Gebrauch gemacht, wo φ, χ, u und v vierkomponentige Spinoren sind und den Zuständen φ = Elektron, χ = Neutrino, u = Proton und v = Neutron entsprechen.

γ_i sind die vierreihigen relativistischen Dirac-Matrizen. Wir führen die 16reihigen Matrizen ein

$$\left.\begin{array}{ll} \Gamma_i = \begin{pmatrix} \gamma_i & 0 & 0 & 0 \\ 0 & \gamma_i & 0 & 0 \\ 0 & 0 & \gamma_i & 0 \\ 0 & 0 & 0 & \gamma_i \end{pmatrix} & \Delta = \begin{pmatrix} 1 & 0 & 0 & 0 \\ 0 & \mu & 0 & 0 \\ 0 & 0 & \mu' & 0 \\ 0 & 0 & 0 & \mu'' \end{pmatrix} \\ \Lambda = \begin{pmatrix} 1 & 0 & 0 & 0 \\ 0 & 0 & 0 & 0 \\ 0 & 0 & \varepsilon & 0 \\ 0 & 0 & 0 & 0 \end{pmatrix} & \Lambda' = \begin{pmatrix} 0 & 0 & 0 & 0 \\ 0 & 1 & 0 & 0 \\ 0 & 0 & 0 & 0 \\ 0 & 0 & 0 & \varepsilon' \end{pmatrix} \\ \multicolumn{2}{c}{\Lambda_i = \Lambda\Gamma_i = \Gamma_i\Lambda; \quad \Lambda_i' = \Lambda'\Gamma_i = \Gamma_i\Lambda'} \end{array}\right\} \qquad (2.2)$$

mit den Vertauschungsrelationen:

$$\left.\begin{array}{r} \Gamma_i\Gamma_k + \Gamma_k\Gamma_i = \delta_{ik}; \\ \Lambda,\ \Lambda',\ \text{und}\ \Delta\ \text{untereinander und mit}\ \Gamma_i\ \text{vertauschbar.} \end{array}\right\} \qquad (2.3)$$

Da εe die Protonladung bedeutet, so ist, je nachdem ob das positive Proton als Partikel oder als Antipartikel (= „Loch" in den negativen Energiezuständen des negativen Protons), $\varepsilon = +1$ oder $\varepsilon = -1$ zu wählen. Dem Neutrino sprechen wir mit Jordan die Neutrinoladung oder *duale Ladung* e zu, dem Neutron entsprechend $\varepsilon' e$[8]). Dann überzeugt man sich leicht, dass

$$P_i = \psi^+ \Lambda_i \psi, \quad P_i' = \psi^+ \Lambda_i' \psi \qquad (2.4)$$

die Vierervektoren der (in Einheiten e cm^{-3} gemessen) Ladungsdichte der elektrischen und der dualen Ladung sind.

p_i ist der Operator $(h/i)\,\partial/\partial x_i$ ($i = 1, 2, 3, 4$). Ein Pfeil $\overleftarrow{p}_i$ soll jeweils bedeuten, dass die Differentiation nach links wirkt:

$$f(x)\,\overleftarrow{p}_i\,g(x) = \frac{h}{i}\,\frac{\partial f}{\partial x_i}\,g(x).$$

3. Das Programm der Neutrinotheorie. Obwohl die folgenden Überlegungen sich nur mit den Kernkräften befassen, empfiehlt es sich ein versuchsweises Programm der in der Vorbemerkung skizzierten unitären Spinortheorie aufzustellen. Es folgen dann die zu behandelnden Kernkräfte (Punkt 4 des Programms) aus einer einfachen Verallgemeinerung der elektrischen Kräfte. Wir unterscheiden zwei Schritte: I. Die *„klassische" Theorie* betrachtet die Feldstärke-Komponenten ψ als gewöhnliche (Dirac'sche c-) Zahlen. II. Die *Quantentheorie des ψ-Wellenfeldes* sieht die Komponenten als Operatoren an.

I. Programm der klassischen Theorie.

1. Es existiert eine reelle und gegenüber Lorentztransformationen invariante Lagrangefunktionen $L(\psi^+, \psi)$. Die aus dem

Extremalprinzip $\delta \int L dx^4 = 0$ folgenden, Euler'schen Gleichungen, sollen folgende Form haben:

2. Wenn die FERMI'sche Konstante $g = 0$ ist, sollen Diracgleichungen für Elektron und Proton folgen, worin die Potentiale A_i in einer bestimmten Weise durch die dem Neutrino zugeordneten Komponenten dargestellt sind. Für das Neutrino soll eine Diracgleichung folgen, welche in einer bestimmten Form eine Wechselwirkung mit dem Vierervektor P_i der elektrischen Ladung enthält. Aus ihr sollen, durch Bildung der Feldvektoren nach einem verallgemeinerten Verfahren (1.1), die Maxwell'schen Gleichungen folgen. Für die Neutronen soll eine Diracgleichung folgen, deren Wechselwirkung mit dem A_i-Feld Null ist, die aber vielleicht eine (noch unbeobachtete) Wechselwirkung mit der elektrischen Ladungsdichte P_i haben darf.

3. Es soll (auch wenn $g \neq 0$) der Erhaltungssatz der Ladung folgen. Wir werden wegen einer (später näher zu präzisierenden) Symmetrieforderung noch die Forderung nach der Erhaltung der „Neutrinoladung" diskutieren[8]).

4. Wenn $g \neq 0$ sollen die „Kernkräfte" (FERMI'sche Theorie des β-Zerfalles[9]) und HEISENBERG-MAJORANA'sche[10]) Austauschkraft) aus der Theorie folgen.

II. Programm der Quantentheorie des Wellenfeldes.

5. Man führt die zu $\psi_\mu(x)$ konjugierten verallgemeinerten Impulse $\pi_\mu(x) = \partial L / \partial (\partial \psi_\mu / \partial t)$ ein. Dann sollen die JORDAN-WIGNER'schen Vertauschungsrelationen[11]) (FERMI-DIRAC-Statistik) für $x_4 = x_4'$

$$\pi_\mu(x)\, \psi_{\mu'}(x') + \psi_{\mu'}(x')\, \pi_\mu(x) = h/i\, \delta_{\mu\mu'}\, \delta(\vec{x} - \vec{x}') \tag{3.1}$$

gelten. $\delta(\vec{r})$ ist die dreidimensionale Dirac'sche δ-Funktion.

6. Aus den Relationen (3.1) sollen, durch Anwendung des in Programmpunkt 2 gegebenen Verfahrens auf (3.1), die bekannten Vertauschungsrelationen für die elektrischen Feldstärken folgen[12]).

Es sollen keine unendlichen Selbstenergien auftreten. Die reinen Zahlen e^2/hc, Verhältnisse der Proton-, Neutron- und Neutrinomasse (μ, μ', μ'') zur Elektronenmasse*), sollen aus der Theorie folgen. Diesen Punkt wollen wir aber nicht zum Programm zählen, da er ausserhalb des Rahmens der folgenden Theorie fällt.

4. Skizzierung der Theorie ohne Kernkräfte. (Neutrinotheorie des Lichtes. Programmpunkte 1 bis 3.)

Die Lagrangefunktion (L-Funktion)

$$L(\psi^+(x), \psi(x)) = \psi^+(x)\,(c\Gamma_i p_i - imc^2 \Delta)\,\psi(x) - e^2 \int dy^4 K(x,y)\,(\psi^+ \Lambda_i' \psi)(y) \cdot (\psi^+ \Lambda_i \psi)(x) \tag{4.1}$$

*) Sowie einige weitere reine Zahlen (die $\omega_{\chi\varphi}$ der Paragraphen 5 bis 7).

ist invariant und, weil die vorkommenden Matrizen hermiteisch sind (2.2), auch reell. Sie erfüllt Programmpunkt 1. Variation von ψ^+ liefert

$$(c\Gamma_i p_i - imc^2 \Delta - eA_i\Lambda_i - eA_i'\Lambda_i')\,\psi = 0 \tag{4.2}$$

und Variation von ψ

$$\psi^+\,(c\Gamma_i \bar{p}_i + imc^2 \Delta + eA_i\Lambda_i + eA_i'\Lambda_i') = 0 \tag{4.2$^+$}$$

Hierin sind

$$A_i\,(x) = e\int dy^4 K\,(xy)\,P_i'\,(y) = e\int dy^4 K\,(xy)\,(\chi^+\gamma_i\chi + \varepsilon' v^+\gamma_i v) \tag{4.3}$$

und

$$A_i'\,(x) = e\int dy^4 K\,(xy)\,P_i\,(y) = e\int dy^4 K\,(xy)\,(\varphi^+\gamma_i\,\varphi + \varepsilon u^+\gamma_i u) \tag{4.4}$$

wo P_i und P_i' die in (2.4) definierten Dichten sind. $K\,(xy)$ ist eine invariante Funktion des Weltabstandes $\sqrt{\sum_i (x_i - y_i)^2}$ und daher symmetrisch. Sie hat die Dimension (Länge) $^{-2}$.

Benützt man jetzt die gespaltene Schreibweise (2.1) für ψ, so folgen die 4 Gleichungen

$$\begin{aligned}(c\gamma_i p_i - imc^2 - eA_i\gamma_i)\,\varphi &= 0\\ (c\gamma_i p_i - i\mu mc^2 - eA_i'\gamma_i)\,\chi &= 0\\ (c\gamma_i p_i - i\mu' mc^2 - eA_i\gamma_i)\,u &= 0\\ (c\gamma_i p_i - i\mu'' mc^2 - eA_i'\gamma_i)\,v &= 0\end{aligned} \tag{4.5}$$

Die erste und dritte Gleichung sind die Diracgleichungen von Elektron und Proton, wenn die A_i die Potentiale darstellen. Da gemäss der Neutrinotheorie des Lichtes dieselben lediglich durch die Neutrinowellenfunktionen dargestellt werden sollen, so ist dies nur der Fall, wenn ε' in (4.3) Null ist. Falls $\varepsilon' \neq 0$ ist, so erhalten unsere Gleichungen eine zwar bis jetzt unbeobachtete Neutron-Ladung Wechselwirkung die in (4.5) formal in der „elektrischen" Wechselwirkung enthalten ist. Aus Symmetriegründen liegt die Vermutung nahe, dass $\varepsilon' = \pm 1$ zu setzen ist. Aus der zweiten Gleichung (4.5) sollen, durch Bildung von Feldstärkegrössen $G_{ik}\,(= \vec{D}, \vec{H})$, die Maxwell'schen Gleichungen gefolgert werden. Aufgabe der Neutrinotheorie ist es, die Definition des A_i, G_{ik} und F_{ik}, d. h. die Funktion $K\,(x, y)$ zweckmässig zu wählen. Wir stellen uns im folgenden auf den (vorläufig noch ungerechtfertigten) Standpunkt, dass auch dieser Teil von Programmpunkt 2 erfüllt sei*). Dies soll uns zu einem heuristischen Prinzip verhelfen, welches uns den Weg zur Einbeziehung der „Kernkräfte" zeigt.

Programmpunkt 3 (Erhaltung der elektrischen Ladung) ist erfüllt. Durch Multiplikation von (4.2) mit $\psi^+ \Lambda$ von links und

*) Im übrigen soll die Möglichkeit, dass Programmpunkte 1—3 und insbesondere 6 durch einen einfachen Ansatz der Art (4.1) *nicht* befriedigt werden können, durchaus offen gelassen werden.

von (4.2⁺) mit $\Lambda\psi$ von rechts und Addition folgt, wegen der Vertauschungsrelationen (2.3) und der Definitionen des elektrischen Ladungsstromes P_i

$$\frac{\partial}{\partial x_i} P_i = 0. \tag{4.6}$$

Multiplikation mit $\psi^+\Lambda'$ resp. $\Lambda'\psi$ von links resp. rechts liefert

$$\frac{\partial}{\partial x_i} P_i' = 0. \tag{4.6'}$$

Es empfiehlt sich aus Analogie A_i' als „duales elektrisches Potential" zu bezeichnen. Jedoch kommt dieser Bezeichnung nur wenig Bedeutung bei, da, wegen der nicht verschwindenden Masse des Elektrons keiner der beiden Anteile zu Maxwellgleichungen führen wird, während in A_i der erste, vom masselosen Neutrino herrührende, Teil Term $\chi^+\gamma_i\chi$ für das Auftreten der Maxwellgleichungen verantwortlich ist.

Die Lösung der Probleme dieses Paragraphen ist Aufgabe der Neutrinotheorie des Lichtes[6]. Die Ladungsdichten P_i und P_i' müssen im Sinne der DIRAC-HEISENBERG'schen Löchertheorie[3]) abgeändert werden. Für Proton resp. Neutron folgt hier auch, im Gegensatz zum Experiment, das magnetische Moment zu 1 resp. 0 BOHR'sche Proton-Magnetonen. Auf diesen Punkt wird später zurückzukommen sein. Die Existenz von Antiprotonen (=negative Protonen) und Antineutronen scheint mir kein Widerspruch mit dem Experiment zu sein, da sie nur ungeheuer viel schwerer zu erzeugen sind als positive Elektronen, und daher noch nicht beobachtet wurden.

Da die Behandlung der Quantisierung des Wellenfeldes vorerst noch hinausgeschoben wird, soll ein wesentliches und für das Verständnis der folgenden Paragraphen notwendiges Ergebnis der Quantisierung hier vorweggenommen werden:

Es bedeuten ψ_0 und ψ_1*) zwei stationäre Zustände, welche Gleichung (4.2) mit $e = 0$ genügen. Es sind dies, wie man aus der gespaltenen Schreibweise (2.1) ersieht, Zustände, bei welchen eine Partikel jeder Sorte in den durch φ_0, χ_0, u_0 und v_0 resp. φ_1, χ_1, u_1 und v_1 gegebenen Zuständen vorhanden ist. Dann bewirkt die „Störung" (der Term mit e^2 in der L-Funktion) Übergänge von ψ_0 nach ψ_1, deren Wahrscheinlichkeit dem Quadrat des „Matrixelementes" des Störungtermes

$$V(\psi_1^+, \psi_0) = e^2 \int d\bar{x}^3 d y^4 K(xy)\,(\psi_1^+ \Lambda_i' \psi_0)(x) \cdot (\psi_1^+ \Lambda_i \psi_0)(y) \tag{4.7}$$

*) Da im folgenden keine Spinovindizes mehr gebraucht werden, so bedeuten untere Indizes der Funktionen ψ, φ, χ, u, v jetzt Quantenzahlen.

proportional sind. In gespaltener Schreibweise lautet der letzte Teil des Integranden

$$\{(\chi_1^+ \gamma_i \chi_0) + \varepsilon' (v_1^+ \gamma_i v_0)\} (x) \cdot \{(\varphi_1^+ \gamma_i \varphi_0) + \varepsilon (u_1^+ \gamma_i u_0)\} (y). \quad (4.8)$$

Der erste Faktor bedeutet Übergänge:

A I: Neutrino im Zustand 0 ⟶ Neutrino im Zustand 1
A II: Neutron „ „ 0 ⟶ Neutron „ „ 1 } (A)

und der zweite solche der Art:

B I: Elektron im Zustand 0 ⟶ Elektron im Zustand 1
B II: Proton „ „ 0 ⟶ Proton „ „ 1 } (B)

Es kann immer nur ein Übergang der Sorte (A) auftreten, wenn einer der Sorte (B) ihn begleitet.

So bedeutet z. B. A I, begleitet von B I, dass ein Neutrino (z. B. aus einem Zustand negativer Energie χ_0) in einen solchen positiver Energie (χ_1) springt (A I), während gleichzeitig ein Elektron seinen Bewegungszustand von φ_0 in φ_1 abändert (B I). Das entstehende Neutrinopaar (Neutrino im Zustand χ_1 plus Loch, resp. Antineutrino im Zustand χ_0) deutet die Neutrinotheorie als das vom Elektron emittierte Lichtquant.

Die 16komponentige Schreibweise ist daher „trivial", d. h. sie liefert keine neuen Gesichtspunkte, da diese Form der Theorie keine Quantensprünge erlaubt, bei welchem ein Teilchen einer Sorte (z. B. Neutrino) sich in ein solches anderer Sorte (z. B. Proton) verwandelt. Daher ist die Alternative, ob das positive Proton Partikel oder Antipartikel ist, unentscheidbar. Die folgenden Paragraphen, welche Übergänge zwischen den Quantenzuständen zulassen, werden einen Entscheid dieser Alternative ergeben.

5. Die Kernkräfte (Programm 4). Die Kernkräfte folgen aus einer einfachen Verallgemeinerung der elektromagnetischen Kräfte des vorangehenden Paragraphen:

Sieht man nämlich von der e^2 proportionalen Wechselwirkung in (4.1) ab, so ist die L-Funktion bilinear in ψ^+, ψ. Die Wechselwirkung tritt erst als biquadratischer Term auf. Dieser Term ist das Produkt zweier mit hermiteischen Operatoren gebildeten, reellen Faktoren und daher a fortiori reell. Ein biquadratischer Zusatzterm

$$L_g(\psi^+(x), \psi(x)) = -e^2 \int dy^4\, M(x, y)\, (\psi^+ \Omega_i^+ \psi)(y) \cdot (\psi^+ \Omega_i \psi)(x) \quad (5.1)$$

wo $M(xy)$ wieder eine Funktion des Weltabstandes $\sqrt{\Sigma (x_i - y_i)^2}$ ist und die Dimension (Länge) $^{-2}$ hat, und Ω_i nicht hermiteische

Matrizen sind, die jedoch bewirken, dass $\psi^+ \Omega_i \psi$ sich wie Vektoren transformieren, ist die einfachste Verallgemeinerung von (4.1). Dann erfüllt L nach wie vor Programmpunkt 1 (Realität und Invarianz). Die Ω_i dürften auch die Operatoren p_i enthalten, doch soll hiervon vorläufig abgesehen werden[13]). Im allgemeinen werden sogar mehrere solche Zusatzterme existieren mit verschiedenen Ω.

Wie bei der Neutrinotheorie des Lichtes, deren Erfolg oder Misserfolg davon abhängt, ob eine Funktion $K(xy)$ existiert, welche Programmpunkt 2 (Maxwellgleichungen) und namentlich auch Punkt 6 (Bosestatistik der Lichtquanten) erfüllt, so sollte auch hier zuerst eine Diskussion von M folgen.

Da wir aber über die Kernkräfte wenig wissen, so wollen wir die Annahme machen

$$M(x,y) = \lambda^2 \cdot \delta(x-y) \tag{5.2}$$

$\delta(x)$ bedeutet die vierdimensionale DIRAC'sche δ-Funktion, λ ist eine Länge. Wir setzen $g = e^2\lambda^2$. Es wird sich später zeigen, dass g die FERMI'sche[9]) Konstante ist. Ihre Grössenordnung ist 10^{-50}. Als Vergleich sei bemerkt, dass dann λ der Comptonwellenlänge einer Partikel vom Atomgewicht 1000 oder aber dem klassischen Radius einer mit e geladenen Partikel vom Atomgewicht 1/10 entspricht (d. h. grössenordnungsweise entspricht λ dem klassischen Protonradius). Unter Verwendung von (5.2) und (5.3) wird

$$L_g = -g\,(\psi^+ \Omega_i^+ \psi)\,(\psi^+ \Omega_i \psi)\,. \tag{5.3}$$

An Stelle der Gleichung (4.2) und (4.2$^+$) folgen jetzt (wir lassen die „elektrischen" Terme von (4.1) mit den Ladungsmatrizen Λ_i weg)

$$(c\Gamma_i p_i - imc^2\Delta - g\,(\psi^+\Omega_i^+\psi)\,\Omega_i - g\,(\psi^+\Omega_i\psi)\,\Omega_i^+)\,\psi = 0 \tag{5.4}$$

und

$$\psi^+\,(c\Gamma_i \vec{p}_i + imc^2\Delta + g\,(\psi^+\Omega_i^+\psi)\,\Omega_i + g\,(\psi^+\Omega_i\psi)\,\Omega_i^+) = 0 \tag{5.4$^+$}$$

Vollführen wir die gleiche Operation, welche uns im vorhergehenden Paragraphen die Gleichungen (4.6) und (4.6') lieferte, so erhalten wir:

$$\frac{\partial P_i}{\partial x_i} = -\frac{ig}{hc}\{(\psi^+\Omega_i^+\psi)\,(\psi^+(\Omega_i\Lambda - \Lambda\Omega_i)\,\psi) + (\psi^+(\Omega_i^+\Lambda - \Lambda\Omega_i^+)\,\psi)\,(\psi^+\Omega_i\psi)\} \tag{5.5}$$

und eine entsprechende Gleichung mit P_i' und Λ' an Stelle von P_i und Λ.

Die Forderung von Programmpunkt 3 (Erhaltung der Ladung) lässt nur die Möglichkeiten zu:

$$\Omega_i \Lambda - \Lambda \Omega_i = \begin{cases} 0 \\ \pm \Omega_i \end{cases} \tag{5.6}$$

Verlangt man auch die Erhaltung der dualen Ladung, so folgt eine analoge Bedingung mit Λ'.

Der Forderung nach relativistischer Invarianz ist Genüge getan mit:

$$\Omega_i = \Gamma_i \Omega = \Omega \Gamma_i \tag{5.7}$$

Hier ist Ω eine Matrix der Form

$$\Omega = \begin{pmatrix} \omega_{\varphi\varphi} & \omega_{\varphi\chi} & \omega_{\varphi u} & \omega_{\varphi v} \\ \omega_{\chi\varphi} & \omega_{\chi\chi} & \omega_{\chi u} & \omega_{\chi v} \\ \omega_{u\varphi} & \omega_{u\chi} & \omega_{uu} & \omega_{uv} \\ \omega_{v\varphi} & \omega_{v\chi} & \omega_{vu} & \omega_{vv} \end{pmatrix} \tag{5.8}$$

Die $\omega_{\varphi\chi}\ldots$ sind reine (mit der vierzeiligen Einheitsmatrix multipliziert zu denkende) Zahlen. Bedingung (5.6) lautet dann (wegen der Vertauschbarkeit von Γ_i, Λ und Ω):

$$\Omega \Lambda - \Lambda \Omega = \begin{cases} 0 \\ \pm \Omega \end{cases} \tag{5.9}$$

Im Matrixelement des neuen Störungsterms

$$V(\psi_1^+ \psi_0) = g \int d\vec{x}^3 (\psi_1^+ \Omega_i^+ \psi_0)(\psi_i^+ \Omega_i \psi_0) \tag{5.10}$$

nimmt der Integrand die Form an:

$$\Big\{\sum_{\varphi,\chi,u,v} \omega_{u\varphi}^* (\varphi_1^+ \gamma_i u_0)\Big\} \cdot \Big\{\sum_{\varphi,\chi,u,v} \omega_{u\varphi} (u_1^+ \gamma_i \varphi_0)\Big\} \tag{5.11}$$

wo Σ die Summationen über alle Kombinationen $(\varphi_1^+ \gamma_i \varphi_0)$, $(\varphi_1^+ \gamma_i \chi_0)$, $(\varphi_1^+ \gamma_i u_0)$ usw. darstellt. Jedes vorhandene Element von Ω^+ bedeutet, dass ein entsprechender Übergang (z. B. im Fall $\omega_{u\varphi}^* \neq 0$, ein Übergang Proton $(u_0) \longrightarrow$ Elektron (ψ_1^+)) vorkommen kann. Da der zweite Faktor die adjungierte Matrix enthält, so tritt jeder solche Übergang nur auf begleitet von einem, in der gleichen Matrix enthaltenen, Übergang eines anderen Teilchens, aber in umgekehrter Richtung (z. B. im betrachteten Falle der Term $\omega_{u\varphi}$, welcher der Verwandlung eines anderen Teilchens von Elektron $(\varphi_0) \longrightarrow$ Proton (u_1^+) entspricht; oder der Term $\omega_{v\chi}$, welcher einen Übergang Neutrino $\longrightarrow$ Neutron darstellt). Da jeder Übergang, zum Unterschied von Paragraph 4, jetzt immer auch mit seinem inversen verkoppelt ist, so liefert die Theorie jetzt Austauschkräfte.

Tatsächlich können (wegen 5.9) jeweils nur wenige Elemente von Ω nicht verschwinden. Das Bild gestaltet sich naturgemäss vollständig anders, je nachdem man (wir haben das positive Elektron willkürlich als „Partikel" festgesetzt) das positive Proton als „Partikel" ($\varepsilon = +1$) oder als „Antipartikel" ansieht ($\varepsilon = -1$). Das negative Proton ist dann „Partikel". Der erste Fall wird Zerstrahlungsprozesse der schweren Teilchen erlauben, während der zweite diese verbietet, da sich nur „Partikel" in „Partikel" verwandeln kann, und da die Ladung erhalten bleibt.

6. Diskussion der möglichen Kernkräfte im Falle $\varepsilon = 1$. (Proton ist Partikel). Wir bilden

$$\Omega\Lambda - \Lambda\Omega = \begin{pmatrix} 0 & -\omega_{\varphi\chi} & (\varepsilon-1)\,\omega_{\varphi u} & -\omega_{\varphi v} \\ \omega_{\chi\varphi} & 0 & \varepsilon\,\omega_{\chi u} & 0 \\ -(\varepsilon-1)\,\omega_{u\varphi} & -\varepsilon\,\omega_{u\chi} & 0 & -\varepsilon\,\omega_{uv} \\ \omega_{v\varphi} & 0 & \varepsilon\,\omega_{vu} & 0 \end{pmatrix} \tag{6.1}$$

$$\Omega\Lambda' - \Lambda'\Omega = -\begin{pmatrix} 0 & -\omega_{\varphi\chi} & 0 & -\varepsilon'\,\omega_{\varphi v} \\ \omega_{\chi\varphi} & 0 & \omega_{\chi u} & -(\varepsilon'-1)\omega_{\chi v} \\ 0 & -\omega_{u\chi} & 0 & -\varepsilon'\,\omega_{uv} \\ \varepsilon'\,\omega_{v\varphi} & (\varepsilon'-1)\,\omega_{v\chi} & \varepsilon'\,\omega_{vu} & 0 \end{pmatrix} \tag{6.2}$$

und diskutieren zuerst die Matrix, welche

$$\left.\begin{aligned} \Omega\Lambda - \Lambda\Omega &= \Omega \\ \Omega\Lambda' - \Lambda'\Omega &= -\Omega \end{aligned}\right\} \tag{6.3}$$

erfüllt. Sie enthält (mit $\varepsilon - 1 = 0$) nur an den Stellen von Null verschiedene Elemente, wo in (6.1) das entsprechende Element ω_{ab} von Ω mit 1 multipliziert steht, d. h. sie erlaubt nur die ihnen zugeordneten Prozesse. Die Erhaltung der dualen Ladung ist automatisch erfüllt, wenn dem Neutron die Ladung $\varepsilon' = +1$ zugesprochen wird. Es sind dies die Umwandlungen:

C I	$\omega_{\chi\varphi}$:	Neutrino $\rightleftarrows$ pos. Elektron	(C)
C II	$\omega_{\chi u}$:	Neutrino $\rightleftarrows$ pos. Proton	
C III	$\omega_{v\varphi}$:	Neutron $\rightleftarrows$ pos. Elektron	
C IV	ω_{vu}:	Neutron $\rightleftarrows$ pos. Proton	

Jeder Prozess an einem Teilchen ist, nach oben Gesagtem, mit einem inversen, an einem anderen Teilchen gekoppelt. So stellt C IV mit sich selbst gekoppelt, die Majorana'sche Wechselwirkungskraft dar. Man überzeugt sich leicht, dass das zugeordnete Matrixelement, wenn die sicher zu primitive Annahme (5.2) durch eine Kraft

$$M(x-y)\begin{cases} = \lambda^{-2} e^{+s^2/\lambda^2} & \text{für } s^2 = \Sigma\,(x_i - y_i)^2 = \Sigma z_i^2 < 0 \\ = 0 & \text{,,}\quad s^2 > 0 \end{cases} \tag{6.4}$$

ersetzt wird, tatsächlich formal den Majorana'schen Ansatz ergibt (Vernachlässigung der γ_i ($i = 1, 2, 3$,)-Glieder gegen das γ_4-Glied, $u^+\gamma_4 = iu^*$ usw.):

$$\int d\vec{x}^3 d\vec{y}^3 u_1^* (x)\, v_0 (x)\, J (\vec{x} - \vec{y})\, v_1^* (y)\, u_0 (y) \qquad (6.5)$$

mit

$$J (\vec{z}) = \frac{e^2}{\lambda^2}\, \omega_{u\,v}^2 \left\{ \int\limits_{-i}^{-i|z|} + \int\limits_{i|z|}^{\infty\, i} \right\} dz_4\, e^{s^2/\lambda^2 - z_4 (E_{u_0} - E_{v_1})/hc} \qquad (6.6)$$

Es stellt eine Austauschkraft zwischen Proton und Neutron dar. Die Wechselwirkungsenergie hängt zum Unterschied gegen Majorana noch von der Energiedifferenz $E_{u_0} - E_{v_1}$ der beiden Zustände u_0 und v_1 ab.

Kombination von C I mit C IV gibt zu dem Fermi'schen Matrixelement Anlass:

$$g^2 \omega_{v\,u} \omega_{\chi\,\varphi}^* \int d\vec{x}^3\, (v_1^+ \gamma_i u_0)\, (\varphi_1^+ \gamma_i\, \chi_0)\,.\, {}^{*}) \qquad (6.7)$$

Es verwandelt sich ein Proton (im Atomkern) aus dem Zustand u_0 in ein Neutron im Zustand v_1 (C IV), während gleichzeitig eine andere Partikel aus dem Neutrinozustande (z. B. negative Energie) χ_0 nach dem (z. B. positiven Energie-) Zustande φ_1 springt, d. h. es entsteht ein Antineutrino und ein positives Elektron (C I). Dies ist aber gerade die Erscheinung der β^+-Radioaktivität.

Die Übergänge C I und C III geben noch zu weiteren Austauschenergien Anlass. Sie erlauben aber auch Prozesse, in welchen ein {Neutron / Proton} zu einem {pos. Elektron / Neutrino} wird, und gleichzeitig ein „Paar", {Neutrino / Antineutrino} und {neg. Elektron / pos. Elektron} entsteht. („*Zerstrahlung" zweiter Art der schweren Teilchen.*) Der Prozess pos. Proton wird Neutrino und ein Neutron (neg. Energie) wird pos. Elektron bedingt eine alternative Erklärung der β^+-Radioaktivität, da jetzt aus dem Kern Proton ein Antineutron, Neutrino und pos. Elektron entstanden ist (*Radioaktivität zweiter Art*). Sie hätte aber die Existenz von zweierlei Neutronsorten (Neutron und Antineutron) im Atomkern zur Folge, die unter „Zerstrahlung" rekombinieren könnten.

Die einzige Matrix, welche $\Omega \Lambda - \Lambda \Omega = 0$ für Λ und Λ' erfüllt, ist eine solche, bei welcher erstens alle Diagonalglieder vorhanden sein könnten, und die daher einfach eine Verallgemeinerung der Wechselwirkungen A I, A II, B I, B II des Paragraphen 4 bedeutet. Dies gibt zu neuen Wechselwirkungs-

*) In der von Konopinski und Uhlenbeck[13]) benützten Form.

energien zwischen gleichen Teilchen Anlass. Des weiteren dürfen in diesem Falle aber noch die folgenden Umwandlungen auftreten.

$$\left.\begin{array}{lll} \text{D I} & \omega_{\varphi u}, \omega_{u\varphi} & \text{pos. Elektron} \rightleftharpoons \text{pos. Proton} \\ \text{D II} & \omega_{\chi v}, \omega_{v\chi} & \text{Neutron} \rightleftharpoons \text{Neutrino}. \end{array}\right\} \quad \text{(D)}$$

Sie treten mit sich selbst oder den ,,Diagonal''-Reaktionen A I, A II, A III und A IV gekoppelt auf und erlauben daher beispielsweise Prozesse, in welchen ein $\left\{\begin{matrix}\text{Neutron}\\ \text{Proton}\end{matrix}\right\}$ ein $\left\{\begin{matrix}\text{Neutrino}\\ \text{Elektron}\end{matrix}\right\}$ wird, und gleichzeitig ein Neutrinopaar (A I) auftritt, welches unter Umständen als Lichtquant in Erscheinung tritt *(Zerstrahlung erster Art der schweren Teilchen).*

Da die Lebensdauer der schweren Materie gegenüber den verschiedenen Zerstrahlungsprozessen durch g bestimmt ist und daher*) nur die Grössenordnung der radioaktiven Halbwertszeiten erreicht, so scheint mir der hier diskutierte Fall ,,Proton ist Partikel'' mit der Erfahrung in Widerspruch zu stehen. Ebenso wiederspricht die mögliche ,,Radioaktivität zweiter Art'' der Erfahrung.

7. Diskussion der möglichen Kernkräfte im Fall $\varepsilon = -1$. (Proton ist Antipartikel). Wir diskutieren hier zuerst den Fall $\Omega\Lambda - \Lambda\Omega = 0$ und wollen, um auch Zerstrahlungen der Neutronen auszuschliessen, gleich auch $\varepsilon' = -1$ setzen. Dann können in dieser ersten Matrix Ω nur die Diagonalelemente von Null verschieden sein. Es treten also auch jetzt die zum Paragraph 4 zusätzlichen Kräfte zwischen gleichen Teilen auf. Hingegen fehlen die Zerstrahlungsprozesse erster Art.

Beide Gl. (6.3) sind erfüllbar durch eine Ω-Matrix, welche nur $\omega_{\chi\varphi}$ und ω_{uv} enthält. D. h. sie erlaubt nur Prozesse

$$\left.\begin{array}{c} \text{E I: Neutrino} \rightleftharpoons \text{pos. Elektron} \\ \text{E II: neg. Proton} \rightleftharpoons \text{Neutron} \\ \text{E II kann auch im Sinne der Antipartikel als} \\ \text{E II: Antineutron} \rightleftharpoons \text{pos. Proton} \end{array}\right\} \quad \text{(E)}$$

geschrieben werden. Diese Prozesse entsprechen daher vollkommen C I und C IV des vorhergehenden Paragraphen. Es folgen daher insbesondere wieder die Majoranakraft (6.5) und das Fermi'sche Matrixelement des β^+-Zerfalls (6.7). (Letzteres mit Vertauschung der Buchstaben u und v, da sich jetzt ein Neutron aus den negativen Energiezuständen in ein negatives Proton

*) Falls die von Null verschiedenen reinen Zahlen ω_{ab} (bis auf wenige Zehnerpotenzen z. B. 137 oder 1847) von der Grössenordnung eins sind.

verwandelt, und somit das ursprünglich vorhandene „neg. Protonloch" (= pos. Proton) zum Verschwinden bringt, und dafür ein Antineutron, Antineutrino und pos. Elektron entsteht.)

Das Fehlen der Prozesse C II und C III verunmöglicht auch die Zerstrahlungsprozesse zweiter Art.

Eine zweite Matrix, in welcher nur $\omega_{u\chi}$ und $\omega_{v\varphi}$ von Null verschieden sind, erfüllt nach (6.1) und (6.2) ebenfalls $\Omega\Lambda - \Lambda\Omega = \Omega$ und $\Omega\Lambda' - \Lambda'\Omega = \Omega$. (Die Vorzeichenumkehr in der Λ'-Gleichung ändert nichts an dem Erhaltungssatz der dualen Ladung.) Sie gibt Anlass zu Prozessen.

F I: neg. Proton ⇄ Neutrino
F II: Neutron ⇄ pos. Elektron
Den ersten schreiben wir wieder für Antipartikel:
F I: Antineutrino ⇄ pos. Proton
(F)

Sie entsprechen vollständig den Prozessen C II und C III. Auch sie geben wieder zu den neuen Austauschkräften zwischen leichten und schweren Teilchen Anlass. Auch erscheint wieder die „Radioaktivität zweiter Art": pos. Proton wird Antineutrino und ein Neutron (neg. Energie) wird pos. Elektron, d. h. es entsteht aus pos. Proton ein Antineutron, ein Antineutrino und ein pos. Elektron. Nur ist ihr Resultat, zum Unterschied gegenüber dem vorhergehenden Paragraphen, identisch mit demjenigen der (Fermi'schen) Radioaktivität erster Art.

Da aber die E-Vorgänge und die F-Vorgänge nur je unter sich gekoppelt werden dürfen, so treten die Zerstrahlungsprozesse zweiter Art ebenfalls nicht auf.

8. Zusammenfassung. Aus der Annahme, dass Elektron, Proton, Neutron und Neutrino verschiedene Quantenzustände eines einzigen Teilchens bedeuten, verbunden mit der Forderung nach Erhaltung der elektrischen Ladung und der dualen (Neutrino-) Ladung folgt die Möglichkeit folgender Prozessgruppen:

Teilchen ⇄ gleiches Teilchen (A)

(C)
Neutrino ⇄ pos. Elektron
(Anti-) Neutron ⇄ pos. Proton
(E)
(Anti-) Neutrino ⇄ pos. Proton
Neutron ⇄ pos. Elektron
(F)

pos. Proton ⇄ pos. Elektron
Neutron ⇄ Neutrino
(D)

Ein Prozess an einem Teilchen verläuft nur dann, wenn gleichzeitig ein anderes Teilchen einen Prozess derselben Gruppe in umgekehrter Richtung durchläuft.

(A) enthält nach der Vorstellung der Neutrinotheorie des Lichts die Licht-Materiewechselwirkung sowie zusätzliche Kräfte zwischen gleichen und zwischen verschiedenen Teilchen, (E) und (F) die Austauschkräfte zwischen verschiedenen Teilchen und die Fermi'sche Erklärung der β-Radioaktivität. (D) sind Zerstrahlungsprozesse der schweren Teilchen.

Wird das positive Elektron als Partikel definiert (im Gegensatz zur Antipartikel), so kann das positive Proton entweder als Partikel ($\varepsilon = 1$) oder Antipartikel ($\varepsilon = -1$) betrachtet werden. Die erste Betrachtungsweise erlaubt die Zerstrahlungsprozesse (D) sowie weitere „Zerstrahlungsprozesse zweiter Art" durch Kombinationen von (E) mit (F), da in diesem Falle (E) und (F) zusammen nur eine einzige Gruppe (C) bilden. (Das eingeklammerte (Anti-) in (E) und (F) fällt weg.)

Die zweite Betrachtungsweise verbietet sowohl die Zerstrahlungsprozesse (D) und auch Kombinationen unter den jetzt verschiedenen Gruppen (E) und (F), somit also auch „Zerstrahlungsprozesse zweiter Art". Die in Atomkernen vorkommenden schweren Teilchen sind dann gegenüber dem positiven Elektron als Antipartikel anzusprechen. Wir ziehen daher diese zweite Festsetzung vor.

Sieht man von den relativ wenigen positiven Elektronen ab, so besteht unsere Welt im wesentlichen nur aus Antipartikeln oder nur aus Partikeln*) (negative Elektronen, Protonen, Neutronen) mit von Null verschiedener Ruhemasse. Eine Zerstrahlung der Welt ist ausgeschlossen. Trotzdem ermöglicht die vorgeschlagene einheitliche Auffassung der Materie Austauschkräfte zwischen allen vorhandenen Partikeln.

Quantitative Folgerungen erlaubt die Theorie erst, wenn die Kräfte experimentell und theoretisch exakter behandelt werden können. Es ist dies aufs Engste verknüpft mit der Entwicklung der Neutrinotheorie des Lichtes. Trotzdem können die vorstehenden Ergebnisse unabhängig von der letztgenannten Theorie angesehen werden. Die Einführung der Neutrinotheorie des Lichtes wurde lediglich als heuristisches Prinzip benützt. Die jetzt $4 \times 4 = 16$ komponentige ψ-Funktion bedeutet eine unitäre Feldtheorie der Materie. Der Einschluss der Neutrinotheorie des Lichtes führt dann zur unitären Feldtheorie im Sinne Born's, welche elektromagnetisches und materielles Feld vereinigt.

*) Da die Theorie ja in Bezug auf Partikel und Antipartikel symmetrisch ist.

Die, wahrscheinlich im Vergleich zu denen von Elektron und Neutrino, grossen Wechselwirkungskräfte von Proton und Neutron mit den virtuellen schweren Partikeln und Antipartikeln (Wechselwirkung mit aufgefüllten Zuständen negativer Energie) lassen vielleicht eine Erklärung der beobachteten Abweichung des magnetischen Momentes vom, aus der störungsfreien Dirac-Gleichung folgenden, Betrag zu.[14])

Den Herren Prof. W. Pauli und G. Wentzel (Zürich), und J. Weigle (Genf) bin ich für manche wertvolle Ratschläge verpflichtet. Insbesondere danke ich Herrn Weigle für viele interessante Diskussionen und Herrn Wentzel für den Hinweis auf die duale Ladung.

Institut de Physique, Université de Genève.

Literatur.

1) G. Mie, Ann. d. Phys. **37**, 512, **39**, 1 und **40**, 1 (1912—1913); H. Weyl, Raum – Zeit – Materie 1920. A. Einstein und W. Mayer, Sitz. Ber. der Preuss. Akad. d. Wiss. 1931.

2) W. Pauli und V. Weisskopf, Helv. Physica Acta **7**, 709 (1934).

3) W. Heisenberg, Zs. f. Phys. **90**, 209 (1934).

4) B. L. van der Warden, Gruppentheoret. Methode i. d. Quantenmechanik, Berlin 1932.

5) L. de Broglie, Une nouvelle conception de la lumière, Act. Scint. Hermann, Paris 1934.

6) G. Wentzel, Zs. f. Phys. **92**, 337 (1934); P. Jordan, Zs. f. Phys. **93**, 464 (1935), **98**, 709 und 759 (1936), **99**, 109 (1936); R. de L. Kronig, Physica **2**, 491, 854, 968 (1935); O. Scherzer, Zs. f. Phys. **97**, 725 (1935).

7) M. Born und L. Infeld, Proc. Roy. Soc. A **144**, 423 (1934) und folgende Arbeiten.

8) P. Jordan, loc. cit., Zs. f. Phys. **98**, 761 u. ff. (1936).

9) E. Fermi, Zs. f. Phys. **88**, 161 (1934).

10) E. Majorana, Zs. f. Phys. **82**, 137 (1933).

11) P. Jordan und E. Wigner, Zs. f. Phys. **47**, 631 (1928), cf. auch W. Heisenberg, Physikalische Prinzipien der Quantenmechanik, Leipzig 1930.

12) Siehe z. B. Heisenberg, loc. cit. 11), Seite 109 und 113.

13) E. J. Konopinski und G. E. Uhlenbeck, Phys. Rev. **48**, 7 und 107 (1935).

14) Vergl. dazu W. Heisenberg, Zeemanfestschrift, Haag 1935.

Radioactive β-decay and nuclear exchange forces as a consequence of a unitary field theory [27]

1032 NATURE JUNE 20, 1936

Radioactive β-Decay and Nuclear Exchange Force as a Consequence of a Unitary Field Theory

THE hypothesis is put forward that positive electron, neutrino, positive proton and neutron are four different quantum states of one elementary particle. Such an assumption would be trivial unless transitions between the different states occur. It is required that Dirac's equation follows from the theory, and that the conservation law of electric charge holds, so only a small number of transitions are allowed. If in addition we satisfy a certain symmetry condition (corresponding to the conservation law of Jordan's neutrino charge[1]) the number of possible processes is further reduced. The permitted transitions are:

(I)	positive electron	⇄	neutrino
(II)	positive electron	⇄	neutron
(III)	positive proton	⇄	neutrino
(IV)	positive proton	⇄	neutron.

Any one of these transmutations can occur only if another one of them takes place in the reverse direction.

Process (IV) (from right to left) and (I) (from left to right) give rise to a transmutation of a neutron into a proton, while simultaneously a positive electron (being in one of the negative energy states of Dirac's theory) becomes a neutrino. Thus a 'Dirac hole' (to be identified with a negative electron) and a neutrino are produced. An explicit calculation according to this theory gives the Fermi[2] formula for radioactive β-decay.

If the transformation (IV) of one particle is coupled with the same transformation (of a second particle) in the reverse direction, the corresponding interaction energy is the one postulated by Majorana[3] in order to explain nuclear constitution (exchange force neutron–proton).

Combining I, II, III and IV, there is a number of further reactions possible which will be discussed elsewhere together with the complete theory[4]. As soon as the neutrino theory of light[5] can be formulated in a satisfactory way, we have a unitary field theory, its field variable being a spinor of 16 components.

ERNEST C. G. STUECKELBERG.
Institut de Physique,
Université de Genève.
May 7.

[1] Jordan, *Z. Phys.*, **98**, 759 (1936).
[2] E. Fermi, *Z. Phys.*, **88**, 161 (1934); Konopinsky and Uhlenbeck, *Phys. Rev.*, **48**, 7 and 107 (1936).
[3] Majorana, *Z. Phys.*, **82**, 137 (1933).
[4] To be published in the *Helv. Phys. Acta*. *Note added in proof:* It seems worth while to point out that combinations between (I) and (III) or between (II) and (IV) which lead to destruction of heavy particles, cannot occur if the negative electron and the positive proton are both considered as *true particles* (being the opposite of the 'holes' or antiparticles in Dirac's theory).
[5] L. de Broglie, "Une nouvelle conception de la lumière", Actualité scient., Hermann, Paris (1934). Further progress has been accomplished by Wentzel, Jordan, Kronig and Scherzer; for references compare Jordan (ref. 1) and *Z. Phys.*, **99**, 112 (1936). See also Kronig, NATURE, **137**, 149 (1936).

Artificial radioactivity giving continuous γ-radiation [30]

Reprinted by permission from Macmillan Publishers Ltd: Nature, vol. 137, pp. 1070–1071, copyright (1936)

Artificial Radioactivity giving Continuous γ-Radiation

IN many cases of nuclear transformations, unstable nuclei are obtained which break down emitting positive electrons ($\beta+$-decay). Because this nucleus is surrounded by orbital negative electrons, the positive electron produced can recombine with one of the negatives giving rise to a γ-ray quantum[1]. The force responsible for recombination is proportional to e (electronic charge). The probability of such a recombination thus involves the factor e^2. Therefore the ratio of the probability of observing a γ-ray quantum to that of finding a positive electron must be of the order of magnitude $e^2/hc \sim 10^{-2}$ (e^2/hc being the only dimensionless number which can be formed from e, the Planck constant h and the velocity of light c).

We conclude that any $\beta+$-emission must be accompanied by a weak γ-radiation. In analogy to the theory of optical dispersion, we can consider the decayed nucleus plus the (positive and negative) electrons as an *intermediate state*. If a transition from the *initial state* (non-decayed nucleus plus orbital electrons) to the *final state* (decayed nucleus plus a γ-ray quantum) is energetically possible, such a transition can occur even if the energy is too small to reach the *intermediate state* (decayed nucleus plus positive electron plus orbital electrons).

Artificial $\beta+$-activity occurs, if the energy difference between the decayed and the undecayed nucleus is greater than mc^2. Emission of a continuous γ-ray spectrum (not accompanied by $\beta+$-emission) should be observed if this energy difference is less than mc^2, but greater than $-mc^2$. The mean life of such a γ-active nucleus is about a thousand times longer than that of a $\beta+$-active nucleus. It seems worth while to look for this new form of radioactivity.

Note added in proof: Only in one of 137 cases of decay is a γ-quantum emitted. In all other cases an unobservable and monochromatic neutrino radiation appears. Prof. W. Pauli remarked in a discussion that this radiation must be followed by emission of the characteristic X-ray spectrum of the atom preceding the unstable element in the periodic system. A complete account will be published in the *Helv. Phys. Acta.*

E. C. G. STUECKELBERG.

Institute of Physics,
University of Geneva.

May 7.

[1] Bloch and Møller, NATURE, **136**, 911 (1935); M. Fierz, *Helv. Phys. Acta*, **9**, 245 (1936); G. Rumer, *Sow. Phys.*, **9**, 317 (1936).

Über die Methode der physikalischen Naturbeschreibung [33]

Naturforschende Gesellschaft in Basel, Verhandlungen, vol. 47 (1936), pp. 181–205

Über die Methode der physikalischen Naturbeschreibung.

Von

E. C. G. Stueckelberg, Genf.

Wenn es zu den Aufgaben der Philosophie und der wissenschaftlicher Theologie gehört, das Warum und Wie unserer Existenz und ihrer Umwelt zu ergründen, so muss als ihr Ausgangspunkt eine vollständige und möglichst übersichtliche Beschreibung unserer Empfindungen vorliegen. Einen Teil dieser Empfindungen fassen wir unter dem Namen Sinnesempfindungen zusammen. Ob ein solcher Ausschnitt genau abgegrenzt werden kann, soll hier nicht untersucht werden. Es kann sogar mit gewisser Berechtigung behauptet werden, dass eine Unterteilung gar nicht streng durchführbar ist. Der schon in der Philosophie der Antike uns entgegentretende Gedanke, dass die Welt nur in ihrer Gesamtheit verständlich sein kann, kommt sicherlich der Wahrheit näher als die philosophischen Systeme des verflossenen Jahrhunderts, welche glaubten, dass alles aus unseren materialistischen Erkenntnissen heraus erklärt werden könne. Die Entstehung dieser Systeme, ich denke z. B. an den Monismus Haeckels, erklärt sich aus der Hybris, welche die Reaktion der Philosophie auf die bedeutenden Erfolge der exakten Naturwissenschaften war. Die ebenfalls nicht zu unterschätzenden Fortschritte unserer heutigen Kenntnisse auf diesem Gebiete zwingen die Naturforschung eher zu einer immer steigenden Bescheidenheit den „Welträtseln“ gegenüber, deren letzte Konsequenz vielleicht darin liegen wird, dass sie die eingangs der Philosophie und Theologie zugewiesene Aufgabe als unlösbar bezeichnen muss.

Wir wollen eine zweite Unterteilung unserer Sinnesempfindungen vornehmen in solche, welche von der belebten und in solche, welche von der unbelebten Umwelt hervorgerufen werden. Hier ist die Trennung vielleicht noch problematischer als die bereits vorgenommene, da der Begriff der letzteren Kategorie von Sinnesempfindungen die Person des Beobachters, der sicher der belebten Welt angehört, und die unbelebte Umwelt in sich schliesst. Eine gewisse Rechtfertigung zu dieser Trennung besteht m. E. nur darin, dass einige Erfahrungstatsachen sich durch die „*unbelebte Materie*“ beschreiben lassen. Die Aufgabe der Physik — oder besser ausgedrückt der physikalischen Naturbeschreibung — ist es dann, ein möglichst übersichtliches System von Aussagen aufzufinden, aus welchen mit Hilfe logischer Deduktion diese Sinneseindrücke beschrieben und insbesondere auch vorausgesagt werden können. Die Physik ist somit ein sowohl in bezug auf die gestellte Aufgabe wie auch in bezug auf das ihr zugewiesene Gebiet ausserordentlich beschränkter Ausschnitt aus der Philosophie.

Das Suchen nach dieser Form der Naturbeschreibung blickt auf eine Jahrtausende alte Entwicklung zurück. Wenn wir trotzdem auch heute in diesem begrenzten Aufgabenkreis noch weit von einer befriedigenden Lösung sind, so ist die eingangs erwähnte Bescheidenheit sicher die einzige richtige Einstellung der Forschung.

Vielleicht darf ich hier eine zur Zeit oft diskutierte Frage der Erziehung des exakten Naturwissenschaftlers streifen. Da die Voraussetzung jedes Umschreibens die Beherrschung des logischen Denkens und der Sprache ist, so möchte ich persönlich das humanistische Gymnasium in seiner jetzigen Form als die beste Vorbereitungsschule bezeichnen. Seine, durch das Wort „Gymnasion“ definierte Aufgabe der turnerischen Schulung des Geistes durch Umgang mit den Sprachen und der Philosophie der Antike halte ich für wesentlich wichtiger als eine vielleicht quantitativ grössere Kenntnis von Erfahrungstatsachen der Naturwissenschaften der Neuzeit. In der Mathematik sehe ich eine organische Weiterentwicklung der Philologie und Logik, welcher wir daher in ihrer zeitlichen Folge, und wegen zeitökonomischen Gründen auch in bezug auf ihre Stundenzahl nur die zweite Stelle im Lehrplan der Schule zuweisen können. — Der Physikunterricht des verstorbenen Herrn H. Veillon am Basler humanistischen Gymnasium, dem eine Anzahl von Baslern die Anregung zum Studium der Physik verdankt, vermittelte uns Schülern durch seine Beschränkung auf einige wenige einfache Versuche und deren theoretische Beschreibung das Wesen des naturwissenschaftlichen Denkens in vorbildlicher Weise.

Kehren wir nunmehr zur Definition des Begriffes „übersichtliche Beschreibung“ zurück. Als Synonym zu diesem Wort empfinde ich den Ausdruck „Theorie“.

Den Begriff der „Beschreibung“ erfüllt sicher jede z. B. chronologisch geordnete Aufzeichnung von Sinnesempfindungen oder Beobachtungen. Für die Gesamtheit aller dieser Aufzeichnungen wollen wir den Begriff der „Erfahrung“ einführen.

Die „übersichtliche Beschreibung oder Theorie“ besteht demgegenüber aus einer Reihe von Sätzen. Sie ist umso übersichtlicher, je weniger Sätze sie enthält. Die Kurzschrift der mathematischen Formel ermöglicht uns Sätze in eine übersichtliche Form zu kleiden, welche, allein mit sprachlicher Ausdrucksweise geschrieben, mehrere Druckseiten füllen würden. Mit Hilfe der logischen Operationen, die wir als Gegebenheiten anerkennen wollen, können wir aus diesen Sätzen weitere Sätze ableiten bis wir zu einer Aussage kommen, deren Objekte mehr oder weniger direkt unsere Sinnesempfindungen sind. Logik und Syntax, welche ja zum Aufgabenkreis der Philologie gehören, erscheinen mir darum als Voraussetzung zur Mathematik, wenn nicht sogar als mit ihr identische Begriffe. Wenn nämlich der Satz anstatt in der Form einer syntaktischen Konstruktion als Formel geschrieben ist, so tritt die mathematische Deduktion an Stelle der logischen Operation.

An eine Theorie können wir eine Frage richten. Z. B. an eine Theorie der Schwere die Frage: „Was geschieht, wenn ich einen Stein in die Luft werfe“. Mit Hilfe der erwähnten Operationen liefert sie mir einen Antwortsatz, den ich kurz die aus der Theorie folgende Antwort nenne. Als Kriterium einer Theorie wollen wir ihre Richtigkeit bezeichnen. Sie ist richtig, wenn die Antwort auf die gestellte Frage mit der Erfahrung übereinstimmt. Sie ist falsch, wenn sich die Antwort nicht mit unserer Beobachtung deckt. Auch den Begriff der Frage müssen wir näher umschreiben. Man darf der Theorie nur sinnvolle Fragen stellen. Den Begriff der sinnvollen Frage umschreiben wir dadurch, dass auf ihre Antwort das Kriterium „*Richtig!*“ oder „*Falsch!*“ anwendbar ist. Eine richtige Frage braucht hingegen nicht unbedingt mit unserer Erfahrung vergleichbar zu sein. Die Frage z. B. wie gross ist der Abstand der Erde von der Sonne am Mittag des 2. Januar 1940 ist sinnvoll, wenn wir eine Operation angeben können, welche uns den Längenbegriff an diesem Datum zu messen gestattet. Im oben angeführten Beispiel ist die Antwort auf die Frage der Bewegung des geworfenen

Steines: „Er beschreibt eine durch Anfangsrichtung und Anfangsgeschwindigkeit bestimmte Parabel" richtig, solange wir den Stein nicht zu weit werfen. Fliegt er nämlich sehr weit, so wird seine Bahn zu einer der Planetenbewegung analogen Ellipse. Eine Schweretheorie, welche sowohl Planetenbewegung wie Wurf umfasst, hat somit einen weiteren Gültigkeitsbereich. Jeder der beiden Theorien ist aber in ihrem Gültigkeitsbereich richtig.

Als weitere Eigenschaft muss eine Beschreibung die innere Widerspruchslosigkeit besitzen. Folgen nämlich durch verschiedene Operationen sich gegenseitig widersprechende Sätze welche Antworten auf sinnvolle Fragen darstellen, so ist die Theorie keine richtige Naturbeschreibung.

1. Die Eigenschaft der Kovarianz einer Theorie.

Als Ursachen unserer Empfindungen führen wir den Begriff des Ereignisses ein. Ein solches findet statt an einem Ort, welchen wir durch die Operation der Messung dreier Zahlen *(Coordinaten)* ermitteln. Solche sind zum Beispiel die Abstände des Ortes vom Fussboden und von zwei zueinander senkrechten Wänden eines rechtwinkligen Zimmers. Die drei Zahlen wollen wir mit x_1, x_2, x_3 bezeichnen. Den Ort desselben Ereignisses können wir aber auch durch die Abstände von drei anderen zu einander senkrechten Ebenen (z. B. denen eines anderen Zimmers) festlegen. Dabei können diese neuen Ebenen (neues Coordinatensystem) beliebig gegen das alte verschoben und verdreht sein. Bezeichnen wir die alten Coordinaten des Ortes in ihrer Gesamtheit durch das Symbol x_i ($i = 1, 2, 3$) und die neuen durch x_i' ($i = 1, 2, 3$), so gibt es eine Rechenoperation der Geometrie, welche uns gestattet die x_i' durch die x_i auszudrücken. Diese Operation nennen wir *Coordinatentransformation.* Wir wollen sie durch das Symbol

$$x_i' = a\,(x_i) \tag{1. 1}$$

darstellen. Zur Festlegung des Ereignisses brauchen wir ferner die Zeit, welche wir durch das Symbol t darstellen.

Dasselbe Ereignis kann aber auch etwa von drei Ebenen eines bewegten Fahrzeuges betrachtet werden. Auch hier haben wir drei Zahlen, welche den Ort beschreiben. Um sie aber aus den x_i zu berechnen, brauchen wir zu jeder Zeit eine andere Operation. Wir drücken das symbolisch durch

$$x_i' = b\,(x_i,\ t) \tag{1. 2}$$

wo b die entsprechende Operation bedeutet.

Eine Theorie ist *covariant in bezug auf die Gruppe einer Coordinatentransformation*, wenn ihre Sätze in allen Systemen, welche durch eine Transformation der betrachteten Mannigfaltigkeit von Operationen auseinander hervorgehen, die gleiche Form haben.

Als Beispiel wollen wir wieder das eingangs erwähnte Gesetz der Schwere betrachten. Die jeweilige Geschwindigkeit des Steines oder besser gesagt des Massenpunktes wird durch drei Geschwindigkeitskomponenten v_1, v_2, v_3 festgelegt, welche wir wieder als Vergrösserung des Abstandes des Punktes vom Fussboden und zwei zueinander senkrechten Wänden beobachten. Dann lautet das Gesetz der Schwere für diesen Massenpunkt: Die Änderung von v_1 pro sec. ist immer gleich gross. Bezeichnen wir sie mit $\dot{v}_1$, so lautet dieser Satz

$$\dot{v}_1 = - g_1$$

g_1 ist eine feste Zahl. Die Änderung pro sec. $\dot{v}_2$ der Geschwindigkeitskomponente v_2 ist Null, oder v_2 ändert sich nicht. Dasselbe gilt für v_3. Beschreiben wir aber das Schweregesetz von einem verdrehten Zimmer aus, dessen Fussboden einen Winkel mit der horizontalen Ebene bildet, so lautet das Gesetz: Die Änderung jedes einzelnen der v_i' per sec. ist gleich drei verschiedenen Konstanten $-g_i'$.

$$\dot{v}_i' = -g_i' \tag{1.3}$$

Aus geometrischen Gründen wissen wir, dass die v_i' und die $\dot{v}_i'$ sich durch dieselbe[1]) Operation (1. 1) aus den v_i resp. $\dot{v}_i$ errechnen lassen wie die x_i' aus den x_i. Damit also (1. 3) formal in jedem Coordinatensystem, welches durch eine Gruppe der Transformationen (1. 1) aus dem ursprünglichen Coordinatensystem hervorgeht, seine Gültigkeit behält, muss sich auch g_i' aus g_i mit Hilfe der Operation a errechnen lassen.

$$g_i' = a\,(g_i) \tag{1.4}$$

Andererseits gibt es Zahlen, wie die Zeit, die Masse des Steines, oder der Abstand zwischen zwei Punkten, welche ihren Wert beim Übergang von einem System zum andern behalten. Wir nennen diese Grössen *Skalare*. Komplexe von Grössen wie (g_1, g_2, g_3) oder (v_1, v_2, v_3), welche in jedem Koordinatensystem verschiedene Werte besitzen, und die durch die Operation (1. 4) auseinander hervorgehen, heissen *Vektoren*. Der Vollständigkeit halber sei noch erwähnt, dass es kompliziertere Gebilde gibt, welche den Namen *Tensoren* tragen und

[1]) bis auf additive, die Parallelverschiebung, darstellende Konstanten.

die aus Zahlengruppen p_{ik}, d. h. Komplexen $p_{11}, p_{12}, \ldots p_{33}$ bestehen. Aus ähnlichen Betrachtungen folgt auch ein Gesetz der Umrechnung der p_{ik} im ersten Zimmer auf die p_{ik}' eines neuen Koordinatensystems. Diese Operation bezeichnen wir symbolisch durch aa, so dass wir die folgenden Umrechnungsgesetze oder Transformationen für Skalare, Vektoren und Tensoren hinschreiben können:

$$\begin{aligned} &\text{Skalar} \ldots\ldots & m' &= m \\ &\text{Vektor} \ldots\ldots & g_i' &= a\,(g_i) \\ &\text{Tensor} \ldots\ldots & p_{ik}' &= aa\,(p_{ik}) \end{aligned} \tag{1. 5}$$

Die Theorie der Schwere (1. 3) lautet in jedem raumfesten System gleich, weil sie Vektoren gleich Vektoren setzt und daher ihre Form und Aussagenmannigfaltigkeit beibehält, wenn auf beiden Seiten der Gleichung dieselbe logische Operation angewendet wird. Wir drücken diese Tatsache aus durch den Satz: *Die Theorie ist kovariant in bezug auf die Transformationsgruppe a.*

Diese Aussage über die Kovarianzeigenschaften einer Theorie gehört mit zu ihrem wichtigsten Inhalt. Eine bestimmte Erfahrung, nämlich dass die Beobachtungen an physikalischen Körpern in jedem Zimmer übereinstimmen, beschreibt die Theorie durch diese Kovarianzeigenschaft.

2. Die Mechanik der physikalischen Körper.

Als Ursache unserer in dieses Kapitel der Physik gehörenden Sinnesempfindungen ordnen wir jedem Teil des Raumes den Begriff der Materie zu. Die Materiequantität, welche wir pro cm^3 antreffen, drücken wir durch eine Zahl, die Materiedichte, aus. Sie ist offenbar ein Skalar. Die Mechanik ist die Beschreibung des Ablaufes der Vorgänge, welche eintreten, wenn ich diese Materie durch einen Stoss, Druck oder Zug beeinflusse. Diese Geschehnisse äussern sich durch eine Verschiebung der Materiepunkte. Ich muss also jedem Materiepunkt, dessen Bewegung ich verfolgen kann, eine Verschiebung, eine Geschwindigkeit und eine Beschleunigung (= Geschwindigkeitsänderung per sec.) zuordnen. Diese letzten drei Grössen haben nach den Betrachtungen des letzten Paragraphen offenbar Vektorcharakter. Wir bezeichnen sie mit s_i, $\dot{s}_i$ und $\ddot{s}_i$ oder auch alle drei zusammen durch s_i. Den Skalar der Dichte wollen wir k nennen. Der Druck und Zug, welcher auf ein einzelnes Materiestück einwirkt und den wir in dem Wort Spannung zusammenfassen, ist ein Tensor p_{ik}.

Nun liefert uns die Punktmechanik, deren Gültigkeit wir am Wurf und an der Bewegung der Himmelskörper geprüft haben, ein erstes System kovarianter Beziehungen, welches wir symbolisch durch

$$M\ (k, s_i, p_{ik}) = 0 \tag{2. 1}$$

darstellen. Dass auch solch eine Beziehung zwischen Skalaren, Vektoren und Tensoren kovariant sein kann, beruht auf gewissen Eigenschaften des Kovarianzbegriffes. Man kann durch bestimmte Operationen aus Skalaren, Vektoren und Tensoren bilden und umgekehrt.

Dieses erste System von Sätzen hat universelle Gültigkeit für jede Art von Materie. Es genügt aber noch nicht zur Beschreibung der Erscheinungen. Wir müssen ein zweites Aggregat von Beziehungen haben, welches einen neuen Skalar, die *Temperatur* T braucht, und welches für jeden Stoff (Luft, Wasser, Eisen, Schwefel etc.) verschieden lautet. Ein solches, für die verschiedenen Materieerscheinungen verschieden lautendes Gesetz, wollen wir als ***phänomenologische Theorie*** bezeichnen und ihre Sätze durch das Symbol

$$MP\ (k, T, s_i, p_{ik}) = 0 \tag{2. 2}$$

darstellen.

Die Kovarianz gegenüber der Gruppe der räumlichen Koordinatentransformationen (1. 5) ist durch die tensorielle Schreibweise gewährleistet. Da auch in einem mit gleichförmiger Geschwindigkeit bewegten Fahrzeug die Beobachtungen der Mechanik denselben Verlauf zeigen, so müssen die Formeln auch gegenüber der Gruppe dieser Transformationen, welche einen speziellen Anteil oder Untergruppe der Operationen b (1. 2) bilden, kovariant sein. Man bezeichnet diese Gruppe als diejenige der ***Galileitransformationen.***

3. Die Elektrodynamik.

Auch die elektrischen Erscheinungen äussern sich in der Beobachtung der Bewegung von Materieteilen, welchen wir hiezu noch eine weitere skalare Eigenschaft, die elektrische Ladungsdichte ϱ und eine vektorielle Eigenschaft, die elektrische Stromdichte J_i zuordnen müssen. Entsprechend den mechanischen Spannungen müssen wir auch gewisse weitere vektorielle Grössen einführen, die wir in vielen Fällen durch einen Skalar Φ und einen Vektor A_i darstellen können. Die mechanische Grundgleichung (2. 1) und die mechanisch phänomenologische Gleichung (2. 2) enthalten dann neben den mechanischen Symbolen noch diese neuen Zeichen, so dass wir als mechanische Beziehung

$$M\ (k,\ s_i,\ p_{ik},\ \varrho,\ J_i,\ \Phi,\ A_i) = 0 \tag{3. 3}$$

und als mechanisch phänomenologische Beziehung

$$MP\ (k,\ T,\ s_i,\ p_{ik},\ \varrho,\ J_i,\ \Phi,\ A_i) = 0 \tag{3. 4}$$

schreiben müssen. Zur vollständigen Beschreibung brauchen wir noch zwei weitere Satzsysteme, deren erstes wieder, wie die mechanische Beziehung, universelle Gültigkeit besitzt, und „Maxwellsche Gleichungen" heisst. Ein zweites Aggregat hat wieder den phänomenologischen Charakter. Wir stellen diese beiden Systeme durch

$$E\ (\varrho,\ J_i,\ \Phi,\ A_i) = 0 \tag{3. 5}$$

und

$$EP\ (k,\ T,\ s_i,\ p_{ik},\ \varrho,\ J_i,\ \Phi,\ A_i) = 0 \tag{3. 6}$$

dar.[1]) Das Licht und die Hertzschen Wellen lassen sich als Ausbreitung von Wellen A_i und Φ im Kontinuum des Raumes beschreiben, wie der Schall durch s_i und ϱ in der Materie darstellbar ist.

Das rein mechanische System und das rein elektrodynamische System von Sätzen haben nun eine gewisse Analogie. In beiden sind die auftretenden Grössen Raum und Zeitfunktionen, d. h. sie haben an jedem Ort zu jeder Zeit bestimmte Werte. Wir nennen solche Grössen *Felder*. Die rein mechanischen Grössen, sowie auch die elektrischen Eigenschaften der Materie ϱ, und J_i haben nur an den Orten von Null verschiedene Werte, wo sich Materie befindet. Den e. m.[2]) Grössen hingegen müssen wir auch in den materielosen Gebieten von Null verschiedene Werte zusprechen. Auch bestehen gewisse Analogien der mathematischen Form: Tensorcharakter, Differentialgleichung, Nahwirkung u. a. m. Es entsprang daher einem Bedürfnis der Vereinfachung unserer Beschreibung, die beiden Phänomene Materie und e. m. Feld unter einem Gesichtspunkt zu vereinigen. So entstanden im letzten Jahrhundert Äthertheorien.

Wie wir dem als Materie in Erscheinung tretenden Kontinuum oder Feld den Namen Materie geben, so gab man dem

[1]) In Wirklichkeit gibt eine Beschreibung des e. m. Feldes durch A_i und Φ nicht den Zerfall in ein allgemeines ((3. 3) und (3. 5)) und in ein phänomenologisches (3. 4) und (3. 6) System von Beziehungen. Nur die Beschreibung des e. m. Feldes durch vier Vektoren E_i, B_i, D_i und H_i erlaubt diese Trennung. Die Formeln ((3. 5) und (3. 6)) haben also lediglich symbolische Bedeutung.

[2]) e. m. steht im folgenden als Abkürzung für „elektromagnetisch".

Kontinuum des materiefreien Raumes, dessen Äusserungen das Feld Φ und A_i ist, den Namen Äther. Eine Reihe von Beschreibungsversuchen setzte es sich zur Aufgabe, die Äthereigenschaften aus den Materieeigenschaften zu erklären. Wir dürfen diese Theorien heute als aussichtslos bezeichnen, hauptsächlich da die Materieeigenschaften sicher nicht vollständig durch die Beziehungen *M* und *MP* gegeben sind. Andere Bestrebungen, welche uns auch heute noch beschäftigen, wollen die Materieeigenschaften aus den Äthereigenschaften erklären. Auf diese Ideen werden wir zurückkommen.

Betrachten wir die Kovarianz der Gleichungen, so sind, in Übereinstimmung mit der Erfahrung, sowohl die mechanischen wie die e. m. Sätze kovariant in bezug auf die räumlichen Koordinatentransformationen. Sieht man aber die Erde als ein Fahrzeug an, das bei Betrachtung kleiner Raumbereiche und kurzer Zeitabschnitte (z. B. ein Laboratoriumsraum während einiger Sekunden) sicher eine gleichförmige Bewegung besitzt, die aber andererseits um Mitternacht gewiss eine andere Richtung hat als am Mittag, so zeigen uns die elektrischen wie die mechanischen Messungen, dass diese Bewegung nicht nachweisbar ist. Es folgt daraus, dass in gleichmässig bewegten Systemen der Ablauf der Naturgesetze unabhängig vom Bewegungszustand ist. Die Physik im gleichmässig bewegten Fahrzeug unterscheidet sich nicht von der im ruhenden Zimmer. Die Theorie muss daher kovariant gegenüber den Galileitransformationen sein. Während die rein mechanischen Sätze diese Eigenschaft haben, fehlt sie der e. m. Beschreibung vollständig. Albert Einstein schlug daher vor, in Weiterentwicklung von Ideen von H. A. Lorentz, den Fehler der Beschreibung in der Form der Galileischen Transformation anstatt in der Form der e. m. Theorie zu suchen. Eine etwas anders geartete Umrechungsvorschrift, die *Lorentztransformation* ergibt Kovarianz der e. m. Sätze. Die Kovarianz der mechanischen Beschreibung lässt sich durch kleine Abänderungen ihrer Grundgesetze erreichen. Die Unterschiede zwischen den Antworten der galileikovarianten und der lorentzkovarianten Mechanik sind so gering, dass sie sich meist unseren Beobachtungen entziehen. In den wenigen Fällen, wo sie eine Rolle spielen, entscheidet die Beobachtung tatsächlich für die Richtigkeit der lorentzkovarianten Beschreibung.

Wir erwähnen, dass diese spezielle Relativitätstheorie den Vektor x_i mit dem Skalar t, den Vektor A_i mit dem Skalar Φ und alle andern Vektoren mit Skalaren zu Vierervektoren (x_1, x_2, x_3, $x_0 = ct$), (A_1, A_2, A_3, $A_0 = \Phi$) zusammenfasst, welche dann ihrerseits beim Übergang von einem System zu einem gleichförmig bewegten System einer Gruppe von Transformationen unterworfen sind, die eine zu (1. 5) sehr analoge Form hat. Auch in diesem vierdimensionalen Raum-Zeit-Kontinuum kann man dann wieder Skalare, Vektoren und Tensoren definieren.

4. Die Atomistik.

Als philosophische These, ohne jeden Versuch physikalische Erscheinungen zu beschreiben, wurde die Theorie des Aufbaues der Materie aus kleinsten, unteilbaren Teilchen, schon im Altertum von Demokritos ausgesprochen. Das Gesetz der konstanten Proportionen der chemischen Verbindungen konnte durch Dalton (1808) vermittels dieser Atomtheorie beschrieben werden. Die Atome Daltons sind die kleinsten Bestandteile der chemischen Elemente. Denken wir uns bestimmte Gruppierungen aus verschiedenen Atomen, welche irgendwie zusammengehalten werden, so haben wir ein *Molekül* vor uns. Eine homogene Materie, welche nicht chemisches Element ist, denkt sich dann Dalton aus vielen gleichartigen Molekülen aufgebaut in gleicher Weise, wie eine Elementsubstanz aus vielen gleichen Atomen. Aus dieser Beschreibung folgt das Gesetz der konstanten Proportionen.

Die atomistische Beschreibung wurde von Clausius, Maxwell und Bolzmann im letzten Jahrhundert dazu verwendet, um die phänomenologischen Beziehungen der Mechanik (2. 2) für gewisse Erscheinungsformen der Materie aus den allgemeinen Prinzipien der Punktmechanik abzuleiten unter Hinzuziehung der Wahrscheinlichkeitsrechnung und der Statistik. Sie ist unter dem Namen *kinetische Gastheorie* eine der fruchtbarsten Beschreibungsmethoden der physikalischen Körper geworden. Im Moment ihrer Aufstellung war ihr Zweck lediglich der Ersatz der vielen phänomenologischen Beziehungen durch ein allgemein gültiges System von Sätzen. Die phänomenologischen Grössen, wie Dichte, Temperatur, Volumen, Materieverschiebungen und Spannungen etc. erklärten sich durch statistische Mittelbildungen über mechanische Grössen der als Massenpunkte oder starre Kugeln beschriebenen Moleküle. Sobald aber die Physik die Möglichkeit hatte, die von gasförmiger Materie verursachte Bewegung kleiner fester Körper (Stäubchen) zu beobachten, so lieferte die Atomistik die richtige Beschreibung dieser neuen Erscheinung, während die phänomenologischen

Beziehungen zu falschen Resultaten führten. (Brown'sche Bewegung.)

Von H. A. LORENTZ wurde das analoge Programm der Zurückführung der phänomenologischen Beziehungen der Elektrodynamik auf die Bewegung von Elektrizitätsatomen durchgeführt. Auch hier zeigte sich bald der Vorteil der atomaren Beschreibung, da Erscheinungen wie die Kathodenstrahlen und Kanalstrahlen Wirkungen erzeugten, welche wir unbedingt den einzelnen Elektrizitätsatomen zuschreiben müssen, wie wir die Bewegung von Stäubchen in Gasen als Wirkung der Materieatome beschreiben.

Beobachtungen und Beschreibungsversuche von RUTHERFORD, BOHR und SOMMERFELD klärten die Beziehung zwischen den Materieatomen und den Elektrizitätsatomen: Wir müssen uns nämlich die Atome der Materie als aus Elektrizitätsatomen zusammengesetzt denken. Ein Materieatom besteht aus einem positiv elektrisch geladenen *Kern*, der praktisch die gesamte Masse des Atoms enthält und dessen Dimensionen rund 100 000 mal kleiner sind als die Atomdimensionen. Die *Elektronen*, wie wir die negativen Elektrizitätsatome heissen, haben ebenfalls einen so kleinen Durchmesser und eine rund einige tausende mal kleinere Masse. Durch ihre Bewegungen erfüllen sie den Raum des Atoms in ähnlicher Weise wie die Gasmoleküle den Raum eines Gefässes. Wenn hingegen die kinetische Gastheorie die phänomenologischen Beziehungen der Mechanik der Gase auf Grund der Punktmechanik ohne Hinzuziehung weiterer Hypothesen ableiten kann, so scheiterten die analogen Versuche von BOHR und SOMMERFELD, aus einer kinetischen Gastheorie der Elektronen das Verhalten der Atome in ihren gegenseitigen Wechselwirkungen und ihren Einwirkungen auf das e. m. Feld zu beschreiben. Die Grundlagen zu einer solchen Beschreibung mussten ja, mangels anderer Kenntnisse, der Äthermechanik der Maxwellschen Gleichungen und den Sätzen der Punktmechanik entnommen werden. Nur einer mit Widersprüchen behafteten *Quantentheorie* gelang der Versuch einigermassen.

Nachdem die Kontinuumsphysik der Materie durch das Atom und dieses durch die Theorie der positiven Atomkerne und des Elektrons ersetzt war, zeigte die Beobachtung der Temperaturstrahlung und des lichtelektrischen Effektes Erscheinungen, welche sich nicht aus der Kontinuumstheorie des elektrischen Feldes erklären liessen. Beschreiben wir hingegen die e. m. Strahlung und somit auch die Lichtstrahlung als die geradlinigen Bahnen von Lichtatomen, so können diese neuen Experimente richtig wiedergegeben werden. Nun steht die Kontinuumsbeschreibung der Materie, welche zu den Schallwellen der Akustik führt, in keinem Widerspruch zum atomaren Aufbau der Körper. Man könnte also denken, dass eine kinetische Theorie der Lichtatome (*Photonen* genannt) die Maxwellschen Kontinuumssätze ergibt. Dass diese Analogie nicht möglich ist, zeigt folgende Uebcrlegnng:

Bei der Ausbreitung der Schallwellen führen die Materieatome kleine Schwingungen um eine raumfeste Gleichgewichtslage aus. Diesen Schwingungen entsprechen in einer makroskopischen Kontinuumsbeschreibung die fortschreitenden Schallwellenzüge der Materie. Die Theorie der Temperaturstrahlung und des lichtelektrischen Effektes, welche auf Max Planck und Albert Einstein zurückgeht, beschreibt hingegen die Lichtausbreitung durch die geradlinige Bewegung von Lichtatomen. Die Erklärung des Lichtes durch solche Photonen bedeutet also einen vollständigen Verzicht auf die Kontinuumstheorie der Lichtwellen der Maxwellschen Gleichungen und steht somit im Widerspruch mit denjenigen Erfahrungstatsachen, welche uns die Wellennatur von Licht und Schall vor Augen führen. Diese Beugungsexperimente gehören aber mit zu den am sichersten fundierten Naturbeobachtungen.

In diesem Entwicklungsstadium unserer Naturbeschreibung müssen wir also eine duale Theorie aufstellen: Gewisse Eigenschaften des e. m. Feldes, z. B. seine Ausbreitung, verlangen die Wellentheorie oder Kontinuumsbeschreibung .Andere Beobachtungen, wie seine Entstehung und Vernichtung in Berührung mit der Materie, zwingen uns zu einer atomistischen Formulierung.

Im nächsten Paragraphen werden wir zeigen, dass auch für die Materie eine solche duale Beschreibung nötig ist, um gewisse Erscheinungen richtig wiederzugeben.

5. Die Wellenmechanik.

Die am Ende des vorhergehenden Paragraphen angedeutete Dualität Atom-Kontinuum, sowie die Schwierigkeiten der kinetischen Theorie des Elektronengases im Atom, veranlassten de Broglie und Schroedinger, nach verbesserten Theorien des Atombaues zu suchen. Schroedinger konnte viele Eigenschaften des Wasserstoffatoms, insbesondere die scharf definierten Wellenlängen des von ihm ausgesandten Lichtes durch eine Wellentheorie des Elektrons erklären. Es sei hier bemerkt, dass der Basler BALMER, der in Gedankenaustausch mit dem damaligen Leiter der Physikalischen Anstalt der Universität Basel, EDUARD HAGENBACH, stand, als erster die zahlentheoretische Fassung der beobachteten Lichtwellenlängen des Wasserstoffes entdeckte. Schrödingers Erklärung dieser Zahlen hatte gegenüber der Bohr-Sommerfeldschen Beschreibung den Vorteil geringerer innerer Widersprüche.

Aus der Kontinuumsvorstellung von DE BROGLIE und SCHROEDINGER folgte konsequenterweise die Beobachtungsmöglichkeit von Beugungserscheinungen der Elektronenstrahlen, deren atomare Natur zu dieser Zeit über jeden Zweifel erhaben schien. Die Kathodenstrahlen mussten also demgemäss neben ihren korpuskularen Eigenschaften auch die für die Wellennatur charakteristischen Beugungserscheinungen zeigen. Die Aufstellung dieser Theorie fällt in die Jahre 1925/26. Ende 1926 entdeckten DAVISSON und GERMER tatsächlich diese Beugungserscheinungen, welche qualitativ genau denjenigen entsprechen, welche kurzwellige e. m. Strahlung (Röntgenstrahlen) beim Durchgang durch Materie zeigt.

Wir haben hier eines der unzähligen Beispiele für den heuristischen Wert der Methoden der physikalischen Naturbeschreibung vor uns: Die zur Erklärung des Atombaues aufgestellte Wellenmechanik liefert uns Antworten, deren Fragen wir noch gar nicht gestellt haben. Sie führt uns somit zu neuen Fragestellungen, deren Antworten, auf ihre Richtigkeit geprüft, uns ein weiteres Kriterium der Gültigkeit der Theorie geben und uns gleichzeitig neue Wege zur Naturbeobachtung liefern.

Die Schrödingersche Theorie präzisierte auch durch gewisse Vorschriften die Abgrenzung von Kontinuumsbeschreibung und atomarer Beschreibung. Jedoch bleibt die Dualität der Theorien auch hier noch ein Element, das leicht zu Widersprüchen führt.

6. Die Quantenmechanik.

Noch vor der Aufstellung der Wellenmechanik hatte W. HEISENBERG eine Beschreibung der atomaren Phänomene gezeigt, deren Resultate in weiten Grenzen denjenigen der Schrödingerschen Theorie entspricht. SCHROEDINGER wies daraufhin auch die Identität beider Methoden nach bei ihrer Anwendung auf bestimmte Fragestellungen. Die Quantenmechanik Heisenbergs gestattete aber eine widerspruchslose Beschreibung der dualen Eigenschaften: Wellen—Korpuskeln. Die atomare

Seite und die Kontinuumsform der physikalischen Erscheinungen folgen nämlich beide aus der von HEISENBERG, PAULI, JORDAN und WIGNER entwickelten Weiterbildung der quantenmechanischen Methode, welche den Namen „*Quantentheorie der Wellenfelder*" oder „*zweite Quantisierung*" trägt. Diese Theorie enthält also **keine wahre Dualität** mehr. Auf eine nähere Erläuterung muss hier verzichtet werden. Es sei lediglich bemerkt, dass sie die elektrischen Feldgrössen A_i und die Feldgrösse der Schrödingerschen Theorie der Elektronen ψ nicht mehr als Zahlen, sondern als Operatoren der Theorie der Funktionalen, eines leider noch wenig bekannten Gebietes der mathematischen Analysis, betrachtet. Ich erwähne nur, dass das Produkt zweier Feldgrössen ψ und φ zum Beispiel nicht immer die komutative Eigenschaft der gewöhnlichen Zahlen besitzt. Es ist also $\psi\varphi$ ein von $\varphi\psi$ verschiedener Begriff. Die Eigenschaft der gewöhnlichen Zahlen, dass 3 mal 4 gleich 4 mal 3 gleich 12 ist, fehlt also den physikalischen Zahlen φ und ψ.

7. Die lorentzkovariante Form der Quantenmechanik.

Da die mechanischen und e. m. Erscheinungen auch in ihrer atomaren Form unabhängig vom Bewegungszustand des Koordinatensystemes sind, solange wir gleichmässig bewegte Systeme wählen, müssen wir auch die Kovarianz der Schrödingerschen Wellenmechanik gegenüber Lorentztransformationen verlangen. Die Beziehung

$$S(\psi, \Phi, A_i) = 0 \qquad (7\,1)$$

welche die Schrödingersche Beziehung aus Paragraph 5 versinnbildlichen soll und die elektromagnetischen Formeln, welche jetzt in der Form

$$E(\psi, \Phi, A_i) = 0 \qquad (7\,2).$$

geschrieben werden müssen, da die der Materie zukommenden Eigenschaften der elektrischen Ladungs- und Stromdichte durch das ψ-Feld ausgedrückt sind, haben in ihrer ursprünglichen Schrödingerschen Form nicht die lorentzkovariante Form. Es gelang hingegen SCHROEDINGER und GORDON leicht, eine kovariante Form zu finden. Die kovariante Form hatte aber die damals als Nachteil empfundene Eigenschaft, ausser den Elektronenwellen negativer Ladung solche positiver elektrischer Ladung zu ergeben. Das 1933 entdeckte positive Elektron, für dessen Entdeckung C. D. ANDERSON einen Teil des diesjährigen Nobelpreises für Physik erhielt, war also schon 1927 von SCHROEDINGER und GORDON gewissermassen vorausgesagt. Damals wurden aber die umgekehrt geladenen Wellen als Widerspruch mit der Erfahrung empfunden. Ein weiterer Mangel der Beschreibung war das Fehlen eines Eigendrehmomentes des Elektrons und einer damit verbundenen magnetischen Doppelladung. P. A. M. DIRAC stellte aus diesem Grunde eine ebenfalls kovariante Wellengleichung auf, welche Drehmoment und magnetische Doppelladung richtig beschrieb. Eine genauere Untersuchung seiner Beziehungen zeigte ihm aber, dass auch er die Existenz umgekehrt geladener Wellen verlangen musste, welche er schon 1931 als *positive Elektronen* erklärte. Die Eigenschaften des später von ANDERSON entdeckten Teilchens und die Vorgänge, welche zu seinem Entstehen und Verschwinden führen, lassen sich auch tatsächlich quantitativ aus der Dirac'schen Theorie errechnen.

Wir bemerken noch, dass das Diracsche ψ-Feld vier Komponenten besitzt, die wir mit ψ_α $(\alpha = 1, 2, 3, 4)$ bezeichnen. Sie transformieren sich aber bei Lorentztransformationen anders als die damals eingeführten Vierervektoren A_i $(i = 0, 1, 2, 3)$. Ihre Transformationseigenschaft bezeichnen wir durch das Wort *Spinoren*. Wir wollen uns hier den Satz merken, dass zwar durch Multiplikation zweier Spinoren Skalare, Vierervektoren und Vierertensoren entstehen können, dass es aber keine Möglichkeit gibt, aus der durch die Worte Skalare, Vektoren und Tensoren definierten Mannigfaltigkeit von Feldgrössen Spinoren zu erzeugen. Diese Tatsache wird uns später eine heuristische Seite der Theorie zeigen.

Die hiermit zu einem vorläufigen Abschluss gebrachte Beschreibung gestattet uns im Prinzip alle Erscheinungen der Physik und Chemie vorherzusagen. Ausser der sogenannten makroskopischen Physik, deren Beschreibung durch die Zurückführungsmöglichkeit der phänomenologischen Beziehungen der Mechanik und Elektrodynamik auf die Quantenmechanik gewährleistet ist, erwähne ich hier im speziellen die Beschreibung der Lichtemission der Elemente (Atome) und der chemischen Verbindungen, zu deren experimenteller Kenntnis u. a. die spektroskopischen Arbeiten von Herrn A. HAGENBACH und im letzten Jahrzehnt auch diejenigen seiner Mitarbeiter, der Herren M. WEHLI, E. MIESCHER und K. WIELAND in Basel sehr vieles beigetragen haben.

Die technische Schwierigkeit der Durchführung der logischen Operationen, welche die Sätze der Theorie mit der beobachteten Materieeigenschaften in Beziehung setzen, gestattet uns allerdings, nur in sehr einfachen Fällen quantitative Resultate vorauszusagen. Da aber diese, wie auch die in komplizierteren Fällen erhaltenen qualitativen Phänomene mit der Erfahrung übereinstimmen, dürfen wir sagen, dass man heute für die Chemie und für diesen Teil der Physik über eine adäquate Beschreibung verfügt.

Der Vollständigkeit halber soll hier doch ein gewisser Widerspruch erwähnt werden, der leider trotz allen Versuchen noch nicht behoben werden konnte: Für die einerseits mit ihrem gemessenen Zahlenwert eingesetzte Masse des Elektrons folgt andererseits durch die Anwendung gewisser logischer Operationen auf die e. m. Gesetze ein unendlich grosser Betrag. Jedoch stört dieser Widerspruch für viele Phänomene der Materie und Lichtausbreitung nicht. Ihn zu beheben ist aber eine der Hauptaufgaben der heutigen theoretischen Physik.

8. Der Aufbau der Atomkerne.

Da die Masse der Atome im wesentlichen in ihrem Kerne konzentriert ist und die Massen der verschiedenen Atomkerne annähernd ganzzahlige Vielfache derjenigen des Wasserstoffatomkerns oder Protons sind, so wurde schon früh die Hypothese geäussert, dass die Kerne selbst nicht nur punktförmige Gebilde bedeuten. Die α-Radioaktivität zeigte auch direkt, dass ein Atomkern in zwei zerfallen kann. Die Gesetze des α-Zerfalls, bei welchem sich ein Radiumatom zum Beispiel in ein Atom des Edelgases Radiumemanation und in ein Atom des Edelgases Helium verwandelt, lassen sich auch leicht qualitativ aus den Methoden der quantenmechanischen Beschreibung ableiten, da die Wellenkorpuskelbeschreibung auch für Protonen und beliebige Atomkerne in dieser Annäherung gilt.

Die künstliche Umwandlung der chemischen Elemente (Alchimie), welche Lord Rutherford 1921 gelang, verlangte dringend nach einer vollständigen quantitativen Beschreibung des Atomkernes. Bei solchen ultrachemischen Reaktionen entdeckte Chadwick ein ungeladenes Teilchen von der Masse des Protons, das sich aber schon durch seine 100 000mal kleinere Dimension vom Wasserstoffatom unterscheidet. Auch für diesen neuen atomaren Bestandteil, *Neutron* genannt, gilt die Wellen-Korpuskelbeschreibung. Nimmt man nun zwischen den verschiedenen Protonen und Neutronen, deren Gesamtmasse der Kernmasse entspricht und deren, von den Protonen allein herrührende, Gesamtladung der positiven Ladung des

Kernes gleich ist, passend gewählte Wechselwirkungen (in Analogie zu den e. m. Wechselwirkungen zwischen Kern und Elektronen) an, so lässt sich nach HEISENBERG und seinen Schülern MAJORANA und v. WEIZSAECKER der Kernaufbau nach quantenmechanischen Gesetzen erklären. Während aber die für den Atombau verantwortlichen Wechselwirkungen zwischen den Elektronen und dem Kern durch die bekannten Sätze über das e. m. Feld beschrieben werden, müssen wir zur Erklärung der Kräfte zwischen Neutron und Proton ein uns unbekanntes neues Feld heranziehen. Hat dieses Feld nicht vielleicht auch andere Äusserungen?

Darauf gibt uns eventuell eine andere Erscheinung der Radioaktivität Auskunft. Es gibt nämlich auch Atomkerne, welche sich unter Aussendung eines negativen oder positiven Elektrons in einen andern Kern verwandeln. Dieser Zerfall lässt sich nun durch keine unserer vielen Beschreibungsmethoden erfassen, da er einem fundamentalen Satz unserer bisherigen Theorie widerspricht: Der Satz von der Erhaltung der Energie und des Schwerpunktes, welcher allen Kontinuums- und Atomtheorien als gemeinsames Element angehört, scheint hier seine Gültigkeit zu verlieren. Für eine Beschreibung, welcher dieser Satz fremd ist, fehlen uns aber jegliche Anhaltspunkte. Verlangen wir aber, dass gleichzeitig mit der Elektronenstrahlung eine andere, bisher unserer Beobachtung entgangene Strahlung auftritt, so können wir deren Ausbreitung so bestimmen, dass Schwerpunkt und Energie erhalten bleiben. PAULI und FERMI bezeichnen diese Strahlung als Neutrinostrahlung, das in der Atomdarstellung ihr entsprechende Teilchen als *Neutrino*. Die Beschreibung dieses Zerfalles (β-Zerfalls) lautet also in der atomaren Theorie: „Ein Neutron verwandelt sich in ein Proton; gleichzeitig entsteht ein negatives Elektron und ein Neutrino. Die konstante Zerfallsenergie verteilt sich nach einem statistischen Gesetz auf die beiden aus dem Kern herausgeschleuderten Partikel Neutrino und Elektron." Während also beim α-Zerfall der entstehende Heliumatomkern die ganze konstante Zerfallsenergie in seiner lebendigen Kraft enthält, erhält das Elektron als einziges beobachtetes Zerfallsprodukt des β-Zerfalles bei jedem einzelnen Zerfall eine andere Energie. So lautet auch die Beobachtung: Alle He-Atome, welche das Ra ausschleudert, haben dieselbe Geschwindigkeit, während die Elektronen, welche aus einem β-Strahler fliegen, eine kontinuierliche Energieverteilung zeigen.

Aber auch aus Kovarianzgründen ist das Neutrinofeld notwendig: Proton, Neutron und Elektron werden durch Spinorfelder u_α, v_α und ψ_α beschrieben. Die Wechselwirkung muss aber aus Kovarianzgründen Tensorcharakter haben. Tensoren können aber nur aus einer geraden Zahl von Spinoren aufgebaut werden. Zweimalige Verwendung eines der drei Spinoren u_α, v_α oder ψ_α würde aber Prozessen entsprechen, bei welchen zwei beobachtbare Teilchen (Proton, Neutron oder Elektron) ausgesandt würden. Da nur das Elektron beobachtet ist, müssen wir einen weiteren Spinor φ_α einführen, welcher einem unbeobachteten Feld (Neutrinofeld) entspricht.

Diese von Fermi aufgestellte Theorie des β-Zerfalls ergibt nun auch ein Feld, welches eine Wechselwirkung zwischen den einzelnen Protonen und Neutronen der Atomkerne liefert und somit die Rolle des e. m. Feldes im Atombau für den Kernbau übernehmen kann. Die Coulombsche Kraft der elektrischen Anziehung erklärt die Atomistik dadurch, dass das eine Ladungszentrum (zum Beispiel der Atomkern) ein Photon aussendet, welches das andere Ladungszentrum (zum Beispiel das Elektron) absorbiert. Die Kernkräfte erklären sich dann durch die Aussendung eines Paares „Elektron-Neutrino" durch das Neutron, welches Paar wiederum von einem Proton absorbiert wird.

9. Die Objekte der physikalischen Naturbeschreibung.

Die Objekte von Aussagen über unsere Sinnesempfindungen entsprechen diesen Empfindungen selbst. Als Objekte einer Theorie hingegen bezeichnen wir diejenigen Begriffe, welche in ihren Sätzen auftreten. Die Objekte der Kontinuumstheorien waren bis zum Beginn von Einsteins Relativitätstheorie die Skalare k, T, ϱ, Φ, die Vektoren s_i, J_i, A_i und Tensoren p_{ik} im dreidimensionalen Raume. Die von Einstein und Lorentz entdeckte Kovarianzeigenschaft änderte die damals vorhandenen Kontinuumstheorien dahin ab, dass sie an Stelle der Begriffe der dreidimensionalen euklidischen Geometrie solche einer *vierdimensionalen pseudoeuklidischen* Geometrie stellte! Skalare, Vierervektoren und Vierertensoren. Eine konsequente Weiterentwicklung dieser speziellen Relativitätstheorie verdanken wir ebenfalls Einstein in der Form seiner allgemeinen Relativitätstheorie: sie beschreibt die Schwerkraft durch ein dem elektrischen Vektorenfelde A_i gewissermassen analoges Tensorfeld g_{ik} im vierdimensionalen

Raume. Das Auftreten des Tensors g_{ik} bewirkt eine Abweichung der Geometrie von ihrer pseudoeuklidischen Form. Aus dieser Veränderung der Geometrie errechnen sich dann die Bahnen von Planeten und von geworfenen Körpern.

Die vier Felder der Elektronenbewegung, Neutrinostrahlung, Protonen- und Neutronenbewegung werden durch vier je vierkomponentige Spinorfelder beschrieben, so dass die gesamten Objekte der Kontinuumstheorie durch die 6 Symbole

$$\psi_\alpha,\ \varphi_\alpha,\ u_\alpha,\ v_\alpha;\ A_i,\ g_{ik}\ ^{1})$$

darstellbar sind.

Die duale Beschreibung führt entsprechend 6 Elementarteilchen ein: Elektron, Neutrino, Proton, Neutron; Photon (Lichtquant), Graviton (Schwerequant).

Die für unsere Theorie so wichtigen positiven Elektronen haben wir in der Aufzählung der Elementarpartikel gar nicht erwähnt. Der Objektbegriff „Partikel der Materie" bedingt nämlich in der Quantenmechanik ein Spinorfeld, welches wiederum die Existenz eines Objektes „*Antipartikel*" verlangt, dessen Eigenschaften beinahe vollständig mit denjenigen der „*Partikel*" übereinstimmen. Das Antipartikel der geladenen Teilchen, Elektron und Proton, unterscheidet sich von dem eigentlichen Teilchen nur durch seine umgekehrte elektrische Ladung. Den ungeladenen Partikeln Neutrino und Neutron kann man auch eine Grösse, welche Ladungscharakter hat, zuschreiben. Ihre Parallele mit der elektrischen Ladung besteht darin, dass auch sie bei allen Materieumwandlungen erhalten bleibt. Wir nennen sie *Neutrinoladung*. Das Antiteilchen der ungeladenen Partikel hat dann umgekehrte Neutrinoladung. Beim Elektron haben wir die Antipartikel in der Form des entdeckten positiven Elektrons vor uns. Wir können aber geradesogut das positive Teilchen als Partikel ansehen und das negative Elektron als Antipartikel. Über das Neutrino können wir experimentell noch nichts aussagen, da weder die Partikel noch ihre Antipartikel direkt beobachtet wurden. Die Einwirkungen der Neutrinostrahlung auf die Materie lassen sich jedoch abschätzen. Ihre Effekte sind aber so schwach, dass es erst einer weiteren Verfeinerung unserer physikalischen Messtechnik bedarf, um sie aufzufinden. Beim Proton und Neutron wurden die Antipartikel nie beobachtet. Das ist aber nicht weiter verwunderlich, weil

¹) Wegen der Fussnote ¹) auf Seite 203 wahrscheinlich eher

$$\psi_\alpha,\ \varphi_\alpha,\ u_\alpha,\ v_\alpha;\ F_{ik},\ g_{ik}$$

zu ihrer Erzeugung unverhältnismässig viel energiereichere Strahlung gebraucht wird, als wir sie auf der Erde erzeugen können. Es ist aber nicht ausgeschlossen, dass negative Protonen einen Teil der kosmischen Strahlung darstellen.

Neben dieser Einteilung in geladene und ungeladene Partikel können wir auch eine solche in schwere und leichte vornehmen. Die Masse des Neutrinos folgt nämlich aus der Theorie des β-Zerfalls: Sie ist exakt Null, oder doch viel kleiner als die des Elektrons. Neutrino und Elektron sind also leichte, Neutron und Proton schwere Elementarteilchen. Diese Eigenschaft können wir analog der elektrischen und neutralen Ladung durch die *leichte* und *schwere Ladung* beschreiben, da auch diese beiden Grössen dem Erhaltungssatz genügen. Elektron und Neutrino haben also leichte Ladung = 1, Proton und Neutron schwere Ladung = 1. Den Antiteilchen ist auch hier die Ladung mit umgekehrtem Vorzeichen zuzusprechen.

Die Massen und elektrischen Ladungen der Teilchen des Vektor-Tensorfeldes sind Null: Wenn wir alle vier definierten Ladungen dieser Teilchen als Null annehmen, so genügen alle beobachteten Umwandlungsprozesse den Erhaltungssätzen der vier Ladungen (siehe Tabelle). Bei Lichtemission und Gravitationswechselwirkungen, welche die Korpuskulartheorie durch Schaffung und Vernichtung der Elementarteilchen Photon und Graviton beschreibt, sieht man die Gültigkeit ohne weiteres. Bei der Erzeugung eines Paares von umgekehrt geladenen Elektronen bei gleichzeitigem Verschwinden eines oder mehrerer Photonen bleibt die neutrale und die schwere Ladung Null, während die elektrische und die leichte Ladung des negativen Elektrons (Antiteilchens) —1 und gleichzeitig diejenigen des positiven Elektrons +1 entstehen. Beim β-Zerfall entsteht aus einem Neutron, dessen elektrische, neutrale, leichte und schwere Ladungen durch das Zahlenquadrupel (0, 1, 0, 1) gegeben ist, ein Proton mit den entsprechenden Zahlen (1, 0, 0, 1), ein negatives Elektron (—1, 0, —1, 0) und ein Neutrino (0, 1, 1, 0). Man sieht, dass auch hier die Summe jeder der entstehenden Ladungen gleich der anfänglich vorhandenen ist.

Eine weitere interessante Grösse ist die Ausbreitung von Störungen (Gruppengeschwindigkeit) der einzelnen Felder. Sie entspricht der Geschwindigkeit der Partikel in der atomistischen Beschreibung und ist ebenfalls in der Tabelle vermerkt.

Auch das Eigendrehmoment *(Spin)* folgt aus der Theorie in Übereinstimmung mit der Erfahrung. Spinorfeld-partikel haben den Spin ½ und Vektor-Tensorfeldteilchen die ganzzahligen Beträge 0, 1, 2, etc. Die Photonen haben insbesondere nur 0 und 1. (Diese Beträge sind in Einheiten der Planckschen Konstanten $h/2\pi$ gemessen.)

Die Spinorteilchen und die Vektortensorteilchen gehorchen auch zwei verschiedenen statistischen Gesetzen, welche den wesentlichen Inhalt der Quantenmechanik der Wellenfelder ausmachen. Spinoren entspricht die *Fermi-Dirac'sche* und Vektortensorfeldern die *Bose-Einsteinsche Statistik.*

Alle diese Eigenschaften sind in der Tabelle, Seite 202, übersichtlich zusammengestellt.

Aus einer Anzahl von universellen Beziehungen (zum Beispiel von der Form (7. 1) und (7. 2)), welche zwischen den sechs Feldern u_α, v_α, ψ_α, φ_α, A_i und g_{ik} herrschen, sollen sich dann die gesamten Physikalischen Gesetze ableiten.

Betrachten wir Proton, Neutron, Elektron und Neutrino als die Bausteine der Materie im engeren Sinne, so dürfen wir die Gesamtheit der Spinorfelder als das Materiefeld oder kurz als Materie bezeichnen. Da wir den naiven Ätherbegriff der Äthertheorien des letzten Jahrhunderts endgültig als erledigt betrachten dürfen, so steht es uns frei, für die Gesamtheit des Vektor-Tensorfeldes der A_i und g_{ik} das Wort Ätherfeld oder kurz Äther in diesem abstrakteren Sinne neu zu definieren.

Wenn wir so den naiven Äthertheorien das Todesurteil gesprochen haben, so ist ihr Leitgedanke dennoch nicht verschwunden. Die naiven Beschreibungsversuche wollten die e. m. Beziehungen auf die mechanischen Kontinuumsbeschreibungen M und MP des Paragraphen 2 zurückführen. Sie wollten also die A_i oder von ihnen abgeleitete Grössen als mechanische Verschiebungen s_i der Aethersubstanz darstellen. Das Ziel, zwei verschiedene Erscheinungen als Äusserungen ein und desselben Objektes verstehen zu wollen, ist aber jederzeit berechtigt, da es sicher eine Verminderung der primären Objekte der Theorie bedeutet und somit zu einer einfacheren und daher besseren Beschreibung führt. Wenn man also nach Wegen sucht, das Materiefeld aus dem Ätherfeld (beides im oben neu definierten Sinne verstanden) zu konstruieren oder umgekehrt, so verfolgt man tatsächlich den Grundgedanken der alten Äthertheorien. MIE, EINSTEIN und MAYER, EDDINGTON, WEYL und in neuester Zeit BORN suchten allgemeine Feldtheorien aufzustellen, in welchen alle oder nur einige der eingeführten Felder durch andere erklärt werden.

Die Objekte der quantenmechanischen Beschreibung.

Feld	Symbol	Transformationscharakter	Partikel	Antipartikel	Masse	Ladung der Partikel: elektr.	neutr.	leichte	schwere	Geschwindigkeit	Drehmoment $\times 2\pi h^{-1}$	Statistik
Materie	ψ_α	Spinor	positives Elektron (+)	negatives Elektron (−)	1	+1	0	+1	0	$<c$	½	Fermi-Dirac
	φ_α		Neutrino (+)	Antineutrino (−)	0	0	+1	+1	0	$=c$	½	
	u_α		positives Proton (+)	negatives Proton (−)	1847	+1	0	0	+1	$<c$	½	
	v_α		Neutron (+)	Antineutron (−)	1848	0	+1	0	+1	$<c$	½	
Äther	A_i	Vektor	Photon	—	0	0	0	0	0	$=c$	0, 1	Einstein-Bose
	g_{ik}	Tensor	Graviton	—	0	0	0	0	0	$=c$	0, 1, 2	

— bedeutet: „keine Antipartikel"
$\alpha = 1, 2, 3, 4$
$i, k = 0, 1, 2, 3$

Die Antipartikel haben in diesen Kolonnen −1 statt +1 stehen (Ladung der Partikel: elektr., neutr., leichte, schwere)

Betrachten wir unsere Tabelle und erinnern uns des Satzes:

„Aus dem Produkt von zwei oder einer beliebigen geraden Zahl von Spinoren kann ich Skalare, Vektoren und Tensoren im vierdimensionalen Raum bilden",

so kann ich höchstens das Vektor-Tensorfeld auf das Spinorfeld zurückführen. Die beiden damit verwandten Sätze:

„Drehmomente von Systemen, welche aus einer geraden Anzahl von Elementarteilchen mit Spin ½ bestehen, sind immer ganzzahlige algebraische Summen der Drehmomente der einzelnen Elementarteilchen",

und

„Systeme, welche aus einer geraden Zahl von Elementarteilchen zusammengesetzt sind, die der Fermi-Diracschen Statistik genügen, befolgen ihrerseits die Regeln der Bose-Einsteinschen Statistik",

bestärken diese Vermutung. Tatsächlich gelang es JORDAN und KRONIG in ihrer Neutrinotheorie des Lichtes, das e. m. Feld durch die φ_α darzustellen. Das Lichtquant ist gewissermassen aus einem Neutrino und einem Antineutrino zusammengesetzt. Seine Ausbreitungsgeschwindigkeit ist daher tatsächlich die Lichtgeschwindigkeit, mit welcher sich die masselosen Neutrinos bewegen. Sein Drehimpuls beträgt $\frac{1}{2}-\frac{1}{2}=0$ oder $\frac{1}{2}+\frac{1}{2}=1$. Seine sämtlichen vier Ladungen sind Null, da es sich ja aus Teilchen und entsprechenden Antiteilchen zusammensetzt.[1])

Wenn wir zwar hoffen können, dass die Photonen sich aus den Spinteilchen erklären lassen, so liegen aber bis heute noch keine Versuche vor, die Gravitonen aus Neutrinos zu erklären.

10. Schlusswort.

Die erkenntnistheoretische Einstellung der heutigen Physik geht im wesentlichen auf ERNST MACH (1838—1916) zurück. Ihre Eigenheit gegenüber früheren Auffassungen ist der Verzicht auf jegliche Metaphysik, welchen ich durch die Betonung des Wortes Beschreibung hervorzuheben versuchte. Fragen, wie: Existieren die Atome wirklich? Gibt es einen leeren Raum?

---

[1]) Es scheint, dass man nicht das Viererveklorfeld A_i, sondern das aus den in der Anm. [1]) auf Seite 188 eingeführten Dreiervektoren E_i und B_i hervorgehende Vierertensorfeld F_{ik} durch die φ_α darstellen soll.

u. a. m. werden als sinnlose Fragen bezeichnet, da es keine Beobachtung gibt, welche eine solche metaphysische Antwort auf ihre Richtigkeit prüfen könnte. Die Relativitätstheorie Einsteins zeigte uns zum ersten Male die Notwendigkeit dieser Auffassung. Die Begriffe Länge und Zeit, welche wir als metaphysische Gegebenheiten ansahen, mussten revidiert werden, um eine richtige Beschreibung der Phänomene zu erlangen. Die Quantenmechanik Heisenbergs zwang uns zu einem weiteren Verzicht auf Metaphysik. Der scheinbare Dualismus Welle-Korpuskel erklärte sich dadurch, dass der physikalische Begriff der Partikel jeden Sinn verliert im Moment, wo wir über keine Methode verfügen, die beiden Eigenschaften eines Teilchens, Ort und Geschwindigkeit, genau zu messen. Der physikalische Begriff Korpuskel ist aber gerade durch diese Eigenschaften definiert. Zur Erkenntnis, dass wir keine Beobachtungsmethode der vollständigen korpuskularen Natur der Partikel besitzen, kam Heisenberg auf Grund der Theorie der Lichtkorpuskeln: Ein Verkleinern der Lichtwellenlänge zur Beobachtung immer kleinerer Objekte bedingt eine Erhöhung des beim Stoss übertragenen Impulsbetrages und somit der Geschwindigkeitsänderung des beobachteten Objektes, so dass eine Kenntnis seiner korpuskularen Eigenschaft nie in aller Schärfe möglich sein wird.

Zum Schlusse sei noch erwähnt, dass P. Jordan, aus dessen Beiträgen zur Neutrinotheorie des Lichtes der Verfasser die Anregung zu diesem Vortrag erhalten hat, eine sehr klare Formulierung des physikalischen Positivismus, wie diese metaphysikfreie Richtung der Physik heisst, in Buchform herausgegeben hat. Das Werk „Anschauliche Quantentheorie" richtet sich in erster Linie an den Fachmann, jedoch hat er diese Gedanken auch in einer weniger technischen Form in zwei kleineren Monographien „Physikalisches Denken in der neuen Zeit" und „Die Physik des 20. Jahrhunderts" einem weiteren Leserkreis zugänglich gemacht.

Genf, Institut de Physique de l'Université,
20. November 1936.

Literatur.

1. Allgemeines.

SIR A. EDDINGTON, New Pathways in Science. Cambridge 1935.
PASCUAL JORDAN, Die Physik des 20. Jahrhunderts. Vieweg 1936.
— Physikalisches Denken in der neuern Zeit (Hanseat. Verlagsanst., Hamburg 1936).

2. Quantentheorie und Quantenmechanik.

HANDBUCH DER PHYSIK (herausgeg. v. Geiger u. Scheel), Bd. 24 (2. Auflage). Springer 1933. (Vergl. insbes. Artikel von W. PAULI.)
PASCUAL JORDAN, Anschauliche Quantentheorie. Springer 1936.

3. Theorie der Atomkerne.

H. A. BETHE und R. F. BACHER, Rev. Mod. Physics **8**, 82 (1936).
C. F. v. WEIZSÄCKER, Die Atomkerne. Akad. Verlagsges. Leipzig 1937.

4. Allgemeine Feldtheorie (ohne Einschluss der Gravitation)

M. BORN, Unitary Theory of Field and Matter. Proc. Ind. Acad. Sc. **3**, 8 und 85 (1936).
P. JORDAN und R. DE L. KRONIG, Zs. f. Phys., Bde. **93—102**. Genaue Literaturangaben vergl. letzte Arbeit, Bd. **102**, 243 (1936). Eine neuere Arbeit von KRONIG lag dem Verf. als Manuskript vor. Physica **3**, 1120 (1936). Er ist Herrn Kronig zu grossem Dank für diese Zusendung des MS. verpflichtet, da erst in dieser letzten Arbeit das statistische Problem (Zusammensetzung zweier Fermi-Dirac-Teilchen zu Bose-Einstein-Teilchen) und das geometrische Problem (Zusammensetzung von Spinoren zu Tensoren, welche den Maxwellschen e. m. Beziehungen für das Vacuum genügen) zusammen behandelt werden.
E. C. G. STUECKELBERG, Helv. Physica Acta **9**, 389 und 533 (1936), Nature **139**, (Februar 1937) (im Druck).

5. Gravitation und e. m. Feld (ohne Einschluss des Spinorfeldes).

W. PAULI, Relativitätstheorie. Teubner 1921.
H. WEYL, Raum, Zeit und Materie. Springer 1920.
A. EINSTEIN und M. MAYER, Sitz.-Ber. der Preuss. Akad. d. Wiss. 1931. Siehe auch SIR A. EDDINGTON: Relativity Theory of Electrons and Protons, Cambridge 1936, welcher zwar die Spinorteilchen mit einschliesst, aber eine grundsätzlich andere Fragestellung behandelt als die hier aufgeworfene (Erklärung der Zahlen 137 und 1847, welche in allen anderen Theorien als Gegebenheiten auftreten!).

Manuskript eingegangen am 1. Dezember 1936.

On the existence of heavy electrons [38]

Reprinted with permission from *Physical Review*, vol. 52 (1937), pp. 41–42.

On the Existence of Heavy Electrons

Different observers[1] believe they have found evidence for the existence of charged particles whose mass amounts probably to about fifty times the electron mass. Furthermore these particles seem to behave according to the Bethe-Heitler theory.[2]

The writer wishes to call attention to an explanation of the nuclear forces, given as early as 1934, by Yukawa,[3] which predicts particles of this sort.

Independently of Yukawa the writer arrived at the same conclusion: Kemmer has shown recently[4] that the formal conception of field theory proposed by the author[5] leads to great difficulties, if the proposed interaction energies have singular character. Interaction energies of nonsingular character can most easily be introduced by assuming the existence of some tensor field differing from that of ordinary radiation.

We describe *matter* by a 16 component spinor ψ, whose first four components refer to the *electron state*, the second four functions to the *neutrino state*, the third group to the *proton state* and the last four components to the *neutron state of matter*. Each of these states is defined by the values of a set of four indices (*charge numbers*) $(\alpha, \beta, \gamma, \delta)$ which have respectively the values $(1, 0, 0, 1)$, $(0, 1, 0, 1)$, $(1, 0, 1, 0)$ and $(0, 1, 1, 0)$. Quantum theory associates *particles* with this field which we denote by the same symbol ψ. They have the spin 1/2 and obey Fermi *statistics*. Their *antiparticles* are to be described as holes according to Dirac's idea and designated by $-\psi$. They have the same charge numbers as the particles but with reversed sign. If e represents the elementary charge $e\alpha$ is the *electric charge* of the particle in the state given by $\psi(\alpha, \beta, \gamma, \delta)$. $\gamma=1$ or 0 indicates whether the particle has *heavy mass* ($\sim$2000) or light mass ($\sim$0 to 1). Matter satisfies evidently the relation $\alpha+\beta=\gamma+\delta=1$. The indices β and δ can be called *neutrino charge* and *light mass number*.

The known form of radiation is described by a tensor field A of four components (the vector potential). We attribute to its particles (photons), designated analogously by $A(\alpha, \beta, \gamma, \delta)$ the charge numbers 0. Emission of a photon by matter is due to an interaction term proportional to $e\alpha$ and can be described by the reaction:

$$\psi(\alpha, \beta, \gamma, \delta) \rightarrow \psi(\alpha, \beta, \gamma, \delta) + A(0, 0, 0, 0). \qquad (1)$$

The components of the A field satisfy Poisson's equation, the charge density being expressed by a suitably chosen generalized Dirac matrix $P=e\psi^{+}\Lambda\psi$ (cf. reference 5). We generalize Poisson's equation, introducing the fundamental length λ in the form:

$$\left(\Delta-\frac{1}{c^2}\frac{\partial^2}{\partial t^2}-\Sigma\frac{1}{\lambda^2}\right)A=-P. \qquad (2)$$

A is now a tensor of more than four components. Σ is a matrix operating on the tensor indices analogously to the way Dirac's matrices act on the spin indices of ψ. We assume for simplicity A to have five components. Furthermore let Σ be of such a form that the four first components which represent a four vector satisfy the ordinary Poisson equation ($\Sigma=0$), while the fifth component (a scalar) satisfies Eq. (2) with $\Sigma=1$. In a nonrelativistic approximation the four-vector part gives the Coulomb potential, while the scalar part gives a static interaction term between particles of the form $(fe^2/r)\exp(-r/\lambda)$. f is a numerical factor, depending on the choice of the Λ matrix. A suitable choice (see reference 5) of the generalized Dirac matrices Λ gives the electrostatic interaction between charged matter particles plus the Heisenberg, Majorana, Wigner and Bartlett interactions between heavy matter particles[6] and the different interactions between heavy and light matter particles (β-decay, etc.) discussed by the author. The Heisenberg interaction seems to demand a second order tensor field A.

The *particles* associated with this *generalized radiation* field have integer spins and obey Bose statistics. Those components for which $\Sigma\neq0$ have a rest mass $m=hc/\lambda\neq0$.

We generalize the *conservation law* of charge numbers expressed in Eq. (1): Then the A particle which appears in nuclear reactions has the charge numbers $(1, -1, 0, 0)$. Radiation satisfies evidently the relations $\alpha+\beta=\gamma+\delta=0$. *Antiparticles* have of course once more the charge numbers with reversed sign. There are naturally no antiphotons as the charge numbers of the four-vector field are identically 0. β^{+} decay can be written down as the result of two successive reactions:

$$\psi(1, 0, 1, 0)\rightarrow\psi'(0, 1, 1, 0)+A(1, -1, 0, 0), \qquad (3)$$

$$A(1, -1, 0, 0)\rightarrow\psi''(1, 0, 0, 1)+(-\psi'''(0, 1, 0, 1)). \qquad (4)$$

A proton ψ decomposes into a neutron ψ' and a positively charged *Bose electron* A, which in turn decomposes into an ordinary (positive) electron ψ'' and an antineutrino—ψ'''. As, even in the β-spectra of highest known energy ($24\ mc^2$), those Bose electrons have never been observed, their mass must be greater than 24 electron masses. Yukawa from other considerations estimates about fifty electron masses.

If the corresponding field component has scalar character, the Pauli Weisskopf theory[7] can be applied. The interaction between this field and the four-vector field (electromagnetic field) leads to a formula differing but little from the Bethe-Heitler theory.[2]

It seems highly probable that Street and Stevenson, and Neddermeyer and Anderson [1] have actually discovered a *new elementary particle*, which has been prediced by theory. This particle is unstable and can only be of secondary origin, its mass being greater than the sum of the masses of electron plus neutrino. There exist very probably also other particles for example: $A(0, 0, 1, -1)$.

It is interesting to note that we have a nonlinear field theory, the field having tensorial and spinorial components. The one set is generalized Maxwell equations (2), which are quadratic in the A's if the electromagnetic interaction is included in (2) and quadratic in the ψ's. The other set has the form of generalized Dirac equations, proposed by the author, and contains linear terms in ψ and bilinear ones in the ψ's and A's.

The writer is indebted to his colleagues Professor J. Weigle (Genève) and G. Wentzel (Zuerich)[8] for many a helpful discussion.

E. C. G. STUECKELBERG

Institut de Physique,
Genève, Switzerland,
June 6, 1937.

[1] J. C. Street and E. C. Stevenson, Bull. Am. Phys. Soc. **12**, 13 (1937); S. H. Neddermeyer and C. D. Anderson, Phys. Rev. **51**, 886 (1937).

[2] H. Bethe and W. Heitler, Proc. Roy. Soc. **A146**, 83 (1934).

[3] H. Yukawa, Proc. Phys. Math. Soc. Japan **17**, 48 (1937).

[4] N. Kemmer, Helv. Phys. Acta **10**, 47 (1937).

[5] E. C. G. Stueckelberg, Nature **137**, 1032 (1936); CR. de la Soc. phys. et sc. nat. Genève **53**, 64 (1936); Helv. Phys. Acta **9**, 389 and 533 (1936). The matrices Λ in this note are the Λ and the Ω matrices in the former theory. Instead of the conservation law of charge numbers the author introduced in these former publications a certain relationship between particles and their antiparticles.

[6] See, for example, the different publications of the Leipzig Institute: S. Fluegge, Zeits. f. Physik **105**, 522 (1937); H. Volz, Zeits. f. Physik **105**, 537 (1937); H. Euler, Zeits. f. Physik **105**, 553 (1937).

[7] W. Pauli and V. Weisskopf, Helv. Phys. Acta **7**, 709 (1934); W. Pauli, Ann. Inst. H. Poincaré p. 137 (1936).

[8] G. Wentzel, Zeits. f. Physik **104**, 34 (1936) has also introduced new particles, but he attributed to them a mass near that of the proton (Bose protons and Bose neutrons).

Die Wechselwirkungs Kräfte in der Elektrodynamik und in der Feldtheorie der Kernkraefte (I) [39]

Helvetica Physica Acta, vol. 11 (1938), pp. 225–244

Die Wechselwirkungskräfte in der Elektrodynamik und in der Feldtheorie der Kernkräfte. (Teil I)

von **E. C. G. Stueckelberg.**

(21. II. 38.)

Inhalt: Es wird gezeigt, dass die Quantentheorie der Wellenfelder auf die gleichen Ausdrücke für die Wechselwirkung zwischen Ladungen führt wie die klassische Behandlung der retardierten Potentiale.

Der Wechselwirkungsoperator hat folgende Form: Retardiertes oder avanciertes Potential der einen Ladung am Orte der zweiten mal zweite Ladung. Kann eine dieser beiden Ladungen in erster Näherung schon Strahlung aussenden, so muss das retardierte Potential dieser Ladung oder aber das avancierte Potential der andern Ladung gewählt werden.

Der vorliegende erste Teil enthält die vollständige Diskussion eines skalaren Feldes. Die Verallgemeinerung auf ein Vierervektorfeld ist nur kurz gestreift und wird in einem zweiten Teile behandelt werden.

Einleitung.

Obwohl die Quantentheorie der Wellenfelder grosse innere Widersprüche enthält, ist sie zur Zeit dennoch das einzige brauchbare Mittel, um die korpuskulare Natur der Strahlung und der Materie zu beschreiben.

So folgt aus der Quantenelektrodynamik einerseits die Existenz diskreter *Lichtquanten*, d. h. das Ergebnis, dass Strahlung der Frequenz $k_0 c$ nur in Beträgen $h k_0 c$ emittiert oder absorbiert werden kann.

Andererseits lässt sich aus ihr das *klassische Ergebnis* der *retardierten Wechselwirkung* zwischen zwei Ladungen ableiten.

In vielen Fällen interessiert uns nur die zweite Eigenschaft des Feldes. Da diese retardierten Wechselwirkungen die Planck'sche Konstante nicht enthalten, so muss es möglich sein zu zeigen, dass alle aus einer Quantentheorie der Wellenfelder folgenden Wechselwirkungsgesetze identisch sind mit den entsprechenden klassischen Ergebnissen.

Wir wollen dies an einem skalaren Felde A beweisen, dessen Feldgleichung die Form hat:

$$(\Box - l^2) A = -4 \pi J . \tag{0.1}$$

$A(x)$ nennen wir in Anlehnung an die Elektrodynamik *Potential* und $J(x)$ die *Ladung*. x ist der Vierervektor des Ortes mit den Komponenten $x_0 = ct$, und x_1, x_2, x_3, welche drei letzteren wir mit $\vec{x}$ bezeichnen. Die Wellengleichung (0.1), deren statische Lösung für eine ruhende Punktladung am Koordinatenanfangspunkt

$$A = \frac{e^{-lr}}{r} \tag{0.2}$$

lautet, bildete den Ausgangspunkt der Feldtheorie YUKAWAS[1]), welche die Kernkräfte aus der Existenz eines neuen Feldes erklärt, dessen Partikel (welche ihm die Quantentheorie der Wellenfelder zuordnet) die Masse $hl|c$ haben. l ist also die reziproke Comptonwellenlänge dieser neuen Partikel. Aus der Reichweite der Kernkräfte ergibt sich ihre Masse grössenordnungsweise zu 100 Elektronenmassen. Wir werden im folgenden sehen, dass die Existenz geladener und ungeladener Partikel gefordert werden muss. Die geladenen Partikel wurden von verschiedenen Autoren in der kosmischen Strahlung beobachtet. Näheres hierüber findet sich in Notizen von YUKAWA[1]), KEMMER, BHABHA und vom Verfasser[2]).

1. Das retardierte und avancierte Potential.

Wir suchen eine Lösung von (0.1). Dazu entwickeln wir $J(x)$ in ein Fourierintegral mit dem Integranden $J(k)$, wobei k den Vierervektor mit der (reellen) Zeitkomponente k_0 und den Raumkomponenten $\vec{k} = (k_1, k_2, k_3)$ bedeutet. dx^4, dk^4 und $d\vec{x}^3$, $d\vec{k}^3$ bedeuten die vier- und dreidimensionalen Volumelemente im Raum resp. im Raum der Wellenvektoren. Es sei also

$$J(x) = \int dk^4 e^{i(k,x)} J(k) \tag{1.1}$$

Integrale ohne Grenzen sind immer vom $-\infty$ bis $+\infty$ zu erstrecken.

Entwickelt man $A(x)$ in analoger Weise, so folgt durch Koeffizientenvergleich für die Koeffizienten $A(k)$ die Beziehung:

$$A(k) = \frac{4\pi J(k)}{(k,k) + l^2} \tag{1.2}$$

(k,x) und (k,k) sind skalare Produkte von Vierervektoren.

Wir bezeichnen als *Eigenvektoren des Feldes* solche Vektoren, deren Zeitkomponente der Beziehung

$$k_0 = \pm k_0(\vec{k}) = \pm \sqrt{(\vec{k},\vec{k}) + l^2} \tag{1.3}$$

gehorcht. Dabei ist $(\vec{k},\vec{k})$ das skalare Produkt des Raumanteils von k mit sich selbst.

Wir setzen voraus, dass $J(k)$ für reelle k-Werte keine Singularitäten besitze. Dann besitzt der Integrand $A(k)$ im Ausdruck für $A(x)$ auf der reellen k_0-Axe je eine Singularität bei $k_0 = \pm \bar{k}_0(\vec{k})$. Da $J(k)$ voraussetzungsgemäss keine Singularitäten in unmittelbarer Nähe der reellen k_0-Axe besitzt, so können wir den Integrationsweg über k_0 von $-\infty$ bis $+\infty$ in (1.1) schon vor dem Koeffizientenvergleich deformieren. Bezeichnen wir diesen deformierten Weg durch (..), so können wir als Lösung von (0.1) schreiben:

$$\begin{aligned} A^{(..)}(y) &= \int d\vec{k}^3 \int_{(..)} dk_0\, A(k)\, e^{i(k,y)} \\ &= \int dx^4 J(x)\, D^{(..)}(y-x)\,. \end{aligned} \tag{1.4}$$

Dabei bedeutet $D^{(..)}(z)$ eine Funktion, welche durch das Fourierintegral

$$D^{(..)}(z) = \frac{2}{(2\pi)^3} \int_{(..)} e^{i(k,z)} \frac{1}{(k,k)+l^2}\, dk^4 \tag{1.5}$$

definiert ist.

Zur Ableitung der zweiten Identität in (1.4) ist von der vierdimensionalen δ-Funktion Gebrauch gemacht

$$\delta(k) = (2\pi)^{-4} \int dx^4\, e^{i(k,x)} \tag{1.6}$$

mit der Eigenschaft

$$\int_K dk^4 f(k)\,\delta(k) = f(0) \text{ oder } = 0\,, \tag{1.7}$$

je nachdem ob der Punkt $k_i = 0$ innerhalb oder ausserhalb des Integrationsbereiches K von (2.7) liegt.

Wählt man als Integrationswege die beiden durch (+) und (—) bezeichneten Wege, die ganz in der positiven resp. negativen Halbebene verlaufen (Fig. 1), und berücksichtigt, dass

$$\begin{aligned} \int_{(\pm)} dk_0 \frac{e^{-ik_0x_0}}{k_0} &= \lim_{\alpha=0}\left(-2i\int_\alpha^\infty dk_0 \frac{\sin k_0 x_0}{k_0} + \int_{-\alpha\,(\pm)}^{+\alpha} \frac{dk_0}{k_0}\right) \\ &= -i\pi\left(\frac{x_0}{|x_0|} \pm 1\right);\ \left(\frac{x_0}{|x_0|} = 0 \ \text{ für } \ x_0 = 0\right) \end{aligned}$$

(α ist hiebei eine positive kleine Grösse)

ist, so erhält man folgenden Ausdruck für $D^{(..)}$:

$$\begin{aligned} D^{(..)}(x) = g^{(..)}(x_0)\frac{1}{|\vec{x}|}\frac{1}{\pi}\int_0^\infty d|\vec{k}|\frac{|\vec{k}|}{\bar{k}_0}(\cos(|\vec{k}|\,|\vec{x}| - \bar{k}_0x_0) \\ -\cos(|\vec{k}|\,|\vec{x}| + \bar{k}_0 x_0)) \end{aligned} \tag{1.8}$$

mit

$$g^{(+)} = \begin{cases} 1 \\ 1/2, \\ 0 \end{cases} \quad g^{(-)} = \begin{cases} 0 \\ -1/2 \\ -1 \end{cases} \quad \text{für } x_0 \gtreqless 0$$

In Formel (1.4) ist daher $A^{(+)}$ nur durch die Ladungsverteilung der Vergangenheit und $A^{(-)}$ nur durch diejenige der Zukunft bestimmt. Sie werden, in Verallgemeinerung der Elektrodynamik, als *retardiertes* und *avanciertes* Potential bezeichnet. $D^{(+)}$ und $D^{(-)}$ selbst sind das retardierte resp. avancierte Potential einer nur zur Zeit $x_0 = 0$ von null verschiedenen Punktladung am Koordinatenanfangspunkt, deren Raum-Zeitintegral den Wert eins hat. Ihrer Ableitung gemäss sind $D^{(+)}$ und $D^{(-)}$ in variante Funktionen.

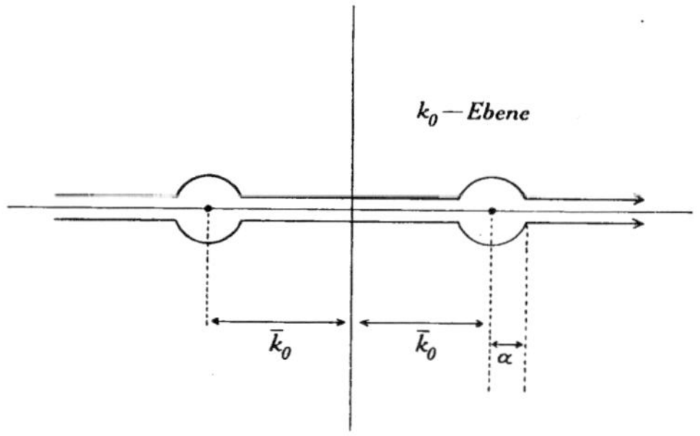

Fig. 1.

Für $l = 0$ (elektrodynamischer Fall) wird

$$D^{(\pm)}(x) = \frac{1}{|\vec{x}|}\,\delta(|\vec{x}| \mp x_0) \qquad (\delta = \text{eindimensionale } \delta\text{-Funktion})$$

Wir definieren noch die ebenfalls invariante Funktion

$$D(x) = D^{(-)}(x) - D^{(+)}(x)$$

$$= \frac{1}{|\vec{x}|}\frac{1}{\pi}\int_0^\infty d\,|\vec{k}|\,\frac{|\vec{k}|}{\bar{k}_0}\cos\left(|\vec{k}|\,|\vec{x}| + \bar{k}_0 x_0\right) - \cos\left(|\vec{k}|\,|\vec{x}| - \bar{k}_0 x_0\right)\Big) \quad (1.9)$$

welche für $l = 0$ in die Heisenberg-Pauli'sche invariante δ-Funktion übergeht. Mit ihrer Hilfe schreiben sich retardiertes und avanciertes Potential in der Form:

$$A^{(\pm)}(y) = \int d\vec{x}^3 \int_{y_0}^{\mp\infty} d\,x_0 D(y - x)\,J(x). \quad (1.10)$$

Im allgemeinen sind also retardiertes und avanciertes Potential zwei verschiedene Ausdrücke. Es gibt aber spezielle Ladungsverteilungen, deren Fourierkoeffizienten $J(k)$ für solche k verschwinden, welche Eigenvektoren des Feldes sind. In diesem Falle ist ein Ausweichen des Integrationsweges in (1.4) nicht notwendig, d. h. beide Integrationswege geben denselben Wert: Das avancierte Potential dieser speziellen Ladungsverteilungen ist gleich dem retardierten.

Møller[3]) zeigte, dass die elektrodynamische Wechselwirkung zweier Elektronen in der Quantentheorie durch folgenden Wechselwirkungsansatz berücksichtigt werden kann: Man berechne das retardierte oder avancierte Potential des ersten Elektrons am Orte des zweiten und multipliziere es mit der Ladung des zweiten Elektrons. Für freie Elektronen führt diese Überlegung tatsächlich auf einen symetrischen Ausdruck, da die Matrixelemente des Stromes in ihrer Fourieranalyse keine Eigenvektoren des Feldes enthalten.

Aus dem Møller'schen Ansatz lässt sich alsdann die Breit'sche Wechselwirkung[4]) durch Entwicklung nach $1/c$ gewinnen, deren Reduktion auf die grossen Komponenten der Diracfunktionen tatsächlich die richtige Spin-Spin- und Spin-Bahn-Wechselwirkung der Pauli'schen Spintheorie liefert.

Erinnert man sich der Vorschrift, dass dieser Wechselwirkungsansatz nur als Störung erster Näherung bei Eigenwertberechnungen verwendet werden darf, so folgt, dass nur statische Ladungsverteilungen in Betracht kommen und somit retardiertes und avanciertes Potential tatsächlich gleich werden.

Hulme[5]) hat später gezeigt, dass diese Møller'sche Vorschrift jeweils dann durch eine strengere zu ersetzen ist, wenn das eine Teilchen strahlen kann: Dann muss nämlich das retardierte Potential dieses strahlenden Teilchens am Orte des zweiten Teilchens multipliziert mit der Ladung des zweiten Teilchens gewählt werden, oder aber das avancierte Potential des nichtstrahlenden Teilchens am Orte des strahlenden multipliziert mit dessen Ladung.

Wir werden im folgenden diese Hulme'sche Verschärfung der Møller'schen Vorschrift als Resultat allgemeiner quantentheoretischer Überlegungen wiederfinden. Ihr klassisches Analogon ist eben dieses Auftreten von Eigenvektoren des Feldes in der „gemischten Ladungsdichte" des einen Teilchens.

*

2. Lagrange und Hamilton Funktion des Feldes.

Das Feld und die Ladung sollen, der Allgemeinheit halber, als komplex angesehen werden. Dann folgt aus der Lagrangefunktion

$$L = \int d\vec{x}\, \mathfrak{L}$$

mit der Lagrangefunktionsdichte

$$\mathfrak{L} = -\frac{1}{8\pi}\left(\left(\frac{\partial A^*}{\partial x}, \frac{\partial A}{\partial x}\right) + l^2 A^* A\right) + \frac{1}{2}\left(A^* J + \left(\frac{\partial A^*}{\partial x}, S\right) + \text{conj.}\right) \tag{2.1}$$

(conj. bedeutet den konjugiert komplexen Ausdruck der Klammerterme) in der üblichen Weise (wenn man A und A^* als zwei unabhängige Funktionen betrachtet) die Formel (0.1). An Stelle von J steht der Ausdruck

$$J_{\text{eff}} = J - \left(\frac{\partial}{\partial x}, S.\right) \tag{2.2}$$

Bezeichnet man J mit *Ladung*, so ist (aus Dimensionsgründen) der Viererverktor S als *Polarisation* aufzufassen. (2.2) ist also die *effective Ladung* in Gl. (0.1).

Die konjugierten Momente folgen in der üblichen Weise zu

$$P = \frac{\delta L}{\delta \dot{A}} = \frac{\partial \mathfrak{L}}{\delta \dot{A}} - \left(\frac{\partial}{\partial \vec{x}}, \frac{\partial \mathfrak{L}}{\partial \frac{\partial A}{\partial \vec{x}}}\right) = \frac{1}{8\pi c}\frac{\partial A^*}{\partial x_0} + \frac{1}{2c} S_0^* \tag{2.3}$$

und entsprechend für P^*. Die Hamiltonfunktion errechnet sich nach der Beziehung:

$$H = -L + \int d\vec{x}^3 (\dot{A} P + \dot{A}^* P^*) = W + V. \tag{2.4}$$

Dabei sind W und V die Volumintegrale von Energiedichten $\mathfrak{W}$ und $\mathfrak{V}$, welche ihrerseits die Form

$$\mathfrak{W} = \frac{1}{8\pi}\left(\left(\frac{\partial A^*}{\partial \vec{x}}, \frac{\partial A}{\partial \vec{x}}\right) + l^2 A^* A\right) + 8\pi c^2 P^* P \tag{2.5}$$

und

$$\mathfrak{V} = -\frac{1}{2}\left(A^* J + \left(\frac{\partial A^*}{\partial \vec{x}}, \vec{S}\right) + \text{conj.}\right) - 4\pi c\,(P S_0 + \text{conj.}) \tag{2.6}$$

haben*).

*) Formel (2.1) ergibt ausser dem angeschriebenen W und V noch einen, nur von der Materie abhängigen (und daher für die Feldgleichungen belanglosen) Zusatzterm der Form

$$+ 2\pi \int d\vec{x}^3 S_0^* S_0 \tag{2.6a}$$

Dieser Term ist aber, wie im zweiten Teile gezeigt werden wird, notwendig, damit die Bewegungsgleichungen der Materie sinnvoll werden.

Die kanonischen Gleichungen folgen durch die, im zweiten und dritten Glied der Gl. (2.3) definierte, funktionelle Differentiation des Hamiltonfunktionals nach A^* resp. P zu

$$\dot{A} = \frac{\delta H}{\delta P} = 8\pi c^2 P^* - 4\pi c S_0 \tag{2.7}$$

und

$$\dot{P}^* = -\frac{\delta H}{\delta A^*} = \frac{1}{8\pi}(\Delta - l^2)A + \frac{1}{2}\left(J - \left(\frac{\partial}{\partial \vec{x}}, \vec{S}\right)\right) \tag{2.8}$$

Differentiation nach der Zeit von (2.7) und Elimination von $\dot{P}^*$ aus (2.8) gibt dann tatsächlich die Gl. (0.1) mit der in (2.2) definierten effektiven Ladung an Stelle von J.

3. Quantisierung der Feldtheorie.

Die Quantisierung der skalaren Feldtheorie wurde von PAULI und WEISSKOPF[6]) durchgeführt. Die Operatoridentitäten

$$\begin{aligned} \dot{P} &= (i/h)[H,P] = -\delta H/\delta A \\ \dot{A} &= (i/h)[H,A] = \delta H/\delta P \end{aligned} \tag{3.1}$$

folgen für jedes analytische Funktional H der Operatoren P und A, wenn gilt

$$[P(\vec{x}), A(\vec{x}')] = (h/i)\,\delta(\vec{x} - \vec{x}') \tag{3.2}$$

und wenn gesternte Operatoren A^* und P^* mit nichtgesternten A und P vertauschbar sind. Dabei bedeutet eine eckige Klammer

$$[a,b] = ab - ba\,.$$

Somit folgen die Bewegungsgleichungen (2.7) und (2.8) auch für die quantentheoretischen Operatoren, wenn wir für H den Operator verwenden, welcher aus den klassischen Gleichungen (2.4), (2.5) und (2.6) folgt, nachdem man darin die P^*, A^*, P und A durch Operatoren ersetzt, welche (3.2) erfüllen.

Dasselbe gilt noch, wenn wir statt V einen Operator K verwenden, für welchen gilt

$$\begin{aligned} (i/h)[K,P^*] &= -\delta K/\delta A^* = \frac{1}{2}\left(J - \left(\frac{\partial}{\partial \vec{x}}, \vec{S}\right)\right) \\ (i/h)[K,A] &= \delta K/\delta P \quad = -4\pi c S_0 \end{aligned} \tag{3.3}$$

Diese Gleichung kann auch als Definition der Ladungsgrössen J und S angesehen werden.

J und S sind Funktionen der kanonischen Variablen, welche den Zustand der Materie beschreiben. Bezeichnen wir diese mit p und q, so sollen aus

$$\begin{aligned} \dot{p} &= (i/h)[K,p] \\ \dot{q} &= (i/h)[K,q] \end{aligned} \tag{3.4}$$

die (klassischen) Bewegungsgleichungen für die Materie folgen.

Das quantentheoretische Problem besteht in der Lösung der Schroedinger-Gleichung

$$\left(H + \frac{h}{i}\frac{\partial}{\partial t}\right)\Psi(t) = 0\,. \tag{3.5}$$

Wir folgen einer von DIRAC, FOCK und PODOLSKI[7]) vorgeschlagenen Methode:

Es wird eine Funktion resp. Funktional $\Psi'(T,t)$ eingeführt, für welche gilt

$$\Psi'(t,t) = \Psi(t) \tag{3.6}$$

und ein Operator $K'(T,t)$, der ebenfalls der Beziehung

$$K'(t,t) = K \tag{3.7}$$

genügt. In (3.7) wird angenommen, dass der Operator K die Zeit t nicht explizit enthalte. K' wird dann im allgemeinen explizit von T und t abhängen. Genügt nun die Funktion Ψ' simultan den beiden Gleichungen:

$$C_T\Psi' = \left(W + \frac{h}{i}\frac{\partial}{\partial T}\right)\Psi'(T,t) = 0 \tag{3.8}$$

$$C_t\Psi' = \left(K'(T,t) + \frac{h}{i}\frac{\partial}{\partial t}\right)\Psi'(T,t) = 0 \tag{3.9}$$

so folgt daraus die Schroedingergleichung für das durch (3.6) definierte einzeitige Ψ mit $H = W + K$.

Damit nun die beiden Gl. (3.8) und (3.9) simultan lösbar sind, müssen die Operatoren C_T und C_t vertauschbar sein. Diese Bedingung

$$[C_T, C_t] = [W, K'(T,t)] + \frac{h}{i}\frac{\partial K'(T,t)}{\partial T} = 0 \tag{3.10}$$

stellt eine Differentialgleichung für die Abhängigkeit des Operators K' von T dar. Ihre Lösung unter Berücksichtigung der Anfangsbedingung (3.7) ist

$$K'(T,t) = e^{iW(t-T)/h}\,K\,e^{-iW(t-T)/h} \tag{3.11}$$

Die Gl. (3.8) lässt sich formal lösen durch

$$\Psi'(T,t) = e^{-iWT/h}\,\psi(t)\,. \tag{3.12}$$

Für die nur von einer Zeit abhängige Funktion $\psi(t)$ gilt dann die aus (3.9), (3.11) und (3.12) folgende Schroedingergleichung

$$\left(K''(t) + \frac{h}{i}\frac{\partial}{\partial t}\right)\psi(t) = 0 \tag{3.13}$$

wo $K''(t)$ den Operator

$$K''(t) = e^{i\,Wt/h}\,K\,e^{-i\,Wt/h} \tag{3.14}$$

bedeutet.

Der Hamiltonoperator K'' in Gl. (3.13) ist also der ursprüngliche Operator K, in welchem die Operatoren $A(\vec{x})$ durch Operatoren $A''(\vec{x},t)$ ersetzt sind, die sich durch dieselbe untäre Transformation (3.14) aus den $A(\vec{x})$ errechnen lassen wie $K''(t)$ aus K. Sieht man vom Operatorcharakter der A ab, so ist (3.13) die Schroedingergleichung der Materie unter dem Einfluss eines gegebenen, explizit von der Zeit t abhängigen Potentialfeldes $A''(\vec{x},t)$.

Dasselbe gilt für P. Um eine explizite Form für die neuen Operatoren A'' und P'' zu haben, entwickeln wir A'' in die Reihe

$$A''(x,t) = A''(\vec{x},0) + t\frac{\partial A''}{\partial t}(\vec{x},0) + \frac{t^2}{2}\frac{\partial^2 A''}{\partial t^2}(\vec{x},0) + \ldots$$

Dabei gelten folgende Beziehungen, welche aus (3.14) folgen:

$$\begin{aligned} A''(\vec{x},0) &= A(\vec{x}) \\ \frac{\partial A''}{\partial t}(\vec{x},0) &= (i/h)[W, A(\vec{x})] = \frac{\delta W}{\delta P} = 8\,\pi c^2 P^*(\vec{x}) \\ \frac{\partial^2 A''}{\partial t^2}(\vec{x},0) &= (i/h)^2[W,[W,A]] = 8\,\pi c^2 (i/h)[W,P^*] \\ &= -8\,\pi c^2 \frac{\delta W}{\delta A^*} = c^2(\Delta - l^2)\,A(\vec{x}) \end{aligned}$$

usw.

Führt man den, nur auf den Parameter $\vec{x}$ wirkenden, symbolischen Operator

$$b(\vec{x}) = \sqrt{-\Delta + l^2} \tag{3.15}$$

ein, so erhält man allgemein:

$$\begin{aligned} \frac{\partial^{2n} A''}{\partial t^{2n}}(\vec{x},0) &= (icb(\vec{x}))^{2n} A(\vec{x}) \\ \frac{\partial^{2n+1} A''}{\partial t^{2n+1}}(\vec{x},0) &= 8\,\pi c^2 (icb)^{-1}(icb)^{2n+1} P^*(\vec{x}). \end{aligned} \tag{3.16}$$

Zur Abkürzung der Reihenentwicklungen des symbolischen Operators b führen wir die Ausdrücke $\cos(bct) = \cos(bx_0)$ und $\sin(bct) = \sin(bx_0)$ ein. Dann kann man schreiben:

$$A''(\vec{x},t) = A(x) = \cos(b(\vec{x})x_0)A(\vec{x}) + 8\pi c^2 b(\vec{x})^{-1}\sin(b(\vec{x})x_0)P^*(\vec{x}). \quad (3.17)$$

Wir werden im folgenden jeweils für $A''(\vec{x},t)$ das Symbol $A(x)$ verwenden, wo x wieder den Weltvector bedeutet.

Diese vom Weltvector x abhängigen Potentiale gehorchen bestimmten Vertauschungsrelationen. Seien x und y zwei Ereignisse, so gilt

$$[A(x), A(y)] = [A(x)^*, A(y)^*] = 0 \quad (3.18)$$

da $A(\vec{x})$ und $P(\vec{x})^*$ in (3.17) vertauschbar sind.

Benützt man die Vertauschungsrelation (3.2) und die dazu konjugierte, und schreibt man die δ-Funktion in Form ihrer Fourierentwicklung (analog (1.6))

$$\delta(\vec{x}) = (2\pi)^{-3}\int d\vec{k}^3 e^{i(\vec{k},\vec{x})}$$

so folgt bei Berücksichtigung von

$$b(\vec{x})e^{i(\vec{k},\vec{x})} = \bar{k}_0(\vec{k})e^{i(\vec{k},\vec{x})} \quad (3.19)$$

(wo $k_0(\vec{k})$ wieder die (positive) Quadratwurzel aus (1.3) bedeutet) die Vertauschungsrelation

$$[A(x)^*, A(y)] = -2\frac{hc}{i}D(x-y). \quad (3.20)$$

Dabei ist D wieder die, durch ihre Fourierdarstellung (1.9) definierte invariante Funktion.

Das in K eventuell vorkommende $P''(\vec{x},t)$ kann durch eine (3.17) analoge Reihe ausgedrückt werden. Man findet dann, dass

$$P''(\vec{x},t) = P(x) = \frac{1}{8\pi c^2}\frac{\partial A''(\vec{x},t)^*}{\partial t} = (8\pi c)^{-1}\frac{\partial A(x)^*}{\partial x_0}. \quad (3.21)$$

Hat der Wechselwirkungsanteil Feld-Materie von K speziell die Form V, deren Dichte durch (2.6) gegeben ist, so ist die entsprechende Dichte $\mathfrak{V}''(\vec{x},t)$ durch den invarianten Ausdruck

$$\mathfrak{V}''(\vec{x},t) = -\frac{1}{2}\left(A(x)^*J(\vec{x}) + \left(\frac{\partial A(x)^*}{\partial x}, S(\vec{x})\right) + \text{conj.}\right) \quad (3.22)$$

gegeben. $J(\vec{x})$ und $S(\vec{x})$ sind die (skalare) Ladung und der Vierervektor ihrer Polarisation am Orte $\vec{x}$. Falls sie, wie in der Diracschen Theorie, nicht von den Potentialen A abhängen, so tritt in ihnen t nach wie vor nicht explizit auf.

4. Ableitung des Wechselwirkungsoperators im Konfigurationenraum.

Beschreibt man die Materie nicht durch ein Materiefeld, sondern durch die Koordinaten q^s $(s = 1, 2, \ldots n)$ von n Teilchen (Konfigurationenraum), so sind die Schroedingerfunktionen Ψ, Ψ' und ψ, welche wir im Paragraphen 3 verwendeten, Funktionale der Funktionen $A(\vec{x})$ und Funktionen der q^s. Der Hamiltonoperator der Materie K'' wird sich dann in eine Summe ΣK^s zerlegen, wo K^s (neben Feldoperatoren) nur auf q^s wirkende Operatoren enthält. Jedes K^s selbst zerfällt in einen Term $R^s(q^s)$, welcher nur auf q^s wirkende Operatoren enthält und in einen Term

$$V^s(q^s, A''(\vec{x}, t)) = V^s(t),$$

welcher ausser auf q^s wirkenden Operatoren, noch Feldoperatoren enthält, die gemäss dem vorhergehenden Paragraphen die Zeit t explizit enthalten.

Die Schroedingergleichung (3.13)

$$\left(K'' + \frac{h}{i}\frac{\partial}{\partial t}\right)\psi = \left(R + V + \frac{h}{i}\frac{\partial}{\partial t}\right)\psi = 0 \tag{4.1}$$

(worin K'' die Summe der K^s und R resp. V die Summen aus den R^s resp. V^s bedeuten) lösen wir durch ein Näherungsverfahren. Dazu denken wir uns den *Störungsterm* V proportional einer kleinen Zahl, nach der wir entwickeln.

Es sei also

$$\psi = \psi^0 + \psi^1 + \psi^2 + \ldots \tag{4.2}$$

wo ψ^m der m-ten Potenz dieser kleinen Zahl proportional sei. Setzt man die Faktoren der einzelnen Potenzen null, so erhält man aus (4.1) das folgende Gleichungssystem:

$$\begin{aligned} &\left(R + \frac{h}{i}\frac{\partial}{\partial t}\right)\psi^0 = 0 \\ &\left(R + \frac{h}{i}\frac{\partial}{\partial t}\right)\psi^1 + V\psi^0 = 0 \\ &\left(R + \frac{h}{i}\frac{\partial}{\partial t}\right)\psi^2 + V\psi^0 = 0. \\ &\ldots\ldots\ldots\ldots\ldots\ldots \end{aligned} \tag{4.3}$$

Die Lösung $\psi^1 + \psi^1 + \psi^2$ der drei angeschriebenen Gln. (4.3) ergibt somit die *zweite Näherung* der Lösung von (4.1). Sie enthält neben der Einwirkung von Materie auf Feld, bereits die Rückwirkung des durch die Materie erzeugten Feldes auf diese selbst. Somit ist die Wechselwirkung zweier Teilchen aufeinander in *ihrer ersten Näherung* enthalten. Gelingt es uns, eine Relation

$$V^s(t)\,\psi^1 = \sum_r U^{sr}\,\psi^0 \qquad (4.4)$$

aufzufinden, so kann man die beiden letzten Gl. des ausgeschriebenen Systems (4.3) durch Addition in die Gleichung

$$R + \frac{h}{i}\frac{\partial}{\partial t}(\psi^1 + \psi^2) + (V + \sum_s\sum_r U^{rs})\,\psi^0 = 0 \qquad (4.5)$$

zusammenziehen.

Entwickelt man die Lösung der Gl.

$$\left(R + V + \sum_r\sum_s U^{rs} + \frac{h}{i}\frac{\partial}{\partial t}\right)\psi = 0\,, \qquad (4.6)$$

wo man sich V und $\Sigma\Sigma U^{rs}$ mit *demselben* kleinen Parameter multipliziert denkt, so erhält man als erste Gleichung (für ψ^0) die erste Gl. von (4.3), während man für die *erste Näherung* ψ^{I} (wo die römischen Indices die Entwicklung nach dem Störungsterm $V + \Sigma\Sigma U^{rs}$ in (4.6) bedeuten) die Gl. (4.5) mit ψ^{I} statt $\psi^1 + \psi^2$ findet. Somit ist die *erste Näherung* von (4.6) identisch mit der *zweiten Näherung* der richtigen Gl. (4.1). Die höheren Näherungen von (4.6) führen aber selbstverständlich zu falschen Ausdrücken für ψ, d. h. solchen, welche der Schroedingergleichung des Problems nicht genügen.

Der Term

$$U^{rs} + U^{sr}, \quad r \neq s \qquad (4.7)$$

stellt dann offenbar die *Wechselwirkungsenergie erster Näherung* zwischen den Teilchen r und s dar.

Zur Auffindung einer Relation (4.4) betrachten wir das Gleichungssystem der n-Gleichungen:

$$\left(R^s + V^s(t^s) + \frac{h}{i}\frac{\partial}{\partial t^s}\right)\psi(t^1..t^s..t^n) = 0\,. \qquad (4.8)$$

Finden wir die Lösung dieses mehrzeitigen Systems, so stellt die Funktion

$$\psi(t,t,,,t) = \psi(t) \qquad (4.9)$$

eine Lösung des Problems (4.1) dar. t^s ist die „Partikelzeit" der s-ten Partikel. Tatsächlich sind aber diese Gln. (4.8) nicht simultan lösbar. Näheres hierüber findet sich bei Bloch[8]). Wir werden aber sehen, dass sie in erster Näherung lösbar sind und uns tatsächlich eine Relation (4.4) vermitteln werden.

Dazu muss die Existenz eines Lorenzsystems gefordert werden, in welchem R^s keine explizit zeitabhängigen Kräfte enthält. Dies ist bei freien Teilchen natürlich der Fall. Ebenso im Ruhsystem des Atomkernes eines gebundenen Elektrons. Es existieren dann Eigenlösungen, für welche gilt (wenn $f(z)$ eine beliebige, durch Reihenentwicklung darstellbare Funktion von z ist)

$$f(R^s)\, u_{\nu_1 \ldots \nu_n} = u_{\nu_1 \ldots \nu_n} \cdot f(h\, \nu_s)\,. \tag{4.10}$$

Die u_{ν_s} sind von den t^s unabhängige Funktionen. Die zeitabhängigen Funktionen

$$v_{\nu_1 \ldots \nu_n} = u_{\nu_1 \ldots \nu_n}\, e^{-i \Sigma \nu_s t^s} \tag{4.11}$$

erfüllen dann die (4.8) in nullter Näherung.

Wir betrachten jetzt den Ausdruck

$$f\left(R^s + \frac{h}{i} \frac{\partial}{\partial t^s}\right) w(t^1 \ldots t^n)\,, \tag{4.12}$$

wo w eine beliebige Funktion der q^s und t^s und Funktional der Feldvariablen ist, die in bezug auf ihre Zeitabhängigkeit in der Form

$$w = \sum_{\omega_1} \ldots \sum_{\omega_n} e^{-i \Sigma \omega_s t^s}\, w_{\omega_1 \ldots \omega_n} \tag{4.12a}$$

darstellbar sei.

Dann wird (4.12) die Summe

$$\sum_{\omega_1} \ldots \sum_{\omega_n} e^{-\Sigma \omega_s t^s} f(R^s - h\, \omega_s)\, w_{\omega_1 \ldots \omega_n}\,.$$

Das Integral hat dann folgenden Sinn:

$$\int d q^s \ldots = \int d q^s v^*_{\nu_1 \ldots \nu_s} f\left(R^s + \frac{h}{i} \frac{\partial}{\partial t^s}\right) w =$$
$$= \int d q^s \sum_{\omega_1} \ldots \sum_{\omega_n} e^{-i \sum_r (\nu_r - \omega_r) t^r} \left(f(R^s - h\, \omega_s)^* u_{\nu_1 \ldots \nu_n}\right)^* w_{\omega_1 \ldots \omega_n}\,.$$

Dabei ist f^* der zu f hermiteisch konjugierte Operator. Ist $f(z)$ eine reelle Funktion, so ist $f^* = f(R^s - h\, \omega_s^*)$, da R^s ein hermiteischer Operator ist. Somit kann für das Integral geschrieben werden:

$$\int d q^s \ldots = \int d q^s \sum_{\omega_1} \ldots \sum_{\omega_n} f(h(\nu_s - \omega_s))\, e^{i \Sigma (\nu_r - \omega_r) t^r}\, u^*_{\nu_1 \ldots \nu_n} w_{\omega_1 \ldots \omega_n} \tag{4.13}$$

Da hiemit jede, durch Reihenentwicklung darstellbare, Funktion des Operators $R^s + h\partial / i \partial t^s$ definiert ist, können die Gleichungen des zu (4.3) analogen Systems, welches durch Entwickeln von ψ in (4.8) erhalten wird, in erster Näherung durch

$$\psi^1 = -\sum \left(R^r + \frac{h}{i}\frac{\partial}{\partial t^r}\right)^{-1} V^r(t^r)\,\psi^0 \tag{4.14}$$

gelöst werden.

Wir haben andererseits die Beziehung

$$\psi^0(t^1\,..\,t^n) = e^{-i\,\Sigma R^r (t^r - t)}\,\psi^0(t) \tag{4.15}$$

Sie ist leicht zu verifizieren, wenn man in Betracht zieht, dass ψ^0 eine Linearkombination von Eigenfunktionen $u_{\nu_1\,..\,\nu_n}$ sein muss, deren Koeffizienten die Zeitabhängigkeit $\exp(-i\,\Sigma\,\nu_r t)$ haben.

Der Operator

$$U^{sr} = -\left\{\left(R^r + \frac{h}{i}\frac{\partial}{\partial t^r}\right)^{-1} V^s(t^s)\,V^r(t^r)\,e^{-i\,\Sigma R^m (t^m - t)}\right\}_{t^1 = t^2 = \ldots = t} \tag{4.16}$$

hat also tatsächlich die Eigenschaft (4.4).

Gemäss der Ableitung des Operators U^{rs} dürfen wir diesen nur in seiner ersten Näherung verwenden. Somit interessieren uns nur solche Matrixelemente, welche zeitunabhängig sind. Berechnet man das vorerst mehrzeitige Matrixelement*)

$$U^{sr}_{\nu',\nu} = \int dq^1 .. \int dq^n \int dA\, v^*_{\nu_1'\,..\,\nu_n'} \left(R^r + \frac{h}{i}\frac{\partial}{\partial t^r}\right)^{-1} V^s(t^s)\,V^r(t^r)\,v_{\nu_1\,..\,\nu_n} \tag{4.17}$$

und verwendet man die Beziehung (4.13), so enthält es den Faktor

$$\frac{i}{h}\,\frac{e^{i(\nu_r' - \omega_r)\,t^r} \cdot e^{i(\nu_s' - \omega_s)\,t^s}}{i(\nu_r' - \omega_r)}. \tag{4.18}$$

Die $\omega_s t^s$ kommen von der Entwicklung von

$$w = V^s(t^s)\,V^r(t^r)\,v_{\nu_1\,..\,\nu_n}(t^1\,..\,t^n)$$

gemäss (4.12a).

Mit Hilfe des convergement gemachten Ausdruckes

$$\lim_{\gamma=0} \frac{e^{i(\nu \pm i\gamma)\,t}}{i(\nu \pm i\gamma)} = -\lim_{\gamma=0} \int_t^{\pm\infty} dt\, e^{i(\nu \pm i\gamma)\,t}$$

kann dieser Faktor in ein Integral umgeformt werden und man erhält für das (mehrzeitige) Matrixelement die Form

$$U^{sr}_{\nu',\nu}(t^1\,..\,t^n) = \int_{t^r}^{+\infty} dt^r \int dq^1 .. \int dq^n \int dA\, v^*_{\nu_1'\,..\,\nu_n'}\,\frac{i}{h}\,V^s(t^s)\,V^r(t^r)\,v_{\nu_1\,..\,\nu_n} \tag{4.19}$$

*) $\int dA$ bedeutet Integration über den Funktionenraum $A(\vec{x})$ (= z. B. Raum der Lichtquanten).

Berücksichtigen wir noch die Bedingung, dass nur zeitunabhängige Matrixelemente eine Rolle spielen, so bedeutet das, dass nur Terme deren Exponenten der Beziehung

$$(\nu_s' - \omega_s) + (\nu_r' - \omega_r) + \sum_{m \neq s, r} (\nu_m' - \nu_m) = 0 \qquad (4.19a)$$

gehorchen, verwendet werden. Wählt man die $v_{\nu_1 .. \nu_n}$ in der üblichen Weise als orthogonales Funktionensystem, so muss $\nu_m' = \nu_m$ sein, da für $m \neq r, s$ der Operator keine auf q^m wirkenden Operatoren enthält. Daher kann $(\nu_r' - \omega_r)$ überall durch $-(\nu_s' - \omega_s)$ ersetzt werden. Das bedeutet, in Gl. (4.18), dass das Integral über dt^r in (4.19) durch

$$-\int_{t^s}^{\pm\infty} dt^s$$

ersetzt werden kann. Addiert man also zum Matrixelement von U^{sr} dasjenige von U^{rs} um dasjenige des Wechselwirkungsoperators (4.7) zu erhalten, und macht beim zweiten von dieser Alternative Gebrauch, so findet man, dass die zeitfreien Matrixelemente des Wechselwirkungsoperators den Matrixelementen des Operators

$$\int_t^{\pm\infty} dt^r e^{i R^r (t^r - t)} \frac{i}{h} [V^s(t), V^r(t^r)] e^{-i R^r (t^r - t)} \qquad (4.20)$$

gleich sind. Hat nun V^s die Form

$$V^s = -\frac{1}{2} \int d\vec{x}^3 (A(x)^* J^s(\vec{x})) + \left(\frac{\partial A(x)^*}{\partial x}, S^s(\vec{x})\right) + \text{conj.} + \text{Terme in } A^2$$

worin J^s und S^s die Ladungs- und Polarisationsoperatoren bei *Abwesenheit des Feldes* bedeuten (ihre Abänderung bei Anwesenheit eines Feldes ist durch „+ Terme in A^2" berücksichtigt), so genügt es in (4.20) nur die in A linearen Terme zu berücksichtigen. Die höheren Terme mitzunehmen, wäre inkorrekt, da wir bei unserem Näherungsverfahren die kleine Grösse, nach der wir entwickeln, nur bis zur zweiten Potenz berücksichtigen wollen. Die Ausdrücke J^s und S^s enthalten also keine Feldoperatoren mehr und sind daher mit den Feldoperatoren vertauschbar.

Wir führen den Operator

$$J^s(y) = e^{i R^r (y_0 - ct)/ch} J^s(\vec{y}) e^{-i R^r (y_0 - ct)/ch} \qquad (4.21)$$

ein, dessen Erwartungswert den Erwartungswert der Ladung des Teilchens s am Orte $\vec{y}$ zu der (im allgemeinen von der in der Schroedingergleichung vorkommenden Zeit t verschiedenen) Zeit y_0/c ergibt (*Operator der retardierten Ladung*). Der Wechselwirkungsoperator darf dann wegen der Vertauschungsrelation (3.18) und (3.20) in der Form geschrieben werden

$$U^{rs} + U^{sr} = -\tfrac{1}{2} \int d\vec{x}^{\,3} \left\{ J^s(\vec{x})\, A^r(x)^* + \left(S^s(\vec{x}),\, \frac{\partial A^r(x)^*}{\partial x} \right) + \text{conj.} \right\}_{x_0 = ct} \tag{4.22}$$

mit*)

$$A^r(x) = \int_{x_0}^{\pm\infty} d y_0 \int d\vec{y}^{\,3} \left(J^r(y)\, D(x-y) + \left(S^r(y),\, \frac{\partial D(x-y)}{\partial y} \right) \right).$$

Den zweiten Term des Integranden kann man durch partielle Integration noch umformen, da die D-Funktion auf der Oberfläche des vierdimensionalen Halbraumes verschwindet, und erhält

$$A^r(x) = \int_{x_0}^{\pm\infty} d y_0 \int d\vec{y}^{\,3} J^r_{\text{eff}}(y)\, D(x-y), \tag{4.23}$$

wo

$$J^r_{\text{eff}}(y) = J^r(y) - \left(\frac{\partial}{\partial y},\, S^r(y) \right) \tag{4.24}$$

die in (2.2) definierte effektive Ladung des Teilchens r am Orte $\vec{y}$ zur Zeit y_0/c bedeutet.

(4.23) ist aber nichts anderes als die klassische Formel (1.10) für das retardierte oder avancierte Potential in operatorieller Form.

Der Operator (4.22) enthält keine Feldoperatoren mehr und führt ohne weiteres auf die Møller'sche Vorschrift. Dabei ist offenbar retardiertes und avanciertes Potential gleich.

Kann aber das Teilchen r strahlen, so bedeutet das, dass die Funktion ψ^1 in (4.14) unendlich wird, d. h., dass im Nenner von (4.18) Glieder mit $\nu_r' = \omega_r$ vorkommen. Das ist immer dann der Fall, wenn die Fourieranalyse des Matrixelementes des Operators $J^r(x)$ für gewisse Übergänge die im Paragraphen 1 definierten Eigenvektoren des Feldes enthält. Die erste Näherung

---

*) Im Zeitanteil des skalaren Produktes muss eigentlich stehen:

$$-\tfrac{1}{2} \int d\vec{x}^3\, S_0{}^s(\vec{x}) \int_{x_0}^{\pm\infty} d y_0 \frac{\partial}{\partial x_0} \cdot \int d\vec{y}^3 (J^r(y)^* D \ldots\ldots)$$

d. h. $\partial/\partial x_0$ steht *unter dem Integralzeichen* nach $d y_0$. Setzt man es, wie in Formel (4.22), *vor das Integralzeichen*, so erhält man wegen der Beziehung:

$$D(x)_{x_0=0} = 0\,; \quad \left\{ \frac{\partial D(x)}{\partial x_0} \right\}_{x_0=0} = -4\pi\delta(\vec{x})\,; \qquad \left\{ \frac{\partial D(x)}{\partial x_i} \right\}_{\substack{x_0=0 \\ i \neq 0}} = 0$$

einen Zusatzterm von der Form

$$-2\pi \int d x^3 (S^s{}_0(\vec{x})\, S_0{}^r(\vec{x})^* + \text{conj.}),$$

welcher sich gegen den in der Anm. zu S. 230 (2.6a) erwähnten Term forthebt.

gibt einen wesentlichen Beitrag zur Störung der Wellenfunktion ψ^0. Der Anfangszustand wird exponentiell abklingen. Trägt man dem schon in der nullten Näherung dadurch Rechnung, dass man an Stelle von (4.11)

$$\left.\begin{aligned} v_{\nu_1 \ldots \nu_n} &= u_{\nu_1 \ldots \nu_n} e^{-i \sum_{m \neq r} \nu_m t^m} \cdot F(-\nu_r + i\gamma, t^r) \\ \text{mit} \quad F(\nu, t) &= \begin{cases} 0 \text{ für } t < 0 \\ e^{i\nu t} \text{ für } t > 0 \end{cases} \end{aligned}\right\} \tag{4.25}$$

schreibt, so lautet, wie man sich durch Fourierdarstellung (oder Integration nach der Zeit in der üblichen Weise) leicht überzeugt (4.18)

$$-\frac{i}{h} \int_{t^r}^{-\infty} dt\, F(\nu_r' - \omega_r + i\gamma, t) \cdot e^{i(\nu_s' - \omega_s) t^s} \tag{4.18a}$$

Formel (4.19) bleibt also zu Recht bestehen, wenn man nur das Vorzeichen $-\infty$ als obere Integrationsgrenze behält.

Die Formel (4.17) für U^{rs} enthält nach wie vor den Faktor (4.18) mit r und s vertauscht. Fügt man jetzt den Term

$$\frac{i}{h} \cdot \frac{-e^{i(\nu_s' - \omega_s) t^s}}{i(\nu_s' - \omega_s)}$$

zu diesem Faktor hinzu, so überzeugt man sich, dass dieser Zusatz *keinen Beitrag* zu zeitunabhängigen Matrixelementen liefert, wenn das s-te Teilchen nicht strahlen kann. Da γ eine, im Vergleich zu $\nu_s' - \omega_s$ verschwindend kleine Grösse bedeutet, so gilt nach wie vor die Energierelation (4.19a), welche wieder gestattet, das mit Zusatzterm versehene (s und r vertauscht) (4.18) als Integral über dt^r von t^r bis $-\infty$ auszudrücken.

Der in (4.22) auftretende Operator (4.23) enthält also nur die Grenze $-\infty$ und ist, gemäss der Definition (1.10) der Operator des retardierten Potentials des r-ten (strahlen könnenden) Teilchens [5]).

5. Ableitung des Wechselwirkungsoperators im gewöhnlichen Raum.

Die Darstellung der Materie durch die Koordinaten von n Raumpunkten q^s ist nur dann möglich, wenn die Teilchenzahl der Materiequanten erhalten bleibt. Das ist aber, wie das Experiment und die Theorie zeigen, nicht der Fall (Paarerzeugung). Man griff daher zu der Felddarstellung der Materie, deren einfachste Form die Darstellung durch einen Schroedinger'schen Feldskalar $q(\bar{x})$ ist. Seine Quantisierung ist von Pauli und Weisskopf[6]) angegeben worden. Eine andere Möglichkeit (die einzige, welche den Halbzahligen Spin der Materiequanten vermittelt) ist die Dirac'sche Darstellung durch einen vierkomponentigen Feldspinor $q^\alpha(\vec{x})$ $(\alpha = 1, 2, 3, 4)$. Die Quantisierung des Diracfeldes bringt gewisse Schwierigkeiten mit sich (Löchertheorie). In einer, von MAJORANA[9]) vorgeschlagenen, Quantisierungsmethode tritt diese Löcherdarstellung nicht explizit in Erscheinung.

Die Schroedingergleichung (3.13) kann auch bei Felddarstellung in der Form (4.1) zerlegt werden. Nur sind R und V nicht mehr Summen von n Termen R^s und V^s, sondern Raumintegrale von gewissen Dichten

$$R = \int d\vec{x}^3 \mathfrak{R}(\vec{x}) \qquad V = \int d\vec{x}^3 \mathfrak{V}(\vec{x})\, t. \tag{5.1}$$

Die Überlegungen bis zu Formel (4.4) behalten ihre Gültigkeit*). (4.4) selbst ist hingegen durch die Forderung zu ersetzen, dass ein Operator $\mathfrak{U}(\vec{x}, \vec{y})$ existiert, für den gilt

$$\mathfrak{V}(\vec{x}, t)\, \psi^1 = \int d\vec{y}^3\, \mathfrak{U}(\vec{x}, \vec{y})\, \psi^0 \tag{5.2}$$

Gl. (4.6) enthält dann an Stelle der Doppelsumme über U^{rs} ein Doppelintegral über den Operator $\mathfrak{U}(\vec{x}, \vec{y})$. Entsprechend der Funktion $\psi(t^1 \ldots t^n)$ führt man ein Funktional $\psi(t(\vec{x}))$ ein, welches der Gleichung:

$$\left(\int d\vec{x}^3 \delta(\vec{x} - y)\, \mathfrak{R}(\vec{x}) + \int d\vec{x}^3 \delta(\vec{x} - \vec{y})\, \mathfrak{V}(\vec{x}, t(\vec{y})) + \frac{h}{i} \frac{\partial}{\partial t(\vec{y})}\right) \psi = 0 \tag{5.3}$$

genügt. Dabei soll $\delta(\vec{x})$ vorerst noch einen endlichen Bereich ausfüllen. Die zeitfreien Eigenfunktionen nullter Näherung u der Gleichung (5.3) sind dann die Eigenfunktionen des Operators, welcher der in einem durch die δ-Funktion um den Ort $\vec{y}$ definierte Volumen enthaltenen Energie der Materie entspricht. $t(\vec{y})$ ist, analog der „Partikelzeit", als „Lokalzeit" zu bezeichnen. Man kann nun leicht die weiteren Rechnungen in voller Analogie zum vorhergehenden Paragraphen durchführen und erhält nach dem Übergang zur δ-Funktion für den Wechselwirkungsoperator statt (4.22)

$$U = -\frac{1}{4} \int d\vec{x}^3 \left\{ J(\vec{x})\, A(x) + \left(S(\vec{x}), \frac{\partial A(x)}{\partial x} \right) + \text{conj.} \right\}_{x_0 = ct} \tag{5.4}$$

mit

$$A(x) = \int_{x_0}^{+\infty} d y_0 \int d\vec{y}^3\, J_{\text{eff}}(y)\, D(x - y)\,. \tag{5.5}$$

*) Die Schroedingerfunktion ψ ist natürlich jetzt Funktional von $A(\vec{x})$ und $q(\vec{x})$.

Dabei sind $J(y)$ und $S(y)$ wieder die retardierten Ladungsoperatoren nullter Näherung ($A = 0$), welche sich aus den Operatoren $J(\vec{y})$ und $S(\vec{y})$ durch

$$J(y) = e^{iR(y_0 - ct)/hc} J(\vec{y}) e^{-iR(y_0 - ct)/hc} \tag{5.6}$$

ergeben. Der Faktor 1/4 in (5.4) kommt davon her, dass $\vec{x}$ und $\vec{y}$ über den ganzen Raum integriert werden und somit jedes Volumelement doppelt gezählt worden ist. Eine Absonderung des Selbstenergieterms, die in (4.22) durch Auslassen der Terme $r = s$ geschieht, wird hier natürlich unmöglich. Ausser (5.4) erhält man, wegen der Nichtvertauschbarkeit der Ladungsoperatoren miteinander, noch Glieder, welche dem Quadrat der Feldoperatoren proportional sind, und welche zu Selbstenergie, Doppelabsorption und -emission und Comptoneffekt beitragen.

6. Verallgemeinerung auf mehrkomponentige Felder.

Die Verallgemeinerung auf mehrkomponentige Tensorfelder kann formal einfach dadurch erhalten werden, dass A, J und S_i weitere Tensorindices erhalten, z. B. für ein Vierervektorfeld A_i, J_i und S_{ik}.

Dann wird aber die Energie W ihren positiv definiten Charakter verlieren. Dem kann nur dadurch abgeholfen werden, dass man Nebenbedingungen einführt.

Für den Fall $l = 0$ lautet diese Nebenbedingung im materiefreien Raum bekanntlich[7])

$$\left(\frac{\partial}{\partial x}, A\right) \psi = 0. \tag{6.1}$$

Sie ist mit sich selbst und mit ihrer konjugiert komplexen (d.h. für zwei Ereignisse x und y) verträglich.

Für den Fall $l \neq 0$ ist es nicht möglich, eine Nebenbedingung zu finden. Führt man aber, neben dem Feld A_i noch ein skalares Feld B ein, das einer Feldgleichung vom Typus (0.1) mit demselben l genügt, so ergibt

$$\left(\left(\frac{\partial}{\partial x}, A\right) + lB\right) \psi = 0 \tag{6.2}$$

eine positiv definite Energie und ist mit sich selbst und mit seiner konjugiert komplexen vertauschbar.

Bei Anwesenheit von Materie sind die Nebenbedingungen in bestimmter Weise abzuändern, damit sie im Laufe der Zeit t erhalten bleiben.

Aus (6.1) folgen in bekannter Weise die Maxwell'schen Gleichungen. Sie besagen u. a., dass ein Photon neben seinem Impuls einen weiteren Freiheitsgrad besitzt: die Polarisation. Diese ist nur zweier Werte fähig (z. B. links- und rechtszirkular), resp. das Photon hat nur zwei Einstellmöglichkeiten des Spins. Aus (6.2) folgen gewisse Maxwell-ähnliche und von PROCA[10]) diskutierte Gleichungen: Das Feld kennt drei linear unabhängige Polarisationsformen, welche einem Teilchen mit drei Einstellmöglichkeiten des Spins entsprechen. Die Eigenwerte des Spins sind natürlich die ganzzahligen Vielfachen h, 0 und $-h$ von h.

In einem zweiten Teile soll das Procafeld als Anwendung der vorgeschlagenen Methode behandelt werden.

Wie mir Herr Dr. N. KEMMER freundlicherweise mitteilt, sind von ihm selbst und von anderen Autoren[11]) verschiedene Arbeiten im Erscheinen, welche das Procafeld und seine Anwendung auf die Wechselwirkung zwischen Neutronen und Protonen behandeln.

Der Vorteil der vorliegenden Methode liegt darin, dass sie, ohne explizites Eingehen auf die Quantenstruktur der Felder und ohne spezielle Annahmen über die Darstellung der Materie, die Berechnung der retardierten und natürlich auch der nicht-retardierten Wechselwirkungsterme gestattet.

In Übereinstimmung mit Kemmer scheint es mir möglich, dass ein komplexes Procafeld (dessen Teilchen „schwere Elektronen" und „schwere Antielektronen" sind) und ein Procafeld (dessen Teilchen ungeladen sind) die Kernkräfte vollständig beschreibt.

Institut de Physique, Université de Genève.

Literatur.

1) YUKAWA, Proc. Phys.-Math. Soc. Japan **17**, 48 (1935) und **19**, 1084 (1937).

2) STUECKELBERG, Phys. Rev. **52**, 41 (1937). — OPPENHEIMER und SERBER, Phys. Rev. **51**, 1113 (1937). — KEMMER, Nature **141**, 116 (1938). — BHABHA, Nature **141**, 117 (1938).

3) MØLLER, Zs. für Phys. **70**, 786 (1931). Quantentheoretische Ableitungen: BETHE und FERMI, Zs. f. Phys. **77**, 296 (1932).

4) BREIT, Phys. Rev. **39**, 616 (1932).

5) HULME, Proc. Roy. Soc. **145**, 487 (1936).

6) PAULI und WEISSKOPF, Helv. Phys. Acta **7**, 709 (1934).

7) DIRAC, FOCK und PODOLSKI, Sow. Phys. **2**, 468 (1932).

8) BLOCH, Sow. Phys. **5**, 301 (1934).

9) MAJORANA, Nuovo Cim. **14** (No. 4) (1937).

10) PROCA, Journ. Phys. **7**, 347 (1936).

11) KEMMER, Proc. Roy. Soc. im Erscheinen. — KEMMER, FROEHLICH und HEITLER, Proc. Roy. Soc. im Ersch. — BHABHA, Proc. Roy. Soc. im Ersch.

Die Wechselwirkungskräfte in der Elektrodynamik und in der Feldtheorie der Kernkräfte (Teil II und III) [40]

Helvetica Physica Acta, vol. 11 (1938) 299–328

Die Wechselwirkungskräfte in der Elektrodynamik und in der Feldtheorie der Kernkräfte. (Teil II und III)

von **E. C. G. Stueckelberg.**

(6. IV. 38.)

Inhalt.

Teil II. Die in Teil I angegebene Methode zur Berechnung der Wechselwirkung zwischen zwei Ladungen wird auf ein Viererpotential verallgemeinert. Eine positiv definite Feldenergie kann auch für ein Feld, dessen Teilchen eine nicht verschwindende Ruhemasse besitzen, durch eine Nebenbedingung erzeugt werden. Es wird die allgemeine Form der durch dieses Feld vermittelten Wechselwirkung zwischen zwei Spinorteilchen gegeben.

Teil III. Die Bewegungsgleichung des Kernkraftfeldes und des Spinorfeldes der Materie werden quantenmechanisch aus einem Hamiltonoperator abgeleitet. Es zeigt sich, dass Operatoren existieren, welche der Kontinuitätsgleichung genügen. Verlangt man die Erhaltung der elektrischen Ladung und die Erhaltung der Dichte der schweren Teilchen, so sind im wesentlichen vier verschiedene Felder möglich. Ihre Teilchen sind: geladene und ungeladene leichte Teilchen mit einer Masse, deren Comptonwellenlänge der Reichweite der Kräfte zwischen schweren Teilchen entspricht, und geladene und ungeladene schwere Teilchen, deren Masse grösser als Proton resp. Neutronmasse ist.

Die empirische Form der Kräfte zwischen Neutron und Proton ergibt sich nur dann, wenn man auch für die ungeladenen leichten Teilchen die Existenz zweier Teilchensorten annimmt (Antiteilchen). Hingegen bestätigt sich die Vermutung, dass eine Theorie ohne Antineutrino im Sinne Majoranas möglich ist.

TEIL II.

7. Verallgemeinerung der Theorie auf ein Viererpotential.

In einem ersten Teile[12]) wurde gezeigt, dass die gegenseitigen Störungen zwischen zwei Materiepartikeln in erster Näherung aus einer Hamiltonfunktion berechnet werden können, in welcher ein Teil der Wechselwirkung Feld-Materie durch gewisse Wechselwirkungsterme Materie-Materie ersetzt werden. Diese Terme hatten folgende Form: Operator der retardierten Ladung des einen Teilchens am Orte des andern mal Ladung des anderen Teilchens.

In der Ableitung beschränkten wir uns auf den skalaren Fall.

Ein solches skalares Feld gibt aber eine Wechselwirkung zwischen den Kernbestandteilen (Protonen und Neutronen), welche ein falsches Vorzeichen und falsche Spinabhängigkeit besitzt: Der skalare Anteil von (4.22) ist positiv, gibt also Abstossung.

Es soll daher als Verallgemeinerung das Feld eines Viererpotentials behandelt werden.

Im vorliegenden zweiten Teil soll daher zuerst die Frage des Vorzeichens der Feldenergie diskutiert werden und nachher sollen die retardierten Potentiale berechnet werden.

Formal geschieht die Verallgemeinerung einfach dadurch, dass den Grössen A (Potential), J (Ladung) und S_k (Polarisation) ein Index i $(i = 0,1,2,3)$ angehängt wird: A_i, J_i, S_{ik}.

Die Formeln von Teil I gelten wörtlich weiter, wenn man die in den A, P, J und S bilinearen Terme durch entsprechende Summen über i (von 0 bis 3) ersetzt.

So zum Beispiel:

$$A^*A \text{ durch } \sum_i \varepsilon_i A_i^* A_i = (A, A)$$

(und analog für P^*P)

$$PS_o \text{ durch } \sum_i \varepsilon_i P_i S_{io}$$

$$\left(\frac{\partial A^*}{\partial \vec{x}}, \frac{\partial A}{\partial \vec{x}}\right) \text{ durch } \sum_k{}' \sum_i \varepsilon_i \frac{\partial A_i^*}{\partial x_k} \frac{\partial A_i}{\partial x_k} \quad \text{usw.}$$

Dabei bedeutet Σ' eine nur über 1,2 und 3 erstreckte Summe. ε_i hat für $i = 1, 2, 3$ den Wert $+1$ und für $i = o$ den Wert -1.

Die Vertauschungsrelationen (3.2) sind durch

$$[P_i(\vec{x}), A_{i'}(\vec{x}')] = \delta_{ii'}(h/i)\,\delta(\vec{x} - \vec{x}') \tag{7.1}$$

zu ersetzen. Die Gleichungen (2.3) resp. (2.7) und (2.8) sind, wegen des Auftretens ε_i, durch

$$\dot{A}_i = \frac{\delta H}{\delta P_i} = \varepsilon_i (8\pi c^2 P_i^* - 4\pi c S_{io}) \tag{7.2}$$

$$\dot{P}_i^* = -\frac{\delta H}{\delta A_i^*} = \varepsilon_i \left(\frac{1}{8\pi}(\Delta - l^2) A_i + \tfrac{1}{2}\left(J_i - \sum_k{}' \frac{\partial S_{ik}}{\partial x_k}\right)\right) \tag{7.3}$$

Formel (3.17) erhält deshalb im zweiten $(P(\vec{x})^*-)$ Term ebenfalls den Faktor ε_i. Dieser hat zur Folge, dass die Vertauschungsrelationen für die explizit zeitabhängigen Operatoren $A_i(x)$ die Form erhalten:

$$[A_i(x)^*, A_k(y)] = -2\frac{hc}{i}\varepsilon_i\delta_{ik}D(x-y). \tag{7.4}$$

Die endgültige Form für den Wechselwirkungsoperator (4.22), (4.23) und (4.24) ändert sich nur insofern, als er durch eine Summe über i (mit ε_i) zu ersetzen ist.

8. Erzeugung positiv definiter Energiedichte durch eine Nebenbedingung.

Den Operator der Energiedichte des Strahlungsfeldes (2.5) formen wir ebenfalls durch die unitäre Transformation (3.14) um. Das bedeutet, dass in (2.5) die $A_i(\vec{x})$ durch die explizit zeitabhängigen Operatoren $A_i(x)$ und die $P_i(\vec{x})$ durch die zeitlichen Ableitungen der $A_i(x)$ ersetzt. (Gleichung (3.21) enthält wegen (7.2) den Faktor ε_i.) Man kann dann die Energiedichte als Summe der Energiedichten einzelner Potentialkomponenten schreiben:

$$\mathfrak{W} = \sum_i \varepsilon_i \mathfrak{W}(A_i) \tag{8.1}$$

mit

$$\mathfrak{W}(A) = \frac{1}{8\pi}\left(\sum_k \frac{\partial A^*}{\partial x_k}\frac{\partial A}{\partial x_k} + l^2 A^* A\right). \tag{8.2}$$

Der Ausdruck (8.2) ist stets positiv, da der Faktor ε_k *nicht* auftritt (die Summe über k ist also *kein* skalares Produkt). Die Energiedichte (8.1) hingegen enthält für $i=0$ einen negativen Summanden. In der Elektrodynamik kann die positiv definite Energie durch die homogene Nebenbedingung (6.1) erzeugt werden. Im Falle $l = o$ hingegen ist diese Nebenbedingung nicht mehr mit ihrer konjugiert komplexen vertauschbar. Führen wir aber neben den vier Potentialkomponenten A_i noch eine skalare Komponente B ein, die ebenfalls einer Wellengleichung (1.1) mit demselben l genügt, so ist die Nebenbedingung (6.2) am Orte y mit ihrer konjugiert komplexen am Orte x vertauschbar. Man findet

$$\begin{aligned}[(6.2)^*, (6.2)] &= \sum_i \varepsilon_i \frac{\partial}{\partial x_i}\frac{\partial}{\partial y} D(x-y) + l^2 D(x-y)\\ &= -(\square - l^2)D(x-y) = 0.\end{aligned}$$

Die letzte Gleichsetzung erfolgte, weil die D-funktion ihrer

*

Herkunft nach (Differenz zwischen avanciertem und retardiertem Potential) der homogenen Wellengleichung genügt.

Die Nebenbedingung kann auch in folgender Form geschrieben werden:

$$\frac{\partial A_0}{\partial x_0}\psi = (-\operatorname{div}\vec{A} - l B)\,\psi. \tag{8.3}$$

Übt man auf beide Seiten der Gleichung die Operation $\frac{\partial A_0}{\partial x_0}$ aus und berücksichtigt, dass sie mit dem Operator der rechten Seite vertauschbar ist, so folgt aus (8.3) und aus der conj. compl. Bedingung (6.2) die Identität:

$$-\frac{\partial A_0^*}{\partial x_0}\frac{\partial A_0}{\partial x_0}\psi = (-\operatorname{div}\vec{A}^*\operatorname{div}\vec{A} - l\,(B^*\operatorname{div}\vec{A} + \operatorname{div}\vec{A}^*.\,B) - l^2 B^* B)\,\psi, \tag{8.4}$$

welche einen der negativen Terme von (8.1) eliminiert.

Für den Term $-\operatorname{grad} A_0^* \operatorname{grad} A_0 - l^2 A_0^* A_0 = f$ schreiben wir $-f + 2f$ und formen den Term $2f$ durch partielle Integration um

$$\int d\vec{x}^3\, 2f = \int d\vec{x}^3\,(A_0^*\,(\Delta - l^2)\,A_0 + A_0^*\,(\Delta - l^2)\,A_0).$$

Berücksichtigt man, dass A_0 der homogenen Wellengleichung genügt, so folgt aus der Nebenbedingung:

$$A_0^*\,(\Delta - l^2)\,A_0\,\psi = A_0^*\frac{\partial^2 A_0}{\partial x_0^2}\psi = -A_0^*\left(\operatorname{div}\frac{\partial\vec{A}}{\partial x_0} + \frac{\partial B}{\partial x_0}\right)\psi = 0.$$

Die Terme, welche div linear enthalten, formen wir noch durch partielle Integration um. Dann kann das Integral der Energiedichte mit dem Intergranden $\mathfrak{W}'$ geschrieben werden:

$$\int d\vec{x}^3\left(\sum_i \varepsilon_i\,\mathfrak{W}(A_i) + \mathfrak{W}(B)\right)\psi = \int d\vec{x}^3\,\mathfrak{W}'\,\psi$$

mit

$$\mathfrak{W}' = \frac{1}{8\pi}\Bigg((\operatorname{rot}\vec{A}^*, \operatorname{rot}\vec{A}) + \left(\operatorname{grad} A_0^* + \frac{\partial\vec{A}^*}{\partial x_0}, \operatorname{grad} A_0 + \frac{\partial\vec{A}}{\partial x_0}\right) + \left(l A_0^* - \frac{\partial B^*}{\partial x_0}\right)\left(l A_0 - \frac{\partial B}{\partial x_0}\right) + (l\vec{A}^* + \operatorname{grad} B^*, l\vec{A} + \operatorname{grad} B)\Bigg).$$

Die Nebenbedingung (6.2) (oder (8.3)) ergibt also eine stets positive Energiedichte.

Führen wir jetzt den neuen Vierervektor des Potentials

$$\Phi_i = A_i + \varepsilon_i\, l^{-1} \frac{\partial B}{\partial x_i} \tag{8.6}$$

ein und den antisymetrischen Feldstärkentensor

$$F_{ik} = \varepsilon_i \frac{\partial \Phi_k}{\partial x_i} - \varepsilon_k \frac{\partial \Phi_i}{\partial x_k} = \varepsilon_i \frac{\partial A_k}{\partial x_i} - \varepsilon_k \frac{\partial A_i}{\partial x_k} \tag{8.7}$$

so kann die Energiedichte unter Verwendung des dreidimensionalen Vektors

$\vec{F}$ mit Komponenten (F_{01}, F_{02}, F_{03})

und des dreidimensionalen Pseudovektors

$\overleftarrow{F}$ mit Komponenten (F_{23}, F_{31}, F_{12})

die Form

$$\mathfrak{W}' = \frac{1}{8\pi}\left((\vec{F}^*, \vec{F}) + (\overleftarrow{F}^*, \overleftarrow{F}) + l^2 (\vec{\Phi}^*, \vec{\Phi}) + l^2\, \Phi_0^*\, \Phi_0\right) \tag{8.8}$$

gebracht werden. Die Energiedichte ist also positiv definit und geht für $l = o$ (B verschwindet identisch) bei reellen Feldstärken in den Energieausdruck der Maxwellschen Elektrodynamik über*).

9. Die Nebenbedingung bei Anwesenheit von Ladungen.

Bei Anwesenheit von Ladungen muss das Funktional ψ nicht nur der Nebenbedingung, sondern auch der Schroedingergleichung (3.13) (wir schreiben im folgenden stets K für K'')

$$\left(K + \frac{h}{i}\frac{\partial}{\partial t}\right)\psi = 0 \tag{9.1}$$

genügen. Hier ist also K der Hamiltonoperator der Materie, in welchem die explizit zeitabhängigen Potentiale auftreten.

Schreibt man K in Form eines Integrals über $d\vec{x}^3$ (der Wechselwirkungsanteil Feld—Materie habe zum Beispiel die Form des Integrals über den Ausdruck (3.22)) und führt als Schrödingerzeit $x_0 = ct$ ein, so errechnet sich aus den Vertauschungsrelationen (7.4)

$$\left[K + \frac{hc}{i}\frac{\partial}{\partial x_0}, A_i(y)\right] = -2\frac{hc}{i}\int d\vec{x}^3 \left(\frac{\delta K}{\delta A_i^*} D(x-y) + \frac{\delta K}{\delta \frac{\partial A_i^*}{\partial x_0}} \frac{\partial D(x-y)}{\partial x_0}\right)$$

*) $\mathfrak{W}$ ist auch tatsächlich die $0-0$ Komponente des Tensors

$$\frac{1}{8\Pi}\left(\sum_m \varepsilon_m F_{im}^* F_{km} + l^2\, \Phi_i^*\, \Phi_k + \text{conj.}\right) - \varepsilon_i\, \delta_{ik}\, \mathfrak{L},$$

wo $\mathfrak{L}$ die Proca'sche Lagrangefunktionsdichte (11.8) bedeutet.

($\partial/\partial x_0$ ist natürlich mit $A_i(y)$ vertauschbar.) Ein analoger Ausdruck folgt für B. Die Ladungs- und Polarisationsgrössen ($\mathfrak{K}$ ist der Integrand von K und hat z. B. die Form (3.22)) werden folgendermassen definiert:

$$\begin{aligned} -2\frac{\partial \mathfrak{K}}{\partial B^*} &= J^B & -2\frac{\partial \mathfrak{K}}{\partial \frac{\partial B^*}{\partial x_i}} &= S_i^B \\ -2\frac{\partial \mathfrak{K}}{\partial A_i^*} &= J_i\,\varepsilon_i & -2\frac{\partial \mathfrak{K}}{\partial \frac{\partial A_i^*}{\partial x_k}} &= S_{ik}\,\varepsilon_i \end{aligned} \tag{9.2}$$

Wir erhalten folgende Vertauschungsrelation

$$\left[K+\frac{hc}{i}\frac{\partial}{\partial x_0},\left(\frac{\partial}{\partial y},A\right)+l\,B\right]=\frac{hc}{i}\int d\vec{x}^3\Big\{\left(\mathrm{div}\,(\vec{J}-l\,\vec{S}^B)\right.$$
$$\left.+l\,(J^B-l\,T)\right)D(x-y)-(J_0-l\,S_0^B)\frac{\partial D(x-y)}{\partial x_0}\Big\} \tag{9.3}$$

Das Argument der Ladungs- und Polarisationsoperatoren ist $\vec{x}$. Für den Tensor S_{ik} wurde die einschränkende Annahme gemacht

$$S_{ik}=\varepsilon_i\delta_{ik}\,T+S'_{ik},\quad S'_{ik}=-S'_{ki}. \tag{9.4}$$

Damit das Funktional ψ gleichzeitig die Nebenbedingung und die Schrödingergleichung erfüllt, müssen die beiden Operatoren: „Nebenbedingung und $K+h\,\partial/i\,\partial t$" vertauschbar sein. Das ist aber gemäss (9.3) nicht der Fall.

Wir addieren darum zur Nebenbedingung noch einen inhomogenen Term, d. h. wir schreiben

$$\left(\left(\frac{\partial}{\partial y},A\right)+l\,B+\int d\vec{x}^3 J_0'(\vec{x})\,D(x-y)\right)\psi=0 \tag{9.5}$$

wo J_0' die 0-Komponente eines combinierten Ladungsvektors ist.

$$J_i'=J_i-l\,S_i^B \tag{9.6}$$

Dann gilt:

$$\left[K+\frac{hc}{i}\frac{\partial}{\partial x_0},\int d\vec{x}^3 J_0'\,D(x-y)\right]$$
$$=\frac{hc}{i}\int d\vec{x}^3\left((-\mathrm{div}\,\vec{J}'+R')\,D(x-y)+J_0'\frac{\partial D(x-y)}{\partial x_0}\right). \tag{9.7}$$

Der Skalar R' ist als Viererdivergenz von J_i' definiert:

$$\frac{i}{hc}[K,J_0']+\mathrm{div}\,\vec{J}'=R'=\frac{1}{c}\dot{J}_0'+\mathrm{div}\,\vec{J}'. \tag{9.8}$$

Vergleich von (9.3) und (9.7) zeigt, dass die inhomogene Nebenbedingung (9.5) im Laufe der Zeit erhalten bleibt, wenn die Operatoridentität

$$\dot{J}_0' + c \operatorname{div} \vec{J}' = l\,(J^B - l\,T) \tag{9.9}$$

identisch erfüllt ist.

Ferner muss J_0' mit den Potentialoperatoren und mit $J_0'^*$ vertauschbar sein, damit die inhomogene Nebenbedingung mit sich selbst und mit ihrer konjugiert komplexen verträglich bleibt.

In der Elektrodynamik verschwinden B und l. Ferner verschwindet auch die Viererdivergenz des elektrischen Stromes. (9.9) ist also erfüllt und die Nebenbedingung (9.5) ist möglich. Sie führt bekanntlich auf die Maxwellschen Gleichungen.

Bei den Kernkräften wird sich zeigen, dass J_0' nicht mit $J_0'^*$ vertauschbar ist. Eine Nebenbedingung in inhomogener Form ist daher nicht möglich. Die einzige Lösung, welche (9.9) erfüllt, besteht darin, dass der Vierervektor J_i' und damit auch seine Viererdivergenz verschwinden, und dass $J^B = lT$.

Aus dem identischen Verschwinden der beiden Seiten der Gleichung (9.9) und aus der Definition der Ladungs- und Polarisationsgrössen (9.2) (9.4) (9.6) und (9.8) folgt dann, dass $\mathfrak{K}$ nur von den folgenden Verbindungen des skalaren Potentials B und des Viererpotentials A_i abhängen kann:

1. Vom Skalar

$$\left(\frac{\partial}{\partial x}, A\right) + l\,B = \left(\frac{\partial}{\partial x}, \Phi\right).$$

2. Von den in (8.6) definierten Potentalen Φ_i.

3. Wegen der Antisymetrie des Tensors S'_{ik} (9.4), von den Feldstärken F_{ik}.

Wegen der, nunmehr homogenen und mit K vertauschbaren, Nebenbedingung verschwindet die unter 1. erwähnte skalare Abhängigkeit. (Natürlich kann widerspruchsfrei eine weitere Abhängigkeit von einem weiteren Skalarfeld C, welches unabhängig von dem zur Erzeugung der Φ_i verwendeten B ist, eingeführt werden.)

Wir schreiben noch die Vertauschungsrelationen dieser neuen Grössen:

$$\begin{aligned}[\Phi_i^*(x), \Phi_k(y)] &= -2\,\frac{hc}{i}\,\varepsilon_i\left(\delta_{ik} - \varepsilon_k\,\frac{\partial^2}{l^2\,\partial x_i\,\partial x_k}\right) D(x-y)\\ [F_{ik}^*(x), \Phi_l(y)] &= -2\,\frac{hc}{i}\,\varepsilon_i\,\varepsilon_k\left(\delta_{kl}\,\frac{\partial}{\partial x_i} - \delta_{il}\,\frac{\partial}{\partial x_k}\right) D(x-y)\,.\end{aligned} \tag{9.10}$$

Gesternte Grössen sind nach wie vor mit ungesternten vertauschbar.

Die Nebenbedingung nimmt, wegen der Definition der Φ_i, die an die Vacuumelektrodynamik erinnernde Form an

$$\left(\frac{\partial}{\partial x}, \Phi\right) \psi = 0. \tag{9.11}$$

Da die Φ_i wie die A_i und die B der homogenen Wellengleichung

$$(\Box - l^2)\, \Phi_i = 0 \tag{9.12}$$

genügen, so folgen für die in (8.7) definierten Feldstärken wegen der Nebenbedingung (9.11) die *Proca*'schen Gleichungen[10])

$$\left(\sum_k \frac{\partial F_{ki}}{\partial x_k} - l^2\, \Phi_i\right) \psi = 0. \tag{9.13}$$

Für $l = 0$ gehen sie in die Maxwell'schen Gleichungen des Vacuums über.

Mit genau gleichem Recht, wie wir (9.11) als Nebenbedingung behandelten und daraus die vier Gleichungen (9.13) herleiteten, können wir eine der Gleichungen (9.13) als Nebenbedingung betrachten und daraus die drei anderen Gleichungen (und die Gleichung (9.11)) entwickeln.

In der Elektrodynamik liess sich durch die *Elimination der Nebenbedingung* die Coulomb'sche Wechselwirkung einführen. Das Feld hatte dann nur noch zwei transversale Komponenten. Eine solche Elimination ist bei nicht verschwindender Ruhmasse($l \neq 0$) unmöglich. Hingegen kann die Nebenbedingung durch eine *Definition der Operatoren* identisch befriedigt werden:

Man wählt die Gleichung (9.13) für $i = 0$ als Nebenbedingung und betrachtet Φ_1, Φ_2 und Φ_3 als *unabhängige Operatoren*, welche den Vertauschungsrelationen (9.10) genügen, und die F_{ik} $(i, k = 1, 2, 3)$ als daraus *abgeleitete Operatoren*. Andererseits sieht man die drei Operatoren

$$\Pi_i = \frac{1}{8\pi c} F^*_{i0} \quad (i = 1, 2, 3) \tag{9.14}$$

als weitere *unabhängige Grössen* an. Für $x_0 = y_0$ gilt nach (9.10):

$$[\Pi_i(x), \Phi_k(y)] = \frac{h}{i}\, \delta_{ik}\, \delta(\vec{x} - \vec{y}) \quad (x_0 = y_0)\,. \tag{9.15}$$

Die Operatoren F_{k0} lassen sich also jetzt gemäss (9.14)* durch die $\Pi_i{}^*$ ausdrücken. Definiert man jetzt Φ_0 ebenfalls als *abgeleiteten Operator* in der Form

$$\Phi_0 = 8\,\pi\, c\, l^{-2} \operatorname{div} \vec{\Pi}^* \tag{9.16}$$

so ist die, als Nebenbedingung betrachtete, letzte Gleichung ($i = 0$) von (9.13) tatsächlich identisch erfüllt.

Unter Verwendung der unabhängigen Operatoren $\vec{\Phi}$ und $\vec{\Pi}$ schreibt sich die Energiedichte (8.8)

$$\mathfrak{W}' = \frac{1}{8\,\pi}\left(l^2\,(\vec{\Phi}^*, \vec{\Phi}) + (\operatorname{rot}\vec{\Phi}^*, \operatorname{rot}\vec{\Phi})\right)$$
$$+ 8\,\pi\, c^2\left((\vec{\Pi}^*, \vec{\Pi}) + l^{-2}\operatorname{div}\vec{\Pi}^* \cdot \operatorname{div}\vec{\Pi}\right). \tag{9.17}$$

Zur Ableitung der Feldgleichungen können zwei Wege eingeschlagen werden:

1. Übergang zur „einzeitigen" Theorie, d. h. Rückgängigmachen des Formalismus, welcher auf Gleichung (3.5) folgte. Die Hamiltonfunktion in (3.5) enthält also dann wieder einen Feldanteil. Dieser ist nichts anderes als (9.17), wo jetzt wieder sämtliche explizit zeitabhängigen Operatoren $F(x)$ ($= F''(\vec{x}, x_0)$) durch die vermittels der Transformation (3.14) verbundenen, nicht explizit zeitabhängigen Operatoren $F(\vec{x})$ zu ersetzen sind. Das geschieht formal einfach dadurch, dass man überall $x_0 = 0$ setzt. Dann sind die drei Φ_i und ihre Ableitungen alle untereinander vertauschbar. Dasselbe gilt für die $\Pi_i{}^*$ und ihre Ableitungen. Π_i und Φ_i hingegen gehorchen der Relation (9.15), d. h. sie sind kanonisch konjugiert. Der Materieanteil K bleibt derselbe, nur sind auch hier die Feldgrössen $F(x)$ durch $F(\vec{x})$ zu ersetzen. Alles dies entspricht genau dem Formalismus KEMMERS[11]). Näheres hierüber in Paragraph 11.

2. Aus der vorliegenden mehrzeitigen Theorie auf Grund der Beziehung

$$\frac{1}{c}\,\dot{F}(y) = \left(\frac{\partial F(y)}{\partial y_0} + \frac{i}{h\,c}\,[K, F(y)]\right)_{y_0 = x_0}. \tag{9.18}$$

Wir lassen die Nebenbedingung in der ursprünglichen Form, d. h. betrachten *alle vier Φ_i als unabhängige Operatoren.* Man muss dann an Stelle des Φ_0 ein $\bar{\Phi}_0$ definieren:

$$\bar{\Phi}_0 = A_0 - \frac{1}{c}\,\dot{B}. \tag{9.18a}$$

Die Nebenbedingung lautet dann in $\bar{\Phi}_0$:

$$\left(\operatorname{div} \vec{\Phi} + \frac{1}{c}\dot{\bar{\Phi}}_0 - 4\pi l^{-2}\left(\operatorname{div} \vec{J} + \frac{1}{c}\dot{J}_0\right)\right)\psi = 0. \qquad (9.19)$$

Die Feldgleichungen für die Komponenten Φ_1, Φ_2 und Φ_3 lauten nach zweimaliger Anwendung der Regel (9.18):

$$(\Delta - l^2)\,\Phi_i - \frac{1}{c^2}\ddot{\Phi}_i = -4\pi\left(J_i - \sum_k{}' \frac{\partial S'_{ik}}{\partial x_k} - \frac{1}{c}\dot{S}'_{i0} - \frac{\partial}{\partial x_i} l^{-2}\left(\operatorname{div} \vec{J} + \frac{1}{c}\dot{J}_0\right)\right). \qquad (9.20)$$

Für das in (9.18a) definierte $\bar{\Phi}_0$ erhält man eine analoge Gleichung. Nur ist $\partial/\partial x_0$ durch $-1/c$ mal die zeitliche Ableitung $(\dot{\ })$ der nachfolgenden Grössen zu ersetzen.

Führt man noch die entsprechenden Feldstärken ein

$$\begin{aligned} F_{ik} &= \frac{\partial \Phi_k}{\partial x_i} - \frac{\partial \Phi_i}{\partial x_k} \qquad (i, k = 1, 2, 3) \\ \bar{F}_{i0} &= \frac{\partial \bar{\Phi}_0}{\partial x_i} + \frac{1}{c}\dot{\Phi}_i \end{aligned} \qquad (9.21)$$

so lassen sich, unter Berücksichtigung der Nebenbedingung (9.19) die Gleichungen (9.20) schreiben

$$\left(\sum_k{}' \frac{\partial \bar{F}_{ki}}{\partial x_k} + \frac{1}{c}\dot{\bar{F}}_{0i} - l^2 \bar{\Phi}_i + 4\pi\left(J_i - \sum_k{}' \frac{\partial S'_{ik}}{\partial x_k} - \frac{1}{c}\dot{S}'_{i0}\right)\right)\psi = 0. \qquad (9.22)$$

Sie entsprechen für $l = 0$ den Maxwell'schen Gleichungen für die Anwesenheit von Ladungen.

10. Die Wechselwirkungsterme des Viererpotentials.

Zur Ableitung der Wechselwirkungsterme kann man entweder explizit die Methode des Paragraphen 4 (Teil I) verwenden, oder aber sich erinnern, dass die Φ_i durch die A_i und B ausdrückbar sind (8.6). Da für diese der wiederholt erwähnte klassische Ausdruck „Retardiertes Potential des ersten Teilchens am Orte des zweiten mal Ladung des zweiten" gilt, so gilt er auch für die Φ_i. Dabei sind dann allerdings als Ladungen die Ausdrücke der rechten Seite der Feldgleichungen (9.20) zu wählen.

Wir wollen uns auf den statischen Fall beschränken. Er sei dadurch definiert, dass

1. alle Grössen $\partial/\partial x_0$ (oder in (9.20) die $(\cdot)$) vernachlässigt werden (Vernachlässigung der Retardierung).

2. Ebenso sollen die J_i $(i \neq 0)$ und die S'_{0k} vernachlässigt werden (Vernachlässigung der Bewegung).

Dann wird gemäss (0.2) und (9.20)

$$\begin{aligned} \Phi^r(x)_0 &= \int d\vec{y}^3 J_0^r(\vec{y})\, v(\vec{x}-\vec{y}), \quad v(\vec{x}) = \frac{e^{-l|\vec{x}|}}{|\vec{x}|} \\ \Phi^r(x)_i &= \int d\vec{y}^3 \sum_k{}' S^r_{ik} \frac{\partial}{\partial y_k} v(|\vec{x}-\vec{y}|) \end{aligned} \tag{10.1}$$

der Vierervektor des Potentials, welches das r-te Teilchen am Orte $\vec{x}$ zur Schrödingerzeit $t = x_0/c$ erzeugt. Wir schreiben im folgenden S_{ik} für S'_{ik}. (Es sei noch bemerkt, dass in dieser Näherung die Potentiale A_i mit den Φ_i identisch werden.)

Die Wechselwirkungsausdrücke werden gemäss (4.22) (man berücksichtige auch die Anmerkung).

$$U^{rs} + U^{sr} = \frac{1}{2} \int d\vec{x}^3 d\vec{y}^3 \Big\{ \Big(J_0^{s*}(\vec{x})\, J_0^r(\vec{y}) + \sum_i{}' \sum_m{}' \sum_k{}' S^{s*}_{im}(\vec{x}) S^r_{ik} \frac{\partial^2}{\partial x_m \partial x_k} \Big) v(\vec{x}-\vec{y}) + \text{konj.} \Big\}$$

Unter Einführung des dreidimensionalen Pseudovektors $\vec{S}$ (S_{23}, S_{31}, S_{12}) und des Operatorvektors $\vec{\nabla}$ lässt sich der letzte Term in die Form

$$(\vec{S}^s \times \vec{\nabla}, \vec{S}^r \times \vec{\nabla}) = (\vec{S}^s, \vec{S}^r)\, \Delta - (\vec{S}^s, \vec{\nabla})(\vec{S}^r, \vec{\nabla}) \tag{10.2}$$

umformen.

Beschreibt man die Ladungen (Materie) durch eine Dirac'sche Theorie, so ist

$$J_i^r(\vec{x}) = f e \tau^r \alpha_i^r \delta(\vec{x} - \vec{q}^r) \tag{10.3}$$

wo f eine Konstante der Dimension einer Zahl, e die elektrische Elementarladung, α_i^r die Dirac'schen Geschwindigkeitsoperatoren (Matrizen) des r-ten Teilchens, τ^r gewisse (im allgemeinen nicht-hermiteische) Matrixoperatoren (isotopic Spin), die mit den α_i^r vertauschbar sind, und $\vec{q}^r$ der Ortsvektor des r-ten Teilchens sind.

Entsprechend wird der antisymetrische Tensor

$$\begin{aligned} &\text{Für } i, k \neq 0: S^r_{ik}(\vec{x}) = + i g e \frac{1}{l} \tau^r \beta^r \alpha^r_i \alpha^r_k \delta(\vec{x} - \vec{q}^r) \\ &\text{Für } i = 0: \quad S^r_{0k}(\vec{x}) = - i g e \frac{1}{l} \tau^r \beta^r \alpha^r_k \delta(\vec{x} - \vec{q}). \end{aligned} \tag{10.4}$$

Hier ist g ebenfalls eine Konstante von der Dimension einer Zahl. β^r ist die Dirac'sche β-Matrix des r-ten Teilchens. Wählt man die Spinoren so, dass $\alpha_0 = 1$ und

$$\beta = \begin{pmatrix} 1 & 0 & 0 & 0 \\ 0 & 1 & 0 & 0 \\ 0 & 0 & -1 & 0 \\ 0 & 0 & 0 & -1 \end{pmatrix}$$

wird, so sieht man, dass nur α_0 und $\beta \alpha_i \alpha_k$ „diagonale" Matrizen sind. Bei Reduktion auf die „grossen Komponenten" der Dirac-Funktion tragen daher die nichtdiagonalen Matrizen erst in der Näherung „Kinetische Energie durch Ruhmasse mal c^2" bei. Somit sind die Vernachlässigungen unter 2. gerechtfertigt. Man kann dann noch (für positive Energien) β durch 1 ersetzen und für $i \beta \alpha_i \alpha_k$ die Matrizen σ_{ik} einführen. Dann wird die Wechselwirkung

$$U^{rs} + U^{sr} = \frac{e^2}{2} (\tau^r \tau^{s*} + \tau^{r*} \tau^s) \Big(|f|^2 + |g|^2 (\vec{\sigma}^r, \vec{\sigma}^s) - |g|^2 (\vec{\sigma}^r, \vec{\nabla}) (\vec{\sigma}^s, \vec{\nabla}) \Big) v(|\vec{q}^r - \vec{q}^s|).\text{*)} \tag{10.5}$$

Hierbei wurde, um den Operator $\triangle$ zu eliminieren, von der Relation

$$(\triangle - l^2)\, v(\vec{x}) = -4\pi\, \delta(\vec{x}) \tag{10.6}$$

Gebrauch gemacht.

Es tritt also, strenggenommen, neben den Termen (10.5) noch ein „Nahwirkungsterm"

$$-4\pi |g|^2 (\vec{\sigma}^r, \vec{\sigma}^s)\, l^{-2}\, \delta(\vec{q}^r - \vec{q}^s) \tag{10.7}$$

innerhalb der letzten Klammer auf.

Dieser Term tritt immer (auch in der Elektrodynamik) auf, wenn man die Umformung (10.2) vollzieht. Er wird aber leicht

*) $\vec{\nabla}$ bedeutet in *beiden* Faktoren die Gradientbildung bezüglich $\vec{q}^r$ (oder beidemale bezüglich $\vec{q}^s$).

übersehen, wenn man die Operation $\vec{\nabla}$ auf $v(\vec{x}-\vec{y})$ zuerst ausführt, d. h. wenn man schreibt

$$(\vec{S}^s\times\vec{\nabla},\vec{S}^r\times\vec{\nabla})\,v(|\vec{z}|)=(\vec{S}^s\times\vec{z},\vec{S}^r\times\vec{z})\frac{1}{|\vec{z}|}\frac{\partial}{\partial|\vec{z}|}\left(\frac{1}{|\vec{z}|}v'(|\vec{z}|)\right)$$
$$+2\,(\vec{S}^r,\vec{S}^s)\frac{1}{|\vec{z}|}v'(|\vec{z}|).$$

Hier bedeutet $v'(|\vec{z}|)$ die Ableitung von v nach $|\vec{z}|$. Formt man jetzt den Vektorproduktterm nach der Formel

$$(\vec{S}^s\times\vec{z},\vec{S}^r\times\vec{z})=(\vec{S}^r,\vec{S}^s)|\vec{z}|^2-(\vec{S}^r,\vec{z})(\vec{S}^s,\vec{z}) \tag{10.8}$$

um, so erhält man genau (10.5) *ohne* den störenden Ausdruck (10.7). Das beruht aber nur darauf, dass wir bei der Umformung des Vektorproduktes einen Term der Ordnung $|\vec{z}|^2/|\vec{z}|^5$ dazuzählen, welcher für $\vec{z}=0$ singulär wird.

Auch bei der Berechnung der Spin-Spinwechselwirkung zweier Elektronen tritt der gleiche Term auf:

Gehen wir nämlich in der üblichen Weise vor: Berechnung des Breit'schen Wechselwirkungsterms durch Entwicklung nach $1/c^2$ der Moeller'schen Wechselwirkung und Reduktion der Diracgleichung auf die „grossen Komponenten", so tritt die Spinwechselwirkung tatsächlich in einer Form $(\vec{\sigma}^s\times\vec{\nabla},\vec{\sigma}^r\times\vec{\nabla})|\vec{z}|^{-1}$ auf. In der Literatur wird nun, der Einfachheit halber, spätestens an dieser Stelle die Umformung (10.8) verwendet, so dass der Zusatzterm vergessen wird.

Erinnert man sich der Tatsache, dass die ganzen so erhaltenen Wechselwirkungsterme (mit Ausnahme des Coulomb'schen Terms)*) nur als *Störung erster Ordnung* verwendet werden dürfen, so tritt der Zusatzterm nur als eine kleine weitere Aufspaltung proportional e^4 zwischen Singlet und Triplet in Erscheinung. Wollte man ihn aber bei der strengen Lösung in Berücksichtigung ziehen, so würde er im anziehenden Falle zu unendlich tiefen Termen führen.

Wir müssen daher bei der Anwendung der so errechneten Wechselwirkungsterme uns stets bewusst bleiben, dass wir sie, zum Unterschiede gegen den in der Elektrodynamik auftretenden Coulombterm, nur als *Störungen* betrachten dürfen.

Tatsächlich brauchen wir aber zur Lösung der Kernprobleme d. h. zur Auffindung der stationären Zustände die *strenge Wechselwirkung*. Wollen wir also die empirischen Wechselwirkungsansätze

*) Siehe Seite 306.

mit den hier erhaltenen Resultaten vergleichen, so müssen wir auf alle Fälle diesen zusätzlichen Nahwirkungsterm fortlassen.

Formel (10.5) hat dann bis auf den grad-Term tatsächlich das richtige Vorzeichen und die richtige Spinabhängigkeit für die Kräfte zwischen Neutron und Proton. Dass auch dem „isotopic spin" Faktor die gewünschte Form gegeben werden kann, soll im § 12 (Teil III) gezeigt werden.

III. TEIL.

11. Bewegungsgleichung und Hamiltonoperator.

Das Kernkraftfeld werde durch mehrere Vierervektoren Φ_i^s beschrieben. Der obere Index s unterscheidet hier, im Gegensatz zu den vorhergehenden Paragraphen nicht mehr die einzelnen Teilchen, sondern eine Anzahl verschiedener Procafelder, deren Operatoren untereinander vertauschbar sind. Die daraus abgeleiteten Sechservektoren F_{ik}^s und die Φ_i^s selbst entsprechen natürlich den auf Gleichung (9.18) folgenden überstrichenen Grössen.

Das Spinorfeld der Materie beschreiben wir in der vom Verfasser vorgeschlagenen Form durch ein 16-komponentiges Spinorfeld[2]) [13]) φ_μ^ν, wo jeder der beiden Indices von 1 bis 4 geht. Die Matrices α_i, β der Dirac'schen Theorie und die von den „Paulitermen" herrührenden Matrices σ_{ik} sollen auf den unteren Index μ wirken, während die Matrices τ (und μ), welche im Paragraphen 10. eingeführt wurden, auf den oberen Index ν auszuübende lineare Operationen darstellen. Sie sind daher mit den Dirac'schen Operatoren vertauschbar. (In den zitierten früheren Arbeiten wurden sie mit Ω und Δ bezeichnet.)

Dann lauten die Bewegungsgleichungen des Feldes:

$$\sum_k \frac{\partial F_{ki}^s}{\partial x_k} - l_s^2 \Phi_i^s + 4\pi \left(J_i^s - \sum_k \frac{\partial S_{ik}^s}{\partial x_k} \right) = 0. \qquad (11.1)$$

Übt man die Operation $\partial/\partial x_i$ auf die Gleichungen aus und addiert, so folgt

$$\left(\frac{\partial}{\partial x}, \Phi^s \right) - 4\pi l^{-2} \left(\frac{\partial}{\partial x}, J^s \right) = 0. \qquad (11.2)$$

Die Bewegungsgleichungen der Materie lauten

$$\Big(-ihc\left(\alpha, \frac{\partial}{\partial x}\right) + mc^2 \beta\mu - \sum_r \frac{1}{2} \big((j^{*r}, \Phi^r) + (s^{r*}, F^r) + \text{conj.} \big) \Big) \varphi = 0. \qquad (11.3)$$

Dabei sind J^r und Sr Abkürzungen für die folgenden, aus φ gebildeten Vektoren und Tensoren:

$$J_i^r = \varphi^* \, j_i^r \, \varphi$$
$$S_{ik}^r = \varphi^* \, s_{ik}^r \, \varphi . \tag{11.4}$$

Die Grössen j^r und s^r werden aus den numerischen Faktoren f^r und g^r, dem elektrischen Elementarquant e und den auf die Spinorindices wirkenden Matrizen in folgender Weise gebildet:

$$j_i^r = f^r \, e \, \alpha_i \, \tau^r, \ \alpha_0 = 1$$
$$s_{ik}^r = g^r \, e \, \frac{1}{l_r} \, \sigma_{ik} \, \tau^r \tag{11.5}$$
$$\sigma_{ik} = i \, \beta \, \alpha_i \, \alpha_k, \ \sigma_{0k} = - i \, \beta \, \alpha_k$$

(s^{r*}, F^r) ist das skalare Produkt der beiden Sechservektoren (d. h. $= \frac{1}{2} \sum_i \sum_k \varepsilon_i \varepsilon_k \ldots$), μ ist eine Matrix, deren Eigenwerte die Massen von Elektron, Neutrino, Proton und Neutron sind (gemessen als Vielfache der Elektronenmasse m). l_r sind für jedes Feld charakteristische reciproke Längen (= Masse der dem Felde r zugeordneten Partikel mal c/h).

In einer klassischen Feldtheorie erhält man die Feldgleichungen für Kernfeld und Materie aus der Extremumsforderung des Raumzeitintegrals einer Lagrangefunktionsdichte $\mathfrak{L}$. Ihr Materieanteil hat die Form:

$$\mathfrak{L}(\varphi) = - \varphi^* \text{ mal Ausdruck (11.3)} \tag{11.6}$$

Für den Feldanteil kann man entweder schreiben

$$\mathfrak{L}(\Phi) = \sum_s \sum_i \varepsilon_i \, \mathfrak{L}(A_i^s)_s + \mathfrak{L}(B^s)_s \tag{11.7}$$

(mit $\Phi_i^s = A_i^s + l_s^{-1} \, \varepsilon_i \, \partial B / \partial x_i$), wo die Summanden Ausdrücke der Form (2.1) darstellen (mit angehängten Indices i und s und ohne den Materieanteil, der ja in (11.6) schon steht), oder aber den Proca'schen Ausdruck

$$\mathfrak{L}(\Phi^s)_s = - \frac{1}{8\pi} \left((F^{s*}, F^s) + l_s^2 \, (\Phi^{s*}, \Phi^s) \right). \tag{11.8}$$

Bei Verwendung von (11.7) muss man die Gleichung (11.2) als Nebenbedingung betrachten.

Der Übergang zur Hamiltonfunktion geschieht in der üblichen Weise (vgl. z. B. Paragraph 2). Allerdings kann nur die Form (11.7) verwendet werden, da in (11.8) die zeitlichen Ableitungen von Φ_0 nicht auftreten. Verwendet man (11.7) so treten in der Hamiltonfunktion die A_i^r, B^r und ihre konjugierten Momente auf.

Der in den Paragraphen 8 und 9 entwickelte Formalismus (die „zweizeitige Formulierung" ist natürlich nicht wesentlich) gestattet (gemäss Formel (9.17) und nachfolgender Bemerkung 1)) einen Hamiltonoperator zu schreiben, welcher nur von je drei Feldgrössen Φ_1^r, Φ_2^r, Φ_3^r, ihren konjugierten Impulsen Π_i^r sowie von ihren konjugiert komplexen Operatoren abhängt.

Die Hamiltonfunktionsdichte lautet:

$$\begin{aligned}\mathfrak{H} = {} & \frac{1}{8\pi}\sum_r \left(l_r^2\,(\vec{\Phi}^{r*}, \vec{\Phi}^r) + (\operatorname{rot}\vec{\Phi}^{r*}, \operatorname{rot}\vec{\Phi}^r)\right) \\ & + 8\pi c^2 \sum_r \left((\vec{\Pi}^{r*}, \vec{\Pi}^r) + l_r^{-2}\operatorname{div}\vec{\Pi}^{r*}\operatorname{div}\vec{\Pi}^r\right) \\ & + \varphi^*\left(-ihc\left(\vec{\alpha}, \frac{\partial}{\partial\vec{x}}\right) + mc^2\beta\right)\varphi \\ & + \frac{1}{2}\sum_r \Bigg(-(\vec{J}^r, \vec{\Phi}^{r*}) + J_0^r\, 8\pi c\, l_r^{-2}\operatorname{div}\vec{\Pi}^r + \text{konj.} \\ & \qquad - \sum_i{}'\sum_k{}' S_{ik}^r \frac{\partial \Phi_i^{r*}}{\partial x_k} + \sum_k S_{0k}^r\, 8\pi c\, \Pi_k^r + \text{konj.}\Bigg) \\ & + \pi\sum_r\left(l_r^{-2}\,(J_0^{r*}J_0^r + J_0^r J_0^{r*}) + \sum_k (S_{0k}^{r*}S_{0k}^r + S_{0k}^r S_{0k}^{r*})\right). \end{aligned} \tag{11.9}$$

Die zu den φ_μ^ν konjugierten Impulse sind natürlich gemäss (11.6) die konjugiert komplexen $\varphi_\mu^{\nu *}$ mal ih. Die Hamiltonfunktion ist in den Kern- und Materiefeldgrössen bilinear bis auf die letzte Linie, welche die (symetrisierten) Terme der Anmerkung (2.6a) enthält. Diese Terme sind biquadratisch in den φ.

Die Bewegungsgleichungen erhält man klassisch und quantentheoretisch aus den kanonischen Gleichungen:

$$\frac{\partial \Phi_i^r}{\partial x_0} = \frac{1}{c}\,\dot{\Phi}_i^r = \frac{\delta H}{c\,\delta \Pi_i^r} = \frac{i}{hc}\,[H, \Pi_i^r] \quad i = 1, 2, 3 \tag{11.10}$$

und einer analogen Gleichung, wo Φ_i^r mit Π_i^r vertauscht ist und wo im dritten Gleichungsglied ein — steht. $\delta H/\delta \Pi_i^r$ bedeutet funktionelle Differentiation des Funktionals H (= Volumintegral von $\mathfrak{H}$) nach der Funktion Π_i^r. Für φ gilt die analoge Beziehung

$$-ihc\frac{\partial\varphi}{\partial x_0} = -ih\dot{\varphi} = -\frac{\delta H}{\delta\varphi^*} = [H, \varphi]. \tag{11.11}$$

Bei der letzten Gleichsetzung in (11.11) ist bei der Differentiation auf die Reihenfolge der Glieder zu achten, da J_0^{r*} mit J_0^r nicht vertauschbar ist.

Die letzten Identitäten (11.10) und (11.11), welche das Korrespondenzprinzip ausdrücken, gelten, wenn das Kernfeld symmetrisch gequantelt wird

$$[\Pi_i^r(\vec{x}), \Phi_k^s(\vec{y})] = \frac{h}{i}\,\delta_{rs}\,\delta_{ik}\,\delta(\vec{x}-\vec{y}) \tag{11.12}$$

und wenn für das Materiefeld die symmetrische (—) oder antisymmetrische (+) Quantisierung gilt:

$$\begin{aligned} &\varphi_\mu^\nu(\vec{x})\,\varphi_\lambda^\varrho(\vec{y}) \pm \varphi_\lambda^\varrho(\vec{y})\,\varphi_\mu^\nu(\vec{x}) = 0 \\ &\varphi_\mu^{\nu *}(\vec{x})\,\varphi_\lambda^\varrho(\vec{y}) \pm \varphi_\lambda^\varrho(\vec{y})\,\varphi_\mu^{\nu *}(\vec{x}) = \delta_{\mu\lambda}\,\delta_{\nu\varrho}\,\delta(\vec{x}-\vec{y}). \end{aligned} \tag{11.13}$$

Alle anderen Operatoren sind miteinander vertauschbar. Da Φ_0^r nicht auftritt, muss die quantentheoretische Ableitung der Feldgleichungen kurz skizziert werden:

1. Differentiation nach der Zeit von (11.10) und Elimination von $\dot{\Pi}_i^r$ aus der kanonisch konjugierten Gleichung führt auf die Gleichungen (9.20) für $i = 1, 2, 3$.

2. *Definiert* man den Operator

$$\Phi_0^r = 8\,\pi\,c\,l_r^{-2}\,\mathrm{div}\,\vec{\Pi}^{r*} + 4\,\pi\,l_r^{-2}\,J_0 \tag{11.14}$$

so folgt aus der zeitlichen Ableitung der kanonisch konjugierten Gleichung (11.10) für $\dot{\vec{\Pi}}^r$ (und Elimination von $\dot{\Phi}_s^r$ durch (11.10) selbst) die vierte Gleichung (9.20).

3. Aus der zu (11.10) kanonisch konjugierten Gleichung ergibt sich durch Divergenzbildung und Verwendung der Definition (11.14) die Beziehung (11.2).

4. Mit Hilfe des so erhaltenen (11.2) eliminiert man die Viererdivergenz des Stromes auf der rechten Seite der Gleichungen (9.20) und erhält die Feldgleichungen in der Form (11.1)

Wir bemerken dazu folgendes:

Die Operatoren

$$\Pi_i^r,\ \Phi_i^r \text{ und } F_{ik}^r = \frac{\partial \Phi_k^r}{\partial x_i} - \frac{\partial \Phi_i^r}{\partial x_k} \quad (i, k = 1, 2, 3)$$

sind *reine Feldoperatoren* und daher mit den Materieoperatoren φ vertauschbar.

Die Operatoren Φ_0^r (definiert durch 11.14) und die Operatoren $F_{i0}^r = \partial\Phi_0^r/\partial x_i + \dot{\Phi}_i^r/c$ sind gemischte Operatoren. Sie sind mit den Materieoperatoren φ *nicht vertauschbar.* Aus den Gleichungen (11.10) folgt direkt die Beziehung

$$F^r_{i0} = 8\,\pi\,c\,\Pi^{r*}_i + 4\,\pi\,S_{0i} \tag{11.15}$$

als *Definition von* F^r_{i0} in Analogie zu (11.14).

Die aus (11.11) folgende Bewegungsgleichung der Materie hat folgende Form:

$$\begin{aligned}&\Big(-i\,h\,c\Big(\alpha, \frac{\partial}{\partial x}\Big) + m\,c^2\,\beta\,\mu + \frac{1}{2}\sum_r \Big(-(\vec{j}^{r*}, \vec{\Phi}^r) + j_0^{r*}\;8\,\pi\,c^2\,l_r^{-2}\,\mathrm{div}\,\vec{\Pi}^{r*}\\ &-\sum_i{}'\sum_k{}' s^{r*}_{ik}\frac{\partial \Phi^r_i}{\partial x_k} + \sum_k{}' s^{r*}_{0k}\,8\,\pi\,c\,\Pi^{r*}_k + \text{konj.}\Big)\Big)\,\varphi\\ &+\frac{1}{2}\sum_v\Big(j_0^{r*}\,2\,\pi\,l_r^{-2}\,(J^r_0\,\varphi + \varphi\,J^r_0) + \sum_k s^{r*}_{0k}\,2\,\pi\,(S^r_{0k}\,\varphi + \varphi\,S^r_{0k}) + \text{konj.}\Big) = 0\,.\end{aligned} \tag{11.16}$$

Wären also die J^r_0 und S^r_{0k} mit φ vertauschbar, so würde (11.16) nach Einsetzen der Definitionen (11.14) und (11.15) tatsächlich identisch mit der klassischen Bewegungsgleichung (11.3). (11.16) ist eine in φ nicht lineare Diracgleichung. Die Nichtlinearität rührt vom Auftreten von Ableitungen der Potentiale in der Lagrangefunktion her, wenn man die A^r_i und B^r als primäre Grössen ansieht (siehe Anm. Teil I, Formel (2.6a)).

12. Die Kontinuitätsgleichung der elektrischen und der schweren Ladung und die explicite Form der Wechselwirkungskräfte im Kern.

Im allgemeinen Formalismus von Teil II ist die *Elektrodynamik* mitenthalten, wenn man für eines der Felder ($r = 0$) $l_0 = 0$ setzt. Dann existiert kein B^0 und man hat $\Phi^0_i = A_i{}^0$ und in (9.9) $J_i{}' = J_i$. Ausser dem trivialen Fall $J^0_i = 0$ ist dann nur die Möglichkeit noch offen, dass J^0_0 mit J^{0*}_0 vertauschbar ist. Zerlegt man jetzt in Real- und Imaginärteil, so teilt sich die Beschreibung in zwei unabhängige reelle Felder auf, die je mit einem unabhängigen reellen Strom in Wechselwirkung stehen. Beide Stromanteile müssen einzeln der Kontinuitätsgleichung genügen. Die Kontinuitätsgleichung und die Realität des Feldes sind somit Konsequenzen von $l_0 = 0$.

Der Formalismus vom vorhergehenden Paragraphen ist hingegen noch nicht allgemein genug um die Elektrodynamik zu beschreiben: Der aus den φ mit Hilfe reeller τ^0 gebildete Strom (11.4) und (11.5) genügt nämlich bei Anwesenheit anderer Felder Φ^r, deren τ^r mit τ^0 nicht vertauschbar sind, nicht der Kontinuitätsgleichung. Man muss daher zum Stromausdruck noch einen aus den Φ^r gebildeten Vierervektor addieren, d. h. die Felder Φ^r müssen Ladungsträger sein.

Ausser diesem, durch die Maxwell'sche Theorie bedingten, *Erhaltungssatz der elektrischen Ladung*, gibt es aber offenbar noch einen weiteren Erhaltungssatz: Bei allen beobachteten Umwandlungen der Materie, wurden noch keine Umwandlungen von schweren Partikeln (Neutron und Proton) in leichte Partikel (Elektron und Neutrino) beobachtet. Wir wollen daher einen *Erhaltungssatz der schweren Ladung* fordern.

Die Matrizen ($\tau^0 = \lambda$)

$$\lambda = \begin{pmatrix} 1 & 0 & 0 & 0 \\ 0 & 0 & 0 & 0 \\ 0 & 0 & 1 & 0 \\ 0 & 0 & 0 & 0 \end{pmatrix}; \text{ und } \lambda' = \begin{pmatrix} 0 & 0 & 0 & 0 \\ 0 & 0 & 0 & 0 \\ 0 & 0 & 1 & 0 \\ 0 & 0 & 0 & 1 \end{pmatrix} \tag{12.1}$$

welche auf den oberen Index von φ wirken, erlauben die vom Spinorfeld getragene elektrische resp. schwere Ladungsdichte in der Form $\varphi^* \alpha_i \lambda \varphi$ zu schreiben. Sind λ^ν die Diagonalelemente der Matrix λ, so hat die 0-Komponente die Form $\sum_\nu \lambda^\nu \varphi^{\nu *} \varphi^\nu$. Die Eigenwerte des Volumintegrals von $\varphi^{\nu *} \varphi^\nu$ sind, bei Verwendung der Löchertheorie und der antisymetrischen Quantelung (vgl. auch Majorana loc. cit. 12)) ganze positive oder negative Zahlen. $\lambda^\nu = 0$ oder 1 ist also die Ladung der Partikel des ν-ten Spinorfeldes. Die Antipartikel haben die Ladung $-\lambda^\nu$.

Wir berechnen jetzt die Viererdivergenz des durch die Matrizen λ geformten Stromes:

Dazu multiplizieren wir (11.16) mit $\varphi^* \lambda$ von links und die konjugiert komplexe Gleichung mit $\lambda \varphi$ von rechts, und subtrahieren die beiden Gleichungen voneinander. Die Viererdivergenz verschwindet nun im allgemeinen nicht, sondern wird ein relativ komplizierter Ausdruck. Er vereinfacht sich sehr, wenn die Matrix λ den folgenden Vertauschungsrelationen genügt:

$$\begin{aligned} [\lambda, \mu] &= 0 \\ [\lambda, \tau^r] &= \Lambda^r \tau^r \\ [\lambda, \tau^{r*}] &= -\Lambda^{r*} \tau^{r*} \\ \Lambda^r &= \text{Vielfaches der Einh. Matrix.} \end{aligned} \tag{12.2}$$

Daraus folgt, dass λ hermiteisch und Λ^r eine reelle Zahl sein muss. Die Divergenzgleichung nimmt dann die Form an

$$\sum_i \frac{\partial}{\partial x_i} (\varphi^* \alpha_i \lambda \varphi) = \frac{i}{2 h c} \sum_r \Lambda^r \Big((\vec{\Phi}^{r*}, \vec{J}^r) - 8 \pi c l^{-2} \operatorname{div} \vec{\Pi}^r \cdot J^r_0 + \sum_i{}' \sum_k{}' \frac{\partial \Phi^{r*}_i}{\partial x_k} S^r_{ik} - \sum_k{}' 8 \pi c \Pi^r_k S^r_{0k} - \text{konj.} \Big) \tag{12.3}$$

*

Dass sich die Terme vierter Ordnung in φ fortheben, folgt aus der Relationen (11.13) und aus der aus (12.2) folgenden Beziehung

$$[\lambda \tau^r, \tau^{r*}] + [\tau^{r*} \lambda, \tau^r] = 0 .$$

Um zu zeigen, dass die rechte Seite die Divergenz eines weiteren Viererstromes ist, multiplizieren wir die Feldgleichung (11.1) mit $\varepsilon_i \Phi_i^{s*}$ von links und die konjugiert komplexe Gleichung ebenfalls von links mit $\varepsilon_i \Phi_i^s$ und addieren die Summen der beiden Gleichungen über i von 0 bis 3.

Man erhält, unter Berücksichtigung der Nichtvertauschbarkeiten, folgende Viererdivergenz:

$$\sum_i \frac{\partial}{\partial x_i} \left(\frac{i}{8 \pi h c} \sum_k \varepsilon_k (\Phi_k^{r*} F_{ik}^r - \Phi_k^r F_{ik}^{r*}) \right) =$$

$$- \frac{i}{2 h c} \left((\vec{\Phi}^{r*}, \vec{J}^r) - (\vec{J}^{r*}, \vec{\Phi}^r) - \Phi_0^{r*} J_0^r + \Phi_0^r J_0^{r*} - \frac{l^2}{4 \pi} [\Phi_0^{r*}, \Phi_0^r] \right.$$

$$\left. - \Phi_i^{r*} \sum_k \frac{\partial S_{ik}^r}{\partial x_k} - \Phi_i^r \sum_k \frac{\partial S_{ik}^{r*}}{\partial x_k} - \frac{1}{4 \pi} \sum_k [F_{k0}^{r*}, F_{k0}^r] \right) . \qquad (12.4)$$

Wegen der Definition von Φ_0^{r*} (11.14) ergeben der dritte, vierte und fünfte Term in der Klammer der rechten Seite gerade

$$(\ldots . - 8 \pi c l^{-2} \operatorname{div} \vec{\Pi}^r \cdot J_0^r + \text{konj.} \ldots) \qquad (12.5)$$

Der letzte Term in der rechten Klammer, nimmt wegen der Definition der F_{k0}^r (11.15) die Form an:

$$\left(\ldots . - \sum_k{}' 4 \pi S_{0k}^{r*} S_{0k}^r + \sum_k{}' 4 \pi S_{0k}^r S_{0k}^{r*} \right) . \qquad (12.6)$$

Multipliziert man also (12.4) mit der Zahl Λ^r und summiert über r, so kann die Summe von (12.3) und (12.4) bei Berücksichtigung der Definition (11.15) als Kontinuitätsgleichung geschrieben werden:

$$\left(\frac{\partial}{\partial x}, \varrho \right) = 0 \text{ mit den Komponenten}$$

$$\varrho_i = \sum_\nu \lambda^\nu (\varphi^{\nu*} \alpha_i \varphi^\nu) + \sum_r \Lambda^r \left(\frac{i}{8 \pi h c} \sum_k \Phi_k^{r*} (F_{ik}^r - 4 \pi S_{ki}^r) \right.$$

$$\left. - \Phi_k^r (F_{ik}^{r*} - 4 \pi S_{ik}^{r*}) \right) . \qquad (12.7)$$

Der Ladungsanteil des r-ten Feldes ist also der r-te Summand der zweiten Summe, genau wie der ν-te Summand der ersten Summe den Ladungsanteil des ν-ten Materiefeldes darstellt.

Der Ladungsanteil des r-ten Feldes verschwindet insbesondere dann, wenn das Feld reell ist. Dass ein analoger Satz für Spinorfelder existiert, hat MAJORANA gezeigt[9]).

Die 0-Komponente ist die eigentliche Ladungsdichte. Sie lautet unter Verwendung der Definition (11.15):

$$\varrho_0 = \sum_\nu \lambda^\nu (\varphi^{\nu *} \varphi^\nu) + \sum_r \Lambda^r \sum_k{}' \frac{i}{h} (\Pi_k^r \Phi_k^r - \Pi_k^{r*} \Phi_k^{r*}). \qquad (12.8)$$

Wie bereits bemerkt, sind die Eigenwerte eines einzelnen Summanden der ersten Summe, bei Berücksichtigung der Diracschen Löchertheorie, positive und negative Vielfache von λ^ν. Dasselbe gilt, nach der Pauli-Weisskopf'schen Theorie[6]) für jeden einzelnen Summand der zweiten Doppelsumme. Die Zahlen λ^ν und Λ^r stellen somit die Ladung der Partikel des ν-ten Materie-(Spinor)- und des r-ten Kernkraft(Tensor)-Feldes dar. Jedes der Felder Φ_i^r (mit Ausnahme der reellen Felder) hat Partikel und Antipartikel. Letztere haben das umgekehrte Vorzeichen der Ladung*).

Die Gleichungen (12.2), welche die *zur Existenz einer Kontinuitätsgleichung* (12.7) notwendigen Bedingungen darstellen, gestatten eine Bestimmung der möglichen Matrices τ^r.

Zuerst folgt aus der ersten Gleichung, dass λ eine Diagonalmatrix sein muss, da die Eigenwerte von μ alle verschieden sind. Die zweite und dritte Gleichung fordern Hermiteicität für λ und bestimmen damit λ^ν und Λ^r als reelle Zahlen. Die Matrices der Gleichung (12.1), elektrische und schwere Ladung, gehorchen offenbar diesen Anforderungen.

Um die Form der vierreihigen Matrices τ^r zu bestimmen, zerlegen wir die allgemeinste vierreihige Matrix in eine Summe von direkten Produkten von zweireihigen Matrices. Es seien 11, 21, 12 und 22 die Numerierung der vier Zustände Elektron, Neutrino, Proton und Neutron (entsprechend den vier möglichen Werten des oberen Index ν von φ). $\tau_0 = 1$, τ_1, τ_2 und τ_3 seien die Einheit und die drei Pauli'schen Matrices, welche auf den ersten Index von 11, 21, usw. wirken. Ebenso seien die τ_i' ($i = 0, 1, 2, 3$) die entsprechenden auf den zweiten Index wirkenden Matrices.

Die gestrichenen und die ungestrichenen Matrices sind natürlich miteinander vertauschbar. Ferner gilt für beide Matrices die bekannte Regel

$$\tau_i \tau_k = -\tau_k \tau_i = \tfrac{1}{2} [\tau_i, \tau_k] = i \tau_l, \quad i k l = \text{cykl.} \qquad (12.9)$$

*) Die Summe über k von 1 bis 3 bedeutet, dass die Teilchen drei Einstellmöglichkeiten der Spins haben.

Die allgemeinste vierreihige Matrix lautet dann

$$\tau^r = \sum_i \sum_k a^r_{ik} \tau_i \tau_k' \tag{12.10}$$

und die speziellen Matrices (12.1) haben die Form

$$\lambda = \tfrac{1}{2} (\tau_0 + \tau_3) \tau_0' ; \; \lambda' = \tfrac{1}{2} \tau_0 (\tau_0' - \tau_3') . \tag{12.11}$$

Einsetzen dieser Entwicklungen in die zweite Gleichung (12.2) und Koeffizientenvergleich beider Seiten der Gleichung gibt folgende Beziehungen zwischen den a^r_{ik}:

$$\begin{array}{ll} \Lambda^r a^r_{0k} = 0 & \Lambda^r a^r_{2k} = + i a^r_{1k} \\ \Lambda^r a^r_{3k} = 0 & \Lambda^r a^r_{1k} = - i a^r_{2k} . \end{array} \tag{12.12}$$

Für die Erhaltung der schweren Ladung folgt eine analoge Gleichung, für den zweiten Index. Nur steht überall statt Λ^r die Grösse $-\Lambda'^r$.

Die Lösungen von (12.12) sind:

$$\begin{array}{l} \Lambda^r = 0 \text{ mit } a^r_{1k} = a^r_{2k} = 0 \text{ oder} \\ \Lambda^r = \pm 1 \text{ mit } a^r_{0k} = a^r_{3k} = 0 \text{ und } a^r_{2k} = \pm i\, a^r_{1k} \end{array} \tag{12.13}$$

und analoge Gleichungen für den zweiten Index mit Λ'^r.

Es sind demgemäss folgende Fälle möglich:

1. Feld ohne elektrische und schwere Ladung:

$\Lambda^1 = \Lambda'^1 = 0$ und

$$\tau^1 = a^1_{00} + a^1_{10} \tau_1 + a^1_{01} \tau_1' + a^1_{11} \tau_1 \tau_1' . \tag{12.14}$$

Ein solches Feld ist offenbar das elektromagnetische Feld. Diese Felder können insbesondere reell sein, da die Matrices τ hermiteisch sind, und die Konstanten reell gewählt werden können.

Die durch dieses Feld hervorgerufene Wechselwirkung zwischen zwei Materieteilchen im Konfigurationenraum folgt durch Einsetzen von (12.14) in (10.5). Beschränken wir uns auf schwere Teilchen, so kann man $\tau_3' \psi = 1 \psi$ setzen, und der τ enthaltende Faktor von (10.5) lautet einfach*):

$$|a|^2 + |b|^2 \tau^r_3 \tau^s_3 + \tfrac{1}{2} (a\, b^* + a^*\, b) (\tau^r_3 + \tau^s_3) \tag{12.15}$$

dabei sind a und b beliebige komplexe Zahlen. Sind sie insbesondere reell, so ist das Feld reell.

---

*) Die Indices r und s in den Gleichungen (12.15), (12.17) und (12.18) beziehen sich natürlich nicht auf verschiedene Felder, sondern, gemäss der Konfigurationenraumbeschreibung des Paragraphes 10, auf zwei verschiedene schwere Teilchen.

2. Feld mit elektrischer, aber ohne schwere Ladung.

$$\Lambda^2 = -1, \Lambda'^2 = 0 \text{ und}$$
$$\tau^2 = (\tau_1 - i\,\tau_2)\,(a_{10}^2 + a_{13}^2\,\tau_3')\,. \tag{12.16}$$

Die Wechselwirkung zwischen zwei schweren Teilchen ist wieder Formel (10.5), wo der τ enthaltende Faktor

$$|a'|^2\,(\tau_1^r\,\tau_1^s + \tau_2^r\,\tau_2^s) \tag{12.17}$$

lautet. Will man insbesondere Kräfte zwischen Neutron und Proton haben, welche unabhängig von der Ladung und nur abhängig vom Symmetriecharakter der Wellenfunktion im Konfigurationsraum der schweren Teilchen sind, so muss sich die Wechselwirkung in der Form (10.5) mit einem τ Faktor („isotopic spin factor"[14]))

$$\left(|a|^2 + |b|^2 \sum_{i=1}^{i=3} \tau_i^r\,\tau_i^s\right) \tag{12.18}$$

schreiben lassen. Bildet man die Summe von (12.15) und (12.17), indem man für beide Felder 1 und 2 dieselben Konstanten f und g in (10.5) nimmt, so erhält man tatsächlich (12.18), wenn man $a'=a$ und $b=i\,a$ setzt. Das führt allerdings zu der Unschönheit, dass das Feld 1 (Feld ohne elektrische und schwere Ladung) komplex ist und also zwei Teilchensorten (Antiteilchen) enthält. Da die τ Matrices der folgenden Felder nur noch τ_1' und τ_2' enthalten, tragen sie nichts mehr zur Wechselwirkung zwischen schweren Teilchen bei. Der Fall $\Lambda^2 = +1$, $\Lambda^1 = 0$ ist identisch mit dem behandelten (Vertauschung von Teilchen und Antiteilchen).

13. Fortsetzung der Diskussion der möglichen Felder.

Während bei den am Ende des vorhergehenden Paragraphen diskutierten Feldtypen die Darstellung durch Vierervektoren Φ_i^r notwendig war, um Übereinstimmung mit dem Experiment (Anziehung im Grundzustand des Deuterons usw.) zu finden, ist sie für die weiteren Felder nicht mehr notwendig. Diese Felder können also z. B. auch skalaren Charakter haben. Man überzeugt sich aber leicht, dass auch für sie analoge Gesichtspunkte gelten und dass insbesondere die Relationen (12.2) gelten, sowie die daraus abgeleiteten Beziehungen (12.12) und (12.13).

Bezeichnen wir die Spinorpartikel Elektron, Neutrino, Proton und Neutron durch*) e (1,0), n (0,0), P (1,1) und N (0,1), die

*) Die beiden Indices in der auf das Symbol folgenden Klammer beziehen sich auf elektrische und schwere Ladung der Teilchen.

neutralen Partikel von Feld 1 (deren wegen der komplexen Konstanten in (12.15) mindestens zwei existieren müssen) mit $\mathfrak{n}$ (0,0) und die geladenen Partikel des Feldes 2 mit $\mathfrak{e}$ (1,0), so geben die Matrices (12.14) und (12.16) zu folgenden möglichen Umwandlungen Anlass:

Feld 1.

$$\text{Spinorpartikel} \longrightarrow \text{gleiche Spinorpartikel}' + \mathfrak{n}\,(0,0) \tag{13.1}$$

Dabei sind natürlich zur Zeit nur die Reaktionen mit schweren Spinorpartikeln „beobachtet", d. h. ihre Existenz muss zur Erklärung der Kernkräfte zwischen gleichen Teilchen gefordert werden.

Feld 2.

$$P\,(1,1) \rightleftarrows N\,(0,1) + \mathfrak{e}\,(1,0) \tag{13.2}$$

$$e\,(1,0) \rightleftarrows n\,(0,0) + \mathfrak{e}\,(1,0)\,. \tag{13,3}$$

Die sämtlichen Symbole sind als algebraische Grössen zu betrachten (negative Symbole bedeuten die entsprechenden Antiteilchen). Aus (13.2) aus (13.3) folgt beispielsweise

$$(-\mathfrak{e}\,(1,0)) \rightleftarrows (-\,e\,(1,0)) + n\,(0,0)\,. \tag{13.3'}$$

D. h. ein (negativ geladenes) Anti-$\mathfrak{e}$-Teilchen $(-\acute{\mathfrak{e}})$ kann in ein negatives Elektron $(-e)$ und ein Neutrino (n) zerfallen.

Auch hier sind vorerst nur die Reaktionen (13.2) „beobachtet", da aus ihnen die Austauschkräfte zwischen Proton und Neutron resultieren.

Da aber die $\mathfrak{e}$-Teilchen offenbar äusserst selten vorkommen, erklärte (13.3') ihre endliche Lebensdauer.

Ferner gibt (13.3) eine Theorie des β-Zerfalles:

Ein Neutron verwandelt sich in ein Proton + ein Anti-$\mathfrak{e}$-Teilchen gemäss der algebraischen Umschreibung von (13.2):

$$N\,(0,1) \longrightarrow P\,(1,1) + (-\,\mathfrak{e}\,(1,0))\,. \tag{13.2'}$$

Hierauf tritt Reaktion (13.3') ein.

Gemäss dem Formalismus von Teil II kann das so gedeutet werden: Ein positives Elektron in einem Zustande negativer Energie springt unter dem Einfluss des retardierten Potentials eines schweren Teilchens, welches sich aus einem Neutron in ein Proton verwandelt, in einen Neutrinozustand positiver Energie. Da die Bewegung der schweren Teilchen langsam erfolgt, kann die Retardierung gemäss Paragraph 10 vernachlässigt werden und eine Wechselwirkung der Form (10.5) in die Hamiltonfunktion eingesetzt werden.

Da die Reichweite des $\mathfrak{e}$-Feldes aus den heuristischen Ansätzen über Kernkräfte als klein gegen die Wellenlänge der de Broglie-wellen von Elektron und Neutrino erscheint, kann die „Fernwirkung" aus Gleichung (10.5) durch eine Nahwirkung ersetzt werden und es folgt eine der Fierz'schen Verallgemeinerungen[5]) der Fermischen Theorie[16]) des β-Zerfalles.

Nun hat aber auch diese Verallgemeinerung immer noch den Nachteil, eine zu schwache Asymetrie der Energieverteilung im kontinuierlichen β-Spektrum zu liefern.

Wir werden sehen, dass die weiteren möglichen Feldtypen eine alternative und nach Rechnungen von WENTZEL[17]) bessere Beschreibung des β-Zerfalles bieten.

Das Nichteintreten der Reaktion (13.3) resp. das nur relativ unwahrscheinliche Eintreten derselben, ergäbe eine unendlich lange, resp. eine längere als die von BHABHA vorgeschlagene[2]) Lebensdauer der $\mathfrak{e}$-Partikel. Die endliche Lebensdauer wäre dann *nur* durch auftretende Zusammenstösse mit Neutronen in Atomkernen bedingt (resp. für Anti-$\mathfrak{e}$-Teilchen mit Protonen)[21]).

Wir setzen unsere Diskussion der Feldtypen fort:

3. Feld ohne elektrische, aber mit schwerer Ladung.

$\Lambda^3 = 0$, $\Lambda'^3 = -1$ ergibt analog (12.16)

$$\tau^3 = (a^3_{01} + a^3_{31}\,\tau_3)\,(\tau_1' + i\,\tau_2')\,. \tag{13.4}$$

Die Teilchen bezeichnen wir mit $\mathfrak{N}$ (0,1). Sie geben zu folgenden Reaktionen Anlass:

$$N\,(0,1) \rightleftharpoons n\,(0,0) + \mathfrak{N}\,(0,1) \tag{13.5}$$

$$P\,(1,1) \rightleftharpoons e\,(1,0) + \mathfrak{N}\,(0,1)\,. \tag{13.6}$$

Da das Proton sicher eine stabile Partikel ist, so folgt aus (13.6), dass die Masse der $\mathfrak{N}$-Partikel grösser als die Differenz zwischen Protonen- und Elektronenmasse ist. Da aber auch aus den Kernspin und Kernstatistikmessungen hervorgeht, dass im Kern nur Neutronen und Protonen aber keine Partikel mit ganzzahligem Spin vorkommen, ist es wahrscheinlich, dass die $\mathfrak{N}$-Partikel auch grössere Masse als die Neutronen besitzen und daher instabil sind.

Gemäss den Überlegungen über die retardierten Potentiale gibt dieses Feld 3 Anlass zu Austausch-Kräften zwischen leichten und schweren Partikeln von sehr kurzer Reichweite (Comptonwellenlänge des Protons). Diese Austauschkräfte erlauben die alternative Erklärung des β-Zerfalles:

Nach (13.5) entsteht eine $\mathfrak{N}$-Partikel und ein Neutrino. Die $\mathfrak{N}$-Partikel zerfällt nach der algebraisch umgeschriebenen Gleichung (13.6):

$$\mathfrak{N}\,(0,1) \longrightarrow P\,(1,1) + (-e\,(1,0))\,. \tag{13.6'}$$

Anders ausgedrückt lautet das: Eine Spinorpartikel geht aus dem Zustande „kerngebundenes Neutron" in einen Zustand „freies Neutrino" über. Das durch diesen Übergang erzeugte retardierte oder avancierte Potential des Feldes 3 induziert den Quantensprung einer anderen Partikel aus einem Zustand „Elektron negativer Energie" in einen Zustand „kerngebundenes Proton".

Eine Vernachlässigung der Retardierung ist natürlich nicht mehr möglich. Wie WENTZEL[17]) zeigt erhält man eine stärkere Asymetrie als diejenige der Fermi'schen Theorie, wenn das $\mathfrak{N}$-Feld vom Kern beeinflusst wird (d. h. wenn „Zwischenzustände" mit gebundenen $\mathfrak{N}$-Partikeln existieren).

4. Feld mit elektrischer und schwerer Ladung gleichen Vorzeichens.

$\Lambda^4 = \Lambda'^4 = -1$ ergibt

$$\tau^4 = a^4_{11}\,(\tau_1 - i\,\tau_2)\,(\tau_1' + i\,\tau_2')\,. \tag{13.7}$$

Die, mit $\mathfrak{P}$ (1,1) bezeichneten Teilchen, geben nur zu der Reaktion

$$P\,(1,1) \rightleftharpoons n\,(0,0) + \mathfrak{P}\,(1,1) \tag{13.8}$$

Anlass. Damit das Proton stabil erscheint, müssen die $\mathfrak{P}$-Teilchen eine grössere Masse als die des Protons haben.

5. Feld mit elektrischer und schwerer Ladung verschiedenen Vorzeichens.

$\Lambda^5 = +1,\ \Lambda'^5 = -1$ gibt

$$\tau^5 = a^5_{11}\,(\tau_1 + i\,\tau_2)\,(\tau_1' + i\,\tau_2') \tag{13.1 9}$$

und ebenfalls nur die einzige Reaktion

$$N\,(0,1) \rightleftharpoons e\,(1,0) + \mathfrak{P}\,(-1,1)\,. \tag{13,10}$$

14. Erweiterung des Strombegriffes J_i^r.

Die Definitionen (11.4) der in den Proca'schen Gleichungen (11.1) auftretenden Stromgrössen sind noch einer, ebenfalls in φ bilinearen, Erweiterung fähig. Fügt man ihnen die Terme

$$K_i^r = \varphi\, k_i^r\, \varphi; \quad R_{ik}^r = \varphi\, r_{ik}^r\, \varphi \tag{14.1}$$

mit den Matrixoperatoren

$$k_i^r = f'^r\, e\, \delta\, \alpha_i\, \varkappa^r; \quad r_{ik}^r = g'^r\, e\, \frac{1}{l_r}\, \delta\, \sigma_{ik}\, \varkappa^r \tag{14.2}$$

hinzu, wo δ die von FERMI eingeführte Matrix [16]) (s. auch PAULI [18])) bedeutet und wo $\varkappa^r$ wieder auf den oberen Index ν von φ_μ^ν wirkende Operatoren darstellen, so ändert sich an den Bewegungsgleichungen des Feldes (11.1) und an der Divergenzgleichung des Feldes (12.4) nichts, ausser dass J_i^r durch $J_i^r + K_i^r$ ersetzt ist. Um die Bewegungsgleichungen für φ aus der Hamiltonfunktion zu erhalten, müssen wir noch (ausser dem erwähnten Ersetzten) die Terme

$$2\pi \sum_r (l_r^{-2} (J_0^{r*} K_0^r + K_0^{r*} J_0^r + K_0^{r*} K_0^r) + \text{entspr. Terme in } S_{0k}^r \text{ und } R_{0k}^r) \tag{14.3}$$

hinzufügen. Es sei bemerkt, dass diese Terme zwar hermiteisch, aber, im Gegensatz zu den Termen der letzten Linie von (11.9) *nicht symetrisch* sind. Nur mit diesen Termen ist eine Kontinuitätsgleichung möglich.

Bei symmetrischer Quantelung von φ erhält man die klassischen Wellengleichungen (11.3), welche aber durch Terme in φ^* ergänzt sind. Bei antisymmetrischer Quantelung*) zeigt sich ein charakteristischer Vorzeichenunterschied eines dieser Terme gegenüber den klassischen Gleichungen. Die Divergenzgleichung (12.3) behält ihre Form, wenn die Matrices $\varkappa$ den Antivertauschungsrelationen

$$\lambda\, \varkappa^r + \varkappa^r\, \lambda = -\Lambda^r\, \varkappa^r \tag{14.4}$$

genügen. Entwickelt man die Matrices $\varkappa$ wieder nach (12.10) mit Konstanten b_{ik}, so folgen in Analogie mit (12.12) und (12.13) die Beziehungen.

$$\begin{gathered} b_{1k}^r = -\Lambda^r\, b_{1k}^r, \quad b_{2k}^r = -\Lambda^r\, b_{2k}^r \\ b_{0k}^r + b_{3k}^r = -\Lambda^r\, b_{0k}^r, \quad b_{3k}^r + b_{0k}^r = -\Lambda^r\, b_{3k}^r. \end{gathered} \tag{14.5}$$

*) Dann gilt in (11.11) nur die Form $[H, \varphi]$ und nicht mehr $\delta H/\delta\varphi$.

Die Lösungen für Λ^r lauten jetzt $0, -1, -2$ statt $0, -1, +1$ wie im Paragraph 12. Die Lösung -2 (doppelt geladene Elementarteilchen) wollen wir ausschliessen.

Die im vorhergehenden Paragraphen besprochenen Feldtypen geben dann zu folgenden zusätzlichen möglichen Reaktionen Anlass:

1. Feld ohne elektrische und ohne schwere Ladung.

$$\varkappa^1 = b^1_{00}\,(\tau_0 - \tau_3)\,(\tau_0' + \tau_3') \tag{14.6}$$

d. h. die Reaktion:

$$\mathfrak{n}\,(0,0) \rightleftharpoons 2\,n\,(0,0)\,. \tag{14.7}$$

2. Feld mit elektrischer, aber ohne schwere Ladung.

$$\varkappa^2 = (b^2_{10}\,\tau_1 + b^2_{20}\,\tau_2)\,(\tau_0' + \tau_3') \quad (14.8)$$

oder in der Reaktionsschreibweise

$$e\,(1,0) \rightleftharpoons (-\,n\,(0,0)) + \mathfrak{e}\,(1,0)\,. \quad (14.8)$$

3. Feld ohne elektrische, aber mit schwerer Ladung.

$$\varkappa^3 = (\tau_0 - \tau_3)\,(b^3_{01}\,\tau_1' + b^3_{02}\,\tau_2') \tag{14.9}$$

oder

$$N\,(0,1) \rightleftharpoons (-\,n\,(0,0)) + \mathfrak{N}\,(0,1)\,. \tag{14.10}$$

4. Feld mit elektrischer und schwerer Ladung gleichen Vorzeichens.

$$\varkappa^4 = (b^4_{11}\,\tau_1 + b^4_{21}\,\tau_2)'\,\tau_1' + (b^4_{12}\,\tau_1 + b^4_{22}\,\tau_2)\,\tau_2' \tag{14.11}$$

mit den Reaktionen

$$P\,(1,1) \rightleftharpoons (-\,n\,(0,0)) + \mathfrak{P}\,(1,1) \tag{14.12}$$

$$N\,(0,1) \rightleftharpoons (-\,e\,(1,0)) + \mathfrak{P}\,(1,1)\,. \tag{14.13}$$

Das Feld 5 mit schwerer und elektrischer Ladung verschiedenen Vorzeichens gibt nur die Matrix $\varkappa^5 = 0$.

Ausser der letzten Reaktion (14.13) sind alle neuen Reaktionen dieselben wie diejenigen der τ Matrices, nur spielt überall das Antineutrino die Rolle des Neutrinos.

Das Feld 4 gibt eine weitere Möglichkeit des β-Zerfalles: Ein Neutron wird eine $\mathfrak{P}$ (1,1)-Partikel (14.13) und sendet ein negatives Elektron aus. Die $\mathfrak{P}$ (1,1)-Partikel zerfallt hierauf, gemäss der algebraischen Umschreibung von (14.12), in ein Proton und ein Neutrino:

$$\mathfrak{P}\,(1,1) \longrightarrow P\,(1,1) + n\,(0,0)\,. \tag{14.12'}$$

Für (12.12') gilt das anlässlich (13.6') gesagte[17]). Das Auftreten von Neutrino und Antineutrino hat folgende tiefergehende Bedeutung:

a) Neutrino und Antineutrino sind verschiedene Partikel. Dann unterscheiden sie sich durch die sogenannte Neutrinoladung. Fordert man die *Erhaltung der Neutrinoladung*, so muss man auch dem Neutron eine Neutrinoladung zusprechen. Die Matrix

$$\lambda'' = 1 - \lambda \tag{14.12}$$

erlaubt dann die von den Spinorpartikeln getragene Neutrinoladungsdichte zu formen. Aus (12.2) folgt dann

$$[\lambda'', \tau^r] = \Lambda''^r \tau^r = -\Lambda^r \tau^r \tag{14.13}$$

d. h. die Partikel des Kernfeldes haben gleichzeitig elektrische und Neutrinoladung umgekehrten Vorzeichens. Aus (14.14) folgt aber

$$\lambda'' \varkappa^r + \varkappa^r \lambda'' = -\Lambda''^r \varkappa^r = (2 + \Lambda^r)\, \varkappa^r \tag{14.14}$$

(14.14) und (14.13) sind nur miteinander verträglich, wenn entweder $\varkappa^r$ oder τ^r verschwindet, d. h. für ein bestimmtes Feld Φ^r treten nur entweder die Reaktionen des Paragraphen 13 oder die Reaktionen aus diesem Paragraphen auf.

b) Es existiert kein Unterschied zwischen Neutrino und Antineutrino. Dann können die Matrices $\varkappa$ so gewählt werden, dass in den Wechselwirkungstermen Feld-Materie in der Hamiltonfunktion der Spinor des Neutrinofeldes φ^2 nur in der Kombination

$$\overline{\varphi}^2 = \varphi^2 + \delta^* \varphi^{2*} \tag{14.15}$$

auftritt. Wählt man die Matrices α_i und β in der Form, dass die α_i rein reell und β rein imaginär erscheinen, so wird die Matrix δ gleich der Einheitsmatrix und man hat $\overline{\varphi}^2 = \overline{\varphi}^{*2}$. Majorana[9]) hat gezeigt, dass man dann auch den Anteil der freien Spinorpartikel in der Hamiltonfunktion allein unter Verwendung der reellen Funktion $\overline{\varphi}^2$ schreiben kann. Das reelle Spinorfeld kennt also, genau wie das reelle Tensorfeld, keine Antipartikel, d. h. es besteht aus nur einer Partikelart (vgl. dazu auch Racah[19]).

15. Schlussbemerkung.

Nachdem die Existenz einer Kontinuitätsgleichung bei Abwesenheit von elektrischen Feldern gezeigt worden ist, macht die Einführung der Wechselwirkung „Elektrisches Feld mit elektrisch geladenen Φ^r und φ^ν Feldern" keine prinzipiellen Schwierigkeiten mehr. Klassisch ist der Fall von Proca[10]) bereits behandelt.

Es sind dann u. a. folgende interessante Eigenschaften der neuen Partikel zu behandeln (vgl. dazu auch BHABHA[2]):

1. Bremsstrahlung, Comptoneffekt und Paarerzeugung der e-Partikel (für spinlose Partikel wurde die zur Bethe-Heitler'schen analoge Formel bereits von PAULI und WEISSKOPF berechnet[6]).

2. Absorption (und Streuung) eines e-Teilchens (oder eines ungeladenen n-Teilchens) durch ein schweres Teilchen im Atomkern (= Atomzertrümmerung durch e- oder n-Teilchen, da schon die Ruhenergie dieser Teilchen genügt, um einen Kernbestandteil aus seiner Bindung zu lösen)[21].

3. Erzeugung von Paaren von e- oder n-Teilchen, durch Rekombination von Proton, Neutron mit Antiproton und Antineutron. Ausstrahlung einer oder mehrerer e- und n-Partikel durch Bremsung von schnellen Neutronen und Protonen.

Das Entstehen der doch offenbar instabilen e-Partikel (über ihre Instabilität vgl. auch soeben veröffentlichte Beobachtungen von BLAKETT[20]) könnte dann eventuell, ausser durch Paarerzeugung aus einer primären Photonenstrahlung, durch diese Rekombination einer primären, aus schweren Antiteilchen bestehenden, kosmischen Strahlung mit Kernbestandteilen gedeutet werden.

Institut de Physique, Université de Genève.

Literatur.

[1]) bis [11]) siehe unter STUECKELBERG, Teil I[12]).
[12]) STUECKELBERG, Helv. Phys. Acta **11**, 225 (1938).
[13]) STUECKELBERG, Helv. Phys. Acta **9**, 389 (1936).
[14]) BREIT, CONDON and PRESENT, Phys. Rev. **50**, 825 (1936).
[15]) FIERZ, Zeitschr. f. Phys. **104**, 553 (1937).
[16]) FERMI, Zschr. f. Phys. **88**, 161 (1934).
[17]) WENTZEL, Zschr. f. Phys. **104**, 34 (1936) und **105**, 738 (1937).
[18]) PAULI, Ann. Inst. H. Poincaré, 1936, p. 109.
[19]) RACAH, Nuovo Cim. **14** (N° 7) (1937).
[20]) BLACKETT, Proc. Roy. Soc. **165**, 30 (1938).
[21]) STUECKELBERG, Helv. Phys. Acta **11**, 378 (1938).

La signification du temps propre en mécanique ondulatoire [53]

Helvetica Physica Acta, vol. 14 (1941), pp. 322–323

La signification du temps propre en mécanique ondulatoire

par E. C. G. STUECKELBERG (Genève).

La *théorie classique* contenue dans

$$\ddot{q}^{\mu} = e\, B^{\mu\nu}\, \dot{q}_{\nu} \text{ }^{1)} \tag{1}$$

montre que $-\dot{q}_{\mu}\,\dot{q}^{\mu} = (d\,s/d\,\lambda)^2 = m^2$ est une constante d'intégration. Le paramètre λ est donc proportionnel au temps propre s. Suivant que $d\,s = \pm\, m\, d\,\lambda$ (1) représente une particule de masse m et de charge $\pm\, e$. Une particule à charge $+\, e$ resp. $-\, e$ est représentée par une ligne d'univers qui évolue vers le futur, resp. vers le passé, si λ augmente ($\dot{q}^4 > 0$ resp. < 0). La réaction de la particule sur le champ peut être décrite par une densité de charge

$$\varrho\,(\vec{x}, t = x^4) = e \int_{-\infty}^{+\infty} \dot{q}^4\, d\lambda\, \delta\,(x^1 - q^1)\; \delta\,(x^2 - q^2)\; \delta\,(x^3 - q^3)\; \delta\,(x^4 - q^4) \tag{2}$$

Son intégrale spatiale vaut $\pm\, e$, suivant que $\dot{q}^4 \gtrless 0$.

La *théorie quantique* découle d'une Hamiltonienne R (correspondant à la masse, c. à d. $R = -\frac{1}{2}\,\mathrm{m}^2$)

$$R = \tfrac{1}{2}\, \pi_{\mu}\, \pi^{\mu};\; \pi_{\mu} = p_{\mu} - e\, \Phi_{\mu}\,(q)\,. \tag{3}$$

Pour autant que les potentiels Φ_4 ne varient que dans des dimensions grandes par rapport aux longueurs d'ondes contenues dans la fonction $\psi\ (q^1, q^2, q^3, q^4, \lambda)$, le paquet d'ondes (normalisé à $\iiiint (d\,q)^4\, |\psi|^2 = 1$) suit la ligne d'univers classique. Un champ électrique n'existant que pendant une très courte période au temps $t = x^4 = 0$ est décrit par un potentiel $\Phi^{\mu} =$ constant, qui possède une discontinuité sur l'hyperplan $x^4 = t = 0$. Cette discontinuité donne lieu, en plus de la réfraction du paquet d'ondes primaire (correspondant à l'accélération de la particule) à une *réflexion* qui correspond à la ligne d'univers d'une particule de même masse mais de *charge opposée.*

La normalisation de ψ et sa dépendance du "temps propre λ" présente certaines difficultés d'interprétation probabiliste. L'exemple décrit semble exprimer le fait suivant:

"Si j'observe au temps $t > 0$ une particule $(m, -e)$ et si je sais qu'au moment $t = 0$ un champ électrique existait, il y a A) une probabilité W_A que la particule a été accélérée par ce

1) $\dot{F}$ signifie la dérivée de F par rapport au paramètre λ. $q^{\mu}(p = 1, 2, 3, 4)$ sont les coordonnées d'un point. $B^{\mu\nu} = -B^{\nu\mu}$ est le champ électromagnétique.

champ, et B) une probabilité $W_B = 1 - W_A$ que la particule n'existait pas pour $t < 0$ mais qu'elle est la partenaire d'une *paire de particules* $(m, -e)$, $(m, +e)$, créée au moment $t = 0$ par ce champ."[1])

Remarque à propos de la création de paires de particules en théorie de relativité [54]

Helvetica Physica Acta, vol. 14 (1941), pp. 588–594

Remarque à propos de la création de paires de particules en théorie de relativité

par E. C. G. Stueckelberg.

(18. X. 1941.)

Résumé: La mécanique de la théorie de relativité peut être mise sous une forme qui permet de comprendre la création de paires de particules de charges électriques opposées sans faire appel à la théorie des quanta. Le changement apporté par cette modification à la théorie des quanta est discuté.

La théorie de relativité exprime les lois physiques dans une forme covariante par rapport à un certain groupe de transformations. Ce groupe est celui des transformations de LORENTZ, pour autant qu'on néglige l'influence de la gravitation. Le groupe est déterminé par les équations de MAXWELL parce que les observations électromagnétiques (par ex. l'expérience de MICHELSON) ne permettent pas de distinguer entre deux systèmes de coordonnées, dont l'un décrit un mouvement rectiligne et non accéléré par rapport à l'autre. Pour tenir compte des effets de gravitation, EINSTEIN a envisagé un groupe plus général de transformations, qui laissent invariant le carré de la distance spatiotemporelle $(ds)^2$ entre deux événements.

La mécanique du point matériel soumis aux forces électromagnétiques et gravifiques peut être exprimée sous une forme covariante par rapport au groupe de ces transformations générales. Les traits fondamentaux de cette mécanique d'EINSTEIN sont les suivants:

Au cours du temps t, le point matériel suit une *trajectoire*, qui est déterminée par les trois fonctions $x^i = q^i(\tau)$. Elles indiquent les valeurs des trois coordonnées d'espace x^i ($i = 1, 2, 3$) à l'instant $t = \tau$. La théorie de relativité fait intervenir le temps t sous la forme d'une quatrième coordonnée $t = x^4$. A la courbe troisdimensionnelle de la trajectoire correspond ainsi une courbe quadridimensionnelle $x^i = q^i(\tau)$; $x^4 = q^4(\tau) = \tau$, appelée *ligne univers*. Elle est exprimée en termes du paramètre τ. Pourtant la substitution du paramètre τ (= au temps) par un paramètre quelconque λ reste possible. Cette substitution $\tau = \tau(\lambda)$ effectuée dans les équations pour x^μ ($\mu = 1, 2, 3, 4$) donne à la représentation paramétrique de la ligne d'univers une forme plus symétrique par rapport aux coordonnées de l'espace-temps:

$$x^{\mu} = q^{\mu}(\lambda) \tag{1}$$

Discutons les équations fondamentales de la mécanique. Soit $\vec{x}$ ($= x^1, x^2, x^3$) le vecteur de l'endroit et $\vec{q}(\tau)$ ($= q^1, q^2, q^3$) la position de la particule au temps $t = \tau$. Soit $\vec{\Gamma}(\vec{x}, t)$ le vecteur du champ de gravitation, $\vec{E}(\vec{x}, t)$ le vecteur du champ électrique et $\vec{B}(\vec{x}, t)$ celui du champ magnétique. Le mouvement $\vec{x} = \vec{q}(\tau)$ de la particule (et sa ligne d'univers) est, en théorie non relativiste, une solution de l'équation fondamentale de NEWTON-LORENTZ:

$$m \frac{d^2\vec{q}}{(d\tau)^2} = m\,\vec{\Gamma} + e\vec{E} + \frac{e}{c}\frac{d\vec{q}}{d\tau} \times \vec{B} \tag{2}$$

$\vec{a} \times \vec{b}$ est le produit vectoriel de deux vecteurs. $\vec{\Gamma}$, $\vec{E}$ et $\vec{B}$ sont à évaluer pour les valeurs ($\vec{q}$, τ), etc. Si $\vec{\Gamma}(x, t) = \vec{\Gamma}(x, x^4) = \vec{\Gamma}(x)$, etc. est connu pour tout le domaine spatiotemporel x ($= x^1, x^2, x^3, x^4$) intéressant, la solution de (2) détermine les trois fonctions $x^i = q^i(\tau)$ (la trajectoire) et naturellement aussi les quatre fonctions $x^i = q^i(\tau)$; $x^4 = q^4(\tau) = \tau$ de la ligne d'univers. En théorie de relativité, l'équation (2) est un peu modifiée, mais ne contient pas de changements importants.

Un point fondamental de cette théorie habituelle est le suivant: La mécanique d'EINSTEIN s'exprime sous une forme qui n'admet que des lignes d'univers ayant *une seule intersection* avec un hyperplan $t (= x^4) = t_0 =$ const (cf. ligne A en fig. 1). Cette seule intersection, qui se fait au point $x^i = q^i(t_0)$, est l'endroit où se trouve la particule à l'instant $t = t_0$. D'autres lignes, d'une forme plus générale (par ex. la ligne B en fig. 1), qui montrent *deux* intersections pour des plans $t (= x^4) = t_0 < 0$ et aucune intersection pour $t (= x^4) = t_0 > 0$ ne peuvent pas figurer dans la mécanique d'EINSTEIN. Ceci est dû au choix particulier du paramètre s en $x^{\mu} = q^{\mu}(s)$. Il est défini comme la longueur de l'arc, c.-à-d. comme l'intégrale de la distance spatiotemporelle $\sqrt{(ds)^2}$ entre deux événements voisins de la courbe. On l'appelle le temps propre. Or $(ds)^2$ n'est positif que pour deux événements, dont l'un est postérieur à l'autre dans tout système de référence. La ligne B fig. 1 ayant des régions où ce n'est certainement pas le cas, ne peut donc pas être exprimée en termes de ce paramètre s. Si, au moment de l'établissement de la théorie d'EINSTEIN, des lignes de ce dernier type n'ont pas été discutées, c'est probablement parce que le *phénomène de la création et de l'annihilation de paires de particules* n'avaient pas été découverts.

Aujourd'hui, vu la découverte de l'électron positif, les lignes B et C fig. 1 admettent une interprétation bien naturelle: Les deux intersections pour $t (= x^4) = t_0 < 0$ de la ligne B représentent les deux endroits des deux partenaires d'une paire de particules. Cette paire est composée d'un électron positif et d'un électron négatif. Leurs lignes d'univers sont d'une forme telle qu'ils se

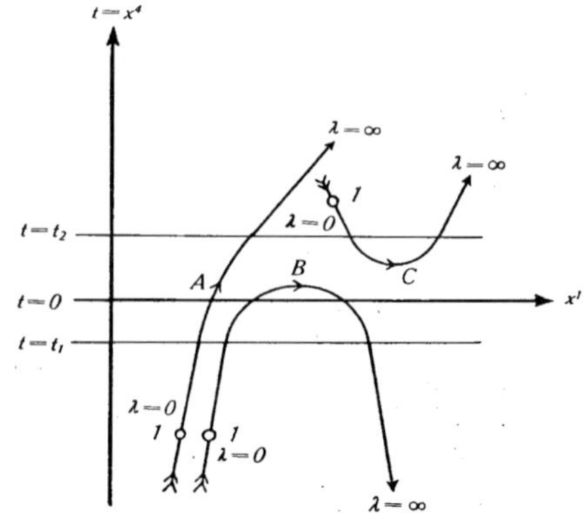

Fig. 1.

Lignes d'univers: A. type habituel (à chaque temps $t = x^4$ correspond *un seul* x^1 représentant l'endroit de la particule); B. type annihilation (à chaque $t = x^4 \ll 0$ correspondent *deux valeurs de* x^1 représentant les endroits d'une paire de particules qui vont s'annihiler pour $t \sim 0$); C. type production de paire (à chaque $t = x^1 \gg 0$ correspondent *deux valeurs de* x^1 etc.).

rencontrent au moment $t \sim 0$, la ligne B décrit ainsi l'anéantissement mutuel des deux corpuscules. On comprend alors pourquoi, pour des temps $t > 0$, il n'existe plus aucune intersection, parce qu'il n'existe plus aucune de ces deux particules. La ligne C est l'illustration spatio-temporelle du phénomène contraire, c.-à-d. de la création d'une paire à l'instant $t \sim 0$.

La question se pose de savoir s'il est possible d'établir une mécanique covariante au sens d'Einstein, qui permette l'existence de telles courbes. Nous nous rappelons que les composantes $E_i = B_{i4}$ $(i = 1, 2, 3)$ du vecteur $\vec{E}$ et les composantes $(\overrightarrow{B})_1 = B_{23}$; $(\overrightarrow{B})_2 = B_{31}$ et $(\overrightarrow{B})_3 = B_{12}$ forment les composantes d'un tenseur covariant et antisymétrique $B_{\mu\nu}$ $(\mu, \nu = 1, 2, 3, 4)$ en quatre dimensions. De même, les composantes $\Gamma_i = -\Gamma^i_{44}$ du vecteur $\vec{\Gamma}$

du champ de gravitation, forment les composantes d'un tenseur affine et mixte $\Gamma^{\mu}_{\nu\varrho}$. Supprimant alors les indices de sommation tensorielle, en $a_\mu b^{\mu\nu} = \sum_\mu a_\mu b^{\mu\nu}$, l'équation de mouvement quadridimensionnelle pour $x^\mu = q^\mu(\lambda)$ deviendra fondamentale de la mécanique

$$\frac{d^2 q^\mu}{(d\lambda)^2} = -\Gamma^{\mu}_{\alpha\beta}\frac{dq^\alpha}{d\lambda}\frac{dq^\beta}{d\lambda} + e B^{\mu\alpha} g_{\alpha\beta}\frac{dq^\beta}{d\lambda} + K^\mu \tag{3}$$

Elle est (au terme K^μ près) la généralisation covariante de (2). ($g_{\alpha\beta}$ est le tenseur fondamental qui relie les composantes co- et contra-variantes $B_{\mu\nu} = g_{\mu\alpha} g_{\mu\beta} B^{\alpha\beta}$). On peut démontrer (vu la structure de $\Gamma^{\mu}_{\nu\varrho}$[1]) que la quantité $m^2 = -g_{\mu\nu}\frac{dq^\mu}{d\lambda}\frac{dq^\nu}{d\lambda}$ est une constante d'intégration si $K^\mu = 0$. Si $m^2 > 0$, les lignes sont du type A prévu par Einstein. La relation entre le paramètre $ds = +\sqrt{ds^2}$ (défini par $ds > 0$ si $d\tau = dq^4 > 0$) et $d\lambda$ est alors, avec $m = +\sqrt{m^2}$:

$$ds = \pm\, m d\lambda \tag{4}$$

La substitution de (4) en (3) réduit (toujours si $K^\mu = 0$) notre formule à la formule d'Einstein, qui prend, à son tour, la forme (2) de la mécanique non relativiste de Newton-Lorentz pour des vitesses $|d\vec{q}/d\tau| \ll c$. m a donc la signification de la masse au repos. Mais il y a une différence très remarquable entre la théorie habituelle et la nôtre. L'ambiguité du signe en (4) a pour conséquence une ambiguité du signe de la charge électrique e dans l'équation fondamentale de la mécanique. ($\pm e$ au lieu de e au 2ème membre de (2)). Notre mécanique (3) a ainsi l'avantage d'être valable à la fois pour les deux charges $\pm e$.

Le terme K^μ en (3) montre la possibilité de faire intervenir dans la théorie des forces nouvelles de nature ni électromagnétique ni gravifique. Si $K^\mu \neq 0$, la quantité m^2 définie plus haut n'est plus une constante d'intégration. Dans la région d'espace temps où ces champs apparaissent, la masse de repos de la particule doit donc varier. La fig. 1 montre l'effet qu'un champ particulier ($K^\mu = 0$ sauf dans l'intervalle $t_1 < t < t_2$ où K^4 diffère de zéro) exerce sur la ligne d'univers. Ce champ accélère la particule (ligne A) pendant l'intervalle $t_2 - t_1$ en diminuant en même temps la masse de repos, la particule gardant la direction de sa trajectoire. Un champ semblable mais plus fort a pour conséquence

[1]) cf. par ex. W. Pauli, Relativitätstheorie. Teubner, Leipzig-Berlin (1921), p. 587, Formule (69).

de déformer la ligne A en la ligne B. Un champ de cette intensité cause donc *l'annihilation d'une paire de particules.* On voit également que les deux partenaires sont de charge électrique opposée parce que le signe $dq^4/d\lambda \lessgtr 0$ détermine le signe de $\pm e$. Un champ semblable et de même intensité mais ayant le signe opposé, déforme une ligne du type A en des courbes du type C et cause ainsi la *création d'une paire de particules.* Une difficulté se présente parce que les particules parcourent une partie de leur existence avec des vitesses supérieures à celle de la lumière. Ceci, et d'autres considérations d'ordre causal, nous semble être un argument important contre l'hypothèse de l'existence de telles forces, malgré la covariance de leur représentation.

Mais, même sans introduire ces champs nouveaux, la mécanique proposée a certains avantages sur celle d'Einstein. La racine carrée qui, en théorie ordinaire, reliait l'énergie à l'impulsion et qui formait le grave obstacle à la quantification de la mécanique relativiste du point matériel a disparu dans notre théorie. Le procédé de quantification de Schroedinger peut alors être mis sous une forme où l'espace et le temps interviennent d'une façon entièrement symétrique. A certains égards, le paramètre λ jouera le rôle de paramètre τ dans l'équation de Schroedinger, tandis que les quatre coordonnées $q (= q^1, q^2, q^3, q^4 = \tau)$ prendront la place des trois $\vec{q} \ (= q^1, q^2, q^3)$ en théorie non relativiste.

L'effet de cette quantification est d'établir une correspondance entre la théorie des rayons $x^\mu = q^\mu(\lambda)$ (optique géométrique) dans l'espace quatridimensionnel avec la propagation des paquets d'ondes $\psi(q, \lambda)$ normalisées à $\iiiint (dq)^4 \mid \psi \mid^2 = 1$ (optique ondulatoire). Nous discutons l'exemple d'un champ électrique $\vec{E}$ à composantes $E_2 = E_3 = 0$ et, pendant l'intervalle très court $0 < t < \delta t$, $E_1 \neq 0$. Dans la limite $\delta t = 0$, $E_1 = \infty$, $E_1 \delta t =$ fini, ce champ est décrit par un potentiel quadrivecteur $\Phi_2 = \Phi_3 = \Phi_4 = 0$; $\Phi_1 = 0$ pour $t > 0$ et $\Phi_1 = E_1 dt$ pour $t < 0$. L'hyperplan $x^4 = t = 0$ est maintenant une surface de discontinuité dans le continu espace-temps. Un rayon incident sur cette surface est réfracté. Fig. 2 montre la réfraction d'un rayon ψ_1 venant d'un point situé dans le demi-espace-temps supérieur $t > 0$. La ligne d'univers correspondant à cette réfraction n'est autre chose que le mouvement d'un *électron négatif* $(dq^4/d\lambda < 0)$ accéléré par le champ $\vec{E}$ pendant l'intervalle δt. En théorie ondulatoire, un rayon réfléchi d'intensité définie et non nulle, est relié à tout rayon réfracté. A cet rayon réfléchi ψ_A correspond un *électron positif*, $(dq^4/d\lambda > 0)$, créé au moment $t = 0$.

L'interprétation probabiliste de ce phénomène a déjà été discutée d'autre part[1]). Elle fera l'objet d'un exposé plus détaillé[2]) de cette modification apportée à la théorie d'EINSTEIN. Remarquons déjà ici que cette théorie permet de prédire les ***espérances mathématiques*** de grandeurs physiques. L'espérance mathématique de la charge électrique $\overline{e}_V(t)$ qu'on observera dans un volume spatial V à un temps donné t et celle de l'énergie $W_V(t)$ dans un tel volume peuvent être calculées. Les prévisions se basent sur une *mesure* faite sur une particule au moment $q^4 \cong t_0 = q_0^4$. La mesure contient la détermination 1⁰ de l'endroit $\vec{q} \cong \vec{q}_0$, 2⁰ de la vitesse $\overline{v} \cong \overline{v}_0$ et 3⁰ dans une expérience de déflection électromagnétique,

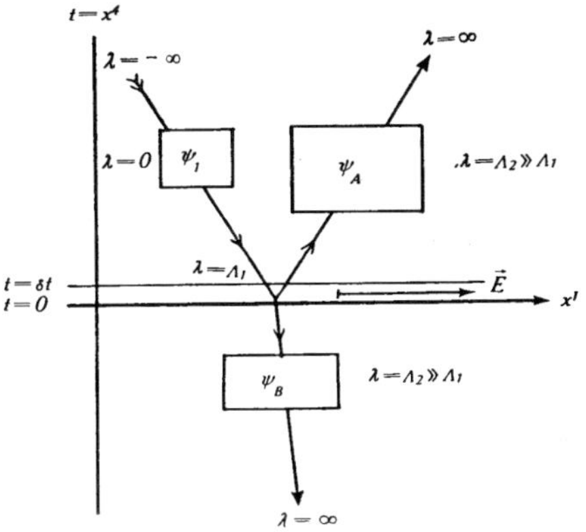

Fig. 2.

Sous l'influence d'une discontinuité de Φ_1, le paquet d'ondes ψ_1 qui se trouve «au moment $\lambda = 0$» à $x^\mu = q^\mu$, se décompose «au moment $\lambda = \Lambda_1$» en un paquet réfracté ψ_B et un paquet réfléchi ψ_A.

de la valeur de e/m. Les résultats sont naturellement soumis à une relation d'incertitude (différente de celle d'HEISENBERG parce que $\Delta t = \Delta q^4 \neq 0$). Dans l'exemple exposé, cette mesure a été exécutée et son résultat est représenté par une particule de charge $-e$ observée pour $t > 0$, qui se trouve sur le rayon marqué par ψ_1. Tout volume entourant ce rayon (pour $t > 0$) fournit l'espérance mathématique $\overline{e}_1 = -e$. Par contre, l'espérance mathématique de la charge contenue dans un volume entourant le rayon réfléchi

[1]) STUECKELBERG. Comm. Soc. Suisse de Phys. Séances des 7 et 8 IX. 1941; Helv. Phys. Acta **14**, 322 (1941); Actes Soc. Helv. des Sci. Nat. **121**, (1941).

[2]) A paraître au prochain numéro des Helv. Phys. Acta.

(marqué par ψ_A) vaut $\bar{e}_A = + eW_A$. W_A est un nombre, contenu entre 0 et 1, déterminé par la théorie. La particule ne pouvant porter que des charges $\pm e$, il y a donc certitude que la particule suit la trajectoire correspondant à celle du paquet d'ondes ψ_1 pour $t > 0$, et la probabilité W_A qu'une antiparticule (de masse d'ailleurs égale à celle de la particule observée) se meut le long de la trajectoire du paquet ψ_A pour $t > 0$.

W_A est donc la probabilité que le champ $\tilde{E}$ a créé une paire dont l'une des partenaires suit la ligne d'Univers ψ_1.

Dans toutes les théories qui ont relié la relativité aux quanta, on a du faire intervenir le phénomène de la création et de l'annihilation de paires de particules. Nous voyons qu'ici, encore une fois, celà est le cas.

Je ne veux pas terminer cet exposé général sans exprimer la grande joie que j'éprouve à publier ces résultats à l'occasion de l'anniversaire de Monsieur le Professeur A. Hagenbach. Je le prie d'accepter cette note comme un signe de ma profonde reconnaissance.

Genève, Institut de Physique de l'Université.
Octobre 1941.

La mécanique du point matériel en théorie de relativité et en théorie des quanta [56]

Helvetica Physica Acta, vol. 15 (1942), pp. 23–37

Reprinted with permission

La mécanique du point matériel en théorie de relativité et en théorie des quanta

par **E. C. G. Stueckelberg.**

(18. X. 41.)

Résumé. Une légère modification de la mécanique d'Einstein (remplaçant l'extrémum de $\int m\,ds = \int m\sqrt{-\dot{q}_\mu \dot{q}^\mu}\,d\lambda$ par celui de $\frac{1}{2}\int \dot{q}_\mu \dot{q}^\mu\,d\lambda$) permet d'établir une nouvelle mécanique relativiste. Ses résultats ne diffèrent pas des résultats obtenus par la forme habituelle, si l'on ne fait intervenir que des champs gravifiques et électromagnétiques. Mais, tout en gardant la covariance de la théorie, on peut introduire des *champs nouveaux* qui ont pour conséquence la *création de particules dans la théorie classique.*

La quantification de la théorie représente l'extension logique de la théorie de Schroedinger aux quatre dimensions de l'espace-temps. Un de ses résultats est la *création de particules par des champs électromagnétiques.*

Exposé général de la théorie[1]).

Théorie classique du mouvement d'un point de masse. — Au cours du temps t, le point matériel décrit une courbe troisdimensionnelle, la *trajectoire.* Celle-ci est déterminée par les trois fonctions $x^i = q^i(\tau)$, qui donnent les valeurs des trois coordonnées $(i = 1, 2, 3)$ x^i au temps $t = \tau$. Si l'on introduit le temps $t = x^4$ comme une quatrième coordonnée, et si, en plus, on considère $t = \tau(\lambda) = q^4(\lambda)$ comme fonction d'un paramètre λ quelconque, la courbe quadridimensionnelle $x^\mu = q^\mu(\lambda)$ $(\mu = 1, 2, 3, 4)$ représente la *ligne d'univers* en termes d'un paramètre quelconque λ.

La théorie d'Einstein (§ 1) donne une loi qui permet de construire ces lignes d'univers, si le *champ de gravitation* $\Gamma^{\lambda}_{\mu\nu}(x) = \Gamma^{\lambda}_{\nu\mu}(x)$) et le *champ électromagnétique* $(B^{\mu\nu}(x) = - B^{\nu\mu}(x))$ sont donnés comme fonctions d'espace-temps $(B^{\mu\nu}(x) = B^{\mu\nu}(x^1, x^2, x^3, x^4))$. Une telle ligne est entièrement déterminée si 1° la position $\vec{x} = \vec{q}$ $(= q^1, q^2, q^3)$ et la vitesse $\vec{v}$ $(= dq^1/d\tau, dq^2/d\tau, dq^3/d\tau)$ sont données pour un certain temps initial $t = q^4$ et si 2° un certain nombre e/m (rapport entre la charge électrique et la masse de repos) qui caractérise le point matériel est connu.

Ne sont admises dans la théorie d'Einstein, que les lignes d'univers ayant *une seule intersection* avec un hyperplan $t = x^4 =$

= const (cf. ligne A en fig. 1)*). En effet, ce ne sont que ces lignes qui correspondent à la conception habituelle de causalité: La seule intersection $x^i = q^i(t = \text{const})$ est l'endroit où l'on trouve la particule au temps t. D'autres lignes (par ex. la ligne B en fig. 1) montrent deux intersections pour des plans $t = x^4 = \text{const} \ll 0$ et aucune intersection pour $t = x^4 = \text{const} \gg 0$. Si, au moment de l'établissement de la théorie d'Einstein, de telles lignes n'étaient pas discutées, c'était parce que le phénomène de la création et de l'annihilation de paires de particules échappait encore aux expérimentateurs et aux théoriciens. Or la ligne B de la fig. 1[1]) décrit une telle annihilation mutuelle de deux particules au moment $t \sim 0$.

Tandis que la mécanique d'Einstein n'admettait donc que des courbes du type A, la mécanique proposée au § 2 se libère de cette restriction. Son résultat sera que, en plus des lignes A, des courbes du type B ou C peuvent apparaître. Les deux partenaires d'une paire ainsi créée ou anéantie ont des charges électriques opposées. Pourtant, les phénomènes B et C ne font apparition que si l'on admet, en plus des champs électromagnétiques et gravifiques, un champ d'un type nouveau $K^\mu(x)$ (§ 3). En l'absence de ce champ, la mécanique proposée ne présente pas de nouveaux phénomènes. Il en sera tout autrement dans la nouvelle mécanique quantifiée.

Théorie quantique du mouvement d'un point de masse. — La mécanique proposée permet une quantification en quatre dimensions analogue à celle introduite par Schroedinger pour les trois dimensions spatiales en théorie non relativiste. Le rapport entre la mécanique classique et la mécanique quantique (§ 5) est alors celui entre l'optique géométrique et l'optique ondulatoire dans le continu quadridimensionnel de l'espace-temps. Pour en donner un exemple, nous considérons le cas où, au temps $t = 0$, un champ électrique homogène très fort E_1 (parallèle à l'axe x^1) apparaît pendant un intervalle très court δt. Cet événement peut être décrit par un potentiel vecteur Φ^μ à une seule composante $\Phi^1 = E_1 \delta t$ pour $t < 0$, qui disparaît pour $t > 0$. L'hypersurface $t = 0$ représente ainsi une surface de discontinuité dans le continu spatiotemporel. Un rayon au sens de l'optique géométrique est *réfracté* sur cette surface (cf. fig. 2). La réfraction n'est pas autre chose que le changement de vitesse dû à l'accélération subie par la particule pendant l'intervalle δt. La ligne d'univers de la mécanique classique correspond à ce rayon réfracté. Mais l'optique ondula-

*) Les figures ont été publiés à l'occasion de l'anniversaire de M. A. Hagenbach dans le numéro précédent. Helv. Phys. Acta **14**, 588 (1941).

toire montre qu'à toute *réfraction* est liée une *réflexion* d'une intensité non nulle. Ce rayon réfléchi est du type C (ou B) de la fig. 1. Notre nouvelle mécanique montre ainsi que, en théorie des quanta, le champ électromagnétique a la propriété de créer et d'annihiler des paires de particules.

Remarquons, pour terminer cette introduction, que la mécanique d'EINSTEIN ne permettait pas de quantification. *L'électron de Dirac* n'est pas la quantification du point matériel d'EINSTEIN, mais celle d'un *système plus complexe* (point de masse avec des degrés de liberté intérieurs)[2]). D'autre part, l'équation de SCHROEDINGER-GORDON n'est pas une quantification du point de masse non plus, mais la *théorie d'un continu scalaire* à deux composantes (= une composante complexe). La *quantification de ce continu,* par PAULI et WEISSKOPF[3]), montre alors le phénomène de la création et de l'annihilation de paires de quanta de charges opposées.

La mécanique ici présentée nous semble donc être la seule mécanique relativiste du point de masse qui permet la quantification directe.

§ 1. La mécanique d'Einstein.

Soit $g_{\mu\nu}(x)$ les composantes covariantes du tenseur fondamental (= potentiel gravifique) et dq^μ les différences de deux événements voisins sur la ligne d'univers $x^\mu = q^\mu(\lambda)$ qui est parcourue par le point matériel. La grandeur $(ds)^2 = -g_{\mu\nu}(q)\,dq^\mu dq^\nu$ peut alors être positive, nulle ou négative. Deux événements sont situés temporellement l'un vis-à-vis de l'autre si $(ds)^2 > 0$. dq^4 ne peut jamais devenir zéro sur une ligne où l'on a partout $(ds)^2 > 0$. Sur une telle ligne, le temps propre est défini par

$$ds = \pm \sqrt{g_{\mu\nu}\,dq^\mu\,dq^\nu} \text{ suivant que } dq^4 \gtrless 0 \tag{1,1}$$

L'équation fondamentale de la mécanique d'EINSTEIN prend alors la forme « masse au repos » par « accélération propre » égale à « force gravifique plus force électromagnétique ». On définit d'abord la quadrivitesse normalisée $w^\mu = dq^\mu/ds$; $w_\mu w^\mu = -1$. La loi, pour une particule de masse de repos m et de charge électrique e s'écrit sous forme

$$m\,\frac{dw^\mu}{ds} = -m\,\Gamma^\mu_{\alpha\beta}\,w^\alpha\,w^\beta + e\,B^{\mu\nu}\,w_\nu \tag{1,3}$$

Cette loi dérive d'un principe de variation $\delta I = 0$, où I se compose de deux parties $I = I_{\text{mat}} + I_{\text{int}}$.

$$I_{\text{mat}} = m\int_{q'}^{q''} ds\ ; \quad I_{\text{int}} = e\int_{q'}^{q''} dq^\mu\,\Phi_\mu(q) \tag{1,4}$$

Sont à varier les lignes d'univers $q^\mu = q^\mu(\lambda)$ reliant l'événement q' à q'' $(q' = (q^{1'}, q^{2'}, q^{3'}, q^{4'}))$. $B_{\mu\nu}$ et $\Gamma^\mu_{\alpha\beta}$ sont les dérivées des potentiels $\Phi_\mu(x)$ et $g_{\mu\nu}(x)$

$$B_{\mu\nu} = \frac{\partial \Phi_\nu}{\partial x^\mu} - \frac{\partial \Phi_\mu}{\partial x^\nu} \; ; \quad \Gamma^\mu_{\alpha\beta} = \tfrac{1}{2} g^{\mu\lambda} \left(\frac{\partial g_{\lambda\alpha}}{\partial x^\beta} + \frac{\partial g_{\lambda\beta}}{\partial x^\alpha} - \frac{\partial g_{\alpha\beta}}{d x^\lambda} \right) \qquad (1,5)$$

évaluées pour $x^\mu = q^\mu$.

§ 2. La mécanique nouvelle.

Nous désignons les dérivées de $q^\mu(\lambda)$ par rapport au paramètre λ par $\dot q^\mu = \frac{dq^\mu}{d\lambda}$. Alors les équations fondamentales sont:

$$\frac{d}{d\lambda} \dot q^\mu = - \Gamma^\mu_{\alpha\beta} \dot q^\alpha \dot q^\beta + e B^{\mu\nu} \dot q_\nu \qquad (2,1)$$

et, pour $\dot q_\mu = g_{\mu\nu} \dot q^\nu$

$$\frac{d}{d\lambda} \dot q_\mu = \tfrac{1}{2} \frac{\partial g_{\alpha\beta}}{\partial x^\mu} \dot q^\alpha \dot q^\beta + e B_{\mu\nu} \dot q^\nu \qquad (2,2)$$

Elles dérivent d'un principe de variation analogue à celui du du § 1. I_{int} a la forme identique à (1,4), mais

$$I_{\text{mat}} = \int_{\lambda'}^{\lambda''} d\lambda \, \tfrac{1}{2} \dot q_\mu \dot q^\mu = \int_{\lambda'}^{\lambda''} d\lambda \, L_{\text{mat}}(q^\mu, \dot q^\mu) \qquad (2,3)$$

On vérifie d'abord que la quantité

$$m^2 = - \dot q_\mu \dot q^\mu \qquad (2,4)$$

est une *constante d'intégration* associée à la ligne d'univers. Donc, pour autant que cette constante est choisie positive $(m^2 > 0)$, la ligne est une succession de points situés temporellement les uns relativement aux autres. A une augmentation $d\lambda$ correspond alors une variation $dq^4 \gtrless 0$ suivant que $\dot q^4 \gtrless 0$. Introduisant le ds de la définition (1,1), on trouve $(m = + \sqrt{m^2})$

$$ds = \pm m \, d\lambda \quad \text{suivant que} \quad \dot q^4 \gtrless 0 \qquad (2,5)$$

La substitution de (2,5) en (2,1) donne l'équation identique à celle d'Einstein (1,3), mais avec les *deux possibilités du signe de e*, soit les équations d'Einstein pour la *particule* (m, e) et pour l'*antiparticule* $(m, -e)$. Notre théorie contient donc les deux charges d'une façon absolument symétrique.

Nous étudions alors les équations décrivant la réaction de la particule sur les champs.

Ce sont les équations de gravitation $G_{\mu\nu} = \varkappa T_{\mu\nu}$ et les équations de MAXWELL. Pour que les premières soient possibles, il faut démontrer l'existence d'un tenseur d'énergie impulsion $T_{\mu\nu}$ satisfaisant à

$$\frac{\partial T^{\mu\nu}}{\partial x^\nu} = 0 \tag{2,6}$$

en vertu de l'équation de mouvement (2,1) et des équations de MAXWELL

$$\frac{\partial [H^{\mu\nu}]}{\partial x^\nu} = 4\pi [J^\mu]\,; \qquad \frac{\partial [J^\mu]}{\partial x^\mu} = 0 \tag{2,7}$$

$[F] = \sqrt{-\|g\|}\,F$ est la densité tensorielle associée à un tenseur F. $\|g\|$ est le déterminant des $g_{\mu\nu}$. La première équation (2,7) dérive d'un principe de variation invariant

$$\delta \iiiint (dx)^4 [\mathfrak{L}] = 0 \tag{2,8}$$

avec

$$\begin{aligned} &\mathfrak{L} = \mathfrak{L}_{\text{él}} = \mathfrak{L}_{\text{maxw}} + \mathfrak{L}_{\text{int}} \\ &\mathfrak{L}_{\text{maxw}} = -\frac{1}{16\pi} B_{\mu\nu} B^{\mu\nu}\,; \quad H^{\mu\nu} = -8\pi \frac{\partial \mathfrak{L}}{\partial B_{\mu\nu}} \\ &\mathfrak{L}_{\text{int}} = J^\mu \Phi_\mu \end{aligned} \tag{2,9}$$

Les fonctions $\Phi_\mu(x)$ sont à varier. Pour définir J^μ nous introduisons la fonction singulière de DIRAC $[\varrho(x)]$[1]) ayant la propriété

$$\underset{\Omega}{\iiiint} (dx)^4 [\varrho(x)]\, f(x) = f(0) \quad \text{ou} \quad = 0 \tag{2,10}$$

suivant que le point $x^\mu = 0$ est contenu en Ω ou non. Si nous ajoutons à $\mathfrak{L}_{\text{él}}$ un $\mathfrak{L}_{\text{mat}}$ défini par

$$[\mathfrak{L}_{\text{mat}}] = \tfrac{1}{2} \int_{-\infty}^{+\infty} d\lambda\, g_{\mu\nu}(x) \dot{q}^\mu \dot{q}^\nu [\varrho(x - q(\lambda))] \tag{2,11}$$

et, si nous définissons en $[\mathfrak{L}_{\text{int}}]$ le courant $[J^\mu]$ par

$$[J^\mu(x)] = e \int_{-\infty}^{+\infty} d\lambda\, \dot{q}^\mu [\varrho(x - q(\lambda))] \tag{2,12}$$

---

*) $[\varrho]$ a en effet les propriétés d'une « densité tensorielle » scalaire. $(dx)^4 = dx^1\, dx^2\, dx^3\, dx^4$.

les équations de la mécanique (2,1) et (2,2) partent du même principe (2,8) avec

$$\mathfrak{L} = \mathfrak{L}_{\text{maxw}} + \mathfrak{L}_{\text{mat}} + \mathfrak{L}_{\text{int}} \tag{2,13}$$

En plus, des fonctions $\Phi_\mu(x)$, les fonctions $q^\mu(\lambda)$ sont à varier. La grandeur

$$T^{\mu\nu} = \frac{2}{\sqrt{-\|g\|}} \left(\frac{\partial\,[\mathfrak{L}]}{\partial\, g_{\mu\nu}}\right)_{\Phi_\mu,\; q^\mu = \text{const}} \tag{2,14}$$

satisfait alors à (2,6). $T^{\mu\nu}$ est, en vertu de l'indépendance de $\mathfrak{L}_{\text{int}}$ des $g_{\mu\nu}$, la somme $T^{\mu\nu}_{\text{maxw}} + T^{\mu\nu}_{\text{mat}}$. $T^{\mu\nu}_{\text{maxw}}$ a la forme habituelle du tenseur d'énergie-impulsion électromagnétique. La densité d'énergie correspondante T^{44}_{maxw} est donc positive. Il en est de même pour

$$[T^{\mu\nu}_{\text{mat}}] = \int_{-\infty}^{+\infty} d\lambda\; \dot q^\mu\, \dot q^\nu\, [\varrho(x - q(\lambda))] \tag{2,15}$$

pour autant que $\dot q^4 \neq 0$. La grandeur

$$W(t) = \iiint_V (d\,x)^3\,[T^{44}(x^1, x^2, x^3, x^4 = t)] = \pm\, \dot q^4 = m w^4 \tag{2,16}$$

$$= \frac{m}{\pm \sqrt{1 - |\vec v\,(t)\,|^2}} \quad \text{suivant que } \dot q^4 \gtrless 0$$

est en effet *toujours* positive. Elle représente l'énergie totale portée par la matière. L'intégrale est à prendre sur un volume spatial V entourant la ligne d'univers à l'instant t.

La charge totale de la particule vaut, à ce même instant:

$$e(t) = \iiint_V (d\,x)^3\,[J^4(x^1, x^2, x^3, x^4 = t)] = \pm\, e \tag{2,17}$$

$$\text{suivant que } \dot q^4 \gtrless 0$$

Remarquons ici une différence fondamentale entre la mécanique habituelle et la nôtre: la particule habituelle est caractérisée par *deux constantes m et e*. Pour prédire la ligne d'univers, il suffit de mesurer, à un certain moment $t = q^4$, les trois coordonnées q^i et les trois composantes de sa vitesse $v^i = \delta q^i/\delta q^4$. Dans notre modèle, la particule est caractérisée par *une seule constante e*. Nous mesurons d'abord, comme dans la théorie habituelle, à un certain moment $t = q^4$, les trois coordonnées q^i et, à un moment plus tard $q^4 + \delta q^4 (\delta q^4 > 0)$, la position $q^i + \delta q^i$. Ensuite, dans une expérience de déflection, nous observons son e/m; e étant donné, $m^2 = -\,\delta q_\mu \delta q^\mu\,|\,\delta\lambda\,|^{-2}$ déterminera la valeur absolue de

$\delta\lambda$. Le signe de $\delta\lambda$ est déterminé par le signe de e/m. De cette manière les q^μ et les $\dot{q}^\mu$ sont déterminés pour une valeur initiale de λ (par ex. $\lambda = 0$) et la ligne d'univers peut être prédite sous sa forme $x^\mu = q^\mu(\lambda)$ en résolvant l'équation différentielle (2,1).

§ 3. La production de paires de particules par des champs non électromagnétiques en mécanique classique.

Si l'on ajoute aux seconds membres de (2,1) un terme $K^\mu(q)$ dû à un champ $K^\mu(x)$, la valeur de $m^2 = -\dot{q}_\mu\dot{q}^\mu$ ne reste plus constante*). Ce nouveau champ a donc pour effet de changer « la masse de repos » de la particule. Si, en particulier, on a $K_\mu = \partial U/\partial x^\mu$, la grandeur

$$R = -\tfrac{1}{2} m^2 + U = \tfrac{1}{2}\dot{q}_\mu q^\mu + U(q) \qquad (3,1)$$

jouera le rôle de *constante d'intégration*. La fig. 1 illustre l'exemple où seule la composante K^4 diffère de zéro dans l'intervalle $t_2 - t_1$ avec $U(t > t_2) = 0$ et $U(t < t_1) = \text{const} \neq 0$.

La courbe A montre un changement de la masse de repos de la particule, accompagné d'une accélération. Si le champ K^4 est plus fort, la ligne se déforme en la forme B. Si le champ est de signe opposé, nous trouverons des lignes du type C. Dans les régions où le champ K_4 disparaît et où U reste constant, les lignes B et C représentent, en vertu de (3.1), des paires de particules de même masse m, mais, comme le démontre (2.17), de charges opposées. Un champ K^μ d'ordre plus général, créera des particules de charges opposées mais de masses différentes. La théorie, complétée par ce nouveau champ K^μ, garde naturellement son invariance relativiste. Mais elle montre des phénomènes qui semblent être *contraires à nos conceptions de causalité*. Pour démontrer ceci, considérons la ligne B de la fig. 1: Une mesure de q^μ et $\dot{q}^\mu$ pour $\lambda = 0$ a été faite, suivant les indications à la fin du § 2. Cette mesure est marquée comme l'événement (1) dans la fig. 1. Ensuite, je produis, au moment t_1, le champ K^4 pendant l'intervalle $t_2 - t_1$. La ligne d'univers, que je peux prédire en me basant sur les résultats de la mesure (1) et sur ma connaissance de l'intensité du champ produit, est du type A ou B. Si elle est du type B, je sais que, pour $t < t_1$, il y a toujours existé une *antiparticule* telle qu'elle rencontrera, à l'instant $t \sim 0$, la particule observée pour s'annihiler avec elle. Ces prévisions me semblent être contraires à nos notions de causalité. Remarquons enfin que les parti-

*) Voir l'éq. (2) de I.

cules peuvent atteindre des vitesses supérieures à celle de la lumière. Par une transformation de LORENTZ, on vérifie que ce dernier phénomène n'est qu'une description alternative de cette même série d'événements.

Ces considérations d'ordre causal nous semblent ainsi interdire l'existence de tels champs.

§ 4. Forme canonique.

L'équation de mouvement sans le champ K^μ (2,1) dérive d'un principe d'HAMILTON $\delta \int d\lambda L = 0$ avec

$$L = \tfrac{1}{2}\, \dot{q}_\mu \dot{q}^\mu + e\,\Phi_\mu(q)\dot{q}^\mu \tag{4,1}$$

Nous introduisons les variables conjuguées $p_\mu = \partial L/\partial \dot{q}^\mu = \dot{q}_\mu + e\,\Phi_\mu$ et définissons l'hamiltonienne par

$$\begin{aligned} R(p\,,q) &= -L + p_\mu \dot{q}^\mu = \tfrac{1}{2}\,\pi_\mu\,\pi^\mu \\ \text{avec} \quad \pi_\mu(p\,,q) &= p_\mu - e\,\Phi_\mu(q) \end{aligned} \tag{4,2}$$

La définition habituelle des parenthèses de POISSON $\{F\,,G\} = (\partial F/\partial p_\mu)(\partial G/\partial q^\mu) - (\partial F/\partial q^\mu)(\partial G/\partial p_\mu)$ fournit les relations

$$\{\pi_\mu\,,\pi_\nu\} = -e\,B_{\mu\nu}(q)\,; \quad \{\pi_\mu\,, f(q)\} = \partial f/\partial q^\mu \tag{4,3}$$

La loi de mouvement

$$\dot{F} = \{R\,,F\} \tag{4,4}$$

permet d'écrire (2,1) sous forme canonique

$$\begin{aligned} \dot{q}^\mu &= \{R\,,q\} = \pi^\mu \\ \dot{\pi}_\mu &= \{R\,,\pi_\mu\} = -\tfrac{1}{2}\,\frac{\partial g^{\alpha\beta}}{\partial q^\mu}\,\pi_\alpha\,\pi_\beta + e\,B_{\mu\nu}\,\pi^\nu \end{aligned} \tag{4,5}$$

L'hamiltonienne elle-même est une constante d'intégration et définit la masse: $2\,R = -m^2$. Le signe de $\pi^4 = \dot{q}^4$ détermine le signe de la charge.

§ 5. La quantification formelle.

En analogie parfaite avec la théorie de SCHROEDINGER, nous introduisons une amplitude de probabilité scalaire et complexe

$$\psi(q^1, q^2, q^3, q^4, \lambda) = \psi(q\,,\lambda) \tag{5,1}$$

satisfaisant à l'équation

$$R(p, q)\,\psi = -\frac{h}{j}\frac{\partial \psi}{\partial \lambda} = h\,j\,\dot{\psi} \tag{5,2}$$

$p_\mu = -jh\partial/\partial q^\mu$ est l'opérateur de différentiation. D'un opérateur hermitéique $F(p, q)$ opérant sur ψ et défini par une série de puissances en p_μ et q^μ, on forme l'espérance mathématique

$$\overline{F}(\lambda) = (\psi, F\psi) = \iiiint (dq)^4\,\psi^*(q, \lambda,)\,(F\psi(q, \lambda)). \tag{5,3}$$

ψ est normalisé à $(\psi, \psi) = 1$.

On vérifie que $\dot{\overline{F}} = \partial \overline{F}(\lambda)/\partial\lambda$ est l'espérance mathématique $\overline{\dot{F}}$ d'un opérateur $\dot{F}$ défini par le commutateur $([A, B] = AB - BA)$:

$$\dot{F} = \frac{j}{h}[R, F] = \{R, F\} \tag{5,4}$$

La dernière égalité suit des lois de commutation de p_μ et q^μ. Elle assure la correspondance entre la mécanique classique et quantique (cf. éq. (4.4)).

Nous nous limitons au cas où les $g_{\mu\nu}$ sont des constantes avec $-\|g\| = 1$. La différence entre $[F]$ et F disparaît. La fonction singulière $\varrho(x - q)$, définie par la limite d'une série de puissances en q^μ, permet alors de définir les opérateurs

$$S^\mu(x) = \frac{e}{2}\left(\pi^\mu \varrho(x - q) + \varrho(x - q)\,\pi^\mu\right) \tag{5,5}$$

et

$$A^{\mu\nu}(x) = \tfrac{1}{2}(\pi^\mu \varrho(x - q)\,\pi^\nu + \pi^\nu \varrho(x - q)\,\pi^\mu) - \tfrac{1}{2}\,g^{\mu\nu}\left(\pi_\alpha\,\varrho(x - q)\,\pi^\alpha - R\,\varrho(x - q) - \varrho(x - q)\,R\right) \tag{5,6}$$

Leurs espérances mathématiques $\overline{S}^\mu(x;\,\lambda)$ et $\overline{A}^{\mu\nu}(x, \lambda)$ formées par des intégrations (5,3) dépendent de λ. On en forme les espérances mathématiques indépendantes de λ:

$$J^\mu(x) = \int_{-\infty}^{+\infty} d\lambda\,\overline{S}^\mu(x, \lambda) \tag{5,7}$$

$$\overline{T}^{\mu\nu}_{\text{mat}}(x) = \int_{-\infty}^{+\infty} d\lambda\,\overline{A}^{\mu\nu}(x, \lambda) \tag{5,8}$$

qui satisfont aux équations:

$$\frac{\partial \overline{J}^\mu(x)}{\partial x^\mu} = 0 \tag{5,9}$$

$$\frac{\partial \overline{T}^{\mu\nu}_{\text{mat}}(x)}{\partial x^\nu} = \tfrac{1}{2}\left(B^{\mu\nu}\,\overline{J}_\nu(x) + \overline{J}_\nu(x)\,B^{\mu\nu}\right) \tag{5,10}$$

Les définitions (5,7) et (5,8), ainsi que les lois (5,9) et (5,10) montrent leur correspondance avec les grandeurs classiques (2,12) (densité du courant électrique) et (2,15) (tenseur d'énergie-impulsion)*).

Remarquons enfin que l'espérance mathématique de la densité de charge

$$\bar{J}^4(x) = \bar{J}^4(\vec{x}, t) = \int_{-\infty}^{+\infty} d\lambda \frac{eh}{2j} \left(\frac{\partial \psi^*}{\partial t} \psi(\vec{x}, t, \lambda) - \psi^* \frac{\partial \psi(\vec{x}, t, \lambda)}{\partial t} \right)$$

$$- e^2 \Phi^4(\vec{x}, t) \int_{-\infty}^{+\infty} d\lambda \mid \psi(\vec{x}, t, \lambda) \mid^2 \qquad (5,11)$$

peut être *positive ou négative*. Elle est partout positive si $\psi(\vec{q}, \tau, \lambda) = u(\vec{q}, \lambda) e^{-j\omega\tau}$ avec $\omega > 0$ et si l'«énergie totale» $h\omega$ est partout plus grande que l'«énergie potentielle» $e\Phi^4$.

L'espérance mathématique de la densité d'énergie matérielle peut être exprimée par

$$T^{44}_{\text{mat}}(\vec{x}, t) = \tfrac{1}{2} \int_{-\infty}^{+\infty} d\lambda \sum_1^4 \mid \pi^\alpha \psi(\vec{x}, t, \lambda) \mid^2 - \int_{-\infty}^{+\infty} \beta d\beta \mid \varphi_\beta(\vec{x}, t) \mid^2 \qquad (5,12)$$

$\varphi_\beta(q) = \varphi_\beta(\vec{q}, q^4)$ est le coefficient de Fourier dans la série

$$\psi(q, \lambda) = \frac{1}{\sqrt{2\pi}} \int_{-\infty}^{+\infty} d\beta \, \varphi_\beta(q) e^{-j\beta\lambda}$$

*) La relation classique $\partial T^{\mu\nu}/\partial x^\nu = 0$ implique en effet que

$$\partial T^{\mu\nu}_{\text{mat}}/\partial x^\nu = - \partial T^{\mu\nu}_{\text{maxw}}/\partial x^\nu = B^{\mu\nu} J_\nu(x).$$

Les relations (5,9) et (5,10) se démontrent de la manière suivant. On a d'abord (a):

$$\frac{\partial S^\mu}{\partial x^\mu} = - e \dot{\varrho} \qquad (a)$$

Si l'espérance mathématique de ϱ disparaît pour $\lambda = \pm \infty$ (ce qui est le cas pour tout événement fini) (5,9) suit immédiatement. Pour démontrer (5,6) on décompose $A^{\mu\nu}$ en deux termes

$$A^{\mu\nu} = A_0^{\mu\nu} + A_1^{\mu\nu} \qquad (b)$$

avec

$$e A_0^{\mu\nu} = \tfrac{1}{2} (\pi^\mu S^\nu + S^\nu \pi^\mu)$$

$$A_1^{\mu\nu} = \frac{h^2}{4} (g^{\mu\nu} g^{\alpha\beta} - g^{\mu\alpha} g^{\nu\beta}) \frac{\partial^2 \varrho}{\partial x^\alpha \partial x^\beta}$$

La divergence $\partial A_1^{\mu\nu} / \partial x_\nu = 0$ disparaît identiquement, tandisque en vertu de (a)

$$\frac{\partial A^{\mu\nu}}{\partial x^\nu} = - \tfrac{1}{2} (\pi^\mu \dot{} + \dot{\varrho} \pi^\mu)$$

L'intégration partielle des espéarnces mathématiques fournit (5,10), si on se sert de la relation

$$\dot{\pi}^\mu = \tfrac{1}{2} e (B_{\mu\nu} \pi^\nu + \pi^\nu B_{\mu\nu}).$$

Pour autant que le paquet d'ondes ψ est composé essentiellement d'ondes correspondant à des valeurs propres de $R = \beta = -\frac{m^2}{2} < 0$, c'est-à-dire à des « masses réelles », l'espérance mathématique de la densité d'énergie est partout positive.

Nous avons ainsi démontré que la théorie quantifiée permet d'évaluer les espérances mathématiques de grandeurs physiques associées à la particule par ex. leur charge et leur énergie. On ne trouvera que des particules à *énergie positive*, mais, en général, des *deux charges* $\pm e$. Au paragraphe suivant, nous étudierons un cas particulier, qui permettra une interprétation probabiliste des résultats formels de ce paragraphe.

§ 6. Production de paires de particules (m, e) et (m, —e) par un champ électrique en mécanique quantique.

Nous voulons démontrer qu'un phénomène, très analogue au phénomène classique de production de paires, étudié au § 3, apparaîtra en théorie quantique, sans qu'on introduise un nouveau champ. Pour pouvoir interpréter les résultats du § 5, nous devons d'abord préciser ce que nous appellons une mesure en théorie quantique; la mesure classique a été exposée au § 2.

Dans une région d'espace-temps, où il n'y a pas de champ électromagnétique, les mesures de q^μ et de $\dot{q}^\mu = \pi^\mu$ peuvent être faites avec la seule limite de précision

$$\Delta q^\mu \, \Delta \dot{q}^\mu \geqslant h. \tag{6,1}$$

Le résultat de la mesure correspond, comme dans le cas classique, au résultat trouvé pour $\lambda = 0$. Il est représenté par un paquet d'ondes

$$\psi(q, 0) = \psi(\vec{q}, \tau, 0) = (2\pi)^{-2} \iiint (dk)^3 \int_{-\infty}^{+\infty} d\omega \, e^{j(\vec{k}\vec{q} - \omega\tau)} \varphi(\vec{k}, \omega) \tag{6,2}$$

avec

$$\iiiint (dq)^4 \, |\psi|^2 = \iiint (dk)^3 \int_{-\infty}^{+\infty} d\omega \, |\varphi|^2 = 1 . \tag{6,3}$$

ψ ne diffère de zéro que dans un petit volume quadridimensionnel entourant le point q_0^μ (c'est le paquet (ψ_1) en fig. 2). De même, $\varphi(k, \omega)$ ne diffère de zéro que dans un petit volume de l'espace impulsion-énergie autour de $\vec{k} = \vec{k}_0$ et $\omega = \omega_0$ ($\omega_0 < 0$ en fig. 2). Ceci correspond en effet à un paquet ψ donnant des espérances mathématiques $\overline{q}^\mu \sim q_0^\mu$; $\overline{\dot{q}}^\mu = (\overline{\dot{\vec{q}}}, \overline{\dot{q}}^4) \sim h(\vec{k}_0, \omega_0)$ avec des incertitudes correspondant à (6,1), qui représente un *électron négatif* si $\omega_0 < 0$.

La fonction $\psi(q, 0)$ étant ainsi trouvée par une première observation, la fonction $\psi(q, \lambda)$ peut être déterminée pour tout λ en résolvant l'équation de SCHROEDINGER (5,2) (l'évolution spatiotemporelle des champs $\Phi_\mu(x)$ étant connue). Ensuite, les espérances mathématiques des grandeurs physiques (par ex. $\overline{J^4}(\vec{x}, t)$) peuvent être prédites.

Regardons de plus près l'exemple illustré par la fig. 2. C'est un champ électrique $\vec{E}$ homogène à composantes $E_1 = B_{14} = -\partial \Phi_1(t)/\partial t$, $E_2 = E_3 = 0$, qui n'existe que pendant l'instant δt. Dans la limite $\delta t = 0$ ce champ peut être décrit par le potentiel discontinu: $\Phi_2 = \Phi_3 = \Phi_4 = 0$; $\Phi_1 = h\gamma/e$ pour $t < 0$, $\Phi_1 = 0$ pour $t > 0$. Les fonctions

$$u_{\vec{k},\omega}(q, \lambda) = \begin{cases} 1\, e^{j(\vec{k}\vec{q} - \omega\tau - \beta\lambda)} + A\, e^{j(\vec{k}\vec{q} + \omega\tau - \beta\lambda)} & \ldots \text{pour } \tau = q^4 > 0 \\ B\, e^{j(\vec{k}\vec{q} - \nu\tau - \beta\lambda)} & \ldots \text{pour } \tau = q^4 < 0 \end{cases} \tag{6,4}$$

$$\omega = \pm\sqrt{|\vec{k}|^2 + 2h^{-1}\beta}\ ; \quad \nu = \pm\sqrt{(k^1 - \gamma)^2 + (k^2)^2 + (k^3)^2 + 2h^{-1}\beta} \tag{6,5}$$

$$A(\vec{k}, \omega) = \frac{\omega - \nu}{\omega + \nu}\ ; \quad B(\vec{k}, \omega) = \frac{2\,\omega}{\omega + \nu}\ ; \quad \frac{\omega}{\nu} > 0 \tag{6,6}$$

satisfont à (5,2). La somme

$$\psi(q, \lambda) = (2\pi)^{-2} \iiint (dk)^3 \int_{-\infty}^{+\infty} d\omega\, u_{\vec{k}\omega}(q, \lambda)\, \varphi(\vec{k}, \omega)$$
$$= \psi_1 + \psi_A + \psi_B \tag{6,7}$$

est la solution de (5,2) avec les conditions initiales correspondant à la mesure $\psi(q, 0) = \psi_1$.(6,7) représente un paquet d'ondes qui suit (à une certaine dispersion près) le rayon spatiotemporel en fig. 2. Au « moment » $\lambda = \Lambda_1$, une partie du paquet (ψ_A) est réfléchi. Son intensité vaut (pour $\lambda \gg \Lambda_1$)

$$\iiiint (dq)^4\, |\psi_A|^2 = W_A = \overline{A^2} \sim A(\vec{k}_0, \omega_0)^2 \tag{6,8 A}$$

Une autre partie suit le rayon réfracté avec l'intensité (pour $\lambda \gg \Lambda_1$)

$$\iiiint (dq)^4\, |\psi_B|^2 = W_B = \overline{B^2 \frac{d\omega}{d\nu}} \sim B(\vec{k}_0, \omega_0)^2 \frac{\nu_0}{\omega_0} \tag{6,8 B}$$

$d\omega/d\nu = \nu/\omega$ et (6,6) montrent que

$$W_A + W_B = 1; \quad W_A, W_B \geqslant 0 \tag{6,9}$$

Passons à l'évaluation de l'espérance mathématique de la charge électrique contenue dans un volume spatial entourant les rayons (les lignes d'univers de fig. 2) et d'une dimension grande par rapport à la dimension spatiale des paquets ψ_1, ψ_A et ψ_B, mais petite par rapport aux distances entre les rayons ψ_1 et ψ_A. On trouvera les valeurs suivantes:

$\bar{e}_1$, l'espérance mathématique de la charge électrique qui suit la ligne d'univers ψ_1 vaut (pour tout $t \gg 0$) $\bar{e}_1 = -e$.

$\bar{e}_B$, l'espérance mathématique de la charge totale qui suit la ligne d'univers ψ_B, vaut (pour tout $t \ll 0$) $\bar{e}_B = -eW_B$.

$\bar{e}_A$, l'espérance mathématique de la charge totale qui suit la ligne d'univers ψ_A, vaut (pour tout $t \gg 0$) $\bar{e}_A = +eW_A$.

Naturellement, l'espérance mathématique de la charge totale pour $t < 0$ est égale à celle pour $t > 0$. L'identité (6,9) n'est pas autre chose que

$$\bar{e}_1 + \bar{e}_A = -e + e\,W_A = -e\,W_B = \bar{e}_B \tag{6,10}$$

L'interprétation physique semble être la suivante:

1⁰ $|\psi|^2 (dq)^4 = dW$ est la probabilité que la particule se trouve, au «moment λ», dans le volume spatiotemporel $(dq)^4$. Il en est de même pour $|\varphi|^2 (dk)^3 d\omega$ dans l'espace d'impulsion-énergie.

2⁰ Notre *première mesure* est exécutée pour une valeur déterminée de λ ($\lambda = 0$). Toute autre mesure, que nous proposons de faire ultérieurement, ne peut évidemment pas être associée à une valeur bien définie de λ. C'est là une différence essentielle entre la théorie classique et la théorie quantifiée.

3⁰ Des mesures de grandeurs physiques (par ex. densité de courant $J^\mu(\vec{x}, t)$, densité d'énergie $T^{\mu\nu}(\vec{x}, t)$) faites à un endroit $\vec{x}$ au temps t ne dépendent pas de λ en théorie classique. Les espérances mathématiques $\bar{J}^\mu$ et $\bar{T}^{\mu\nu}(\vec{x}, t)$ du § 5 expriment donc les espérances mathématiques correspondant à des mesures de ces grandeurs physiques.

Ceci montre que dans notre problème, il y a une probabilité W_A de trouver un *électron positif* sur n'importe quel point (ou plutôt n'importe quelle région) de la ligne d'univers ψ_A et une probabilité W_B de trouver un électron négatif sur la ligne ψ_B. Autrement dit, notre théorie fournit une probabilité W_A que l'électron négatif actuellement observé pour $\lambda = 0$ (ψ_1) soit le partenaire d'une paire (ψ_1 et ψ_A) créée à l'instant $t = 0$ par le champ $\vec{E}$ et une probabilité $W_B = 1 - W_A$ que l'électron négatif observé ait déjà existé au passé ($t < 0$) et ait été accéléré par le champ $\vec{E}$ à l'instant $t = 0$.

Cette interprétation n'est pas opposée à nos notions de causalité. W_A n'est autre chose que *la probabilité que le champ considéré ait produite une paire de particules* qui suivent les trajectoires ψ_1 et ψ_A. Le résultat numérique correspond à celui de PAULI et WEISSKOPF [3)6)]. L'exemple contraire «j'observe un électron positif à un instant $t < 0$» est déjà un peu plus délicat. (Pour son illustration, on a qu'à changer le signe de l'axe t en fig. 2). Il y a alors une probabilité W_A que l'électron positif observé pour $t < 0$ s'anéantisse sous l'influence du champ $\vec{E}$ à l'instant $t = 0$, avec un partenaire ψ_A et la probabilité $W_B = 1 - W_A$ que l'électron observé continue d'exister pour $t > 0$ et poursuive un mouvement accéléré à l'instant $t = 0$ par ce champ $\vec{E}$. Ce second exemple rappelle un peu l'exemple du § 3, où l'observation d'une particule (1) (ligne B en fig. 1) et l'existence d'un champ K^μ à une période future à l'observation (1) nécessitait l'existence d'une antiparticule, qui poursuit un chemin bien défini pour tout $t < 0$. En effet, l'exemple quantique ci-dessus implique une probabilité bien définie (espérance mathématique $\bar{e}_A = -e\, W_A$) qu'une *antiparticule* existait au temps $t < 0$. Cet effet n'est autre chose que la *fluctuation* de la densité de charge dans la théorie du champ quantifié.

§ 7. Conclusions.

Il est possible d'établir une mécanique classique covariante par rapport aux transformations de la théorie de relativité générale, qui ne fait pas intervenir la racine carrée $ds = \sqrt{-\dot{q}_\mu \dot{q}^\mu}\, d\lambda$. Ses résultats sont identiques à ceux de la mécanique habituelle, pour autant que l'on ne fait pas intervenir de nouveau champs (§ 3). La quantification de la théorie introduit une densité de probabilité quadridimensionnelle et invariante $|\psi|^2$ normalisée à $\iiiint |\psi|^2 (dq)^4 = 1$. Une interprétation physique est possible (§ 6) et fait prévoir la production de paires de particules. La difficulté de la théorie de DIRAC (énergies négatives) ne se présente pas, la racine carrée étant éliminée.

La réaction des particules sur le champ a été étudiée, elle aussi, et fera l'objet d'une publication ultérieure. Elle n'est possible en théorie quantifiée qu'en suivant les méthodes de WENTZEL[4]) et de DIRAC[5]). Une difficulté d'interprétation physique apparaît alors: Si le champ produit par la particule (m, e) est le *champ retardé habituel d'une charge* e, celui de l'antiparticule$(m, -e)$ est le *champ avancé d'une charge* $-(-e) = e$. Les deux champs ont pourtant le même effet de freinage.

Institut de Physique de l'Université de Genève.

Littérature.

[1]) cf. Résumé Soc. Suisse de Physique, Helv. Phys. Acta **14**, 51 (1941), et l'exposé général de la théorie, Helv. Phys. Acta **14**, 588 (1941), mentionné comme I.

[2]) Une étude à ce sujet est en préparation pour la publication dans les Helvetica Physica Acta.

[3]) Pauli et Weisskopf, Helv. Phys. Acta **7**, 709 (1934).

[4]) Wentzel, Ztschr. f. Phys. **86**, 479 et 635 (1934).

[5]) Dirac, Ann. Inst. Henri Poincaré (1939).

[6]) Pour démontrer ceci dans la théorie du champ quantifié de Pauli et Weisskopf on doit faire des considérations analogues à telles que fait M. F. Hund (Ztschr. f. Phys. **117**, 1 [1940]).

An unambiguous method of avoiding divergence difficulties in quantum theory [67]

Reprinted by permission from Macmillan Publishers Ltd: Nature, vol. 153, pp. 143–144, copyright (1944).

No. 3874, JANUARY 29, 1944 NATURE 143

AN UNAMBIGUOUS METHOD OF AVOIDING DIVERGENCE DIFFICULTIES IN QUANTUM THEORY

By PROF. E. C. G. STUECKELBERG
University of Geneva

THE classical theory of a point charge and the quantum theory of wave packets contain in their usual form well-known divergences. Dirac[1] has elaborated a classical particle theory which avoids these difficulties, but his results cannot as yet be applied to quantum theory. On the other hand, Heisenberg[2] has recently proposed a new formalism which permits the calculation of collision cross-sections in quantum theory without being disturbed by diverging terms. However, no connecting link such as the correspondence principle has been given in order to apply this quantum formalism to a given problem such as Rutherford scattering, Compton effect or radiation damping.

Dirac's theory being well known, we content ourselves with recalling briefly the general idea of Heisenberg's description of collision phenomena. If no reaction between particles takes place, the behaviour of the system of particles can be described by a plane wave in configuration space $\Psi^{(0)}$. Such a wave can always be decomposed into an incoming $\Psi^{(0-)}$ and an outgoing $\Psi^{(0+)}$ spherical wave. If reactions take place, the number of particles scattered in a given direction (as observed by an observer at infinite distance from the collision region) is determined by a change of phase in the outgoing wave. This effect is described by a unitary phase operator $e^{-i\eta}$ determining the true outgoing spherical wave $\Psi^{(+)} = e^{-i\eta}\Psi^{(0+)}$ at infinity. The operator η must be Hermitic and relativistically invariant. Applying the method to quantized fields $u(\vec{x},t)$, Heisenberg discusses the effects due to η values of the form

$$\eta = \varepsilon \int_{-\infty}^{+\infty} dt \int (dx)^3\, u\left(\vec{x},t\right)^n.$$

If u is written as a series

$u = \Sigma\, (2Vk^4)^{-\frac{1}{2}} (a_k \exp(i(k,x)) + a_k^* \exp(-i(k,x)))$

with $(k,x) = (\vec{k},\vec{x}) - k^4 t$, $k^4 > 0$, the operator η has to be chosen so that all the a^*'s stand left of all the a's. This rule is invariant and prevents singularities from occurring. The space-time integral guarantees the invariance of the formalism and implies the usual conservation laws of energy (k^4) and momentum $\vec{k}$. No description for finite distances is possible in this theory.

We have succeeded in connecting the usual quantum theory to the Heisenberg formalism, and the result is but the logical translation of Dirac's classical theory into quantum language. Our method will best be understood in the following example:

Let $u(x) = u(\vec{x},t)$ be a field of matter (particles) the wave packets of which follow the same world-lines $x^a = q^a(\lambda)$ as the particles do. If $m\,\ddot{q}_a = \varepsilon\partial\varphi/\partial q^a$ is the differential equation of the corresponding particles in a given field of force $\varphi(\vec{x},t)$,

$$(\Box - (m^2 - 2\varepsilon m\varphi))\, u = 0 \qquad (1)$$

is the wave equation. The reaction on the field φ is described by

$$(\Box - \varkappa^2)\,\varphi = -\rho;\ \rho = \varepsilon m u^2;\ (\text{or} = \varepsilon \int d\lambda\, \delta(x - q(\lambda))) \qquad (2)$$

In quantum theory we treat both u and φ as operators, developing them into series with a_k^*, a_k (for u) and b_μ^*, b_μ (for φ). Thus, the explicit time dependence of u and φ is given by $(\Box - m^2)u = (\Box - \varkappa^2)\varphi = 0$, while the true time derivative $\dot{u}$ follows from a canonical formalism

$$\dot{u} = \frac{\partial}{\partial t} u\left(\vec{x},t\right) + i\varepsilon\left[H(t), u\left(\vec{x},t\right)\right];$$

$$\varepsilon H(t) = \varepsilon m \int (dx)^3 \varphi u \left(\vec{x},t\right)^2 \qquad (3)$$

This expression (3) and the explicit time dependence of the operators are equivalent to (1) and (2). In order to apply our theory to an actual problem, we have to express the Schrödinger probability amplitude $\Psi(t)$ in terms of the probability amplitude $\Psi(-T)$ at a time $t = -T$ before the collisions have occurred ($\lim T = \infty$). Quantum theory proceeds as follows: the Schrödinger differential equation

$$\frac{\partial\Psi(t)}{\partial t} = -i\varepsilon H(t)\,\Psi(t) \qquad (4)$$

is solved by successive approximations in ε. We try, however, the solution

$$\Psi(t) = e^{\alpha(t)}\,\Psi(-T). \qquad (5)$$

Substituted into (4), it leads to the integral equation ($\partial\alpha/\partial t \equiv \dot{\alpha}$)

$$-iH(t) = \dot{\alpha}(t) + \frac{\varepsilon}{2!}[\alpha,\dot{\alpha}] + \frac{\varepsilon^2}{3!}[\alpha,[\alpha,\dot{\alpha}]] + \ldots \qquad (6)$$

In order to obtain a solution for the unknown operator $\alpha(t)$ ($\alpha(-T) = 0$), we develop $\varepsilon\alpha$ in the following series $\varepsilon\alpha = \varepsilon\alpha^{(1)} + \varepsilon^2\alpha^{(2)} + \varepsilon^3\alpha^{(3)} + \ldots$

The first terms are

$$\begin{aligned}\varepsilon\dot{\alpha}^{(1)} &= -i\varepsilon H(t)\\ \varepsilon^2\dot{\alpha}^{(2)} &= -\frac{\varepsilon^2}{2}[\alpha^{(1)},\dot{\alpha}^{(1)}] \equiv \varepsilon^2\alpha\overset{(2)}{R} + \varepsilon^2\alpha\overset{(2)}{C}\\ \varepsilon^3\dot{\alpha}^{(3)} &= -\frac{\varepsilon^3}{2}[\alpha^{(2)},\dot{\alpha}^{(1)}] - \frac{\varepsilon^3}{6}[\alpha^{(1)},\dot{\alpha}^{(2)}]\end{aligned} \qquad (7)$$

etc. Integrating from $-T$ to t and applying the usual commutation rules ($[u(x), u(y)] = iD(x-y)$), the first terms are

$$\varepsilon\alpha^{(1)} = -i\varepsilon m \int_{-T}^{t} dt \int (dx)^3 u^2\varphi\left(\vec{x},t\right)$$

$$\varepsilon^2\alpha\overset{(2)}{R} = -i\frac{\varepsilon^2}{2}m^2 \int_{-T}^{t} dt \int (dx)^3 u\left(\vec{x},t\right)^2 \text{ret}_{(\varphi)}(u^2) \qquad (8)$$

$$\varepsilon^2\alpha\overset{(2)}{C} = -i\varepsilon^2 m^2 \int_{-T}^{t} dt \int (dx)^3 u\varphi\left(\vec{x},t\right) \text{ret}_{(u)}(u\varphi)$$

$$\varepsilon^3\alpha^{(3)} = -i\varepsilon^3 m^3 \int_{-T}^{t} dt \int (dx)^3 (u\left(\vec{x},t\right)^2 \text{oper}(uu\varphi) + \ldots).$$

$\text{ret}_{(\varphi)}(\rho)$ is the retarded potential of a charge density ρ in (2), oper (ρ) is another more complicated integral operator. After developing u and φ in terms of the a,a^* and b,b^* operators, we can, as did Heisenberg, interchange their order so as to have all a^*,b^* left of the

144 NATURE JANUARY 29, 1944, VOL. 153

a,b. This change corresponds to an invariant subtraction of all diverging terms; the term $a^*_{k'}a_{k'}\, a^*_{k'}\, a_k$ in $\alpha_C^{(2)}$ contributes to the 'Coulomb' self-energy of the *u*-particles. The other terms of $\varepsilon^n\alpha^{(n)}$ correspond to the transition-probabilities for collisions in which $n + 2$ particles take part (in the sense of chemical reactions). For example, $\varepsilon^2\alpha^{(2)}_R$ contains the term $a^*_{k''}a^*_{k'''}a_{k'}a_k$, where two particles with the momenta k and k' disappear and two particles of momenta k'' and k''' are created. This is Rutherford scattering (if φ is the electromagnetic field) and $\varepsilon^2\alpha^{(2)}_C$ contains the Compton effect $(a^*_{k'}b^*_{\mu'}a_k b_\mu)$, where a φ-quantum of momentum μ collides with a *u*-particle of momentum k. The terms in $\varepsilon^3\alpha^{(3)}$ contain the *Bremstrahlung* $(a^*_{k''}a^*_{k'''}\, b^*_{\mu}a_{k'}a_k)$, where, in addition to the Rutherford scattering, a μ-quantum is created.

Putting $t = + T$ and passing to $T \to \infty$, our theory takes the Heisenberg form. However, we have unambiguously determined the Hermitic Heisenberg operator $\eta = i\varepsilon\alpha(\infty)$ and, furthermore, for any given t, we can describe the quantum mechanical state of the system.

We can arbitrarily change the numerical coefficients of each individual $\varepsilon^n\alpha^{(n)}$ (or of their invariant parts) without destroying the conservation laws or the invariance. For example, a theory which contains only $\varepsilon\alpha = \varepsilon^2\alpha^{(2)}R$ and nothing else shows Rutherford scattering but no radiation effects (no Compton effect and no *Bremstrahlung*, etc.). But such a theory is possible even classically. Consider a system of particles, where the force acting upon any one of them is the mean value between the advanced and retarded effect of all other particles. Such a theory is invariant and conserves energy and momentum, but it is not conformal to our causal representation of phenomena. The same acausal behaviour is contained in the quantum mechanical $\Psi(t)$. There exists now a finite probability that a quantum appears at a certain event $(\vec{x},t)$, without a finite probability that a cause has occurred at a preceding event in the invariant past of $(\vec{x},t)$. In electrodynamic phenomena, the causal behaviour has been experimentally checked. Therefore, our theory is unambiguous if applied to quantum electrodynamics. Applied to nuclear forces, however, we have great liberty in the choice of $\varepsilon\alpha$ (or of Heisenberg's η) if we go back to a causal description at small distances. But one must say that it is not necessary to go back to causal description even for distances of 10^{-13} cm.

We have applied our theory to the case of line width. A classical point particle with an internal (scalar) degree of freedom (τ, ρ, σ) treated according to Dirac's method leads to

$$\ddot{\tau} + \frac{1}{2\pi}\,\varepsilon^2\mu_0\sigma\dot{\tau} + \mu_0{}^2\tau = \varepsilon\sigma\,\varphi\,(q(\lambda))^{\text{inc}};\ \dot{\sigma}\sim 0.$$

Its line broadening due to radiation damping is (if $\sigma = +1$ and $\varkappa = 0$) therefore

$$J(\mu) = \gamma^2\,((\mu - \mu_0)^2 + \gamma^2)^{-1};\ \gamma = \frac{\varepsilon^2}{4\pi}\mu_0. \tag{9}$$

The corresponding quantum mechanical model is given by two fields of matter u and v the rest masses of which differ by $m_v{}^2 - m_u{}^2 \cong 2m\mu_0$. With $\varepsilon H = 2\varepsilon m\int(dx)^3 uv\varphi$, only particles of mass $m_u(<m_v)$ are stable (analogous to $\sigma = +1$ in classical theory). Excitation of the *u*-particle into the *v*-state and subsequent emission of a φ-quantum, or the dispersion of a φ-wave produces φ-quanta of frequency μ with a probability given by

$$J(\mu) = \left(\sin\frac{\gamma}{\mu-\mu_0}\right)^2;\ \gamma = \frac{\varepsilon^2}{4\pi}\mu_0 \tag{10}$$

instead of (9). This is a rigorous solution of the quantum-mechanical problem. The approximate treatment of Wigner and Weisskopf[3] leads to (9). Total intensity of the emitted light and the dispersion for $|\mu - \mu_0| \gg \gamma$ are, however, the same.

[1] Dirac, *Proc. Roy. Soc.*, A, **167**, 148 (1938).
[2] Heisenberg, *Z. Phys.*, **120**, 513 and 673 (1943).
[3] Weisskopf and Wigner, *Z. Phys.*, **63**, 54 (1930).

Une propriété de l'opérateur S en mécanique asymptotique [78]

Helvetica Physica Acta, vol. 19 (1946), pp. 242–243

Une propriété de l'opérateur S en mécanique asymptotique

par E. C. G. Stückelberg.

En mécanique asymptotique, nous introduisons conformément aux idées de Heisenberg, un opérateur unitaire:

$$\boldsymbol{S} = \mathbf{1} - i\,\alpha + \cdots = e^{-i(\alpha + \cdots)} \tag{1}$$

défini par une série unimodulaire en l'opérateur hermitien α, qui s'écrit[1]):

$$\alpha = \varepsilon\,\alpha^{(1)} + \cdots; \qquad \alpha^{(1)} = \varepsilon \int (dx)^4 u^+ u^+ \varphi^+ \varphi \tag{2}$$

u et φ sont des «demi-champs», c'est-à-dire des fonctions complexes des variables $\vec{x}$ et t, mais ne possèdent que des fréquences positives (donc uniquement des opérateurs a^+ et b^+).

Alors l'effet «de première approximation (obtenu en n'utilisant dans S que les deux termes $1 - i\,\alpha^{(1)}$) admet l'interprétation facile suivante: la collision entre un quantum initialement (c'est-à-dire à l'époque $t = -T$) dans l'état k' du champ u et un quantum dans l'état μ' du champ φ. En effet, la probabilité pour que finalement (à l'époque $t = +T$, $T = \infty$) l'on observe un quantum à l'état k'' et μ'' est proportionnelle à la région $(\Delta x)^4$ de l'espace temps où les paquets d'onde k', μ', k'' et μ'' diffèrent simultanément de zéro.

Si, comme l'a fait Heisenberg, on ne considère alors dans la série (1) que les termes en $\alpha^{(1)}$, on démontre facilement qu'en seconde approximation déjà (en posant $S = 1 - i\alpha^{(1)} + \frac{1}{2}\,(\alpha^{(1)})^2$), il faut admettre l'existence de colisions entre trois quanta k' et l' du champ u et μ' du champ φ dont l'explication est en désaccord avec nos notions de causalité, et celà à l'échelle de l'espace-temps macroscopique: un quantum l' du champ u émet «un quantum γ (du champ φ) d'*énergie négative*», pour pouvoir passer à un état l'' en émettant un quantum μ'', tandis que plus tard, au temps t''

---

[1]) Nous renonçons, dans la suite, d'employer des caractères gras pour les opérateurs α, μ^*, ...

$> t'$, les quanta k' et μ' entrent en collision triple avec ce quantum v''' d'ébergie négative.

Ainsi donc, dans notre théorie qui exclut des énergies négatives, il y faudrait pourtant inclure des phénomènes dont la description spatiotemporelle comporterait de tels états.

Il existe un et un seul moyen d'éviter la contradiction précédente: c'est d'écrire S sous la forme d'une série de la forme

$$S = \frac{1 - i/2 \cdot \alpha}{1 + i/2 \cdot \alpha} = \exp\left(- i\, 2\, \mathrm{tg}^{-1} \frac{\alpha^{(1)}}{2}\right)$$

α étant la série engendrée par $\alpha^{(1)}$ et

$$\alpha^{(2)} = \int (dx)^4 \left[u^+ u\, \varphi^+ \,\mathrm{sym}_a\, (\varphi u^+ u) + \varphi^+ \varphi\, u^+ \,\mathrm{sym}_\varphi\, (u\, \varphi^+ \varphi)\right)$$

$$\alpha^{(3)} = \cdots .$$

Donc, sans utiliser le principe de correspondance asymptotique que nous avions énoncé dans un mémoire précédent[2]), nous retrouvons, par simple application du principe de causalité, l'expression de l'opérateur S en fonction de α, telle que nous l'avions obtenue dans les §§ 6 et 8 du mémoire cité.

Diffusion d'électrons par un champ coulombien magnétique

par P. Banderet et M. Fierz, Bâle.

Dirac a, le premier, considéré l'équation de Schroedinger d'un électron dans un champ coulombien magnétique. Il a constaté qu'elle n'avait de solution acceptable du point de vue physique que si le pôle magnétique avait comme intensité un multiple de $\frac{hc}{2e}$. Les fonctions propres ont été établies par Tamm et par Fierz; ce sont en coordonnées polaires des produits d'une fonction de r par des fonctions des angles qui ne sont pas des fonctions sphériques.

Le problème de la diffusion d'électrons par le champ d'un pôle magnétique de Dirac présente de ce fait un caractère particulier. On ne peut pas comme d'habitude, construire une solution représentant asymptotiquement une onde plane avec, comme perturbation, un onde émise par le centre. Mais on peut choisir les coefficients arbitraires de la solution de manière à avoir la superposition d'une onde incidente se comportant pour grands r comme

---

[2]) Stückelberg, Mécanique fonctionnelle H.P.A. **14**, 51 (1943) (réf. I); **16**, 427 (1944) (réf. II); **17**, 3 (1945) (réf. III) **18**, 21 et 195 (réf. III et IV).

$\delta(\cos\Theta)$, δ étant la fonction singulière de DIRAC, et d'un onde émergente. Les coefficients arbitraires sont complètement déterminés par cette méthode. On obtient pour l'onde émergente la superposition d'une diffusion analogue à celle de RUTHERFORD, et d'une perturbation qui n'est importante que dans la direction d'où vient l'onde incidente. Les résultats sont semblables à ceux que donne la mécanique classique.

(Un article plus détaillé paraîtra ultérieurement dans les H. P. A.)

Sur la théorie des antennes de radio — Comparaison avec l'expérience

par J. PATRY, Albiswerk-Zürich-A.G.

Il y a quelques temps, MÜLLER et l'auteur ont publié trois brèves communications sur la théorie des antennes selon la méthode de HALLÈN[1]). Cette méthode est très pratique, car elle permet d'étudier non seulement les antennes linéaires simples, mais aussi des types plus compliqués. Comme elle n'est qu'une approximation, il était intéressant de comparer avec l'expérience les résultats auxquels elle conduit. METZLER, de l'administration des P.T.T. a fait, il y a un certain temps, une série de mesures qui nous ont servi de contrôle. Elles ont été publiées dans la thèse de doctorat de METZLER.

Les courbes projetées par l'auteur (qui paraîtront prochainement dans le Bulletin S. E. V.) montrent une très bonne concordance entre l'expérience et la théorie. Des calculs sont en cours pour comparer les résultats théoriques avec les mesures accomplies par des Anglais[2]). La théorie détaillée, dont la valeur pratique est ainsi démontrée, sera exposée prochainement dans le périodique «Schweizer Archiv».

---

[1]) MÜLLER und PATRY, H.P.A. **17**, 127, 159, 455 (1944).

[2]) ESSEN and OLIVER, Wireless Engineer **22**, 507 (1945).

A convergent expression for the magnetic moment of the neutron [81]

with D. Rivier

Reprinted with permission from *Physical Review*, vol. 74 (1948), p. 218. Erratum, idem 986.

PHYSICAL REVIEW VOLUME 74, NUMBER 2 JULY 15, 1948

A Convergent Expression for the Magnetic Moment of the Neutron

D. RIVIER AND E. C. G. STUECKELBERG
University of Geneva, Geneva, Switzerland
June 2, 1948

CALCULATION of the magnetic moment of the neutron has always yielded a divergent result.[1] We have avoided this difficulty by the use of an invariant perturbation method based upon the S matrix theory.[2] The magnetic moment remains finite but no definite numerical value comes out of the calculations. Our method is quite general and can be applied to all diverging expressions arising from field quantization.

The diverging terms are due to the Dirac δ function appearing in

$$D_\kappa^s(r,t)=(t/2|t|)D_\kappa^0(r,t)$$
$$=(1/4\pi)[\delta(T^2)-\tfrac{1}{2}\theta(T^2)(\kappa/T)J_1(\kappa T)]$$

with

$$T^2=t^2=r^2 \quad \text{and} \quad \theta(T^2)=0 \quad \text{for} \quad T^2\gtrless 0,$$

where $D_\kappa^0(r,t)$ is the antisymmetric invariant function appearing in the commutation relations of a field of quanta of restmass κ. If we define $\delta(T^2)$ by the following limiting process:

$$\delta(T^2)=\lim_{\kappa_i\to\infty}-\tfrac{1}{2}\theta(T^2)\sum_1^N c_i(\kappa_i/T)J_1(\kappa_i T),$$

with

$$\sum_1^N c_i=-1 \qquad \lim_{\kappa_i\to\infty}\sum_1^N (c_i/\kappa_i^2)=0,$$

a calculation then shows that it is always possible to find a set of $c_i=c_i(\kappa_1^2,\cdots\kappa_N^2)$, identical for each individual diverging term, such that each such term gives a finite contribution to the magnetic moment. Our calculations could be described as the generation of the S matrix by the usual Hamiltonian and its correction[3] by (2). Thus corrected our S matrix might not satisfy causality anymore. That it still does can be seen in the following way. All the terms of the S matrix contain space-time integrations upon invariant $D_\kappa(r,t)$ functions. In our calculations these appear only in the form

$$D_\kappa^c=D_\kappa^s+(i/2)D_\kappa^1$$

where D_κ^1 is the symmetric invariant D function involving a T^{-2} singularity. This is exactly the condition that expresses causality[4] or this complex function contains only incoming waves in the past and outgoing waves in the future. This condition reduces to that introduced by Wentzel[5] in his non-relativistic wave optics of the S matrix.

[1] J. M. Jauch, Phys. Rev. **63**, 334 (1943), does obtain a finite value but it seems to us that he has discussed only one of many terms.
[2] E. C. G. Stueckelberg, Helv. Phys. Acta **17**, 3 (1944); *ibid.* **18**, 3 (1945).
[3] In this interpretation (2) could be considered as an "invariant substraction."
[4] E. C. G. Stueckelberg, Helv. Phys. Acta **19**, 242 (1946); Phys. Soc. Cambridge Conference Report 199 (1947).
[5] Wentzel, Helv. Phys. Acta **21**, 49 (1948).

Erratum: A Convergent Expression for the Magnetic Moment of the Neutron

[Phys. Rev. **74**, 2 (1948)]

D. RIVIER AND E. C. G. STUECKLBERG
University of Geneva, Geneva, Switzerland

IN our previous letter two misprints have occurred which make the text impossible to understand: The formulas defining T^2 and $\theta(T^2)$ are

$$T^2 = t^2 - r^2,$$
$$\theta(T^2) = 0, \quad T^2 < 0,$$
$$\theta(T^2) = 1, \quad T^2 > 0.$$

Causalité et structure de la Matrice S [82]

with D. Rivier, *Helvetica Physica Acta*, vol. 23 (1950), pp. 215–222

Causalité et structure de la Matrice *S*

par **E. C. G. Stueckelberg** et **D. Rivier** (Genève).

(19. X. 1949.)

La causalité en théorie des champs quantifiés définit la structure des noyaux intégraux dans les coefficients du développement de la matrice S en fonction des opérateurs de translation dans l'espace des quanta. Cette structure, qui fait apparaître la fonction $D^c(x)$, est celle que l'on obtient en intégrant de manière invariante l'équation d'évolution fonctionnelle, à une certaine indétermination près dans le cas de la méthode proposée. Cette indétermination remplace le problème de l'élimination des divergences par celui de la détermination des termes non définis de la matrice S.

1. La causalité en physique classique et en physique quantique.

Pour fixer les idées, considérons le système *classique* constitué par n particules chargées, soumises aux lois de l'électrodynamique de Maxwell. Une loi physique exprime, par l'intermédiaire d'équations:

$$\xi''_{(k)} = F_{(k)}[\tau'', \tau'; \ldots \pi'_{(i)}, \xi'_{(i)} \ldots]$$

$$\pi''_{(k)} = G_{(k)}[\tau'', \tau'; \ldots \pi'_{(i)}, \xi'_{(i)} \ldots] \qquad \xi''_{(k)} = \frac{1}{m_{(k)}} \int^{\tau''} d\lambda\, \pi_{(k)}(\lambda) \qquad (1.1)$$

les relations fonctionnelles existant entre les quantités de mouvement $\pi''_{(k)}$ et les lieux $\xi''_{(k)}$ de chaque particule (k) au temps τ'' et ce temps τ'' d'une part, et les mêmes grandeurs $\pi'_{(i)}$, $\xi_{(i)}$, τ' à un instant initial τ' d'autre part.

Ces équations s'obtiennent par intégration d'équations différentielles décrivant les processus élémentaires. Dans le cas envisagé celles-ci s'écrivent*):

$$d^{(i)} p_\alpha^{(k)} = e^{(k)} e^{(i)} ds_\beta^{(k)} ds_{[\alpha}^{(i)} \partial^{\beta]} D^{(\mathrm{ret})}(s_{(k)}/s_{(i)}) \qquad (1.2)$$

La fonction $D^{\mathrm{ret}}(x_{(k)}/x_{(i)})$ décrit ce que nous appelons «l'action causale» de (i) sur (k). Remarquons que la fonction $D^{\mathrm{ret}}(x/y)$ est, comme il est nécessaire, invariante par rapport au groupe de Lorentz. Notons aussi que les effets $d^{(i)} p_\alpha^{(k)}$ sur la particule (k) dus aux diverses particules… (i),… s'additionnent les uns aux autres.

*) Pour les notations, voyez le numéro [3]) des références.

Considérons maintenant le système *quantique* correspondant: il s'agit de n quanta, en nombre variable. Les lois physiques décrivant l'évolution du système sont déterminées par un opérateur unitaire $\boldsymbol{S}$, représenté par des matrices de transition:

$$S\left[\tau'' ; \ldots \xi''_{(k)} \ldots \xi''_{(n'')} / \tau' ; \ldots \xi'_{(i)} \ldots \xi'_{(n')}\right] \tag{1.3}$$

fonctionnelles seulement des variables lieux initiales $\xi''_{(k)}$ et finales $\xi'_{(i)}$ et des deux époques τ'' et τ' entre lesquelles le système évolue.

Une matrice de transition s'obtient aussi par intégration d'un élément différentiel:

$$\begin{aligned} dS\left[\ldots \xi''_{(k)} \ldots / \ldots x_{(l)} \ldots x_{(m)} \ldots / \ldots \xi'_{(i)} \ldots\right] \sim \\ \ldots D^{+}(\xi''_{(k)} / x_{(k)})\, \Gamma_{(k)}\, d x_{(k)}\, D^{c}(x_{(k)} / x_{(l)})\, \Gamma_{(l)}\, d x_{(l)} \ldots \times \\ \times \ldots \Gamma_{(m)}\, d x_{(m)}\, D^{c}(x_{(m)} / x_{(i)})\, \Gamma_{(i)}\, d x_{(i)}\, D^{+}(x_{(i)} / \xi'_{(i)}) \ldots \end{aligned} \tag{1.4}$$

par convention, dans $D^{+}(x/y)$ on a $x^4 > y^4$. L'élément différentiel dS est un produit d'amplitudes de probabilité:

$D^{+}(x/x_{(k)}) = u(x)$ représente l'amplitude de probabilité relative au quantum k observée au temps $x^4 = \tau$ (x^1, x^2, x^3), émergeant de $x_{(k)}$ si $x^4 > x^4_{(k)}$ ou convergeant en $x_{(k)}$ si $x^4 < x^4_{(k)}$; en $x_{(k)}$ elle a été localisée avec autant de précision qu'il est possible;

$\Gamma_{(i)}$ est une fonction du lieu $x_{(i)}$ caractérisant l'amplitude de probabilité du processus élémentaire en ce point;

enfin la fonction $D^{c}(x_{(k)}/x_{(l)})$ décrit l'amplitude de probabilité de «l'action causale de (l) sur (k)»: c'est le correspondant quantique de $D^{\mathrm{ret}}(x_{(k)}/x_{(l)})$, qu'il faut déterminer maintenant.

Qu'il soit impossible de maintenir la fonction $D^{\mathrm{ret}}(x_{(k)}/x_{(l)})$ résulte du fait que celle-ci ne représente pas une amplitude de probabilité. Mais nous savons que:

$$D^{c}(x/y) \sim D^{+}(x/y) = D^{1}(x/y) - i D^{0}(x/y) \quad \text{pour } x^4 > y^4.$$

Or, du fait que la fonction $D^{1}(x/y)$, à l'inverse de $D^{0}(x/y)$, ne s'évanouit pas à l'extérieur du cône de lumière, la seule façon de prolonger de manière invariante $D^{c}(x/y)$ pour $x^4 < y^4$ est de l'écrire:

$$D^{c}(x/y) \sim (D^{1} + a D^{0})(x/y) = \left(\frac{1+ai}{2} D^{+} + \frac{1-ia}{2} D^{-}\right)(x/y)$$

mais il est clair que le coefficient de $D^{+}(x/y)$ doit ici être nul, sans quoi des acausalités manifestes*) contribueraient à l'élément différentiel dS. On *doit* poser $a = i$ et:

$$D^{c}(x/y) \sim D^{-}(x/y) \quad \text{pour } x^4 < y^4.$$

*) En particulier, une partie de l'action de (l) sur (k) ne dépendrait pas de la succession temporelle des événements $x_{(l)}$ et $x_{(k)}$.

et la fonction causale seule possible est donc, en introduisant un facteur de normalisation égal à $i/2$:

$$D^c \sim D^s + \frac{i}{2} D^1 \qquad x^4 \mp y^4 \tag{1.5}$$

Dans (1.4), l'interprétation de la contribution:

$$\ldots D^+(\xi''_{(k)}/x_{(k)})\, \Gamma_{(k)}\, dx_{(k)}\, D^c(x_{(k)}/x_{(l)})\, \Gamma_{(l)}\, dx_{(l)} \ldots \text{ pour } x^4_{(k)} < x^4_{(l)} \tag{1.6}$$

s'obtient en inversant en $x_{(k)}$ et en $x_{(l)}$ les phénomènes de cause et d'effet dans les processus décrits par la contribution:

$$\ldots D^+(\xi''_{(k)}/x_{(k)})\, \Gamma_{(k)}\, d\,x_{(k)}\, D^c(x_{(k)}/x_{(l)})\, \Gamma_{(l)}\, d\,x_{(l)} \ldots \text{ pour } x^4_{(k)} > x^4_{(l)}$$

Plus précisément, le processus décrit par (1.6) est l'émission simultanée en $x_{(k)}$ d'un paquet observé en τ'' et d'un paquet participant en $x_{(l)}$ à un nouveau processus. Cela est immédiat si l'on remarque que:

$$D^c(y/x) = \pm D^c(x/y)$$

le signe étant positif ou négatif suivant la statistique qui régit les ensembles de quanta.

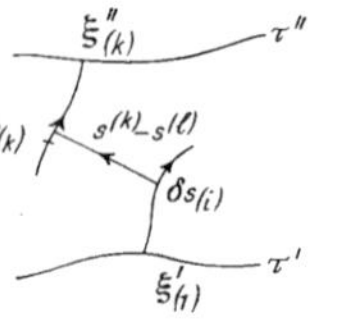

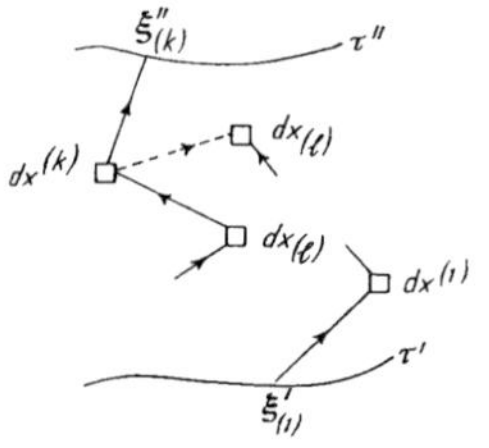

système classique — système quantique

Fig. 1.

Une différence essentielle entre la fonction classique $D^{\text{ret}}(x)$ et la fonction quantique $D^c(x)$ réside dans leur comportement à *l'extérieur du cône de lumière*; par exemple dans le cas scalaire:

$$\left.\begin{aligned} D^{\text{ret}}(x) &= 0 \\ D^c(x) &= \frac{1}{8\pi i}\,\frac{\varkappa}{R}\, H_1^{(1)}(i\varkappa R) \end{aligned}\right| \begin{aligned} &R^2 = x^\alpha x_\alpha > 0 \\ &\varkappa \text{ masse du quantum} \end{aligned} \tag{1.7}$$

2. Généralisation à des variables quelconques.

Le passage à des variables quelconques $u''_{(1)}, u''_{(2)}, \ldots u''_{(n'')}$ définissant l'état des n quanta s'opère comme on le sait par l'intermédiaire d'un opérateur $\boldsymbol{U}$ représenté par une matrice de transformation:

$$U(\xi'_{(1)} \ldots \xi'_{(n')} / u'_{(1)} \ldots u'_{(n')}) \tag{2.1}$$

et possédant un opérateur hermitien conjugué $\boldsymbol{U}^+$ représenté par la matrice $U^+(u''_{(1)} \ldots u''_{(n'')} / \xi''_{(1)} \ldots \xi''_{(n'')})$.

La transformée de S s'écrit:

$$S[\tau'', u''_{(1)} \ldots u''_{(n'')} / \tau', u'_{(1)} \ldots u'_{(n')}] = U^+(u''_{(1)} \ldots u''_{(n'')} / \xi''_{(1)} \ldots \xi''_{(n'')}) \times$$
$$\times S[\tau'' \ldots \xi''_{(k)} \ldots / \tau' \ldots \xi'_{(i)} \ldots] \, U(\xi'_{(1)} \ldots \xi'_{(n')} / u'_{(1)} \ldots u'_{(n')}) \tag{2.2}$$

il est important de remarquer que la multiplication matricielle représente ici une intégration sur tous les points d'une surface temporelle d'élément $d\sigma^\alpha(\xi)$. Ainsi, nous avons explicitement:

$$S[\tau'', u''_{(1)} \ldots u''_{(n'')} / \tau', u'_{(1)} \ldots u'_{(n')}] \sim \ldots \int^{(k)} \int^{(l)} \ldots \int^{(m)} \int^{(i)} \ldots \int d\sigma^{\alpha''}(\xi''_{(k)}) \ldots \times$$
$$\times \ldots \int d\sigma^{\alpha'}(\xi'_{(i)}) \ldots U^+_{\alpha''}(u''_{(1)} \ldots u''_{(n'')} / \ldots \xi''_{(k)} \ldots) \, D^+(\xi''_{(k)} / x_{(k)}) \times$$
$$\times \Gamma_{(k)} \, d x_{(k)} \, D^c(x_{(k)} / x_{(l)}) \, \Gamma_{(l)} \, d x_{(l)} \ldots \Gamma_{(m)} \, d x_{(m)} \, D^c(x_{(m)} / x_{(i)}) \times$$
$$\times \Gamma_{(i)} \, d x_{(i)} \, D^+(x_{(i)} / \xi'_{(i)}) \ldots U_{\alpha'}(\ldots \xi'_{(i)} \ldots / u'_{(1)} \ldots u'_{(n')}); \tag{2.3}$$

là comme dans ce qui suit $\int^{(k)} \ldots$ est mis pour $\int_{\tau'}^{\tau''} (d x^{(k)})^4$.

En utilisant alors comme il est d'usage un système de paquets orthogonal et complet $u'_{(i)}, u''_{(k)}, \ldots$, satisfaisant donc aux relations:

$$\left. \begin{aligned} \int d\sigma^\alpha(x) \, u^{+''}(x) \, \boldsymbol{\Omega}_\alpha \quad u'(x) &= \delta(u''/u') \\ \int dV(u) \, u(x'') \quad u^+(x') &= D^+(x''/x') \end{aligned} \right\} \tag{2.4}$$

où $\boldsymbol{\Omega}_\alpha$ est un opérateur d'orthogonalisation dépendant de la variance des paquets u, on obtient pour S en posant:

$$U_\alpha(\xi''/u') = \boldsymbol{\Omega}_\alpha u'(\xi''): \tag{2.5}$$

$$S[\tau'', u''_{(1)} \ldots u''_{(n'')} / \tau', u'_{(1)} \ldots u'_{(n')}] \sim \ldots \int^{(k)} \int^{(l)} \ldots \int^{(m)} \int^{(i)} \ldots u''^+_{(k)}(x_{(k)}) \, \Gamma_{(k)} \, d x_{(k)} \times$$
$$\times D^c(x_{(k)} / x_{(l)}) \, \Gamma_{(l)} \, d x_{(l)} \ldots \Gamma_{(m)} \, d x_{(m)} \, D^c(x_{(m)} / x_{(i)}) \, \Gamma_{(i)} \, d x_{(i)} \, u'_{(i)}(x_{(i)}) \ldots \tag{2.6}$$

Une nouvelle généralisation permet de considérer des transitions où le système de paquets initial $u'_{(1)} \ldots u'_{(n')}$ et le système final $v''_{(1)} \ldots \ldots v''_{(n'')}$ diffèrent entre eux. On a alors:

$$S\left[\tau'' v''_{(1)} \ldots v''_{(n'')} / \tau' u'_{(1)} \ldots u'_{(n')}\right] = V^{+}\left(v''_{(1)} \ldots v''_{(n'')} / \ldots \xi''_{(k'')} \ldots\right)$$
$$S\left[\ldots \xi''_{(k)} \ldots / \ldots \xi'_{(i)} \ldots\right] U\left(\ldots \xi'_{(i)} \ldots / u'_{(1)} \ldots u'_{(n')}\right) \tag{2.7}$$

où la matrice $V^{+}(v''_{(1)} \ldots v''_{(n'')} / \ldots \xi''_{(k)} \ldots)$ représente l'hermitien conjugué de l'opérateur $\boldsymbol{V}$ de transformation des variables $\xi''_{(k)}$ aux variables $u''_{(l)}$. On obtient donc une représentation *mixte*:

$$S\left[\tau'', v''_{(1)} \ldots v''_{(n'')} / \tau', u'_{(1)} \ldots u'_{(n')}\right] \sim \ldots \int\limits^{(k)}\int\limits^{(l)} \ldots \int\limits^{(m)}\int\limits^{(i)} \ldots v''^{+}_{(k)}\left(x_{(k)}\right) \times$$
$$\times\; \Gamma_{(k)}\, d x_{(k)}\, D^{c}\left(x_{(k)} / x_{(l)}\right) \Gamma_{(l)}\, d x_{(l)} \ldots \Gamma_{(m)}\, d x_{(m)}\, D^{c}\left(x_{(m)} / x_{(i)}\right) \times$$
$$\times\; \Gamma_{(i)}\, d x_{(i)}\, u'_{(i)}\left(x_{(i)}\right) \ldots \tag{2.8}$$

On peut voir que la forme générale de S s'obtient (après intégration multiple sur $\ldots dx^{(k)} \ldots dx^{(i)} \ldots$) en substituant dans (1.4) les paquets particuliers $D^{+}(\xi''_{(k)}/x_{(k)})$ (émergents) et $D^{+}(x_{(i)}/\xi'_{(i)})$ (incidents) par les paquets généraux $u''^{+}(x_{(k)})$ et $u'(x_{(i)})$.

3. Passage à l'espace des quanta.

Il est maintenant indiqué de passer à l'espace des quanta, à l'aide des opérateurs de translation dans cet espace, $\boldsymbol{a}^{+}(u'')$ et $\boldsymbol{a}(u'')$, ... qui permettent l'introduction des opérateurs de champs:

$$\boldsymbol{u}_{(i)}(x) = \frac{1}{\sqrt{2}} \int dV\left(u'_{(i)}\right) \boldsymbol{a}\left(u'_{(i)}\right) u'_{(i)}(x) + \frac{1}{\sqrt{2}} \int dV\left(u'_{(i)}\right) \boldsymbol{a}^{+}\left(u'_{(i)}\right) u^{+}_{(i)}(x) \tag{3.1}$$

Dans cet espace l'opérateur $\boldsymbol{S}\,[\tau'', \tau']$, satisfaisant à la condition d'unitarité:

$$\boldsymbol{S}^{+}\left[\tau'', \tau\right] \boldsymbol{S}\left[\tau'', \tau\right] = \mathbf{1} \tag{3.2}$$

peut naturellement se développer en termes des opérateurs $\boldsymbol{a}^{+}(u')$ et $\boldsymbol{a}(u')$, en posant*):

$$\boldsymbol{S}\left[\tau'', \tau'\right] = \mathbf{1} + \sum_{(n'', n')} \sum_{j=0}^{\infty} \varepsilon^{j}\, \boldsymbol{S}^{(j)}_{\ldots n'' \ldots n'}\left[\tau'', \tau'\right] \tag{3.3}$$

*) Le paramètre numérique ε que nous introduisons ici apparaît lors de la construction explicite de S, basée sur une méthode de perturbation (voyez § 4).

avec:

$$\varepsilon^j \boldsymbol{S}^j_{\ldots n'' \ldots n'}[\tau'', \tau'] = \varepsilon^j \int dV(u''_{(1)}) \ldots \int dV(u''_{(n'')}) \int dV(u'_{(1)}) \ldots \times$$
$$\times \ldots \int dV(u'_{(n')})\, \boldsymbol{a}^+(u''_1) \ldots \boldsymbol{a}^+(u''_{(n'')})\, S^{(j)}[\tau'', u''_{(1)} \ldots u''_{(n'')}/\tau', u'_{(1)} \ldots u'_{(n')}] \times$$
$$\times \cdot \boldsymbol{a}(u'_{(1)}) \ldots \boldsymbol{a}(u'_{(n')}). \tag{3.4}$$

où nous avons ordonné les opérateurs $\boldsymbol{a}^+(u''_{(i)})$ et $\boldsymbol{a}(u'_{(k)})$ dans chacun des termes $S^{(j)}_{\ldots n'' \ldots n'}$ de manière que les opérateurs $\boldsymbol{a}^+(u''_k)$ soient tous à gauche des opérateurs $\boldsymbol{a}(u'_{(i)})$, et en fixant une fois pour toutes l'ordre de succession de ces opérateurs. Cela est toujours possible à l'aide des relations de commutations:

$$[\boldsymbol{a}(u''_{(k)}), \boldsymbol{a}^+(u'_{(i)})]_\pm = \mathbf{1}\, \delta(u''_{(k)}/u'_{(i)})$$
$$[\boldsymbol{a}^+(u''_{(k)}), \boldsymbol{a}^+ \cdot (u'_{(i)})]_\pm = [\boldsymbol{a}(u''_{(k)}), \boldsymbol{a}(u'_{(i)})]_\pm = \mathbf{0} \tag{3.5}$$

Cela fait, il est clair que, de par la signification même des opérateurs $\boldsymbol{a}^+(u''_{(k)})$ et $\boldsymbol{a}(u'_{(i)})$ lorsqu'ils opèrent sur la fonctionnelle de l'espace des quanta, les coefficients $S^{(j)}[\tau'', . u''_{(k)} . /\tau'', . u'_{(i)} \ldots]$ qui figurent dans le développement (3.4) ne peuvent être que les matrices de transition dont la structure a été définie en (2.6): alors seulement l'opérateur $\boldsymbol{S}$ est causal.

Nous voyons donc apparaître, à côté de l'*invariance*, deux propriétés nécessaires de l'opérateur $\boldsymbol{S}$: l'*unitarité*, exprimée par (3.2) et la *causalité* exprimée par la structure des coefficients $S[\tau'' \ldots /\tau' \ldots]$ donnée par (2.6).

4. Construction de la matrice S à partir d'un processus élémentaire donné.

Nous avons trouvé la structure générale de l'opérateur $\boldsymbol{S}$. Mais la tâche essentielle est de le construire à partir d'un processus élémentaire donné.

Une première méthode part de l'équation d'évolution fonctionnelle:

$$-\frac{1}{i} \frac{\delta \psi}{\delta \tau(x)} = \varepsilon\, \boldsymbol{h}(x) \tag{4.1}$$

dont l'intégration invariante donne un opérateur $\boldsymbol{S}$ qui a bien la structure causale (3.4), mais où la fonction $D^c(x/y)$ est donnée par (1.5) pour toute valeur de $x-y$, y compris $x-y=0$. L'intégration «invariante et causale» de (4.1) montre qu'en général $\varepsilon\boldsymbol{h}(x)$ est donné par:

$$\varepsilon \boldsymbol{h}(x) = \varepsilon \boldsymbol{h}^{(1)}(x) + \varepsilon^2 \boldsymbol{h}^{(2)}[x, \tau] \tag{4.2}$$

résultat auquel mène aussi l'application des conditions d'intégrabilité de (4.1)[1].

Cette méthode a été utilisée, sous des formes variées, par divers auteurs dans le cas particulier de l'électrodynamique[1]) d'un champ scalaire[2]) ou de couplages plus généraux[3]). Elle a le désavantage d'introduire dès l'ordre $j = 2$ des termes divergents pour les coefficients $S^{(j)}_{\ldots n'' \ldots n'}[\tau'', \tau']$. Divers procédés ont été proposés pour éliminer ces divergences. Les uns sont appuyés sur des considérations physiques[4]), les autres sont d'allure plus mathématique[3, 5]). Cependant, une forme satisfaisante de la théorie devrait éviter de semblables corrections a posteriori, qui n'assurent pas toujours de manière visible sa causalité, son unitarité, ou même sa cohérence.

Une seconde méthode a l'avantage, en posant d'emblée:

$$\boldsymbol{S} = \exp(-i\boldsymbol{\alpha}) \quad \boldsymbol{\alpha}^{+} = \boldsymbol{\alpha} \quad \boldsymbol{\alpha} = \sum_{1}^{\infty} \varepsilon^{i} \boldsymbol{\alpha}_{(i)} \tag{4.3}$$

d'assurer initialement l'invariance et l'unitarité de $\boldsymbol{S}$. Elle procède ensuite de la manière suivante:

1. On choisit l'opérateur $\boldsymbol{\alpha}_{(1)}$ qui correspond au processus élémentaire envisagé. Celui qui correspond à $\boldsymbol{h}^{(1)}$ de (4.2) s'écrit:*)

$$\varepsilon \boldsymbol{\alpha}_{(1)} = \varepsilon \int (d x)^4 \boldsymbol{h}_{(1)}(x) \qquad \boldsymbol{\alpha}^{+}_{(1)} = \boldsymbol{\alpha}_{(1)} \tag{4.4}$$

2. On développe alors $\boldsymbol{S}$ selon (4.3) et l'on contrôle en comparant avec (3.4) que chacun des coefficients $S^{(j)}[\tau''; u''/\tau'; u']$ ait bien la structure causale (2.6);

$$\boldsymbol{S} = \boldsymbol{1} + (-i\varepsilon) \int (d x)^4 \boldsymbol{h}_{(1)}(x) + \frac{(-i\varepsilon)^2}{2!} \cdot \int (d x')^4 \boldsymbol{h}_{(1)}(x') \times \\ \times \int (d x)^4 \boldsymbol{h}_{(1)}(x) + \ldots \tag{4.5}$$

On voit tout de suite que dès les $S^{(2)}[\,/\,]$ ce n'est pas le cas. Pour rendre causals ces coefficients, on modifie $\boldsymbol{\alpha}$ en lui ajoutant selon (4.3) un terme $\varepsilon^2 \boldsymbol{\alpha}_{(2)}$, hermitien naturellement, ce qui a pour résultat d'ajouter aux coefficients $S^{(2)}[/]$ une «correction causale», dans la mesure où elle leur donne la structure (2.6). Formellement le résultat est alors celui que donne la première méthode, à cela près cependant que les fonctions $D^c(x/y)$ qui apparaissent dans les noyaux ne sont pas définies pour $x = y$.

Les coefficients $S^{(2)}[\ldots/\ldots]$ rendus causals, on développe $\boldsymbol{S}$ selon (4.3) avec $\boldsymbol{\alpha} = \varepsilon\,\boldsymbol{\alpha}_{(1)} + \varepsilon^2\,\boldsymbol{\alpha}_{(2)}$ et l'on contrôle alors la structure causale des termes $S^{(3)}[\ldots/\ldots]$. Cela conduit de nouveau à une correction causale donnant formellement les mêmes coefficients $S^{(3)}[\,/\,]$ que la méthode différentielle, et obtenue en ajoutant à $\boldsymbol{\alpha}$ un terme $\varepsilon^3 \boldsymbol{\alpha}_{(3)}$.

*) Dans (4.4) et (4.5) $\int$ est mis pour $\int_{\tau'}^{\tau''}$.

L'on procède ainsi de suite pour les $S^{(4)}[\ldots/\ldots]$ $S^{(5)}[\ldots/\ldots]\ldots$, en sorte que la série (4.3) est bien déterminée aux arbitraires près signalés tout à l'heure, dus à l'indétermination de la fonction $D^c(x/y)$ pour $x = y$.

Ainsi donc la première méthode, qui est différentielle et la seconde, qui est intégrale, conduisent *formellement* aux mêmes résultats (avant la suppression des expressions divergentes ou indéterminées); mais, tandis que dans le premier cas on se trouve en face d'expressions divergentes en général à partir du deuxième ordre, dans le second ces expressions sont indéterminées. On voit donc qu'au problème de l'élimination des divergences de la matrice S la seconde méthode substitue celui de la détermination de cette matrice à partir d'une expression partiellement non définie. Pour résoudre ce problème, il existe une méthode*) générale qui conduit pour les termes du deuxième ordre à des résultats analogues à ceux de M. Schwinger, en particulier pour la polarisation du vide et pour l'énergie propre des particules élémentaires. Ces résultats, joints à ceux établis jusqu'ici pour le troisième ordre semblent satisfaisants. Un prochain travail exposera cette méthode et son application au problème de la polarisation du vide.

Ces recherches ont bénéficié de l'aide matérielle de la C. S. A.; nous l'en remercions.

Références.

[1]) Principalement Tomonaga et ses collaborateurs; Schwinger, Feynmann et F. J. Dyson (Phys. Rev. **75**, 3, 486; **75**, 11, 1736 (1949)) qui donne une bibliographie très complète des mémoires des auteurs précédents, à laquelle nous renvoyons.

[2]) A. Houriet et A. Kind, H.P.A. **22**, 3, 319 (1949).

[3]) E. C. G. Stueckelberg et D. Rivier, Phys. Rev. **74**, 2, 218 (1948); D. Rivier, H.P.A. **22**, 3, 265 (1949).

[4]) J. Schwinger, Phys. Rev. **74**, 10, 1439 (1948); **75**, 4, 651 (1949).

[5]) D. Rivier, loc. cit.; Pauli et Villars, Rev. Mod. Phys. **21**, 434 (1949). Nous remercions MM. Pauli et Villars de nous avoir donné un manuscrit de leur travail avant sa publication.

---

*) Cette méthode faisait l'objet d'une communication destinée à la Conférence de Physique de Bâle (Septembre 1949).

A propos des divergences en théorie des champs quantifiés [83]

with D. Rivier, *Helvetica Physica Acta*, vol. 23, suppl. 3 (1950), pp. 236–239

A propos des divergences en théorie des champs quantifiés

par **E. C. G. Stueckelberg** et **D. Rivier**, Genève.

Comme nous le montrons ailleurs[1]), la causalité impose à la matrice S qui décrit l'évolution d'un système une structure bien déterminée: lorsqu'on développe celle-ci suivant les opérateurs de translation dans l'espace des quanta, les coefficients $S^{(i)}[\tau''; u''\cdot\cdot/\tau'; u'\cdot\cdot]$ sont des intégrales multiples où n'apparaissent, à côté des champs liés à un seul point de l'espace temps, que les fonctions*):

$$D^c(x/y) = D^s(x/y) + \frac{i}{2} D^1(x/y) \quad x \neq y \tag{1}$$

Formellement, cette structure est aussi celle que l'on obtient par intégration invariante de l'équation différentielle[2]) d'évolution du système à la différence suivante près: tandis que l'intégration conduit à une expression à première vue déterminée pour la matrice S, la construction causale de S laisse une certaine indétermination pour le noyau intégral formé des fonctions $D^c(x/y)$: en effet, de la manière dont ces fonctions sont introduites, il n'est pas possible de leur fixer a priori une valeur au point $x = y$. Cela est essentiel: en effet, si l'on défini aussi en $x = y$ la fonction $D^c(x/y)$ par (1) comme *doit* le faire l'intégration de l'équation différentielle, on est alors conduit à des coefficients $S^{(i)}[\ldots]$ qui divergent en général. Par contre, l'indétermination de la fonction $D^c(x/y)$ au point $x = y$ permet une détermination a posteriori des noyaux intégraux qui conduisaient dans la théorie différentielle à des divergences; cette définition est univoque et donne une valeur finie aux coefficients $S[\ldots]$. Comme nous allons le montrer, il subsiste après cela dans ces coefficients un certain arbitraire; mais celui-ci peut être partiellement éliminé par des considérations physiques.

Le procédé de définition des noyaux intégraux $\Delta^c(x/y)$ est appuyé sur le fait que les divergences proviennent essentiellement des singularités non intégrables des produits de fonctions $D^c(x/y)$, situées aux points $x = y$. Il est alors indiqué de définir les noyaux «vrais» à partir de ceux que donne l'intégration en leur ôtant leurs singularités non intégrables. On y parvient d'abord par l'utilisation

*) Pour les notations, voyez le second travail cité sous [2]).

d'un opérateur $\boldsymbol{\vartheta}$ tel que $\boldsymbol{\vartheta}\, \Delta^c(x/y)$ soit intégrable. Il suffit de prendre pour $\boldsymbol{\vartheta}$ la multiplication par la fonction:

$$\vartheta^{(n)}(iz) = a_{\alpha_1 \dots \alpha_n} (iz^{\alpha_1}) \cdots (iz^{\alpha_n}) \tag{2}$$

où $z^{\alpha_i} = x^{\alpha_i} - y^{\alpha_i}$, avec sommation de 1 à 4 sur les indices vectoriels $\alpha_1 \dots \alpha_n$, et où n dépend de l'acuité de la singularité. Puis, pour conserver au résultat sa signification, il est nécessaire de multiplier la valeur de l'intégrale par un opérateur $\boldsymbol{\vartheta}_s$ qui doit se réduire, dans le domaine où $\Delta^s(x/y)$ est intégrable, à l'inverse de $\boldsymbol{\vartheta}$.

La réalisation de ces opérations est simple dans l'espace de Fourier; si nous écrivons (en prenant pour fixer les idées un terme du deuxième ordre):

$$S^{(2)}(\varphi''/\varphi') = \int (dx'')^4 \varphi''(x'')\, \boldsymbol{\Delta}^c \varphi'(x'') \tag{3}$$

où:

$$\boldsymbol{\Delta}^c \varphi'(x) = \frac{1}{2}\int (dx)^4 \left(D^c(x/y)\right)^2 \varphi'(y)$$

il suffit d'étudier*):

$$\boldsymbol{\Delta}^s \varphi(x) = \int (dy)^4 \Delta^s(x/y)\, \varphi(y) \tag{4}$$

où:

$$\Delta^s(x/y) = \frac{i}{4}\left(D^1 D^s + D^s D^1\right)(x-y)$$

qui s'écrit dans l'espace de Fourier:

$$\boldsymbol{\Delta}^s \varphi(x) = (2\pi)^{-3/2} \int dV(\vec{k})\, \Delta^s(k)\, e^{ikx} \psi(\vec{k}) \tag{5)**}$$

avec:

$$\Delta^s(k) \sim \frac{i}{4}(2\pi)^{-3} \int (dp)^4 \left[\frac{\delta(p^2+\varkappa^2{}_u)}{(k^2-p)^2+\varkappa^2{}_u} + \frac{\delta((k-p)^2+\varkappa^2{}_u)}{p^2+\varkappa^2{}_u}\right] \tag{6}$$

ou encore, en utilisant un algorithme dû à M. Schwinger[3]) et en effectuant la translation[4]) $p^\alpha \to p^\alpha + uk^\alpha$

$$\Delta^s(k) \sim \int (dp)^4 \Delta^s(k^2, p^2) \tag{6a}$$

avec:

$$\Delta^s(k^2, p^2) \sim -\frac{i}{4}(2\pi)^{-3} \int_0^1 du\, \delta'[p^2 + (\varkappa^2 + (u-u^2)k^2)] \tag{7}$$

---

*) Nous nous limitons au cas où τ'' et τ' dans $S^2[\tau''\ \varphi''/\tau'\ \varphi']$ sont des surfaces à l'infini: $\tau'' = -\tau' = x^{4''} = \infty$.

**) $\psi(\vec{k})$ est la composante de Fourier du paquet d'ondes $\varphi(x)$.

En lieu et place de l'intégrale (6a) qui *diverge* en général, comme c'est le cas ici, nous écrivons alors:

$$\Delta^s(k^2) = \underbrace{\int_0^{(k^2)} d(k^2) \ldots \int_0^{k^2} d(k^2)}_{(n)} \int (dp)^4 \cdot \left(\frac{\partial}{\partial (k^2)}\right)^n \Delta^s(k^2, p^2) \tag{8}$$

où $n \geqslant 0$ est le plus petit entier tel que:

$$\int (dp)^4 \left(\frac{\partial}{\partial (k^2)}\right)^n \Delta^s(k^2, p^2) \tag{9}$$

ait un sens.

La vraie valeur du noyau $\Delta^s(k^2)$ est donc:

$$\Delta^s(k^2) = \Delta^s_{\text{déf}}(k^2) + b_0 + b_1 k^2 + \cdots b_{n-1} k^{2(n-1)} \tag{10}$$

où $\Delta^s_{\text{déf}}(k^2)$ est une fonction parfaitement définie de k^2. Il s'introduit donc n constantes arbitraires b_i et dans l'espace x, $\Delta^s(x/y)$ n'est définie qu'à la série:

$$b_0 \delta(x-y) + b_1 \square\, \delta(x-y) + \cdots b_{n-1} \square^{(n-1)} \delta(x-y) \tag{11}$$

près, faisant apparaître des singularités intégrables à l'origine $x - y = 0$. Dans l'exemple choisi (6), on a $n = 1$ et:

$$\Delta^s_{\text{déf}}(k^2) = \frac{1}{32\, i\, \pi^2} \left[\operatorname{Log} \varkappa^2 \left(\frac{\alpha+1}{\alpha-1}\right)^\alpha - 2\right] \qquad \alpha^2 = 1 + \frac{4\varkappa^2}{k^2}$$

Nous avons étudié avec cette méthode les termes de seconde approximation et en partie ceux de la troisième. On peut brièvement résumer les résultats de la manière suivante:

Toutes les divergences dues à la limite supérieure infinie de k (catastrophe ultraviolette) disparaissent. Grâce aux constantes arbitraires b_i, dans l'approximation du *deuxième* ordre le terme appelé énergie propre ou masse propre du photon (ou du méson) peut être annulé (dans le cas du photon, il suffit pour cela de poser nulle une des constantes). Dans la même approximation, il est possible d'éviter une renormalisation de la charge: la charge induite Δe peut être annullée. Notons encore à propos de la deuxième approximation que le courant induit satisfait à l'équation de continuité, à moins que le potentiel inducteur soit lui-même induit par un courant ne satisfaisant pas à l'équation de continuité, cas offrant, semble-t-il, peu d'intérêt.

En *troisième* approximation et dans le cas de l'électrodynamique, l'étude du rapport des coefficients des opérateurs de moment intrinsèque $S^{\alpha\beta}$ et de moment orbital $L^{\alpha\beta}$ dans la matrice S montre que ce rapport, qui vaut $g_0 = 2$ en première approximation, s'écrit*):

$$g = g_0 + 2\,\varkappa_u \frac{\mu}{e}$$

où:

$e = \varepsilon(1 + a\,\varepsilon^2)$ charge de l'électron «renormalisée»

$\varkappa_u =$ masse de l'électron

$$\mu = \frac{1}{2\pi\varkappa_u}\,\varepsilon^3$$

a est arbitraire; on obtient donc:

$$g = 2 + \frac{\varepsilon^2}{\pi}\left[1 - \varepsilon^2(a + \cdots)\right] \tag{13}$$

$\varepsilon^2 a \ll 1$ est nécessaire pour conserver un sens à un développement en ε. Donc, limité à ε^2, notre résultat, qui coïncide avec celui de M. SCHWINGER[5]), est indépendant d'une renormalisation de la charge.

Références.

[1]) E. C. G. STUECKELBERG et D. RIVIER, H. P. A. **23**, 215 (1950).

[2]) Par exemple A. HOURIET et A. KIND, H. P. A. **22**, 319 (1949); D. RIVIER, H. P. A. **22**, 265 (1949).

[3]) J. SCHWINGER, Phys. Rev. **75**, 651 (1949).

[4]) Due aussi à M. SCHWINGER.

[5]) J. SCHWINGER, Phys. Rev. **76**, 790 (1949).

*) Nous remercions M. T. GREEN qui nous a aidé dans l'évaluation du facteur μ.

Relativistic Quantum Theory for Finite Time Intervals [84]

Reprinted with permission from *Physical Review*, vol. 81 (1951), pp. 130–133

PHYSICAL REVIEW VOLUME 81, NUMBER 1 JANUARY 1, 1951

Relativistic Quantum Theory for Finite Time Intervals

E. C. G. STUECKELBERG*
University of Geneva, Geneva, Switzerland

(Received August 11, 1950)

If transition probabilities are evaluated for transitions occuring during a finite time interval, additional divergencies occur different from those commonly encountered for infinite time intervals. The expressions obtained can however be made convergent, if an indeterminacy of time is attributed to each epoch of observation. The method is applied to the emission of a photon by a free electron.

I. INTRODUCTION

THE convergent results in the relativistic quantum theory of elementary particles, which have been recently obtained by different authors,[1] apply only to time periods of infinite duration between two observations. If one tries to evaluate transition probabilities for processes which are localized in space-time by a *sharply defined boundary* (for example two time-like hypersurfaces specifying an initial and final observation), one obtains divergent results. These divergences arise from regions near the boundary, where processes occur without conservation of the momentum-energy component normal to the hypersurface. However, we show here that one can obtain convergent results if *diffuse boundaries* are introduced. We show in Sec. I that this generalization is possible without affecting the unitarity and causality of the array of probability amplitude forming the S-matrix. In Sec. III, we evaluate, in second-order approximation, the time-independent probability for the emission of a photon by an electron.

Time, with these unsharp limits, no longer appears as a *parameter*, t, whose values $t=t'$ and $t=t''$ are fixed for the two limits of the *period of evolution*, $t''-t'=2T$, during which the photon emission takes place. The initial and final epochs themselves, $t\cong t'$ and $t\cong t''$, are now of finite duration, $\Delta t'$ and $\Delta t''$, and must be given in terms of *two probability amplitudes for time*, $f'(t)$ and $f''(t)$, describing the precision with which t' and t'' have been determined. In the probability $dw(\omega)$ that an electron has emitted a photon of frequency between ω and $\omega+d\omega$ during the period considered, the Fourier transforms, $g'(\omega)$ and $g''(\omega)$, of the two probability amplitudes figure as convergence factors for the integral.[2] We have:

$$dw(\omega)=d\omega(|g''|^2+|g'|^2)(\omega)n(\omega) \equiv dw''(\omega)+dw'(\omega). \quad (1)$$

* Work supported by the Swiss Atomic Energy Commission.

[1] S. Tomonaga, Progr. Theor. Phys. **1**, No. 2, 27 (1946); J. Schwinger, Phys. Rev. **74**, 1439 (1948); **75**, 651 (1949); **76**, 790 (1949); R. P. Feynman, Phys. Rev. **76**, 749, 769 (1949); F. Dyson, Phys. Rev. **75**, 486, 1736 (1949).

[2] To $g'(\omega)=\exp(i\omega t')$; $g''(\omega)=\exp(i\omega t'')$ correspond the epochs $f'(t)=\delta(t-t')$ and $f''(t)=\delta(t-t'')$ of sharply determined time values, for which the integral of Eq. (1) diverges.

$g'(\omega)$ and $g''(\omega)$ are normalized to $|g'(0)|=|g''(0)|=1$. Their absolute squares are time independent. We are thus led to think of (1) as the sum of two probabilities for processes which take place *only during the epochs of the initial and of the final observations*. Furthermore, these processes show no conservation of energy: ω is the surplus energy in the final state over that in the initial state. This excess energy can be interpreted as having been furnished by the measuring apparatus during either of the two epochs of observation. Then $|g'(\omega)|^2$ and $|g''(\omega)|^2$ indicate the probability that such an energy is available.

II. THE CONVERGENCE CONDITIONS FOR PROBABILITY AMPLITUDES

We describe a process taking place in a given space-time (x-space) region, V, as the *annihilation of momentum-energy* out of the *incoming* matter or radiation waves and the *creation of momentum-energy* into the *outgoing* waves. We represent the incoming waves of electrons, positrons and photons by the wave packets $u'(x)$, $v'(x)$ and $\varphi'(x)$ with a positive frequency spectrum and the outgoing waves by their conjugate complex $u''^{\dagger}(x)$, $v''^{\dagger}(x)$ and $\varphi''^{\dagger}(x)$ with a negative frequency spectrum. The particular packets

$$u_A'(x)=(2\pi)^{-\frac{3}{2}}\pi_A(k'n')\exp(ik'x) \tag{2}$$

are plane electron waves. A packet (2) represents a quantum of sharply defined *momentum-energy* k', lying in the *momentum-energy space* (p-space) on the hypersurface of rest mass κ, $p=k'$, where

$$k'^4=+(\kappa^2+|\mathbf{k}'|^2)^{\frac{1}{2}} \tag{3}$$

n' numbers the two spin orientations, π^A in (3) is normalized to $\pi_A{}^{\dagger\prime\prime}\pi'^A=2\kappa i\delta(n''/n')$ for electrons. For photons of rest mass[3] μ we write $p=l'$ and number their polarizations by $n'=1, 2, 3$ (normalization $\pi_\alpha{}''^{\dagger}\pi'^\alpha=\delta(n''/n')$). Then the nth order contribution to the probability amplitude of a process is the n-fold space-time integral over V:

$$\epsilon^n S_n[V](u''\cdots\varphi''\cdots/u'\cdots)$$

$$=i\epsilon^n\int dx''V(x'')\cdots\int dy''V(y'')\cdots\int dxV(x)\cdots$$

$$\times\int dx'V(x')\cdots u_A{}''^{\dagger}(x'')\cdots\varphi_\alpha{}''^{\dagger}(y'')\cdots$$

$$\times\Delta_{B\cdots}^{(c)A\cdots\alpha\cdots}(x''-y'',\cdots x''-x,$$

$$\cdots x''-x',\cdots)u'^B(x')\cdots. \tag{4}$$

For sharply defined boundaries, $V(x')$ is a *discontinuous function*, with the two values 0 or 1 for events x' outside or inside V. The causal[4] function, $\Delta^{(c)A\cdots}(x''-y'',\cdots)$ is a covariant function of the $n-1$ relative displacements of the n events. It is contragradient in its indices to the vector, $\alpha\cdots\beta\cdots$, or spinor, $A\cdots B\cdots$, indices of the packets.

Let us now see how the *generalization to a continuous real function* $V(x)$ related to the time uncertainties of the initial and final observation epochs is possible without affecting the unitarity of the theory or changing the causal function $\Delta^{(c)}$. The unitarity of the S-matrix corresponding to (4) implies that the hermitian part of S_n is determined in terms of the S_m for $m<n$. Therefore S_1 is antihermitian and is given in terms of a single space-time integral of the hermitian interaction energy density and the real function $V(x)$. In electrodynamics the typical element is:

$$\epsilon S_1[V](u''\varphi''/u')$$

$$=i\epsilon 2^{-\frac{1}{2}}\int dxV(x)(\varphi_\alpha{}''^{\dagger}u''^{\dagger}\gamma^\alpha u')(x). \tag{5}$$

In terms of this S_1, we *define* the hermitian part of S_2 by means of the unitarity relation. We obtain for this part expressions of the type (4) (with $n=2$), where $i\Delta^{(c)}$ is replaced by non-causal functions $-\frac{1}{4}(\Delta^{(+)}+\Delta^{(-)})$. The anti-Hermitian part of S_2 is then determined in terms of a causality correction $i\Delta^{(s)}$, such that the sum of the hermitian and antihermitian parts gives $i\Delta^{(c)}$. The higher order approximations are obtained in exactly the same way. This procedure shows that the unitarity and the causality of the S-matrix are independent of the particular form given to $V(x)$, as long as it is a real function.

In order to find the convergence conditions of (4), we transform it into p-space. The transformed integral is $n-1$ fold. If m is the total number of incoming and outgoing packets, then

$$\epsilon^n S_n[V](k''n'',\cdots l'',\cdots/k'n',\cdots)$$

$$=i\epsilon^n(2\pi)^{4n-\frac{3}{2}m}\int dp\cdots\int dq\cdots\int dr\cdots$$

$$\times V(k''-p-q-r-\cdots)\cdots V(l''+p)\cdots$$

$$\times V(0+q)\cdots V(-k'+r)\cdots\pi_A{}''^{\dagger}\cdots\pi_\alpha{}''^{\dagger}\cdots$$

$$\times\Delta^{(c)A\cdots\alpha\cdots}{}_{B\cdots}(p,\cdots q,\cdots r,\cdots)\pi'^B,\cdots. \tag{6}$$

This formula involves the Fourier transforms of $V(x)$ and $\Delta^{(c)}$. We shall use the same letters to describe these functions in both x- and p-spaces. We see at once that the existence of the Fourier transform of the causal

[3] In order to avoid the difficulties connected with zero rest mass photons, we suppose the photon to have a finite rest mass, small compared to that of the electron.

[4] By the term "causal" we imply that $\Delta^{(c)}$ in Eq. (4) is a network of causal functions $\Delta^{(c)}(x-y)$ [given in Eq. (9)] describing an outgoing wave at y and an incoming wave at x, if x is later than y (the "creation at y" precedes the "annihilation at x"), and *vice versa* if y is later than x. See Stueckelberg and Rivier, Phys. Rev. 74, 218 (1948); Helv. Phys. Acta 23, 215 (1950); 23, supp. III, 236 (1930).

function $\Delta^{(c)}(p, \cdots)$ is a *necessary condition for the convergence of the amplitude S_n*. For the particular case $V(p)=\delta(p)$ one obtains the usual result:

$$\epsilon^n S_n[\infty](k''n''\cdots l''\cdots/k'n'\cdots)$$
$$=i\epsilon^n(2\pi)^{4n-\frac{3}{2}m}\delta(-k''-l''-\cdots+k'+\cdots)$$
$$\times\pi_A{}''^{\dagger}\cdots\pi_\alpha{}''^{\dagger}\cdots$$
$$\times\Delta^{(c)A\cdots\alpha\cdots}{}_{B\cdots}(-l'', \cdots 0, \cdots k', \cdots)\pi'^{B}\cdots, \quad (7)$$

which is commonly used to define transition probabilities for a region V extending over "infinite" physical space-time.

III. PHOTON EMISSION BY A FREE ELECTRON

In the second approximation of quantum electrodynamics, the probability amplitude of an electron-electron transition inside of V is

$$\epsilon^2 S_2[V](u''/u')$$
$$=\epsilon^2\int dxV(x)\int dyV(y)u''^{\dagger}(x)$$
$$\times(\gamma\partial\Delta_2{}^{(c)}+\kappa\Delta_1{}^{(c)})(x-y)u'(y)$$
$$\equiv\delta(u''/u')(-\tfrac{1}{2}W(u')+\tfrac{1}{2}C(u')-i\Phi(u')). \quad (8)$$

Its Hermitian and anti-Hermitian parts correspond to the separation of the two scalar, causal functions into real and imaginary parts according to:

$$i\Delta^{(c)}=-\tfrac{1}{4}(\Delta^{(+)}+\Delta^{(-)})+i\Delta^{(s)}. \quad (9)$$

$\Delta^{(\pm)}$ are functions involving only definite frequencies larger than the sum of the two rest masses κ and μ.

$$\Delta^{(-)}(-p)=\Delta^{(+)}(p)=\begin{Bmatrix}0\\ 2\Delta^{(1)}(p^2)\end{Bmatrix}\text{ for } p^4\lesseqgtr 0, \quad (10)$$

$$\Delta^{(1)}(p^2)=0\text{ for }-p^2<(\kappa+\mu)^2. \quad (11)$$

$\Delta^{(s)}$ is the causality correction, defined in terms of $\Delta^{(1)}$ by the following differential equation

$$\left(-\frac{d}{d(p^2)}\right)^n\Delta^{(s)}(p^2)=\frac{n!}{2\pi}\int_0^\infty d(z)^2\frac{\Delta^{(1)}(-z^2)}{(p^2+z^2)^{n+1}} \quad (12)$$

for the lowest n, for which the principal value converges.[5]

Let us now recall briefly the physical meaning of the three parts in Eq. (8) due to $\Delta^{(\pm)}$ and $\Delta^{(s)}$ in (9):

(1) $W(u')$ is a *decrease of the probability* of observing only an emerging electron u'', due to the process of photon emission (in which case we should observe the outgoing waves u'' and φ'').

(2) $C(u')$ is an *increase of this probability*, due to the diminution of the probability for the spontaneous process of three quantum creation (photon, electron and positron), when the electron state u' is occupied.

(3) $\Phi(u')$ is a phase change, different for different wave packets, and giving rise to a *change of the dispersion law for electron waves* [i.e., differing from (2)].

We evaluate these quantities for the space-time region V bounded by two time-like hyperplanes $t\cong t'$ and $t\cong t''$ defined by the amplitudes $f'(t)$ and $f''(t)$. In terms of the Fourier transforms, normalized to $|g'(0)|=|g''(0)|=1$, the transform $V(p)$ in (6) (Appendix I) is

$$V(p)=\delta(\mathbf{p})i(2\pi\omega)^{-1}(g'-g'')(\omega),\ \omega=p^4. \quad (13)$$

Omitting the δ-symbol in (8) (because it is evidently diagonal in the momentum and spin space of the electron), we have the following p-space representation of $S_2[V]$ in terms of W, C and Φ:

$$W(k')=\int_{-\infty}^{+\infty}d\omega|g''-g'|^2(\omega)\frac{(2\pi)^3\epsilon^2}{k'^4\omega^2}$$
$$\times(k'p\Delta_2{}^{(+)}+\kappa^2\Delta_1{}^{(+)})(p), \quad (14)$$

$$\mathbf{p}=\mathbf{k}';\quad p^4=k'^4+\omega. \quad (14a)$$

C is the same expression involving $-\Delta^{(-)}$ instead of $\Delta^{(+)}$, and in the phase $\Phi(u')$, $4\Delta^{(s)}$ has to be substituted for $\Delta^{(+)}$. In the limit, where the period $2T=t''-t'$ between the two epochs t'' and t' is long with respect to their indeterminacies $\Delta t''$ and $\Delta t'$, a frequency ω_0 may be defined

$$2T\gg\omega_0{}^{-1}\gg\Delta t'',\ \Delta t' \quad (15)$$

allowing in the integrand of (14) the substitution:

$$\omega^{-2}|g''-g'|^2(\omega)=\begin{cases}\omega^{-2}(2\sin\omega T)^2,\text{ for }\omega^2<\omega_0{}^2 & (16)\\ \omega^{-2}(|g''|^2+|g'|^2)(\omega), & \\ \qquad\text{for }\omega^2>\omega_0{}^2. & (17)\end{cases}$$

If the intergrand is different from zero for $\omega=0$, W, C, and Φ consist of a *time proportional* part plus a *time independent* part.

In our particular case, we obtain a time proportional part for the phase alone. Its form, $2T\Delta\kappa^2(2k'^4)^{-1}$, shows that the dispersion law Eq. (2) is affected in the form of a rest mass change $\Delta\kappa^2$. The value of the invariant $\Delta\kappa^2$ is undetermined on account of the arbitrary constants in the definition of $\Delta^{(s)}$ in (12). The integrand of the time-independent probability (14) decomposes, on account of (17), into the two contributions due to the two limits of the region V. The four vector, p, is the momentum-energy in the electron+photon state. The substitution of $2\Delta^{(1)}$ for $\omega\ll\kappa=k'^4$ (Appendix II) gives the following expression for the low frequency

[5] The integration of (12) introduces a finite series of terms of the type $a_0+a_1p^2+\cdots+a_{n-1}(p^2)^{n-1}$ with n arbitrary constants. A detailed discussion of this arbitrariness will be published in the *Helvetica Physica Acta*.

probability spectrum to find a photon within ω and $\omega+d\omega$ created at $t\cong t'$.

$$dw'(\omega)=d\omega\,|g'(\omega)|^2\frac{\epsilon^2}{(2\pi)^2}\frac{(\omega^2-\mu^2)^{\frac{1}{2}}}{\omega^2\mu^2}. \tag{18}$$

This expression may be compared to the frequency spectrum of the total energy of a spherically symmetric classical radiation field $\chi^\alpha(x)$ of longitudinal photons ($n'=3$). Writing $\omega=l'^4$, the total energy is in such a case:

$$P^4=\int d\sigma(l')l'^4\sum_{n'=1}^{3}(b^\dagger b)(l'n')$$

$$=\int d\omega 4\pi\omega(\omega^2-\mu^2)^{\frac{1}{2}}(b^\dagger b)(l3). \tag{19}$$

$b^\dagger$ and b are the coefficients of the field $\sqrt{2}\chi^\alpha(x)$, developed in terms of plane wave packets $\varphi'^\alpha(x)$ and their conjugates of the form (2) with the three polarizations n' normalized to $\pi_\alpha''^\dagger\pi'^\alpha=\delta(n''/n')$.

We now determine χ^α from the condition at $t=t'$

$$\chi(x')=0;\quad \chi^4(x')=-\frac{\epsilon}{4\pi}\frac{\exp(-\mu|\mathbf{x}'|)}{|\mathbf{x}'|}+\frac{\epsilon}{\mu^2}\delta(\mathbf{x}');$$

$$\partial_4\chi(x')=-\frac{\epsilon}{\mu^2}\operatorname{grad}\delta(\mathbf{x}');\quad \partial_4\chi^4(x')=0. \tag{20}$$

At $t=t'$, this field compensates the static field everywhere except "inside" of the point particle. We find

$$b^\dagger b(l3)=\frac{1}{2}\frac{\epsilon^2}{(2\pi)^3}\frac{\omega^2-\mu^2}{\omega^2\mu^2}. \tag{21}$$

Comparing (18) with (19) and (21), we see that the classical radiation field corresponding to $dw'(\omega)$ compensates the field of the point charge ϵ, except within a sphere with a radius of the order of $\Delta t'$. The energy density of the total field (static+radiation) is strictly zero outside of this sphere.

IV. CONCLUSIONS

These arguments show, that the introduction of a finite period of evolution in current quantum electrodynamics produces no difficulty of convergence, if diffuse time boundaries are used.

It is also interesting to note that the emission of a photon by a free electron has a simple classical analog. The *photon field* expectation values correspond to a *classical* radiation field, which compensates (or "interferes away") the point electron's static field at the epoch of observation $t\cong t'$ everywhere, except within a small sphere with a radius of the order of the indeterminacy $\Delta t'$ of the time measurement.

APPENDIX I

If we wish to interpret the S-matrix as an operator operating on a state vector ψ, the state vector refers not to a time like hypersurface as in the Tomonaga-Schwinger theory, but to a time-like layer. We may describe such a layer by a real function $F(x)$, which has the values one or zero for events x lying in the future or in the past of the layer. The region V will then be given in terms of the two layers

$$V(x)=F'(x)-F''(x) \tag{I.1}$$

and the S-matrix transforms according to

$$\psi[F'']=S[V]\psi[F'] \tag{I.2}$$

the initial state into the final state. The probability amplitude for the time measurement is the gradient of $F(x)$:

$$f_\alpha'(x)=\partial_\alpha F'(x). \tag{I.3}$$

It reduces to a function of time alone in the case considered in Section III.

$$\mathbf{f}'(x)=0;\quad f_4'(x)\equiv f'(t)=(2\pi)^{-1}\int_{-\infty}^{+\infty}d\omega e^{-i\omega t}g'(\omega). \tag{I.4}$$

The Fourier transforms of $V(x)$ is then given by (13).

APPENDIX II

In terms of the invariant functions of given rest mass

$$D_\kappa^{(+)}(x)=(2\pi)^{-3}\int d\sigma(k)e^{ikx};\quad d\sigma(k)=(k^4)^{-1}(dk)^3 \tag{II.1}$$

the $\Delta^{(+)}$-functions in (8) are

$$\Delta_1^{(+)}(x)=-\tfrac{3}{4}D_\kappa^{(+)}(x)D_\mu^{(+)}(x)\equiv-\tfrac{3}{4}\Delta^{(+)}(x), \tag{II.2}$$

$$\partial_\alpha\Delta_2^{(+)}(x)=-\tfrac{1}{4}\partial_\alpha D_\kappa^{(+)}(x)\cdot D_\mu^{(+)}(x)-(1/2\mu^2)\times\partial_\beta D_\kappa^{(+)}(x)\cdot\partial^\beta\partial_\alpha D_\mu^{(+)}(x). \tag{II.3}$$

From the divergence of (II.3), we can explicitly evaluate the Fourier transforms for $p^4>0$ and their limiting values for $\mathbf{p}=0$, $p^4=\kappa+\omega$ and $\omega\ll\kappa$:

$$\Delta^{(+)}(p)=2\Delta^{(1)}(p^2)=(2\pi)^{-5}(-p^2)^{-1}[(p^2+(\kappa+\mu)^2)(p^2+(\kappa-\mu)^2)]^{\frac{1}{2}}\cong(2\pi)^{-5}\kappa^{-1}[2(\omega^2-\mu^2)^{\frac{1}{2}}+\sim\kappa^{-1}]. \tag{II.4}$$

$$\Delta_2^{(+)}(p)=2\Delta_2^{(1)}(p^2)=-\tfrac{1}{8}(-p^2)^{-1}(-p^2+\kappa^2+\mu^{-2}(p^2+\kappa^2)^2-2\mu^2)\Delta^{(1)}(p^2)\cong-(\tfrac{1}{4}+\tfrac{1}{2}\mu^{-2}\omega^2+\sim\kappa^{-1})2\Delta^{(1)}(p^2). \tag{II.5}$$

From these approximations, the expression

$$2(k'p\Delta_2^{(1)}\,|\,\kappa^2\Delta_1^{(1)})(p^2)\cong(2\pi)^{-5}\kappa(\mu^{-2}(\omega^2-\mu^2)^{\frac{1}{2}}+\sim\kappa^{-1}) \tag{II.6}$$

is obtained, leading from (14) to (18).

Elimination des constantes arbitraires dans la théorie relativiste des quanta [85]

with T. A. Green, *Helvetica Physica Acta*, vol. 24 (1951), pp. 153–174

Elimination des constantes arbitraires dans la théorie relativiste des quanta

par **E. C. G. Stueckelberg***) et **T. A. Green****) (Genève).

14. XII. (1950).

Summary: This article shows how the influence of the undetermined constants in the integral theory of collisions[1][2][3][4]) can be avoided. A rule is given by which the probability amplitudes ($\boldsymbol{S}[V]$-matrix) may be calculated in terms of a given *local action*. The procedure of the integral method differs essentially from the differential method employed by TOMONAGA[6]), SCHWINGER[5]), FEYNMAN[7]) and DYSON[8]) in that the two sorts of diverging terms occuring in the formal solution of a SCHROEDINGER equation are avoided. These two divergencies are: 1) the well known «*self energy*» *divergencies* which have been since corrected by methods of regularization (RIVIER[1]), PAULI and VILLARS[9])); 2) the more serious *boundary divergencies* (STUECKELBERG[4])) due to the sharp spatio-temporal limitation of the space-time region of evolution V in which the collisions occur. The convergent parts (anomalous g-factor of the electron and the LAMB-RETHERFORD shift) obtained by SCHWINGER are, in the present theory, the boundary independent amplitudes in fourth approximation. Up to this approximation the rule eliminates the arbitrary constants from all conservative processes.

§ 1. contains an outline of the physical meaning of boundary effects. These effects involve a classical field which describes the actions of the counters necessary to distinguish between the incoming and outgoing particles.

§ 2. the rule is formulated. It is based upon the unitarity and causality conditions developed in previous publications[1][2][3][4]).

In § 3. the limiting process is given which avoids the influences of the boundary. In particular, the limiting process necessary to apply the results to zero mass photons is given.

§ 4., § 5. and § 6. contain the application of the rule to the second, third, and fourth approximations of quantum electrodynamics for finite mass photons. It is shown that a certain ambiguity, related to charge renormalization, occuring in the work of SCHWINGER[5]) and DYSON[8]) is avoided if the explicit influence of the boundary region is taken into account.

---

*) Recherche subventionnée par la Commission Suisse d'Energie Atomique.

**) Actuellement au Radiation Laboratory, Berkeley, California, U.S.A.

*

§ 1. Introduction.

La théorie quantique des particules élémentaires (ou quanta) décrit les résultats des expériences par les amplitudes de probabilité pour la transition d'une distribution des quanta à une autre. La description spatio-temporelle de ces amplitudes distingue entre des *actions locales* ayant lieu à un certain événement x (par exemple l'émission d'un photon par un électron), et des actions à distance ou *actions multilocales* (par exemple, la déflection d'un électron à l'événement x dans le champ (retardé, symétrique, ou causal) produit, à l'événement y, par une autre particule qui subit également un changement de sa quantité de mouvement-énergie).

De telles créations, collisions, et annihilations sont influencées par des champs macroscopiques. Pour étudier les propriétés des particules élémentaires, il faut que les actions locales et multilocales

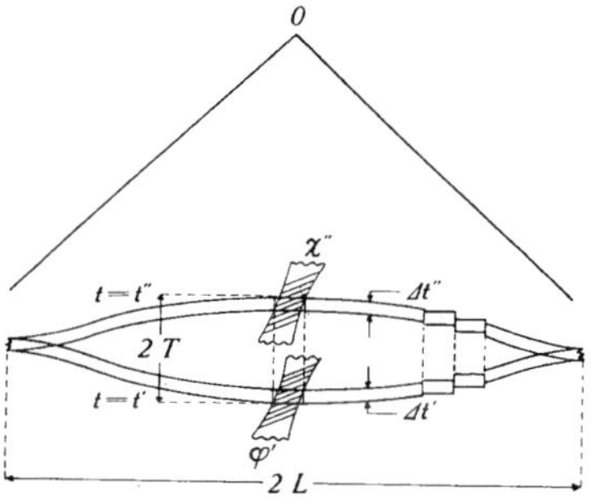

Fig. 1.

La région spatiale de volume $(2\,L)^3$ est remplie de compteurs qui sont mis en marche à deux époques, de durée $\Delta t'$ et $\Delta t''$, séparées par une période $t'' - t' = 2\,T$. Le compteur indiqué est un *compteur idéal*, s'il n'est actionné que par des quanta qui se trouvent dans un paquet d'onde donné:

$$\varphi'(\vec{x}, t') = \chi''(\vec{x}, t'')$$

Les observations sont recueillies à l'origine O.

contribuant à l'amplitude examinée se produisent toutes dans une région spatiale donnée, et pendant une période durant laquelle aucune influence de la matière macroscopique ne se manifeste dans cette région. Nous appelons cette région spatio-temporelle la période d'évolution V.

D'autre part, pour pouvoir observer la distribution des quanta qui entrent dans V *(distribution incidente)* et la distribution des quanta qui sortent de V *(distribution émergente)*, il faut que V soit entouré d'une hypercouche quadridimensionelle de matière macrocroscopique (les compteurs) qui enregistre ces deux distributions. Or, la présence d'un tel champ macroscopique donne lieu à de nombreux processus où il y a échange d'énergie-impulsion entre ce champ (décrit classiquement) et les particules élémentaires. Dans les deux figures, nous avons donné deux réalisations possibles d'un tel champ macroscopique.

Dans la fig. 1, ce champ décrit la matière des compteurs au repos. Cette matière ne présente d'action sur les particules que pendant

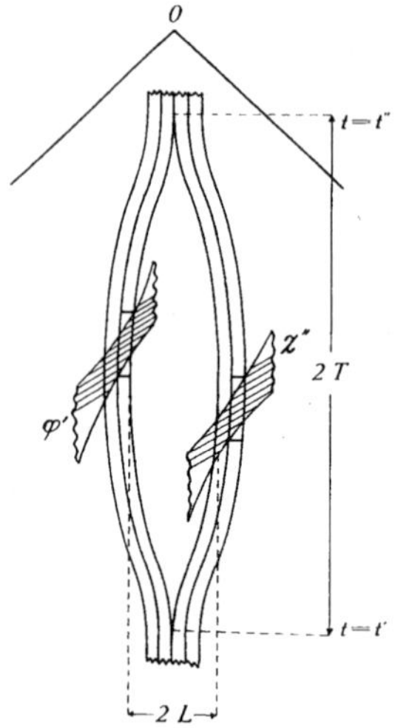

Fig. 2.

A $t < t'$ l'espace est rempli d'une sphère compacte de compteurs élastiques. Elle s'élargit, puis elle se contracte, en créant pendant la période 2 T une région «vide» entourée d'une couche de compteurs. La couche est composée de deux sortes de compteurs. Les uns enregistrent les quanta incidents (paquets φ') et les autres enregistrent les quanta émergents (paquets χ'').

deux courtes époques. De pareils compteurs (dont les éléments de construction sont aussi des particules élémentaires) ne peuvent être construits que si l'on peut varier, à volonté, les constantes de couplage entre leurs éléments et les quanta qu'on désire observer. Il n'est pas en contradiction avec la théorie des quanta de supposer l'existence d'une telle matière pour discuter des expériences idéales (Gedankenexperimente). Ce dispositif a l'avantage de permettre de formuler une théorie limite pour des époques de durée nulle, c'est-à-dire lorsque les deux hypercouches de la fig. 1 tendent vers deux hypersurfaces temporelles. Un pareil dispositif correspond à la réalisation expérimentale de la théorie différentielle.

Dans un article précédent[4]) l'un de nous a montré que cette limite n'existe pas en général. D'autre part, même dans les cas où elle existerait, les effets d'une certitude pareille dans l'observation du temps mettrait à disposition (sous forme du champ classique) une énergie arbitrairement grande, qui donnerait lieu à des probabilités énormes pour des actions localisées près de l'une ou l'autre de ces surfaces.

Sur la fig. 2, un autre arrangement de compteurs est donné. Il peut être réalisé pour autant que des arrangements de compteurs existent qui permettent d'observer la direction et le sens du quantum qui les traverse.

Pour préciser le formalisme, il est utile d'introduire ici la notion du «compteur idéal». C'est un appareil qui précise dans quel paquet d'onde φ un quantum incident ou émergent a été repéré. Soit $M(\varphi)$ une distribution incidente. Elle indique, pour chaque paquet incident $\varphi = \varphi', \varphi'', \ldots$, le nombre de quanta $M(\varphi)$ qui s'y trouvaient. L'amplitude $\Psi'[M()]$, fonctionnelle de $M(\varphi)$, précise ainsi un «*état incident*». De même, $\Psi''[N()]$ précise «*l'état émergent*» sous forme d'une fonctionnelle de la distribution des quanta $N(\chi)$ parmi les paquets émergents $\chi = \chi', \chi'', \ldots$ Alors l'amplitude de transition

$$S\,[V(\,)\,,\ N(\,)/M(\,)] \to \boldsymbol{S}\,[V] \tag{1.1}$$

est fonctionnelle de la région V (décrite par un champ classique $V(x)$), de $N(\chi)$ et de $M(\varphi)$. L'opérateur $\boldsymbol{S}[V]$ est ainsi une matrice unitaire dans l'espace des deux distributions, fonctionnelle d'un champ classique. Elle relie les deux «états» par la relation:

$$\Psi'' = \boldsymbol{S}\,[V]\,\Psi' \tag{1.2}$$

Si, dans le dispositif de la fig. 1, la limite pour des hypersurfaces temporelles existe, on peut considérer $\Psi''[N()] \equiv \Psi''[t''(), N()]$ comme une fonctionnelle de l'hypersurface temporelle $t = t''(\vec{x})$. Cette condition est nécessaire et suffisante pour donner un sens à l'équation fonctionnelle des temps multiples (Stueckelberg[10]), Tomonaga[6]), Schwinger[5])). Cette limite n'existant pas, l'idée d'un «état à l'époque $t''()$» doit être abandonnée. La description des collisions doit donc se faire sous forme d'une théorie explicitement intégrale (1.2).

Le champ classique caractérisant la région d'évolution V est une fonction déterminée.

$$V(x) = \left\{ \begin{matrix} 0 \text{ hors de} \\ 1 \text{ dans} \end{matrix} \right\} \text{la région d'évolution } V \tag{1.3}$$

Elle figure dans les intégrations spatio-temporelles sous forme d'un poids de l'élément de volume, que nous écrivons sous la forme

$$\int V dx \ldots = \int (dx)^4 \, V(x) \ldots \tag{1.4}$$

Les époques d'observation de la fig. 1, resp. la région des compteurs de la fig. 2, sont caractérisées par des champs classiques du type

$$v(x) = (V - V^2)(x) \, ; \;\; \vartheta_{\alpha\beta}(x) = (\partial_\alpha V \cdot \partial_\beta V)(x) \tag{1.5}$$

qui ne sont différents de zéro que dans la région de l'hypersurface.

§ 2. Le développement de l'opérateur S en termes d'une action locale $h(x)$.

Le développement de $\boldsymbol{S}$ en fonction d'un paramètre de couplage ε est

$$\boldsymbol{S} = \boldsymbol{I} + \sum_{n=1}^{\infty} \varepsilon^n \boldsymbol{S}_{(n)} \,. \tag{2.1}$$

La contribution d'ordre n

$$\boldsymbol{S}_{(n)} = \sum_{m=0}^{\infty} \boldsymbol{S}_{(nm)} \equiv \boldsymbol{H}_{(n)} + i \, \boldsymbol{A}_{(n)} \tag{2.2}$$

est une somme *d'expressions multilinéaires* d'ordre m dans les opérateurs de création $\boldsymbol{d}^\dagger(\chi)$ et d'annihilation $\mathbf{c}(\varphi)$ de quanta dans les paquets émergeants $\chi(x)$, resp. incidents $\varphi(x)$. La *condition d'unitarité* établit une relation de récurrence

$$\boldsymbol{H}_{(n)} = -\frac{1}{2} \sum_{n'=1}^{n-1} \boldsymbol{S}^\dagger_{(n')} \boldsymbol{S}_{(n-n')} \tag{2.3}$$

déterminant univoquement la *partie hermitienne* $\boldsymbol{H}_{(n)}$ de $\boldsymbol{S}_{(n)}$ par l'approximation précédente. Toute expression multilinéaire d'ordre $m = m_x + m_y + \ldots. + m_z$ a la forme

$$\begin{aligned} \boldsymbol{S}_{(nm)}[V] &= \sum_\chi \ldots \sum_\varphi \ldots \boldsymbol{d}^\dagger(\chi) \ldots \boldsymbol{I} \, \mathbf{c}(\varphi) \ldots S_{(nm)}(\chi \ldots / \varphi \ldots) \\ &= i \int V dx \int V dy \ldots \int V dz \, (\boldsymbol{\varphi}^{m_x}(x) \, \boldsymbol{\varphi}^{m_y}(y) \ldots \boldsymbol{\varphi}^{m_z}(z))^\sim \\ &\quad \times D^{(c)}_{(nm)}(x - z, \, y - z, \, \ldots) \end{aligned} \tag{2.4}$$

Le symbole $\sim$ indique que l'expression multilinéaire est *ordonnée* (les $\boldsymbol{d}^\dagger$ opérant *après* les $\mathbf{c}$). L'opérateur $\boldsymbol{I}$ fait correspondre à tout état incident $\Psi'[M()]$ l'état émergent $\Psi''[N()]$ qu'on obtient si le système saturé des paquets $\varphi(x)$ est développé en termes du système saturé des $\chi(x)$. Les noyaux $D^{(c)}$ doivent satisfaire à la *condition de causalité*, discutée autre part[2]). Elle exige que pour toute *constellation des événements* $x, y, \ldots, z$, différente de leur coïncidence simultanée $x = y = \ldots = z$, les $D^{(c)}$ se résolvent en une somme de produits de fonctions causales d'un seul déplacement spatio-temporel $D^{(c)}(x-z)$:

$$D^{(c)}(x-z, y-z, \ldots) \doteq D^{(c)}_{(1)}(x-z)\, D^{(c)}_{(2)}(y-z) \ldots D^{(c)}_{(3)}(x-y) \cdots + \cdots \quad (2.5)$$

$$D^{(c)}(x-y) = D^{(\pm)}(x-y) \quad \text{pour} \quad x^4 \gtrless y^4 \quad (2.6)^{*)}$$

A la décomposition de $\boldsymbol{S}_{(nm)}$ en ses parties hermitienne et antihermitienne (2.2) correspond une décomposition des noyaux

$$i\, D^{(c)}_{(nm)} = -\frac{1}{2} D^{(1)}_{(nm)} + i\, D^{(s)}_{(nm)} \quad (2.7)$$

Nous avons évité le signe d'égalité (=) en (2.5) et y avons substitué le signe $\doteq$ exprimant ainsi que (2.5) est indéfini lorsque deux ou plusieurs événements coïncident. Considérons maintenant les conséquences de ces deux conditions. L'unitarité exige que $\boldsymbol{H}_{(1)} = 0$ d'où il résulte que $D^{(1)}_{(1\,m)} = 0$ en (2.7). La causalité et l'hermiticité de $\boldsymbol{A}_{(1)}$ excluent un $D^{(s)}_{(1\,m)}(x-z, \ldots) \neq 0$ sauf pour la coïncidence simultanée de tous les événements et pour leurs constellations infinitésimalement voisines. Un tel $D^{(s)}$ est caractérisé par une *distribution*;

$$D^{(s)}_{(1\,m)}(x-z, \ldots) = \sigma(x-z, y-z \ldots) = a\,\delta(x-z)\,\delta(y-z) \ldots + \ldots \quad (2.8)$$

Elle s'exprime en termes du $\delta(x)$ spatio-temporel et d'un nombre fini de ses dérivées. Par intégration partielle (cf. § 3), le $\boldsymbol{A}_{(1)}$ peut être réduit à une intégrale simple sur une *densité d'action locale* $\boldsymbol{h}_{(1\,m)}(x)$ et une action de surface explicite (cf. § 3).

Le développement de $\boldsymbol{S}$ en termes de cette action locale est obtenu de la manière suivante:

L'*unitarité*, vu la formule de récurrence (2.3), nous fournit $\boldsymbol{H}_{(2)}$ et ainsi la première partie $-\frac{1}{2} D^{(1)}_{(2\,m)}$ dans la décomposition (2.7) des noyaux. La *condition de causalité* détermine ensuite l'autre partie $i\, D^{(s)}_{(2\,m)}$ à des distributions (2.8) près. La partie antihermitienne ainsi déterminée, $\boldsymbol{A}_{(2)}$ (resp. $D^{(s)}_{(2\,m)}$) est appelée le *complément causal* de $\boldsymbol{H}_{(2)}$ (resp. $D^{(1)}_{(2\,m)}$). Avec le $\boldsymbol{S}_{(2)}$ ainsi trouvé, on forme $\boldsymbol{A}_{(3)}$ et y détermine les $D^{(s)}_{(3\,m)}$ à des nouvelles distributions près et ainsi de suite. L'arbitraire contenu dans ces distributions est égal, à des actions

*) $D^{(\pm)}(x)$ sont des fonctions à fréquences définies ($\pm$), cf. Stueckelberg et Rivier[2]).

de surface près, à l'arbitraire qui nous est permis dans le choix de l'action élémentaire $\boldsymbol{h}_{(1\,m)}(x)$. $\boldsymbol{S}$ est développé en termes d'une action élémentaire précisée si une règle détermine univoquement le complément causal $\boldsymbol{A}_{(n)}$ (resp. $D^{(s)}_{(nm)}$) pour tout $\boldsymbol{H}_{(n)}$ (resp. $D^{(1)}_{(nm)}$) obtenu par récurrence (2.3). Une règle pareille serait la «*règle naturelle*» exprimée par le signe d'égalité en (2.5). Elle réduit, par récurrence, d'abord l'arbitraire en (2.5) à l'arbitraire des $D^{(s)}_{(2m)}$ de la deuxième approximation qui doit être défini en termes des potentiels symétriques (voir § 4). Malheureusement, cette règle ne peut pas être appliquée en général, car les produits (2.5) de fonctions intégrables peuvent être non intégrables. Ceci nous oblige à substituer à cette règle un procédé plus général qui équivaut à la «règle naturelle» pour des produits intégrables. Ce procédé fait intervenir les deux étapes suivantes: 1⁰ Si le produit (2.5) n'est pas intégrable, on en construit une *fonction covariante approximative*, ${}_PD^{(c)}(x, y, \ldots)$. Elle possède une représentation de Fourier et ne se distingue du produit que dans les très hautes fréquences (d'ordre P). 2⁰ Un *processus limite* définit alors $D^{(c)}(x, y, \ldots)$ à un minimum de constantes arbitraires près.

1⁰ *La fonction approximative*: Pour trouver ${}_PD^{(c)}(x, y, \ldots)$, on choisit un des facteurs dans (2.5) et on le remplace par une *fonction non covariante*, par exemple

$$D^{(c)}_{(3)}(x) \to {}_PD^{(c)}_{(3)}(x) = (2\,\pi)^{-4} \int_P dp\, e^{ipx}\, D^{(c)}_{(3)}(p) \tag{2.9}$$

P définit l'intérieur d'un cylindre sphérique resp. hyperbolique de rayon P autour d'un axe temporel ou spatial. Le ${}_PD^{(c)}(x-y, \ldots)$ de (2.5) ainsi défini possède un développement de Fourier. Par exemple, la fonction des deux déplacements écrits en (2.5) peut être développée en

$${}_PD^{(c)}(x, y) = (2\,\pi)^{-8} \int dp \int dq\, e^{i(px+qy)}\, {}_PD^{(c)}(p, q) \tag{2.10}$$

en termes de sa transformée de Fourier.

$${}_PD^{(c)}(p, q) = (2\,\pi)^{-4} \int_P dk\, D^{(c)}_{(1)}(p-k)\, D^{(c)}_{(2)}(q-k)\, D^{(c)}_{(3)}(k) \tag{2.11}$$

La fonction ${}_PD^{(c)}(x, y)$ en (2.10) a perdu sa covariance au même titre que (2.9). Pourtant, pour un P arbitrairement grand, il existe un ensemble arbitrairement grand de référentiels de Lorentz dans lesquels les ${}_PD^{(c)}(x, y)$ *non covariants* ne se distinguent que par les contributions à très grande fréquence. On peut alors définir un

${}_P D^{(c)}(x, y)$ *covariant* si on fixe, pour chaque terme (2.11) du développement (2.10), une orientation invariante du cylindre P par rapport aux quadrivecteurs p et q. Cette fonction covariante (2.10) continue à rester l'équivalent du produit en (2.5) sauf pour les contributions à très grande fréquence. En général, $D^{(c)}(x, y, ..)$ est une expression spinorielle et tensorielle. Elle se réduit à différents termes, composé chacun d'un facteur covariant (dépendant des deux vecteurs p et q) qui multiplie une fonction invariante (scalaire) de la longueur des trois vecteurs $\xi = p^2$, $\eta = q^2$, $\zeta = (p-q)^2$

$$ {}_P D(^{(c)}p, q) = {}_P D^{(c)}(p^2, q^2, (p-q)^2) \equiv {}_P D^{(c)}(\xi, \eta, \zeta) \qquad (2.12) $$

2⁰ *Le passage à la limite* $P \to \infty$. Si la limite

$$ D^{(c)}(\xi, \ldots) = \lim_{P=\infty} {}_P D^{(c)}(\xi, \ldots) \qquad (2.13) $$

existe, la fonction est la fonction définie par le produit. Si elle n'existe pas, nous cherchons, parmi les systèmes d'équations aux dérivées partielles qui suffisent à déterminer $D^{(c)}(\xi, \ldots)$,

$$ \partial_\xi^{n_\xi} \partial_\eta^{n_\eta} \ldots D^{(c)}(\xi, \eta, \ldots) = \lim_{P=\infty} \partial_\xi^{n_\xi} \partial_\eta^{n_\eta} \ldots {}_P D^{(c)}(\xi, \eta \ldots) \qquad (2.14) $$

le «*système définissant*» pour lequel a) le deuxième membre existe, et b) la solution générale $D^{(c)}(\xi, \ldots)$ possède le *minimum de constantes* arbitraires. Ce $D^{(c)}$est donné à un polynome près. C'est-à-dire la transformation suivante reste possible:

$$ D'^{(c)}(p, q) = D^{(c)}(p, q) + \sigma(p^2, q^2, (p-q)^2) \qquad (2.15) $$

$\sigma(p^2, \ldots)$ est un polynôme en p^2, q^2, ... contenant ce minimum de constantes arbitraires, soit la représentation de Fourier d'une distribution. Cette règle est la généralisation recherchée de la «règle naturelle». Car, si la limite (2.13) existe, l'équation (2.13), est ellemême le système définissant.

§ 3. Les actions de surface.

1⁰ *Actions de surface implicites*: En choisissant comme paquets des ondes planes, on constate que les intégrales d'espace-temps entrant dans l'expression des amplitudes de transition qui se rapportent à des *processus conservatifs*, sont proportionnelles à l'hypervolume $2\,T\,(2\,L)^3$). En plus de ces processus, il y a les *processus non-conservatifs* qui empruntent de l'énergie et de la quantité de mouvement au champ classique. Un exemple d'un tel processus (l'émission d'un photon par un électron) a été discuté autre part[4]) Les amplitudes sont proportionnelles à $(2\,L)^3$ seulement (aire de l'hypersurface). Nous les appelons *actions de surface implicites.* Dans une approximation donnée, certaines actions seulement contiennent des processus conservatifs. Dans les amplitudes de tels processus, l'influence des compteurs peut être éliminée si l'on donne à la région V des dimensions grandes par rapport aux dimensions des hypercouches. Pour rendre négligable la contribution des processus non conservatifs dans le dispositif de la fig. 1, on doit passer à la limite

$$2\,T \gg \Delta t'',\ \Delta t' \gg \mu^{-1} \tag{3.1}$$

où μ est *la masse de la particule la plus légère.*

2⁰ *Actions de surface explicites*: Outre les actions implicites que nous venons de discuter, l'introduction des *actions à distance infinitésimale* sous forme de distributions (2.8) fait apparaître des actions de surface qui contiennent explicitement les champs macroscopiques (1.5). Par une intégration partielle, on démontre que la différence entre une action locale et une action à distance infinitésimale est une action explicite de surface. Par exemple

$$\begin{aligned}\int V\,dx\,(\boldsymbol{\varphi}\,\boldsymbol{\chi})^{\sim}(x) - \int V\,dx \int V\,dy\,(\boldsymbol{\varphi}(x)\,\boldsymbol{\chi}(y))^{\sim}\,\delta(x-y)\\ = \int v\,dx\,(\boldsymbol{\varphi}\,\boldsymbol{\chi})^{\sim}(x)\end{aligned} \tag{3.2}$$

$$\begin{aligned}\frac{1}{2}\int V\,dx(\boldsymbol{\varphi}\,\Box\,\boldsymbol{\chi} + (\Box\,\boldsymbol{\varphi})\,\boldsymbol{\chi})^{\sim}(x)\\ - \int V\,dx \int V\,dy\,(\boldsymbol{\varphi}(x)\,\boldsymbol{\chi}(y))^{\sim}\,\Box\,\delta(x-y)\\ = \frac{1}{2}\int v\,dx\,(\boldsymbol{\varphi}\,\Box\,\boldsymbol{\chi} + (\Box\,\boldsymbol{\varphi})\,\boldsymbol{\chi})^{\sim}(x) + \int \vartheta_{\varrho}^{\ \varrho}\,dx\,(\boldsymbol{\varphi}\,\boldsymbol{\chi})^{\sim}(x)\end{aligned} \tag{3.3}$$

Les deux champs classiques $v(x)$ et $\vartheta_{\alpha\beta}(x)$ de (1.5) figurent donc explicitement dans les intégrands spatio-temporels. Pour éviter qu'ils contribuent aux amplitudes de probabilité, il faut de nouveau donner aux époques d'observation une durée suffisamment longue (3.1) pour qu'aucun effet ne se produise dans la région des compteurs. On voit en (3.1) que la limite $\mu \rightarrow 0$, c'est-à-dire le passage à l'électrodynamique des photons à masse nulle, fait intervenir des durées infinies pour les époques d'observation.

§ 4. La deuxième approximation en électrodynamique.

1⁰ *Résumé de la théorie.* Nous appliquons la règle établie au § 2 pour développer $\boldsymbol{S}$ en termes de l'action trilinéaire

$$\boldsymbol{h}^J_{(13)}(x) = (\boldsymbol{\varphi}_\alpha \boldsymbol{J}^\alpha)(x)\,; \quad \boldsymbol{J}^\alpha(x) = (\boldsymbol{u}^\dagger \gamma^\alpha \boldsymbol{u})^\sim(x) \tag{4.1}$$

Elle contient un champ tensoriel *(champ des photons de masse non nulle μ)* satisfaisant aux équations d'onde

$$\partial_\alpha \boldsymbol{\varphi}_\beta - \partial_\beta \boldsymbol{\varphi}_\alpha - \boldsymbol{B}_{\alpha\beta} = \boldsymbol{0}\,; \quad \partial_\alpha \boldsymbol{B}^{\alpha\beta} - \mu^2 \boldsymbol{\varphi}^\beta = \boldsymbol{0} \tag{4.2}$$

qui sont équivalentes à

$$(\Box - \mu^2)\, \boldsymbol{\varphi}^\alpha(x) = \boldsymbol{0}\,; \quad \partial_\alpha \boldsymbol{\varphi}^\alpha(x) = \boldsymbol{0} \tag{4.3}$$

et un champ spinoriel *(champ des positrons-électrons de masse $\varkappa$)* ayant les équations d'onde

$$\boldsymbol{u}^\dagger(x)(\gamma\partial - \varkappa) = \boldsymbol{0}\,; \quad (\gamma\partial + \varkappa)\,\boldsymbol{u}(x) = \boldsymbol{0} \tag{4.4}$$

La normalisation des paquets est contenue dans les lois habituelles de commutation, resp. d'anticommutation. Nous n'écrivons ici que les relations

$$i\,[\boldsymbol{\varphi}'_\alpha, \boldsymbol{\varphi}_\beta]_- = (g_{\alpha\beta} - \mu^{-2}\partial_\alpha\partial_\beta)\, D^{(0)}_{(\mu)}(x'-x)\,\boldsymbol{1} \equiv D^{(0)}_{(\mu)\alpha\beta}(x'-x)\,\boldsymbol{1} \tag{4.5}$$

$$[\boldsymbol{u}', \boldsymbol{u}^\dagger]_+ = -(\gamma\partial - \varkappa)\, D^{(0)}_{(\varkappa)}(x'-x)\,\boldsymbol{1} \equiv -\Delta^{(0)}_{(\varkappa)}(x'-x)\,\boldsymbol{1} \tag{4.6}$$

2⁰ *La première approximation*

$$\boldsymbol{S}_{(1)} = i\boldsymbol{A}_{(13)} = i\int V dx\, \boldsymbol{h}^J_{(13)}(x) \tag{4.7}$$

ne contient aucun processus conservatif. Les processus non-conservatifs, qui empruntent de l'énergie au champ superficiel, ont été discutés autre part[4]).

3⁰ La *deuxième approximation*, écrite conformément au développement multilinéaire (2.2) et (2.4), a la forme

$$\begin{aligned}
\boldsymbol{S}_{(2)}[V] &= \boldsymbol{S}_{(20)} + \boldsymbol{S}_{(22)} + \boldsymbol{S}_{(24)} + \boldsymbol{S}_{(26)} = \boldsymbol{H}_{(2)} + i\boldsymbol{A}_{(2)} \\
&= i\int V dx \int V dy \Big(\tfrac{1}{2}\boldsymbol{I}\, D^{(c)}_{(20)}(x-y) \\
&\quad - \tfrac{1}{2}\boldsymbol{\varphi}^\alpha(x)\, D^{(c)}_{(22)\alpha\beta}(x-y)\, \boldsymbol{\varphi}^\beta(y) - i\boldsymbol{u}^\dagger(x)\, \Delta^{(c)}_{(22)}(x-y)\, \boldsymbol{u}(y) \\
&\quad - i\boldsymbol{u}^\dagger(\gamma\boldsymbol{\varphi})(x)\, \Delta^{(c)}_{(24)}(x-y)\, (\gamma\boldsymbol{\varphi})\,\boldsymbol{u}(y) \\
&\quad + \tfrac{1}{2}\boldsymbol{J}^\alpha(x)\, D^{(c)}_{(24)\alpha\beta}(x-y)\, \boldsymbol{J}^\beta(y)\Big)^\sim - \tfrac{1}{2}(\boldsymbol{S}^\dagger_{(1)}\boldsymbol{S}_{(1)})^\sim[V]
\end{aligned} \tag{4.8}$$

Le *terme sixlinéaire* ne contient aucun noyau. Les *deux termes quadrilinéaires* ont des noyaux*)

$$\Delta^{(c)}_{(24)}(x) = (\gamma\,\partial - \varkappa)\, D^{(c)}_{(\varkappa)}(x) \equiv \Delta^{(c)}_{(\varkappa)}(x) \tag{4.9}$$

$$D^{(c)}_{(24)\alpha\beta}(x) = (g_{\alpha\beta} - \mu^{-2}\,\partial_\alpha\,\partial_\beta)\, D^{(c)}_{(\mu)}(x) \equiv D^{(c)}_{(\mu)\alpha\beta}(x) \tag{4.10}$$

dans lesquels le complément causal doit encore être défini. Ce complément introduit dans $\boldsymbol{A}_{(24)}$ *des actions à distance (actions bilocales)*

$$\boldsymbol{h}_{(24)}(x, y) = \frac{1}{2}\, \boldsymbol{J}^\alpha(x)\, D^{(s)}_{(\mu)\alpha\beta}(x-y)\, \boldsymbol{J}^\beta(y) - i\, \boldsymbol{u}^\dagger (\gamma\,\boldsymbol{\varphi})(x)\, \Delta^{(s)}_{(\varkappa)}(x-y)\, (\gamma\,\boldsymbol{\varphi})\, \boldsymbol{u}(y) \tag{4.11}$$

Nous définissons $\Delta^{(s)}$ et $D^{(s)}_{\alpha\beta}$ comme des *pures actions à distance* en les représentant par des *potentiels symétriques*

$$(\Box - \mu^2)\, D^{(s)}_{(\mu)\alpha\beta}(x) = -\,g_{\alpha\beta}\,\delta(x) + \mu^{-2}\,\partial_\alpha\,\partial_\beta\,\delta(x)$$

$$(\gamma\,\partial + \varkappa)\, \Delta^{(s)}_{(\varkappa)}(x) = -\,\delta(x) \tag{4.12}$$

qui sont intégrables. C'est par cette définition que nous marquons l'absence d'actions locales quadrilinéaires. Notre règle donne alors les corrections causales des *termes bilinéaires*

$$\boldsymbol{h}_{(22)}(x, y) = -\frac{1}{2}\,(\boldsymbol{\varphi}^\alpha(x)\,\boldsymbol{\varphi}^\beta(y)^{\sim}\, D^{(s)}_{(22)\alpha\beta}(x-y) - i\,(\boldsymbol{u}^{\dagger A}(x)\,\boldsymbol{u}^B(y)^{\sim}\, \Delta^{(s)}_{(22)AB}(x-y)) \tag{4.13}$$

par les deux produits

$$D^{(s)}_{(22)\alpha\beta}(x) \cdot -\frac{1}{2}\,\mathrm{trace}\,(\gamma_\alpha\, \Delta^{(s)}_{(\varkappa)}(x)\,\gamma_\beta\, \Delta^{(1)}_{(\varkappa)}(-x) + \underset{\cdots}{^{(1)\,(s)}}) \tag{4.14**)}$$

$$\Delta^{(s)}_{(22)}(x) \doteq \frac{1}{2}\,\gamma^\alpha\,(\Delta^{(s)}_{(\varkappa)}(x)\, D^{(1)}_{(\mu)\alpha\beta}(x) + \underset{\cdots}{^{(1)\,(s)}})\,\gamma^\beta \tag{4.15}$$

---

*) *Règle de notations pour les noyaux.* Les lettres D et Δ sont utilisées pour les noyaux tensoriels $D_{\alpha\beta}$, resp. spinoriels Δ_{AB}. Les indices supérieurs $^{(1)}$, $^{(s)}$ et $^{(c)}$ sont employés pour marquer la décomposition des noyaux causals. Deux indices numériques inférieurs $_{(nm)}$ indiquent que ce noyau apparaît dans la n-ième approximation d'un terme m-linéaire. Les indices inférieurs $_{(\varkappa)}$, $_{(\mu)}$, $_{(\varkappa+\mu)}$ indiquent l'étendue spatiale ($\sim \varkappa^{-1}$, μ^{-1}, $(\varkappa+\mu)^{-1}$) des noyaux spatiaux obtenus par intégration sur le temps. Les lettres G et Γ sont des fonctions tensorielles, resp. spinorielles, utilisées pour définir les noyaux. Appliqués à ces fonctions, les indices $^{(s)}$, $_{(\varkappa)}$, etc. gardent leur signification. L'index numérique $_{(l)}$ numérote les différentes fonctions G et Γ d'après leur entrée dans le texte. De tous ces indices, nous omettrons, dans les calculs intermédiaires, ceux qui sont sous-entendus.

**) $\underset{\cdots}{^{(1)\,(s)}}$ est la terme obtenu par permutation des indices $^{(1)}$ et $^{(s)}$.

Le *terme zérolinéaire* (en $\boldsymbol{I}$) contient, dans sa partie hermitienne, une diminution de la probabilité de trouver du vide, diminution due à la production spontanée de trois quanta d'après $\boldsymbol{A}_{(13)}$ de la première approximation. Son complément causal équivaut à un changement de phase arbitraire et indépendant de l'état incident et ne contient donc aucune action observable.

4. *Evaluation des produits*: Evaluons maintenant, en suivant notre règle, les noyaux des termes bilinéaires. En vertu de l'identité de Schwinger

$$\int_0^1 du\, \delta'\big(a + (b-a)u\big) = -\big(a^{-1}\delta(b) + b^{-1}\delta(a)\big) \qquad (4.16)$$

on trouve la transformée approximée

$$_PD^{(s)}_{(22)\alpha\beta}(p) = \frac{2}{(2\pi)^3}\int_P dk \int_0^1 du\, \delta'\big(k^2 + \varkappa^2 + u(p^2 - 2pk)\big)$$
$$\times\left[(p_\alpha k_\beta + p_\beta k_\alpha - 2k_\alpha k_\beta) - g_{\alpha\beta}(pk - k^2 - \varkappa^2)\right] \qquad (4.17)$$

La somme étant convergente pour une région P finie, les intégrations peuvent être interchangées. (4.17) est covariante si P décrit un cylindre de rayon P autour du vecteur p. Cette orientation du cylindre est telle que (4.17) est invariante par rapport à la *translation de l'origine* $k' = k - up$, nécessaire pour mettre en évidence les fonctions scalaires de p^2. Effectuant cette translation, on obtient

$$_PD^{(s)}_{(22)\alpha\beta}(p) = \frac{2}{(2\pi)^3}\int_0^1 du \int_P dk\, \delta'[k^2 + A(p^2, u)]$$
$$\times\left(2(u-u^2)p_\alpha p_\beta - g_{\alpha\beta}\left((u-u^2)p^2 - \frac{k^2}{2} - \varkappa^2\right)\right) \qquad (4.18)$$

$$A(p^2, u) = p^2(u-u^2) + \varkappa^2 \qquad (4.19)$$

Pour simplifier, nous avons déja utilisé les relations

$$\int_P dk\, k^\alpha f(k^2) = 0; \qquad \int_P dk\, k^\alpha k^\beta f(k^2) = \frac{1}{4} g^{\alpha\beta}\int_P dk\, k^2 f(k^2) \qquad (4.20)$$

dont la deuxième ne doit être appliquée qu'aux dérivées des scalaires dans (4.18) qui convergent pour $P = \infty$. La définition de $D^{(s)}_{(22)\alpha\beta}(p)$ se fait à partir de (4.18) en employant la règle générale (2.14). L'intégration en k peut être effectuée sur la première dérivée (par rapport à p^2) des scalaires indépendants de k^2 et sur la deuxième dérivée des termes avec k^2 si l'on utilise les valeurs limites*)

$$\lim_{P=\infty} \int_P dk\, \delta''(k^2+A) = \frac{\pi}{A}$$

$$\lim_{P=\infty} \int_P dk\, k^2\, \delta'''(k^2+A) = -\frac{2\pi}{A} \tag{4.21}$$

On trouve que

$$\begin{aligned} D^{(s)}_{(22)\alpha\beta}(p) &= (p_\alpha p_\beta - g_{\alpha\beta} p^2)\, G(p^2) \\ &\quad + b_0 p_\alpha p_\beta + (b_1 + b_2 p^2)\, g_{\alpha\beta} \end{aligned} \tag{4.22}$$

est déterminé en termes d'une fonction $G(p^2)$ et de *trois constantes arbitraires*. Vu que les noyaux définis en deuxième approximation apparaîtront dans les approximations supérieures, il est avantageux de les écrire dans une forme où l'opérateur d'onde est mis en évidence. On obtient ainsi

$$\begin{aligned} D^{(s)}_{(22)\alpha\beta}(p) &= p_\alpha p_\beta G_{(0)}(p^2) - g_{\alpha\beta} \frac{(p^2+\mu^2)^2}{\varkappa^2} G_{(1)}(p^2) \\ &\quad + g_{\alpha\beta}\big(b_1 \mu^2 - (p^2+\mu^2)\, b_2\big) \end{aligned} \tag{4.23}$$

On se rendra compte que $G_{(0)}$ ne contribue en quatrième approximation qu'aux effets de surface (elle contient une constante arbitraire). $G_{(1)}$ est une fonction définie qui figurera dans le résultat (LAMB shift) de la quatrième approximation.

$$G^{(s)}_{(\varkappa)\,(1)}(p^2) = \frac{1}{16\pi^2} \int_0^1 \frac{dv\left(v^2 - \frac{1}{3}v^4\right)}{\left(1 + \frac{p^2(1-v^2)}{4\varkappa^2}\right)\left(1 - \frac{\mu^2(1-v^2)}{4\varkappa^2}\right)^2} \tag{4.24}$$

Pour évaluer (4.15), il est utile de décomposer ce spineur en

$$\Delta^{(s)}_{(22)}(p) = \Delta^{(s)}_{I\,(22)}(p) + \frac{1}{2\mu^2} \Delta^{(s)}_{II\,(22)}(p) \tag{4.25}$$

correspondant aux deux termes dans (4.5) et (4.10). L'évaluation de $\Delta^{(s)}_I(p)$ est analogue à celle de $D^{(s)}_{(22)\alpha\beta}$. Nous l'écrivons sous une forme analogue à (4.23) en mettant en évidence l'opérateur d'onde.

$$\begin{aligned} \Delta^{(s)}_{I\,(22)}(p) &= (i\gamma p + \varkappa)\, \Gamma_{(1)}(p)\, (i\gamma p + \varkappa) \\ &\quad + b_3 \varkappa + b_4 (i\gamma p + \varkappa) \end{aligned} \tag{4.26}$$

$\Gamma_{(1)}(p)$ est un spineur défini et les deux constantes sont arbitraires.

---

*) L'évaluation invariante en coordonnées polaires hyperboliques donne $0 \cdot \infty$.

Considérons maintenant $\Delta^{(s)}_{II(22)}(p)$. D'après (4.15) et (4.25), on a

$$_P\Delta^{(s)}_{II(22)}(p) = \frac{1}{(2\pi)^3}\int_P dk\,[(\gamma k)\,(i\,(\gamma,\,p-k)-\varkappa)\,(\gamma k)]\big(a^{-1}\,\delta(b)+b^{-1}\delta(a)\big)$$

$$a = k^2+\mu^2; \qquad b = (p-k)^2+\varkappa^2 \tag{4.27}$$

Nous écrivons l'expression spinorielle [...] sous la forme

$$[\ldots] = -(i\gamma p+\varkappa)\big(i\,(\gamma,\,p-k)-\varkappa\big)(i\gamma p+\varkappa) \\ -(i\gamma p+\varkappa)\,b - i\gamma k\,b \tag{4.28}$$

Alors l'identité $b\,\delta\,(b)\equiv 0$ et la règle

$$\lim_{P=\infty}\int_P dk\;\delta(k^2+\mu^2) = \text{const.} \tag{4.29}$$

déterminent (à un terme du type $b_4\;(i\gamma p+\varkappa)$ près)

$$\Delta^{(s)}_{II(22)}(p) = -(i\gamma p+\varkappa)\;\Gamma^{(s)}_{(o)}(p)\;(i\gamma p+\varkappa) \tag{4.30}$$

$\Gamma^{(s)}_{(0)}$ est une fonction à définir en employant la règle pour évaluer le produit

$$_P\Gamma^{(s)}_{(0)}(p) = \frac{1}{(2\pi)^4}\int_P dk\big(D^{(s)}_{(\mu)}(k)\;\Delta^{(1)}_{(\varkappa)}(p-k)+\overset{(1)\,(s)}{\ldots}\big) \tag{4.31}$$

Elle contient deux constantes arbitraires. Nous montrerons que $\Gamma^{(s)}_{(0)}\;(p)$ ne contribue en quatrième approximation qu'aux effets de surface.

5. *Discussion des actions introduites par la deuxième approximation.* La mise en évidence des opérateurs d'onde $p^2+\mu^2$, resp. $i\gamma p+\varkappa$, montre par intégration partielle que, des actions bilinéaires, seule l'action locale

$$\boldsymbol{h}_{(22)}(x) = -\frac{1}{2}\,b_1\,\mu^2(\boldsymbol{\varphi})^{2\sim}(x) - i\,b_3\,\varkappa(\boldsymbol{u}^\dagger\boldsymbol{u})^\sim(x) \tag{4.32}$$

est explicitement indépendante de la surface. Elle contient une contribution proportionnelle au volume $(2\,L)^3\;2\,T$ de la période d'évolution. On peut montrer, qu'à des effets de surface près (4.32) est équivalent à une renormalisation des masses, déjà arbitraires, μ et $\varkappa$ du photon, resp. de l'électron, qu'on a introduites dans la théorie.

Les actions bilocales quadrilinéaires (4. 11) donnent lieu aux processus conservatifs de l'interaction entre deux charges, de l'effet Compton, et de la production de paires.

L'*interaction entre deux charges* contient un terme $\sim 1/\mu^2$. Par intégration partielle, ce terme se réduit à une action à distance entre des événements de surface parce que la charge satisfait à l'équation de continuité. L'influence de ce terme peut être négligée si la limite (3.1) est atteinte de la manière décrite à la fin du § 3. Dans les termes de l'*effet Compton,* on démontre que l'amplitude pour l'émission et l'absorption d'un photon longitudinal devient petite par rapport aux effets transversaux si $\mu \to 0$.

§ 5. La troisième approximation.

Pour ne pas allonger notre exposé, nous ne donnons pas la formule explicite analogue à (4.8), mais nous nous bornons à discuter le terme trilinéaire en $\boldsymbol{u}^\dagger \boldsymbol{u} \boldsymbol{\varphi}\,(x)$*). L'action contenue dans son complément causal est d'abord trilocale. On a

$$\begin{aligned}\boldsymbol{h}_{(33)}(x, y, z) = -\Big[&\boldsymbol{\varphi}^\alpha(x)\, D^{(s)}_{(22)\alpha\beta}(x-y)\, D^{(s)\beta\gamma}_{(\mu)}(y-z)\, \boldsymbol{J}_\gamma(z) - \frac{1}{4}\binom{(1)\,(1)}{\cdot\;\cdot\;\cdot} \\ &+ \boldsymbol{u}^\dagger(x)\, \Delta^{(s)}_{(22)}(x-y)\, \Delta^{(s)}_{(\varkappa)}(y-z)\, (\gamma\boldsymbol{\varphi})\, \boldsymbol{u}(z) - \frac{1}{4}\binom{(1)\,(1)}{\cdot\;\cdot\;\cdot} \\ &+ \boldsymbol{u}^\dagger(\gamma\boldsymbol{\varphi})\,(x)\, \Delta^{(s)}_{(\varkappa)}(x-y)\, \Delta^{(s)}_{(22)}(y-z)\, \boldsymbol{u}(z) - \frac{1}{4}\binom{(1)\,(1)}{\cdot\;\cdot\;\cdot} \\ &+ \boldsymbol{u}^\dagger(x)\, \Delta^{(s)}_{(33)\alpha}(x-y, y-z)\, \boldsymbol{u}(z)\, \boldsymbol{\varphi}^\alpha(y)\Big]^\sim \end{aligned} \tag{5.1}$$

Les trois premiers termes convergent sans autre et s'expriment en termes des fonctions apparues déjà en deuxième approximation. Le noyau nouveau est le produit

$$\begin{aligned}\Delta^{(s)}_{(33)\alpha}(x, y) \doteq \frac{1}{2}\Big(&\gamma^\varrho\, \Delta^{(s)}_{(\varkappa)}(x)\, \gamma_\alpha\, \Delta^{(s)}_{(\varkappa)}(y)\, \gamma^\sigma\, D^{(1)}_{(\mu)\varrho\sigma}(x+y) \\ &+ \overset{(s)\,(1)\,(s)}{\cdot\;\cdot\;\cdot} + \overset{(1)\,(s)\,(s)}{\cdot\;\cdot\;\cdot}\Big) - \frac{1}{8}\binom{(1)\,(1)\,(1)}{\cdot\;\cdot\;\cdot}\end{aligned} \tag{5.2}$$

Le dernier terme $-\frac{1}{8}(\ldots)$ converge sans autre. Puisqu'il ne contribue pas aux processus conservatifs de la quatrième approximation, il sera omis de la discussion au même titre que les $-\frac{1}{4}(\ldots)$ dans (5.1). Par analogie avec (4.25), nous mettons en évidence la contribution due au terme $\sim \mu^{-2}$.

$$\Delta^{(s)\alpha}_{(33)} = \Delta^{(s)\alpha}_{I\,(33)} + \frac{1}{2\,\mu^2}\, \Delta^{(s)\alpha}_{II\,(33)} \tag{5.3}$$

*) Dans les termes de plus haute linéarité, les produits (2.5) convergent. Les termes linéaires et trilinéaires en $\boldsymbol{\varphi}$ sont nuls pour des raisons de symétrie.

La première partie (~ 1) est

$$\Delta^{(s)}_{I\alpha}(p,q) \doteq \frac{1}{2(2\pi)^3}\int_P dk\left((bc)^{-1}\delta(a) + (ca)^{-1}\delta(b) + (ab)^{-1}\delta(c)\right)$$
$$\times g_{\mu\nu}\left(\gamma^\mu(i(\gamma,p-k)-\varkappa)\gamma_\alpha(i(\gamma,q-k)-\varkappa)\gamma^\nu\right)$$
$$a = (p-k)^2+\varkappa^2;\quad b=(q-k)^2+\varkappa^2;\quad c=k^2+\mu^2 \qquad (5.4)$$

Elle sera évaluée analoguement à la première partie de (4.25). L'identité de SCHWINGER à employer est

$$\left((bc)^{-1}\delta(a)+\ldots\right)$$
$$= \frac{1}{2}\int_{-1}^{1} dv\int_0^1 du\, u\, \delta''\!\left(c(1-u)+\frac{1}{2}(a+b)\,u+\frac{1}{2}(a-b)\,uv\right) \qquad (5.5)$$

La convergence de (5.4) nous permet d'interchanger de nouveau les intégrations. Par contre, pour mettre les variables scalaires,

$$\xi = p^2 \quad \eta = q^2 \quad \text{et} \quad \zeta = (p-q)^2 \qquad (5.6)$$

en évidence, une translation

$$k' = k - \frac{u}{2}(p+q+(p-q)\,v) \qquad (5.7)$$

doit être effectuée. Elle laisse invariant le domaine P (cylindre), *seulement si celui-ci est orienté parallèlement à l'axe* $p+q+(p-q)\,v$ *dépendant du scalaire* v. En vertu de (5.5), chaque facteur scalaire dans (5.4) est une somme sur des contributions dépendant de ce scalaire v. Ce changement du domaine, attribuant à chaque contribution un cylindre à orientation différente, ne peut affecter que la partie due aux très hautes fréquences.

Le résultat final est l'expression covariante

$$_P\Delta^{(s)\alpha}_{I(33)}(p,q) = \frac{1}{4(2\pi)^3}\int_{-1}^{1} dv\int_0^1 du\,u\int_P dk\,\delta''\left(k^2+B(\xi,\eta,\zeta,u,v)\right)$$
$$\times\left(\gamma^\varrho(i(\gamma,p-r)-\varkappa)\gamma^\alpha(i(\gamma,q-r)-\varkappa)\gamma_\varrho - k^2\gamma^\alpha\right)$$
$$r = \frac{u}{2}\left(p+q+(p-q)\,v\right) \qquad (5.8)$$

avec

$$B(\xi,\eta,\zeta,u,v) = \mu^2(1-u) + \frac{p^2+\varkappa^2}{2}(u-u^2)(1+v)$$
$$+\frac{(q^2+\varkappa^2)}{2}(u-u^2)(1-v) + u^2\varkappa^2\left(1+\frac{(p-q)^2(1-v^2)}{4\varkappa^2}\right)$$

Le premier terme converge. Au deuxième terme, la règle fait correspondre le système définissant

$$\lim_{P=\infty} \partial_\xi \frac{-1}{4(2\pi)^3} \int_P dk\, k^2 \delta''\left(k^2 + B(\xi, \ldots)\right) = \frac{2\pi}{4(2\pi)^3} \partial_\xi \operatorname{Ln} \frac{B(\xi, \ldots)}{\varkappa^2} \quad (5.9)$$

Elle montre que la fonction logarithmique du deuxième membre est la solution générale avec une constante arbitraire. Mettant en évidence les opérateurs d'onde, on peut écrire:

$$\begin{aligned} \Delta^{(s)\alpha}_{I(33)}(p, q) &= (i\gamma p + \varkappa)\, \Gamma^{\alpha}_{(2)}(p, q) + \Gamma^{\alpha\sim}_{(2)}(-q, -p)\,(i\gamma q + \varkappa) \\ &+ \gamma^\alpha \frac{((p-q)^2 + \mu^2)}{\varkappa^2} G_{(2)}\left((p-q)^2\right) \\ &- i(p-q)_\gamma \frac{\sigma^{\alpha\gamma}}{2\varkappa}\left(\lambda + \frac{(p-q)^2 + \mu^2}{\varkappa^2} G_{(3)}\left((p-q)^2\right)\right) \\ &- b_5 \gamma^\alpha; \quad \sigma^{\alpha\gamma} = \frac{1}{2}[\gamma^\alpha, \gamma^\gamma]_- . \end{aligned} \quad (5.10)^*)$$

Les termes contenant la fonction spinorielle $\Gamma^{\alpha}_{(2)}$ n'apportent pas de contribution dans l'exemple discuté. Nous écrivons par contre les fonctions scalaires qui déterminent la *constante* λ.

$$\begin{aligned} G_{(2)}(\zeta) = &\left[\frac{1}{16\pi^2} \int_0^1 u\, du \int_{-1}^1 dv \left\{\left(1 - u + \frac{u^2}{4}(1 + v^2)\right)\right.\right. \\ &\left.\left. - \frac{\varkappa^2}{\zeta}(2 - 2u - u^2)\, v \frac{\partial}{\partial v}\right\} \frac{\varkappa^2}{B(-\varkappa^2, -\varkappa^2, \zeta, u, v)}\right] \\ &- \frac{\mu^2}{\zeta + \mu^2}\left([\zeta] - [-\mu^2]\right) \end{aligned} \quad (5.11)^{**})$$

$$G_{(4)}(\zeta) = \frac{1}{8\pi^2} \int_0^1 u\, du \int_{-1}^1 dv \frac{\varkappa^2 (u - u^2)}{B(-\varkappa^2, -\varkappa^2, \zeta, u, v)}$$

$$G_{(3)}(\zeta) = \frac{\varkappa^2}{\zeta + \mu^2}\left(G_{(4)}(\zeta) - G_{(4)}(-\mu^2)\right); \qquad \lambda = G_{(4)}(-\mu^2) \quad (5.12)$$

La deuxième partie de (5.3) est

$$_P\Delta^{(s)\alpha}_{II(33)}(p, q) = \frac{1}{(2\pi)^3} \int_P dk\, (\gamma^\mu \ldots^\alpha \ldots \gamma^\nu)\, k_\mu k_\nu \left((bc)^{-1} \delta(a) + \ldots\right) \quad (5.13)$$

Elle contient l'expression spinorielle et les facteurs $\delta(a) \ldots$ de (5.4).

---

*) Dans (5.10) $\Gamma^{\alpha\sim}_{(2)AB} = \Gamma^{\alpha}_{(2)BA}$.

**) $[\zeta]$ et $[-\mu^2]$ sont le premier terme évalué pour ζ et pour $\zeta = -\mu^2$.

*

En analogie parfaite avec (4.27) et (4.28), on réarrange l'expression spinorielle pour obtenir la forme suivante (Λ^α est un spineur):

$$\begin{aligned}(k\gamma)(\,.\overset{\alpha}{.}.\,)(k\gamma) &= (i\gamma p+\varkappa)\,\Lambda^\alpha(p,q,k)\,(i\gamma q+\varkappa)\\ &-(i\gamma p+\varkappa)\big(i(\gamma,p-k)-\varkappa\big)\gamma^\alpha b\\ &-\gamma^\alpha\big(i(\gamma,q-k)-\varkappa)(i\gamma q+\varkappa\big)\,a\\ &-\gamma^\alpha\,a\,b \end{aligned}\qquad (5.14)$$

Au dernier terme, la règle (4.29) fait correspondre une constante, qu'on peut considérer comme déjà contenue dans la constante b_5. Le deuxième terme en (5.14) s'écrit, vu que $b\,\delta(b)\equiv 0$, sous la forme

$$-(i\gamma p+\varkappa)(2\pi)^{-3}\int_P dk\big(i(\gamma,p-k)-\varkappa\big)\gamma^\alpha\big(c^{-1}\delta(a)+a^{-1}\delta(c)\big)\qquad (5.15)$$

qui s'exprime en termes du $\Gamma_{(0)}(p)$ de (4.30). La même transformation peut être faite sur le troisième terme. Donc, le terme $\sim\frac{1}{2}\mu^{-2}$ en (5.3) a la forme

$$\begin{aligned}\Delta^{(s)\alpha}_{II\,(33)}(p,q) &= (i\gamma p+\varkappa)\,\Gamma^\alpha_{(3)}(p,q)\,(i\gamma q+\varkappa)\\ &-(i\gamma p+\varkappa)\,\Gamma_{(0)}(p)\,\gamma^\alpha\\ &-\gamma^\alpha\,\Gamma_{(0)}(q)\,(i\gamma q+\varkappa)\end{aligned}\qquad (5.16)$$

$\Gamma^\alpha_{(3)}(p,q)$ doit être évalué en appliquant notre règle. Il contient une constante, mais n'apporte pas de contribution en quatrième approximation.

Discussion des résultats du terme trilinéaire de la troisième approximation: Elle ne peut être que formelle: La distinction entre des contributions dépendant explicitement de la surface et des contributions de volume n'a pas de sens, car comme dans la première approximation, aucun effet n'est possible sans qu'il n'emprunte de l'énergie au champ macroscopique. Mentionnons toujours, qu'une intégration partielle sépare la contribution proportionnelle à b_2 dans $D^{(s)}_{(22)\alpha\beta}$ en une action de surface explicite et une action de volume. Pour trouver l'action de volume, on utilise (3.3) dans (5.1) avec $\boldsymbol{\varphi}_\alpha(x)$ et $\boldsymbol{\chi}^\alpha(x)=D^{(s)\alpha\beta}_{(\mu)}(x-y)\ldots$ On tient compte du fait que $\boldsymbol{\varphi}_\alpha(x)$ et $D^{(s)\alpha\beta}_{(\mu)}$ satisfont à l'équation d'onde homogène, resp. inhomogène. La même séparation s'applique aux termes avec b_4 dans $\Delta^{(s)}_{(22)}$. On utilise dans (5.1) une relation analogue à (3.3):

$$\begin{aligned}\frac{1}{2}\int Vdx\big(-\boldsymbol{u}(\gamma\partial)\cdot\boldsymbol{w}+\boldsymbol{u}\cdot(\gamma\partial)\boldsymbol{w}\big)^\sim(x)-\int dx\big(V\boldsymbol{u}(x)\cdot\gamma\partial V\boldsymbol{w}(x)\big)^\sim\\ =+\frac{1}{2}\int v\,dx\big(-\boldsymbol{u}(\gamma\partial)\cdot\boldsymbol{w}+\boldsymbol{u}\cdot(\gamma\partial)\boldsymbol{w}\big)^\sim\end{aligned}\qquad (5.17)$$

et on tient compte du fait que $\boldsymbol{u}$ resp. $\Delta^{(s)}_{(\varkappa)}$ satisfont à l'équation d'onde homogène, resp. inhomogène. Le terme avec la constante arbitraire $b_5\,\gamma^\beta$ et le terme avec la constante définie, λ, dans $\Delta^{(s)\alpha}_{(1)(33)}$ sont explicitement indépendants de la surface. Ainsi on trouve comme action de volume une action locale trilinéaire

$$\boldsymbol{h}_{(33)}(x) = \left(\frac{1}{2}\,b_2 + 2\,\frac{1}{2}\,b_4 + b_5\right)\boldsymbol{h}^J_{(13)}(x) + \lambda\,\boldsymbol{h}^N_{(33)}(x) \qquad (5.18)$$

avec

$$\boldsymbol{h}^N_{(33)}(x) = \frac{1}{2}\boldsymbol{B}_{\alpha\beta}\boldsymbol{N}^{\alpha\beta}(x)\,; \quad \boldsymbol{N}^{\alpha\beta}(x) = -\frac{1}{2\,\varkappa}\,(\boldsymbol{u}^\dagger\,\sigma^{\alpha\beta}\boldsymbol{u})^\sim(x) \qquad (5.19)$$

Cette action est équivalente à une renormalisation du paramètre arbitraire ε dans l'*action de charge*

$$\varepsilon \to \varepsilon'\,; \qquad \varepsilon' = \varepsilon\left(1 + \varepsilon^2\left(\frac{1}{2}\,b_2 + b_4 + b_5\right)\right) \qquad (5.20)$$

et à l'introduction d'une nouvelle action trilinéaire, l'*action du moment* (5.19). Elle s'introduit avec un paramètre

$$\eta = \varepsilon^3\,\lambda = \varepsilon^3\,G_{(4)}(-\mu^2) \longrightarrow \frac{\varepsilon^3}{8\,\pi^2} \qquad (5.21)^*)$$

§ 6. La quatrième approximation.

Nous ne discutons que les termes quadrilinéaires**). Considérons d'abord le terme en $\boldsymbol{u}^\dagger\boldsymbol{u}\,\boldsymbol{\varphi}\,\boldsymbol{\varphi}$: Son complément causal est une action quadrilocale.

$$\begin{aligned}\boldsymbol{h}_{(44)}(x,y,z,w) = \Big(&i\,\boldsymbol{\varphi}_\alpha(x)D^{(s)\alpha\beta}_{(22)}(x-y)\,D^{(s)}_{(\mu)\beta\gamma}(y-z)\,\boldsymbol{u}^\dagger(z)\,\gamma^\gamma\Delta^{(s)}_{(\varkappa)}(z-w)(\gamma\boldsymbol{\varphi})\boldsymbol{u}(w)\\ &+ \mathbf{conj} - \frac{1}{4}\,(\overset{(1)}{.}\,\overset{(1)}{.}\,\overset{(s)}{.} + \ldots)\\ &+ i\,\boldsymbol{u}^\dagger\,\Delta^{(s)}_{(22)}(x-y)\,\Delta^{(s)}_{(\varkappa)}(y-z)\,(\gamma\boldsymbol{\varphi})(z)\,\Delta^{(s)}_{(\varkappa)}(z-w)\,(\gamma\boldsymbol{\varphi})\,\boldsymbol{u}(w)\\ &+ \mathbf{conj} - \frac{1}{4}\,(\overset{(1)}{.}\,\overset{(1)}{.}\,\overset{(s)}{.} + \ldots)\\ &+ i\,\boldsymbol{u}^\dagger(\gamma\boldsymbol{\varphi})(x)\,\Delta^{(s)}_{(\varkappa)}(x-y)\,\Delta^{(s)}_{(22)}(y-z)\,\Delta^{(s)}_{(\varkappa)}(z-w)(\gamma\boldsymbol{\varphi})\,\boldsymbol{u}(w) - \frac{1}{4}(\ldots)\\ &+ i\,\boldsymbol{u}^\dagger(x)\,\Delta^{(s)\alpha}_{(33)}(x-y,\,y-z)\,\boldsymbol{\varphi}_\alpha(y)\,\Delta^{(s)}_{(\varkappa)}(z-w)\,(\gamma\boldsymbol{\varphi})\,\boldsymbol{u}(w)\\ &+ \mathbf{conj} - \frac{1}{4}\,(\overset{(1)}{.}\,\overset{(1)}{.}\,.)\\ &+ i\,\boldsymbol{u}^\dagger(x)\,\Delta^{(s)\alpha\beta}_{(44)}(x-y,y-z,z-w)\,\boldsymbol{u}(w)\,\boldsymbol{\varphi}_\alpha(y)\,\boldsymbol{\varphi}_\beta(z)\Big)^\sim\end{aligned} \qquad (6.1)^{***})$$

---

*) Dans (5.21), $\to$ signifie le passage à $\mu^2 \to 0$.

**) Les termes de plus haut degré convergent sans autre. Les processus conservatifs des termes bilinéaires donnent une nouvelle renormalisation arbitraire des masses μ et $\varkappa$.

***) **conj** signifie l'expression conjugée. $\overset{(1)}{.}\,\overset{(1)}{.}\,\overset{(s)}{.} + \ldots$ comprend les trois permutations de $^{(1)\,(1)\,(s)}$.

Elle contient, outre les fonctions provenant de la deuxième et troisième approximation, le noyau

$$\Delta^{(s)\alpha\beta}_{(44)}(x,y,z) = \frac{1}{2}\big(\gamma^{\varrho}\Delta^{(s)}_{(\varkappa)}(x)\,\gamma^{\alpha}\Delta^{(s)}_{(\varkappa)}(y)\,\gamma^{\beta}\Delta^{(s)}_{(\varkappa)}(z)\,\gamma^{\sigma}(g_{\varrho\sigma}-\mu^{-2}\,\partial_{\varrho}\partial_{\sigma})\,D^{(1)}_{(\mu)}(x+y+z)$$

$$+^{(s)\,(s)\,(1)\,(s)} +^{(s)\,(1)\,(s)\,(s)} +^{(1)\,(s)\,(s)\,(s)}\big) - \frac{1}{8}\,\big(^{(1)\,(1)\,(1)\,(s)} + \ldots\big)$$

$$\equiv \Delta^{(s)\alpha\beta}_{I\,(44)} + \frac{1}{2\,\mu^2}\,\Delta^{(s)\alpha\beta}_{II\,(44)} - \frac{1}{8}(\ldots) \qquad (6.2)$$

Comme cette approximation contient des processus conservatifs, on peut omettre le terme $\frac{1}{8}$ $(\ldots)$ en (6.2) et les termes $-\frac{1}{4}$ $(\ldots)$ en (6.1), car ils ne contribuent qu'à des effets implicites de la surface. Le reste de (6.2) est encore une fois séparé en ses deux parties ~ 1 et $\sim \mu^{-2}$. L'évaluation de la première partie utilise une identité de SCHWINGER. Il est explicitement donné par SCHAFROT[11]). Convergeant sans autre, il ne contient aucune constante arbitraire. La partie $\sim \frac{1}{2}\,\mu^{-2}$ peut s'évaluer d'une manière analogue aux deux évaluations précédentes. On montre qu'il se réduit à une expression contenant le $\Gamma_{(0)}$introduit dans les deux approximations précédentes:

$$\Delta^{(s)\alpha\beta}_{II\,(44)} = -\,\delta(x)\,\gamma^{\alpha}\,\Gamma_{(0)}(y)\,\gamma^{\beta}\,\delta(z) \qquad (6.3)$$

Vu qu'une séparation entre des processus conservatifs et non conservatifs est possible, il suffit de démontrer (6.3) par une intégration partielle des dérivées $\partial_{\varrho}\,\partial_{\sigma}\,D^{(1)}_{(\mu)}$ dans le $_P\Delta^{(s)\alpha\beta}_{II\,(44)}$ correspondant à (6.2). Parmi les quatre permutations, seules les permutations $^{ss s 1}$ et $^{s 1 s s}$ apportent une contribution, car, chaque fois qu'on a $\boldsymbol{u}^{\dagger}\gamma^{\varrho}\Delta^{(1)}_{(\varkappa)}$ ou $\Delta^{(1)}_{(\varkappa)}\,\gamma_{\sigma}\boldsymbol{u}$, l'équation de continuité annuelle le terme. Vu (4.31) et vu les équations d'onde (4.12), on obtient (6.3).

Discussion des termes quadrilinéaires et conservatifs de la quatrième approximation: 1. *Le terme quadrilinéaire en* $\boldsymbol{u}^{\dagger}\boldsymbol{u}\,\boldsymbol{\varphi}\,\boldsymbol{\varphi}$: Nous montrons d'abord l'influence des constantes arbitraires b_2, b_4, et b_5 contenues dans les distributions dans $D^{(s)\alpha\beta}_{(22)}$ (4.23), $\Delta^{(s)}_{(22)}$ (4.26), et dans $\Delta^{(s)\alpha}_{(33)}$ (5.10). L'intégration partielle (3.3) appliquée aux deux termes avec $D^{(s)}_{(22)}$ dans (6.1), y substitue (en vertu des équations d'onde pour $\boldsymbol{\varphi}$ et $D^{(s)}_{(\mu)}$) $2\cdot\frac{1}{2}\,b_2$ fois l'action correspondante de la deuxième approximation dans (4.11). L'intégration partielle (5.17) des deux termes contenant $\boldsymbol{u}^{\dagger}\,\Delta^{(s)}_{(22)}$ et $\Delta^{(s)}_{(22)}\,\boldsymbol{u}$ donne (en vertu des équations d'onde pour $\boldsymbol{u}$, $\boldsymbol{u}^{\dagger}$, et $\Delta^{(s)}_{(\varkappa)}$) $2\cdot\frac{1}{2}\,b_4$ fois cette action. Le cinquième terme dans (6.1), contenant $\Delta^{(s)}_{(22)}$ «au milieu», donne $2\cdot\frac{1}{2}\,b_4$ fois cette même action, car dans l'application de (5.17), $\boldsymbol{u}$ et $\boldsymbol{w}$ sont des $\Delta^{(s)}_{(\varkappa)}$ qui satisfont les deux à l'équation inhomogène. Cette manière d'évaluer les contributions élimine l'arbitraire contenu dans la discussion de SCHWINGER*) et de DYSON**). Finalement, la partie arbitraire des deux termes contenant $\Delta^{(s)}_{(33)}$ donne $2\,b_5$ fois cette action. Ains, on obtient dans (6.1), comme contribution due aux constantes arbitraires, une action bilocale

$$\boldsymbol{h}_{(44)}(x,y) = 2\left(\frac{1}{2}\,b_2 + 2\,\frac{1}{2}\,b_4 + b_5\right)\boldsymbol{h}_{(24)}(x,y) \tag{6.4}$$

Cette action équivaut à la deuxième approximation en ε' (de l'action de charge) développée jusqu'à ε^4:

$$\begin{aligned}\varepsilon'\,\boldsymbol{h}^J_{(13)}(x) &= \varepsilon\,\boldsymbol{h}^J_{(13)} + \varepsilon^3\,\boldsymbol{h}^J_{(33)}(x)\\ &= \varepsilon\left(1+\varepsilon^2\left(\frac{1}{2}\,b_2 + b_4 + b_5\right)\right)\boldsymbol{h}^J_{(13)}\end{aligned} \tag{6.5}$$

Il nous reste à discuter l'influence des constantes dans $\Gamma_{(0)}(x)$ figurant dans $\Delta^{(s)}_{II(22)}$ dans $\Delta^{(s)\alpha}_{II(33)}$ et dans $\Delta^{(s)\alpha\beta}_{II(44)}$. On vérifie immédiatement que les opérateurs d'onde contenus dans $\Delta^{(s)\alpha}_{II(33)}$ et $\Delta^{(s)}_{II(22)}$, opérant (par intégration partielle) sur les $\Delta^{(s)}_{(\varkappa)}$ et les $\boldsymbol{u}$, réduisent ces termes à la forme (6.3), et que *leur somme est nulle.*

2⁰ *Le terme quadrilinéaire en* $\boldsymbol{u}$ *et* $\boldsymbol{u}^\dagger$: Il contient la même renormalisation de la charge (6.4) et (6.5). La contribution $\sim\mu^{-2}$ de $\Delta^{(s)}_{(22)}$ et $\Delta^{(s)\alpha}_{(33)}$ n'entre pas dans $\boldsymbol{h}_{(44)}$. Le terme ~ 1, $\Delta^{(s)}_{I(44)}$, est convergent sans ambiguïté. Dans la fonction $\Delta^{(s)}_{II(44)}$, les deux termes avec μ^{-2} se compensent mutuellement. A part l'action tétralocale contenant $\Delta^{(s)}_{I(44)}$, et l'action (6.4), une *nouvelle action bilocale à longue distance* ($\sim\mu^{-2}$) apparaît:

$$\boldsymbol{h}^{NJ}_{(\mu)(44)}(x,y) = -\lambda\left(\partial_\alpha \boldsymbol{N}^{\alpha\beta}(x)\,D^{(s)}_{(\mu)\beta\gamma}(x-y)\,\boldsymbol{J}^\gamma(y)\right)^\sim \tag{6.6}$$

Elle exprime que la charge $\boldsymbol{J}^\alpha$ agit par son potentiel symétrique sur le moment $\boldsymbol{N}^{\alpha\beta}$ (anomalie du facteur g). Une *action à courte distance* ($\sim\varkappa^{-1}$)

$$\begin{aligned}\boldsymbol{h}_{(\varkappa)(44)}(x,y) &= \frac{1}{2\,\varkappa^2}\,\boldsymbol{J}^\alpha(x)\left\{G_{(\varkappa)(1)} - 2\,G_{(\varkappa)(2)}\right\}(x-y)\,\boldsymbol{J}_\alpha(y)\\ &\quad - \frac{1}{\varkappa^2}\,\partial_\alpha \boldsymbol{N}^{\alpha\beta}(x)\,G_{(\varkappa)(3)}(x-y)\,\boldsymbol{J}_\beta(y)\end{aligned} \tag{6.7}$$

entre les deux charges, resp. entre moment et charge, s'ajoute aux actions à longue distance (LAMB-RETHERFORD shift).

---

*) Bibliographie 5, p. 794, Eqs. (1.48)—(1.52).
**) Bibliographie 8, p. 1752, Eq. 98.

3⁰ *Le terme quadrilinéaire en* φ converge avec l'emploi de la règle naturelle, donc sans l'introduction de constantes arbitraires. Il a été explicitement calculé par KARPLUS et NEUMAN[12]).

4⁰ Mentionnons encore que la partie hermitienne de $\boldsymbol{S}_{(4)}$ contient aussi des actions de volume dues au réarrangement de $-\frac{1}{2}(\boldsymbol{A}_{(2)}\boldsymbol{A}_{(2)})^{\sim}$. On constate en particulier que les termes dus à μ^{-2} n'apportent pas de contribution.

Nous avons donc démontré que l'action de charge détermine, en utilisant la règle proposée, les amplitudes des processus conservatifs dans la matrice $\boldsymbol{S}[V]$ *en quatrième approximation**). Celles-ci dépendent uniquement des *deux masses arbitraires* (μ et $\varkappa$) et d'un *paramètre de couplage arbitraire* (ε).

5⁰ Le passage à la limite $\mu \rightarrow 0$ est possible sans autres précautions que celles mentionnées à la fin des §§ 3 et 4, car toute contribution $\sim \mu^{-2}$ a disparu dans les amplitudes qui se rapportent aux processus conservatifs**).

Institut de Physique de l'Université, Genève.

BIBLIOGRAPHIE.

1) D. RIVIER, H.P.A. XXII **3**, 265 (1949).
2) E. C. G. STUECKELBERG et D. RIVIER, H. P. A. XXIII, 1—2, 216 (1950).
3) E. C. G. STUECKELBERG et D. RIVIER, H.P.A. XXIII, Sup. 3, 236 (1950).
4) E. C. G. STUECKELBERG, Phys. Rev. **81**, 130 (1951).
5) J. SCHWINGER, Phys. Rev. **76**, 790 (1949).
6) TOMONAGA, Progress in Theoretical Physics 1 (1946).
7) FEYNMAN, Phys. Rev. **76**, 749 et 769 (1949).
8) DYSON, Phys. Rev. **75**, 486 et 1736 (1949).
9) W. PAULI et F. VILLARS, Rev. Mod. Phys. **21**, Nr. 3, 434 (1949).
10) E. C. G. STUECKELBERG, H. P. A. **11**, 242 (1938).
11) SCHAFROTH, H.P.A. XXII, 501 (1949).
12) KARPLUS and NEUMAN, Phys. Rev. **80**, 381 (1950).

*) *Note ajoutée aux épreuves:* Entre temps, M. A. PETERMANN (Lausanne) a trouvé la démonstration que les contributions provenant des termes $\sim \mu^{-2}$ s'éliminent dans toutes les approximations pour les processus conservatifs. Ce résultat fera l'objet d'une publication prochaine.

**) Dans la fonction $G_{(2)}$ (Eqs. (5.10) et (5.11)) la constante Ln $\varkappa/\mu$, que l'on obtient en effectuant l'intégration en u, correspond à la constante

$$\left(1 + \mathrm{Ln}\,\frac{\varkappa}{2\,k_{\mathrm{min.}}}\right)$$

qui apparaît dans la théorie de SCHWINGER[5]).

Restriction of Possible Interaction in Quantum Electrodynamics [86]

with A. Petermann

PHYSICAL REVIEW VOLUME 82, NUMBER 4 MAY 15, 1951

Restriction of Possible Interactions in Quantum Electrodynamics*

A. PETERMANN AND E. C. G. STUECKELBERG
Lausanne and Geneva, Switzerland
(Received March 27, 1951)

IT can be shown that the integral method of collision theory[1] permits the development of a unitary and causal S-matrix in terms of one or more scalar and hermitian local interaction densities of the type

$$h(x)=\Omega(\partial^{u\dagger},\partial^{u},\cdots\partial^{A\alpha},\cdots)_{AB\cdots\alpha\cdots}(u^{\dagger A}u^{B}\cdots A^{\alpha}\cdots)^{\#}(x). \quad (1)$$

Such a density may involve an arbitrary product of quantized fields $u^{\dagger}$, u, $\cdots A^{\alpha}$ and an arbitrary coupling operator Ω containing derivation operators $\partial^{u\dagger}$ operating upon the different factors. (The symbol $^{\#}$ implies that the creation operators operate after the annihilation operators.) To make our problem specific, let u be the field of charged particles and A^{α} the electromagnetic potential.

To a given term in the development of S, there corresponds a Feynman graph.[2] The derivation operators at a given point x operate upon the free field factors $u^{A}(x)$ and upon the causal functions $\Delta^{(c)AB}(x-y)$, $D^{(c)\alpha\beta}(x-y)$, $\cdots$ corresponding to the external lines and internal lines ending at x. The anti-hermitian parts in each approximation involve a certain number of arbitrary constants. Their contribution to a conservative process can be shown to be equivalent to the introduction of new local interaction densities of the type (1). Therefore, an arbitrary choice of interaction densities (1) will in general introduce an infinite number of such interactions and will thus introduce, for instance, infinite series of derivatives which correspond to nonlocal interactions contradicting causality.

This contradiction can be avoided in the following way. In the

Dyson[3] terminology, to a given order term there corresponds a complicated graph. This can be reduced to a simpler graph by formally substituting into the latter "effective" causal functions and "effective" coupling operators for the true functions and operators. Only if the degree of linearity in momentum space of these effective quantities is (aside from a logarithmic term) of no higher order than the degree of the true quantities will this complicated graph not introduce additional interaction densities. The arbitrary constants reduce, therefore, to a renormalization of the (arbitrary) constants (coupling parameters) of a finite number of interaction densities.

A sufficient condition for this reduction is the following inequality between the degrees of linearity c, $-s$, $-i$, in momentum space of $\Omega(p, \cdots)\rightarrow p^c$, $\Delta^{(c)AB}(p)\rightarrow p^{-s}$, $D^{(c)\alpha\beta}(p)\rightarrow p^{-i}$, and the multilinearities m of the charged field u and $u^\dagger$, and k of A^α in (1):

$$0\leqslant c\leqslant m(s-4)+\tfrac{1}{2}k(i-4)+4. \qquad (2)$$

This condition (2) is also necessary because we can show that if the highest powers of the contributions from two or more different graphs compensate each other, the total contribution is zero identically. This particular property of the graphs has the consequence that the contributions arising from the terms $\sim\mu^{-2}$ in $D^{(c)\alpha\beta}$ for photons of non-zero rest mass μ compensate each other.[4] Therefore, we can put $i=2$ (instead of $i=0$).

If we apply this condition to charged particles of spin 0, $\frac{1}{2}$, 1, we see at once that the method used cannot be applied to the interaction of vector mesons ($s=0$) with the photon field.[5] For Dirac electrons ($s=1$), only the coupling of zero order is possible (charge-potential interaction). The interaction between magnetic moment and field (Pauli term) cannot be introduced. For scalar mesons ($s=2$), the coupling of order one allows the charge-potential interaction. Furthermore, the quadrilinear couplings $u^\dagger uA^\alpha A_\alpha$ (necessary for gauge invariance) and $u^\dagger u^\dagger uu$ (found by Rohrlich[6]) have to be introduced. In every case, the bilinear mass renormalization interactions are necessary and the non-gauge-invariant quadrilinear term $(A^\alpha A_\alpha)^2$ is also[7] compatible with (2). However, no other interactions can be introduced.

* Assisted by the Swiss Atomic Energy Commission.

[1] E. C. G. Stueckelberg and D. Rivier, Helv. Phys. Acta **23**, 216 (1950); **23**, Supp. 3, 236 (1950).
[2] R. P. Feynman, Phys. Rev. **76**, 749, 769 (1949).
[3] F. J. Dyson, Phys. Rev. **75**, 486, 1736 (1949).
[4] E. C. G. Stueckelberg and T. A. Green, Helv. Phys. Acta **24**, No. 2 (1951).
[5] T. Kinoshita and Y. Nambu, Prog. Theor. Phys. **5**, 473, 749 (1950).
[6] F. Rohrlich, Phys. Rev. **80**, 666 (1950).
[7] P. T. Matthews, Phil. Mag. **41**, 185 (1950).

The normalization group in quantum theory [87]

with A. Petermann, *Helvetica Physica Acta*, vol. 24 (1951), pp. 317-319

The normalization group in quantum theory

by E. C. G. STUECKELBERG and A. PETERMANN (Genève)*).

In order to discuss complex interactions, we generalize DYSON's method[1]) in the following way:

Consider a given n-th order contribution to the S-matrix

$$S_{nm}[V](u''\varphi''\ldots/u'\ldots) = \text{const.} \prod_i \left(\int V dx_i\right) u''^{\dagger}(x_1)\varphi''^{\dagger}(x_2)\sigma_{M_1}(x_1\ldots x_i\ldots)\cdot$$
$$\cdot \Delta(x_i x_k)\ldots u'(x_l)\ldots\sigma_{M_2}\ldots\sigma_{M_n}\cdot \tag{1}$$

corresponding to a definite FEYNMAN graph of n specified points. It is given in terms of the wawe packets $\varphi, u, \ldots$ the causal func-

*) Work supported by the Swiss Atomic Energy Commission.

tions $\Delta(xx')$, $D(xx')$,... and the distributions $\sigma_M(xx'\ldots)$ specifying points *MNL*... They depend numerically on the *packet normalizations* Z,... their *rest mass* $\varkappa$... and the *coupling constants* ε_M in the following way:

$$(\partial/\partial \log Z)\,(u, \Delta) = (\tfrac{1}{2}\,u, \Delta);\quad (\partial/\partial \log \varkappa)\,\Delta = -\varkappa\,\Delta\,\Delta^{*})$$
$$(\partial/\partial \log \varepsilon_M)\,\sigma_M = \sigma_M;\ \text{etc.} \tag{2}$$

The particular distributions $\sigma_0(x\,x') = i\,\varepsilon_0\,\varkappa\,\delta(x-x')$ and $\sigma_1(x\,x') = \varepsilon_1(\gamma\,\partial + \varkappa)\,\delta(x-x')$ contribute only to mass and packet renormalization**).

$$(\partial/\partial\,\varepsilon_0)\,S_n = (\partial/\partial \log \varkappa)\,S_n;\quad (\partial/\partial\,\varepsilon_1)\,S_n = (\partial/\partial \log z)\,S_n, \tag{3}$$

S_n is evaluated following the rule given in a previous paper[2]). Therefore it involves a great number of *arbitrary constants* arising from all proper self energy and vertex parts $\Sigma_\alpha(xx'\ldots)$ contained in S_n. Let them be ordered according to their degree of complexity in such a way that Σ_β may need the definition of Σ_α only if $\beta > \alpha$. A change in thear bitrary constants can be expressed in terms of infinitesimal operators.

$$\boldsymbol{P}_{N\beta}\,\Sigma_\beta = \sigma_N;\quad \boldsymbol{P}_{N\beta}\,\Sigma_\alpha = 0 \qquad \text{if } \beta > \alpha\,. \tag{4}$$

Operating upon any S_n, they form a Lie Group:

$$[\boldsymbol{P}_{N\beta}, \boldsymbol{P}_{M\alpha}] = \sum_{\gamma} (\delta_{M,L}\,\boldsymbol{P}_{N\gamma} - \delta_{N,L}\,\boldsymbol{P}_{M\gamma}) \tag{5}$$

γ numbers the vertices Σ_γ arising from overlapping of Σ_α and Σ_β in a single point σ_L.

If $P_{M\alpha}$ operates on the sum $S = \Sigma\,S_n$, the sum being extended over all terms belonging to the same process, the identity

$$\boldsymbol{P}_{M\alpha}\,S = (\partial/\partial \log \varepsilon_M)\,S\,, \tag{6}$$

holds. Thus, the arbitrariness contained in the evaluation of the S_n's is equivalent to a renormalization of the interaction parameters ε_M.

This method shows that a chosen set of local interactions σ_M cannot be introduced without in general adding interactions up to an infinite order in the derivatives of $\delta(x)$ and involving actions between any number of quanta. We have shown[3]) that the only exceptions are: Actions of zero order involving three and four scalar

*) Matrix multiplication in x-space, for u the same equation holds with $\Delta\,u$.

**) To be shown by partial integration according to (3,3) and (5,17) in[2]) and omitting surface contributions.

fields. Actions of zero resp. first order between a vector field and two spinor resp. two scalar fields, if the corresponding charge satisfies continuity. Thus, a theory of the *Proca particle* or of a *Dirac particle with Pauli terms* introduces all derivatives and therefore describes non-local interactions. One can show, that non local interactions of the type $\sigma(xx')$ between the charge and the vector field are possible without contradicting *macroscopic causality*, if the FOURIER representation $\sigma(p^2)$ has only complex singularities. If this "finite extension" is developped in terms of p^2, we find all multipole actions corresponding to the infinite series formed of (4). This theory should be considered as a phenomenological approach to a true description of spin 1 particles, in which they appear as bound states of an even number of elementary particles of spin $1/2$. If the pseudoscalar meson field should be unable to account simultaneously for the apparent magnetic moment of the nucleon and for the nuclear forces, the nucleon would also be a composite particle formed from an odd number of elementary particles of spin $1/2$ and of particles of spin 0.

References:

[1]) DYSON, Phys. Rev. **78**, 1736 (1949).
[2]) STUECKELBERG and GREEN, HPA. **24**, 153 (1951).
[3]) STUECKELBERG and PETERMANN, Phys. Rev. **82**, 548 (1951).

Thermodynamique en Relativité Générale [89]

with Wanders, G, *Helvetica Physica Acta*, vol. 26 (1953), pp. 307–316

Reprinted with permission

Thermodynamique en Relativité Générale

par **E. C. G. Stueckelberg** et **G. Wanders*)** (Genève).

(17 IV 1953.)

Summary. Non-relativistic phenomenological Thermodynamics is extendet to the space-time of General Relativity[1]). One finds an unique energy—momentum tensor, entropy—and substance current and irreversibility.

La *Thermodynamique phénoménologique* et *non-relativiste* est basée sur deux principes exprimant l'impossibilité de perpetuum mobile de première et seconde espèce. *Le premier principe* se traduit par le *principe de conservation de l'énergie* H (au cours de l'évolution temporelle $t' \leqslant t'' \leqslant t''' \leqslant \ldots$) auquel s'ajoutent les principes de conservation des *trois composantes de la quantité de mouvement* π_i et de C *quantités de substances indépendantes* N_A, $(A = 1,2, \ldots, C)$:

$$H' = H'' = H''' = \ldots \tag{1}$$

$$\pi_i' = \pi_i'' = \ldots; \quad i = 1 \text{ à } d\,(= 3) \tag{2}$$

$$N_A' = N_A'' = \ldots; \; A = 1 \text{ à } C \tag{3}$$

Le *deuxième principe localise* d'abord ces quantités dans différents *systèmes*, ou *phases*: I, II, ...:

$$H = H(I) + H(II) + \ldots \tag{4}$$

$$\pi_i = \pi_i(I) + \pi_i(II) + \ldots \tag{5}$$

$$N_A = N_A(I) + N_A(II) + \ldots \tag{6}$$

Il décompose ensuite le transfert d'énergie d'un système à un autre en deux termes: *travail* et *chaleur*. Le postulat de l'impossibilité de transformer une quantité de chaleur retirée d'un système I intégralement en du travail fourni à un autre système II, sans que l'état des autres systèmes III, IV, ... ne soit modifié est la forme

*) Recherche subventionnée par la Commission Suisse de l'Energie Atomique (C. S. A.).

[1]) B. Leaf, Phys. Rev. **84**, 345 (1952), a fait récemment l'analyse correspondante en relativité restreinte.

classique du deuxième principe. On en déduit: d'une part, l'existence de l'*entropie* S, également localisable:

$$S = S(I) + S(II) + \dots \tag{7}$$

qui ne peut qu'augmenter au cours du temps:

$$S' \leqslant S'' \leqslant S''' \leqslant \dots \tag{8}$$

D'autre part, on démontre qu'en tout système et à toute époque doivent exister une fonction d'état positive, la *température* T, et un certain nombre de coefficients positifs ou nuls: les *viscosités transversale* et *longitudinale* η et ξ, la *conductibilité thermique* $\varkappa$ et les *coefficients de diffusion* λ_A, λ_B, ... des différentes substances, qui sont aussi des fonctions d'état, ainsi que les *C potentiels chimiques* μ_A.

En *Théorie de la Relativité*, le transfert s'exprime par le flux d'énergie à travers une surface. Mais, la décomposition de ce flux en travail et chaleur n'est plus univoquement possible. Ceci a amené certains auteurs (Pauli, Tolman) à introduire, en Relativité Restreinte, la notion d'une température d'un corps en mouvement. Quant à nous, il nous semble que la seule manière d'introduire le deuxième principe en Théorie de la Relativité est de remplacer son énoncé classique par sa conséquence (7) et (8), c'est-à-dire, le postulat d'une grandeur S, extensive au même titre que H, π_i et N_A, qui augmente au cours du temps pour tout observateur.

Dans un continu Riemannien quadri-dimensionnel, le correspondant d'un terme des sommes (4), (5), est:

$$dH_\beta = d\sigma_\alpha\, \Theta^\alpha_\beta; \quad dH_i = d\pi_i; \quad dH_4 = -dH \qquad (4')\ (5')$$

où $d\sigma_\alpha$ est l'élément d'une hypersurface tridimensionnelle caractérisant une époque; il n'est pas possible de définir une quantité de mouvement et une énergie totales. Par contre, la quantité totale de substance A et l'entropie totale sont données par:

$$N'_A = \int_{t'} d\sigma_\alpha\, n^\alpha_A \tag{6'}$$

$$S' = \int_{t'} d\sigma_\alpha\, s^\alpha . \tag{7'}$$

Les $(d + C + 1)$ équations d'évolution (1), (2) et (3) se traduisent en:

$$D_\beta\, \Theta^\beta_\alpha = 0 \qquad (1')\ (2')$$

$$D_\beta\, n^\beta_A = 0 \tag{3'}$$

où D_α est le symbole de la dérivée covariante par rapport à x^α. Le deuxième principe prend la forme (i = *irréversibilité*):

$$D_\alpha s^\alpha - i = 0; \quad i \geqslant 0. \tag{8'}$$

Comme il fallait s'y attendre, le remplacement du deuxième principe par une de ses conséquences est insuffisant pour définir la température. Par contre, *deux axiomes* supplémentaires introduisent d'une manière univoque les fonctions $T\mu_A$, ξ, η, $\varkappa$ et λ_A, ainsi que la quadri-(d-) vitesse v^α, comme nous le montrons dans cet article. Ces axiomes sont:

1° Les $(d + C + 2)$ principes (1'), (2'), (3') et (8') réglant l'évolution des courants de quantité de mouvement, d'énergie, des substances chimiques indépendantes et de l'entropie ne sont pas indépendants (en d'autres termes: l'état ne dépend que de $(d + C + 1)$ variables d'état: $f_1, f_2, \ldots, f_{(d+C+1)}$.

2° Les $(d + C + 1)$ variables d'état peuvent être choisies telles que les $(d + C + 2)$ courants Θ^β_α, n^β_A et s^β ne dépendent que linéairement des dérivées $D_\alpha f_1, D_\alpha f_2, \ldots, D_\alpha f_{(d+C+1)}$*).

Cependant, ces axiomes n'imposent que des relations entre les signes des fonctions, sans fixer individuellement le signe de chacune d'elles. En particulier, le caractère positif de la température est perdu.

De l'axiome 1° suit l'existence de $(d + C + 2)$ coefficients homogènes εv^α, μ_A, T reliant les $(d + C + 2)$ principes par:

$$\varepsilon v^\alpha D_\beta \Theta^\beta_\alpha + \sum_A \mu_A D_\beta n^\beta_A + T (D_\beta s^\beta - i) = 0 \tag{9}$$

qui sont univoquement définis (au signe de v^α près) si on normalise la quadri-vitesse v^α à:

$$v^2 = v_\alpha v^\alpha = g_{\alpha\beta} v^\alpha v^\beta = -\varepsilon; \quad \varepsilon^2 = 1. \tag{10}$$

Partant du tenseur $\Theta^{\alpha\beta}$, symétrique, le plus général satisfaisant l'axiome 2°, nous cherchons les restrictions qui doivent lui être imposées pour que (9) soit une identité, et nous trouvons les expressions de n^α_A, s^α, et i. Il s'agit donc de calculer:

$$r = -\varepsilon v^\alpha D_\beta \Theta^\beta_\alpha = D_\alpha u^\alpha + \varepsilon \Theta^{\beta\alpha} v_{\beta\alpha} \tag{11}$$

où

$$u^\alpha = -\varepsilon v_\beta \Theta^\beta_\alpha \tag{12}$$

est la projection sur la quadri-vitesse v^α du courant d'énergie-impulsion, c'est-à-dire, le *flux d'énergie interne*, et:

$$v_{\beta\alpha} = \frac{1}{2} (D_\beta v_\alpha + D_\alpha v_\beta). \tag{13}$$

*) L'axiome 2° exprime que ces courants décrivent le *phénomène du transport*.

Choisissant les champs v^α, μ_A, et T comme variables d'état, nous décomposons $\Theta^{\alpha\beta}$ en une somme de termes ne contenant chacun essentiellement que les dérivées d'un seul champ et nous analysons séparément ces différents termes.

Le fluide parfait. Un premier terme $\Theta^{\alpha\beta}_{(0)}$ ne contient aucune dérivée; sa forme générale est:

$$\Theta^{\alpha\beta}_{(0)} = m v^\alpha v^\beta + \varepsilon p g^{\alpha\beta} \tag{14}$$

où: $m = m(T, \mu_A, \mu_B, \ldots)$ et $p = p(T, \mu_A, \mu_B, \ldots)$.

On trouve:

$$u^\alpha_{(0)} = \varphi v^\alpha \quad \text{avec} \quad \varphi = m - p \tag{15}$$

et:

$$r_{(0)} = \dot\varphi + m D_\alpha v^\alpha \tag{16}$$

$\dot\varphi =$ (dérivée hydrodynamique de φ) $= v^\alpha D_\alpha \varphi$.

En introduisant les variables σ, ν_A, ν_B, ... conjuguées de T, μ_A, μ_B, ...:

où
$$T = \varphi_\sigma\, ; \; \mu_A = \varphi_{\nu_A} \tag{17}$$

$$\varphi_\sigma = \left(\frac{\partial\varphi}{\partial\sigma}\right)_{\nu_A} \qquad \varphi_{\nu_A} = \left(\frac{\partial\varphi}{\partial\nu_A}\right)_{\sigma,\, \nu_{B \neq A}}$$

$$\dot\varphi = T\dot\sigma + \sum_A \mu_A \dot\nu_A$$

et $r_{(0)}$ prend la forme exigée par (9) si:

$$m = T\sigma + \sum_A \mu_A \nu_A \tag{18}$$

en effet, on a alors:

$$r_{(0)} = T D_\alpha(\sigma v^\alpha) + \sum_A \mu_A D_\alpha(\nu_A v^\alpha). \tag{19}$$

Nous montrons dans l'Appendice qu'étant données deux fonctions m et φ de $(C+1)$ variables x_1, x_2, ... x_{C+1} il est toujours possible d'effectuer un changement de variables $x_1, x_2, \ldots \to T, \mu_A, \mu_B, \ldots$ tel que (18) soit vérifiée. Ainsi (18) choisit le système de variables qui doit être identifié au système: température, potentiels chimiques, pour que (9) soit satisfaite.

(19) montre que $\Theta^{\alpha\beta}_{(0)}$ fournit des *contributions au courant d'entropie et aux courants de substances parallèles à la quadri-vitesse:*

$$s^{\alpha}_{(0)} = \sigma v^{\alpha} \qquad n^{\alpha}_{A(0)} = \nu_A v^{\alpha} \tag{20}$$

mais ne donne *pas de contribution à l'irréversibilité* i: $i_{(0)} = 0$.

Si $\Theta^{\alpha\beta}$ se réduit à $\Theta^{\alpha\beta}_{(0)}$, l'équation d'évolution $D_\alpha \Theta^{\alpha\beta} = 0$ devient, en tenant compte de

$$D_\alpha(m v^\alpha) = \dot{p}$$

et en passant au référentiel lorentzien local de repos au point x, dans le cas $d = 3$ ($g_{ii} = 1$, $g_{44} = -1$, $g_{\alpha\beta} = 0$ pour $\alpha \neq \beta$; $v^i = 0$, $v^4 = 1$, au point x, $\varepsilon = 1$):

$$m\,\partial_4 v^i = -m\, G_4{}^i{}_4 - \partial^i p$$

avec

$$G_\mu{}^\varrho{}_\nu = \frac{1}{2} g^{\varrho\varrho'}(\partial_\mu g_{\varrho'\nu} - \partial_{\varrho'} g_{\mu\nu} + \partial_\nu g_{\varrho'\mu})$$

qui est l'équation du mouvement d'un *fluide parfait* dans le *champ gravifique* $G_4{}^i{}_4$, m étant la *densité de masse* (au repos), p la *pression.*

La viscosité. Le terme de $\Theta^{\alpha\beta}$ contenant les dérivées du champ v^α a la forme générale:

$$\Theta^{\alpha\beta}_{(1)} = \Theta^{\alpha\beta}_{(\eta)} + \Theta^{\alpha\beta}_{(\xi)} \tag{21}$$

où:

$$\Theta^{\alpha\beta}_{(\eta)} = -\varepsilon\eta\big(2v^{\alpha\beta} + \varepsilon(v^\alpha \dot{v}^\beta + v^\beta \dot{v}^\alpha)\big) - \zeta(v^\alpha \dot{v}^\beta + v^\beta \dot{v}^\alpha) \tag{22}$$

et:

$$\Theta^{\alpha\beta}_{(\xi)} = -\varepsilon\xi(g^{\alpha\beta} + \varepsilon v^\alpha v^\beta) v_\varrho{}^\varrho - \chi v^\alpha v^\beta v_\varrho{}^\varrho \tag{23}$$

$$\eta = \eta(T, \mu_A, \ldots); \; \zeta = \zeta(T, \mu_A, \ldots) \ldots$$

$\Theta^{\alpha\beta}_{(\eta)}$ donne:

$$u^\alpha_{(\eta)} = \zeta \dot{v}^\alpha$$

$$r_{(\eta)} = D_\alpha(\zeta \dot{v}^\alpha) - \eta(2v^{\alpha\beta} v_{\alpha\beta} + \varepsilon \dot{v}^\alpha \dot{v}_\alpha) - \zeta(\dot{v}^\alpha \dot{v}_\alpha)$$

Le premier terme de $r_{(\eta)}$ est une divergence dont le développement contient des dérivées de T et de μ_A dont est fonction ζ. Tandis que les deuxième et troisième termes sont des *formes définies,* ce premier terme ne peut pas être écrit: ($TD_\alpha s^\alpha_{(\eta)}$ + forme définie), comme l'exige (9). Nous enlevons donc de (22) le terme en ζ. Alors:

$$u^\alpha_{(\eta)} = 0 \tag{24}$$

$$r_{(\eta)} = -\eta\,(2\,v^{\alpha\beta}\,v_{\alpha\beta} + \varepsilon\,\dot{v}^\alpha \dot{v}_\alpha)\,. \tag{25}$$

(24) entraîne $\Theta^{\alpha\beta}_{(\eta)}\,\Theta_{(\eta)\alpha\beta} \geqslant 0$ comme on le voit en passant au référentiel lorentzien local de repos. Or:

$$\Theta^{\alpha\beta}_{(\eta)}\,\Theta_{(\eta)\alpha\beta} = \eta^2\,(2 v^{\alpha\beta}\,v_{\alpha\beta} + \varepsilon\,v^\alpha\,v_\alpha)\,.$$

Ainsi la *contribution à l'irréversibilité due à* $\Theta^{\alpha\beta}_{(\eta)}$

$$\dot{i}_{(\eta)} = -\frac{1}{T}\,r_{(\eta)} \tag{26}$$

est *positive si* η *et* T *sont toujours de même signe.* Plus exactement: $i_{(\eta)}$ n'est une forme définie que si la *métrique est définie* (statique pure) ou *indéfinie avec une seule dimension privilégiée* (temps)*); v^α doit être alors un vecteur temporel.

En passant au référentiel lorentzien local de repos, et dans le cas limite d'un champ gravifique nul:

$$\Theta^{i4}_{(\eta)} = \Theta^{44}_{(\eta)} = 0 \tag{27}$$

$$\Theta^{ij}_{(\eta)} = -\eta\,(\partial^i v^j + \partial^j v^i) = -\tau^{ij}_{(\eta)} \tag{28}$$

on retrouve le tenseur des tensions $\tau^{ij}_{(\eta)}$ dû à la viscosité transversale; η *est donc identifié comme coefficient de viscosité transversale.* (27) exprime que l'existence de la viscosité transversale ne contribue pas à la densité d'énergie de repos, ni au courant d'énergie.

Dans le référentiel lorentzien local de repos la densité d'énergie $\Theta^{44}_{(\xi)}$ est proportionnelle à la divergence v^ϱ_ϱ de la quadri-vitesse v^α. Raisonnant comme plus haut on voit que $\chi = 0$. Alors:

$$u^\alpha_{(\xi)} = 0 \tag{29}$$

$$r_{(\xi)} = -\xi\,(v^\varrho_\varrho)^2\,. \tag{30}$$

On a ainsi une *contribution positive à l'irréversibilité*

$$\dot{i}_{(\xi)} = \frac{\xi}{T}\,(v^\varrho_\varrho)^2 \tag{31}$$

---

*) $\varepsilon = 1$ implique les signatures du $g_{\alpha\beta}$ diagonalisé $(-1, -1, \ldots, -1)$ ou $(1, 1, \ldots, 1, -1)$; $\varepsilon = -1$ les signatures $(1, 1, \ldots 1)$ ou $(1, -1, \ldots, -1, -1)$.

si ξ et T sont toujours de même signe. Dans le référentiel lorentzien local de repos, et si le champ gravifique est nul:

$$\Theta^{i4}_{(\xi)} = \Theta^{44}_{(\xi)} = 0 \tag{32}$$

$$\Theta^{ij}_{(\xi)} = -\xi\,(\partial_\varrho v^\varrho)\,g^{ij} = -\tau^{ij}_{(\xi)}. \tag{33}$$

$\tau^{ij}_{(\xi)}$ est le tenseur des tensions dû à la viscosité longitudinale et *ξ est identifié comme coefficient de viscosité longitudinale.* (32) s'interprète comme (27).

La conduction thermique. Les dérivées du champ T donnent le terme $\Theta^{\alpha\beta}_{(q)}$:

$$\Theta^{\alpha\beta}_{(q)} = -\varepsilon\varkappa\,(v^\alpha\partial^\beta T + v^\beta\partial^\alpha T) - (2\varkappa + \omega)\,v^\alpha v^\beta \dot{T}$$
$$-\varepsilon\psi g^{\alpha\beta}\dot{T} - \varrho T\,(v^\alpha\dot{v}^\beta + v^\beta\dot{v}^\alpha) \tag{34}$$

où le dernier terme est une première partie du terme retranché de $\Theta^{\alpha\beta}_{(\eta)}$ (la seconde partie sera introduite dans les composantes de $\Theta^{\alpha\beta}$ contenant les dérivées du champ μ_A). (34) peut être écrit:

$$\Theta^{\alpha\beta}_{(q)} = q^\alpha v^\beta + q^\beta v^\alpha - \lambda\,(v^\alpha v^\beta + \varepsilon\,\frac{\psi}{\lambda}\,g^{\alpha\beta})\,\dot{T} \tag{35}$$

où:

$$q^\alpha = -\varepsilon\varkappa\,T^\alpha_\perp - \varrho\,T\dot{v}^\alpha;\quad v_\alpha q^\alpha = 0$$

$$T_{\alpha\perp} = (\text{dérivée normale de } T) = \partial_\alpha T + \varepsilon v_\alpha \dot{T};\quad v_\alpha T^\alpha_\perp = 0\,.$$

L'argument qui entraîna $\zeta = \chi = 0$ nécessite ici $\omega = \psi = 0$. On a donc:

$$u^\alpha_{(q)} = q^\alpha \tag{36}$$

et, en posant: $q^\alpha = Ts^\alpha_{(q)}$, $(v_\alpha s^\alpha_{(q)} = 0)$

$$r_{(q)} = T D_\alpha s^\alpha_{(q)} - \frac{1}{T}\frac{\varepsilon}{\varkappa}\,(q^\alpha q_\alpha) + \frac{1}{\varkappa}\,q^\alpha\dot{v}_\alpha\,(\varkappa - \varrho). \tag{37}$$

Pour que *la contribution à l'irréversibilité soit positive, il faut que:* $\varkappa = \varrho$, $\varkappa \geqslant 0$; alors:

$$i_{(q)} = \frac{1}{T^2}\,\frac{\varepsilon}{\varkappa}\,(q^\alpha q_\alpha) \geqslant 0 \text{ car } \varepsilon\, q^\alpha q_\alpha \geqslant 0\,. \tag{38}$$

La *contribution au courant d'entropie $s^\alpha_{(q)}$ est définie plus haut,* elle est *normale à la quadri-vitesse.*

Dans le référentiel lorentzien local de repos:

$$\Theta^{ij}_{(q)} = \Theta^{44}_{(q)} = 0 \tag{39}$$

$$\Theta^{4i}_{(q)} = -\varkappa\,\partial^i T - \varkappa\,T\,\partial_4 v^i - \varkappa\,T\,G^i_{44}\,. \tag{40}$$

Lorsque $\partial_4 v^i = 0$ et en absence de champ gravifique, on retrouve le courant d'énergie dû à la *conduction thermique*, $\varkappa$ étant le *coefficient de conductibilité thermique*. Dans le cas général, les deux derniers termes de (40) ne permettent plus d'identifier $\Theta^{4i}_{(q)}$ à un courant de chaleur; en Thermodynamique relativiste la décomposition du courant d'énergie en chaleur et travail n'a plus de sens. Dans le cas statique ($\partial_4 v^i = 0$ et $\Theta^{4i}_{(q)} = 0$), (40) montre que la température n'est pas uniforme dans un champ gravifique non nul.

La diffusion des substances. La discussion du terme de $\Theta^{\alpha\beta}$ contenant les dérivées du champ μ_A est semblable à la précédente. On trouve:

$$\Theta^{\alpha\beta}_{(A)} = j^{\alpha}_{A} v^{\beta} + j^{\beta}_{A} v^{\alpha} \tag{41}$$

où

$$j^{\alpha}_{A} = -\, \varepsilon\, \lambda_A\, (\mu^{\alpha}_{A\perp} + \varepsilon\, \mu_A\, \dot{v}^{A}) \tag{42}$$

$$u^{\alpha}_{(A)} = j^{\alpha}_{A}\,. \tag{43}$$

$\Theta^{\alpha\beta}_{(A)}$ donne une *contribution au courant de la substance A, normal à la quadri-vitesse:*

$$n^{\alpha}_{(A)} = \frac{1}{\mu_A}\, j^{\alpha}_{A}\,; \quad v_{\alpha}\, n^{\alpha}_{(A)} = 0 \tag{44}$$

et une *contribution à l'irréversibilité:*

$$i_{(A)} = \frac{1}{T}\, \frac{\varepsilon}{\mu_A \lambda_A}\, (j^{\alpha}_{A}\, j_{\alpha A}) \tag{45}$$

positive si $\mu_A\, \lambda_A$ a même signe que T.

Dans le référentiel lorentzien local de repos:

$$\Theta^{ij}_{(A)} = \Theta^{44}_{(A)} = 0 \tag{46}$$

$$\Theta^{4i}_{(A)} = -\, \lambda_A\, \partial^i \mu_A - \lambda_A\, \mu_A\, \partial^4 v^i - \lambda_A\, \mu_A\, G^{\,i}_{4\,4}\,. \tag{47}$$

Si $\partial_4 v^i = 0$ et $G^{\,i}_{4\,4} = 0$, $\Theta^{4i}_{(q)}$ est le courant d'énergie familier dû à la *diffusion de la substance A*, avec *coefficient de diffusion* λ_A. Si l'on tient compte de la diffusion et de la conduction thermique, les potentiels chimiques et la température sont liées au champ gravifique, dans le cas statique, par l'équation:

$$\varkappa\, \partial^i T + \sum_A \lambda_A\, \partial^i \mu_A = -\,(\varkappa\, T + \sum_A \lambda_A\, \mu_A)\; G^{\,i}_{4\,4}\, (v^4)^2. \tag{48) **)$$

Les réactions chimiques. Si les substances A, B, ... peuvent réagir entre elles (3') doit être remplacée par (r_A = "*rate of production*")*):

$$D_{\alpha}\, n^{\alpha}_{A} - r_A = 0 \tag{49}$$

---

*) A ne pas confondre avec le r introduit plus haut.

**) Vu qu'on doit encore satisfaire à $m\, \dot{v}^i = -\, \partial^i p$, la solution de (48) ne peut être que $T/T_0 = \mu_A/\mu_{A\,0} = (g_{44})^{-\frac{1}{2}}$.

où r_A est la vitesse de production ($\gtrless 0$) de la substance A. La chimie relie les r_A par la loi des proportions constantes:

$$\sum_A c_{aA}\, r_A = 0 \quad a = 1, 2, \ldots, c \leqslant C \tag{50}$$

a dénombrant les espèces atomiques.

Dans (9), $D_\alpha n_A^\alpha$ est remplacé par $(D_\alpha n_A^\alpha - r_A)$ et un terme i' s'ajoute à l'irréversibilité:

$$i' = -\sum_A \frac{1}{T}\, \mu_A\, r_A\,. \tag{51}$$

La substance A peut participer à une série de réactions possibles $p, q, \ldots$: $r_A = \sum_p r_{A\,p}$, de *vitesses de réaction* a_p, i. e. $r_{A\,p} = c_{A\,p}\, a_p$ où les $c_{A\,p}$ sont des entiers positifs ou négatifs. Posant ensuite:

$$a_p = -\zeta_p \sum_A c_{A\,p}\, \mu_A \tag{52}$$

il vient:

$$i' = \sum_p i'_p \text{ avec } i'_p = \frac{\zeta_p}{T}\Big(\sum_A c_{A\,p}\, \mu_A\Big)^2 \tag{53}$$

Les réactions chimiques donnent une *contribution positive à l'irréversibilité si* ζ_p et T *ont le même signe.*

En *résumé* (9) est une identité, c'est-à-dire: le deuxième principe est une conséquence des autres principes d'évolution si:

$$\Theta^{\alpha\beta} = \Theta^{\alpha\beta}_{(o)} + \Theta^{\alpha\beta}_{(\eta)} + \Theta^{\alpha\beta}_{(\xi)} + \Theta^{\alpha\beta}_{(q)} + \sum_A \Theta^{\alpha\beta}_{(A)}$$

avec:

$$\Theta^{\alpha\beta}_{(o)} = m\, v^\alpha v^\beta + \varepsilon\, p\, g^{\alpha\beta}$$

$$\Theta^{\alpha\beta}_{(\eta)} = -\varepsilon\,\eta\,(2\, v^{\alpha\beta} + \varepsilon\,(v^\alpha \dot{v}^\beta + v^\beta \dot{v}^\alpha))$$

$$\Theta^{\alpha\beta}_{(\xi)} = -\varepsilon\xi\,(g^{\alpha\beta} + \varepsilon\, v^\alpha v^\beta)\, v_\varrho^\varrho$$

$$\Theta^{\alpha\beta}_{(q)} = v^\alpha q^\beta + v^\beta q^\alpha \qquad q^\alpha = -\varepsilon\,\varkappa\,(T_\perp^\alpha + \varepsilon T \dot{v}^\alpha)$$

$$\Theta^{\alpha\beta}_{(A)} = v^\alpha j_A^\beta + v^\beta j_A^\alpha \qquad j_A^\alpha = -\varepsilon\,\lambda_A\,(\mu_{A\perp}^\alpha + \varepsilon\,\mu_A\,\dot{v}^\alpha)$$

et:

$$m = T\sigma + \Sigma\,\mu_A \nu_A; \qquad T = \varphi_\sigma;\ \mu_A = \varphi_{\nu_A}; \qquad \varphi = m - p$$

$$\eta(T, \mu_A, \ldots)\, T \geqslant 0 \qquad \xi\,(T, \mu_A, \ldots) T \geqslant 0$$

$$\varkappa(T, \mu_A, \ldots) \geqslant 0 \quad \lambda_A(T, \mu_A, \ldots) T \mu_A \geqslant 0 \quad \zeta_p\,(T, \mu_A \ldots)\; T \geqslant 0$$

$$n_A^\alpha = n_{A/\!/}^\alpha + n_{A\perp}^\alpha \qquad n_{A/\!/}^\alpha = \nu_A\, v^\alpha \qquad n_{A\perp}^\alpha = \frac{1}{\mu_A}\, j_A^\alpha\,.$$

Le courant d'entropie vaut alors:

$$s^\alpha = s_{/\!/}^\alpha + s_\perp^\alpha \qquad s_{/\!/}^\alpha = \sigma\, v^\alpha \qquad s_\perp^\alpha = \frac{1}{T}\, q^\alpha$$

et l'irréversibilité est donnée par:

$$i = \frac{\eta}{T}(2\, v^{\alpha\beta} v_{\alpha\beta} + \varepsilon\, \dot{v}^{\alpha} \dot{v}_{\alpha}) + \frac{\xi}{T}(v_p^p)^2 + \frac{1}{T}\frac{\varepsilon}{\varkappa T}(q^{\alpha} q_{\alpha})$$

$$+ \sum_A \frac{1}{T}\frac{\varepsilon}{\mu_A \lambda_A}(j_A^{\alpha} j_{\alpha A}) + \sum_p \frac{\zeta_p}{T}\Big(\sum_A c_{A\,p}\,\mu_A\Big)^2 .$$

Appendice.

Il est possible de définir les variables T, μ_A, μ_B, ... de telle façon que: $m = T\sigma + \sum_A \mu_A \nu_A$.

Soient deux fonctions m et φ données en termes de variables quelconques, σ'', ν_A'', ...:

$$m = m''(\sigma'', \nu_A'') \qquad \varphi = \varphi''(\sigma'', \nu_A'') .$$

Il est possible de définir un changement de variables:

$$\sigma' = \sigma'(\sigma'', \nu_A'') \qquad \nu_A' = \nu_A''$$

tel que:

$$m = \sigma' \frac{\partial \varphi}{\partial \sigma'} .$$

En effet: $\frac{\partial \varphi}{\partial \sigma'} = \frac{\partial \varphi''}{\partial \sigma''}\frac{\partial \sigma''}{\partial \sigma'}$ et la relation exigée s'écrit:

$$\frac{\partial}{\partial \sigma''} \log \sigma' = \frac{1}{m''}\frac{\partial \varphi''}{\partial \sigma''}(\sigma'', \nu_A'')$$

équation différentielle en σ'' dépendant paramétriquement de ν_A'', ... et définissant la fonction: $\sigma' = \sigma'(\sigma'', \nu_A'')$ cherchée.

Opérons le changement de variables:

$$\sigma = \sigma' \qquad \nu_A = \nu_A' \sigma' .$$

Alors:

$$m = \frac{\partial \varphi}{\partial \sigma}\sigma + \sum_A \frac{\partial \varphi}{\partial \nu_A}\nu_A$$

et en définissant T et μ_A comme conjuguées de σ et ν_A: $T = \frac{\partial \varphi}{\partial \sigma}$ $\mu_A = \frac{\partial \varphi}{\partial \nu_A}$, on obtient la relation exigée.

Stueckelberg's Complete List of Publications

[1] Photographische Bestimmung von Kathodentemperaturen im Elektrischen Lichtbogen, Ph.D. thesis, Physikalische Anstalt Basel, October 18, 1927, 38 p; printed in *Helvetica Physica Acta*, vol. 1 (1928), pp. 75–109.

[2] Ionization by Collisions of the Second Kind in Mixtures of Oxygen with the Rare Gases, with Smyth, H. D., *Physical Review*, vol. 32 (1928), pp. 779–783.

[3] The Origin of the Continuous Spectrum of the Hydrogen Molecule, with Winans, J. G., *National Academy of Sciences*, vol. 14 (1928), pp. 867–873.

[4] Diatomic Molecules According to the Wave Mechanics I: Electronic Levels of the Hydrogen Molecular Ion, with Morse, P. M., *Physical Review*, vol. 33 (1929), pp. 932–947.

[5] Simultaneous Ionization and Dissociation of Oxygen and Intensities of the Ultra-Violet 0_2^+ Bands, *Physical Review*, vol. 34 (1929), pp. 65 67.

[6] Primäre und sekundäre Ionen in Sauerstoff und Kohlendioxyd, with Smyth, H. D., *Helvetica Physica Acta*, vol. 2 (1929), pp. 303–304.

[7] Störungsrechnung des Wasserstoffmolekülions und des Wasserstoffmoleküls, with Morse, P. M., *Helvetica Physica Acta*, vol. 2 (1929), pp. 304–306.

[8] Recombination of electron and alpha-particle, with Morse, P. M., *Physical Review*, vol. 35 (1930), pp. 116–117.

[9] Computation of the Effective Cross Section for the Recombination of Electrons with Hydrogen Ions, with Morse, P. M., *Physical Review*, vol. 36 (1930), pp. 16–23.

[10] The Ionization of Carbon Dioxide by Electron Impact, with Smyth, H. D., *Physical Review*, vol. 36 (1930), pp. 472–477.

[11] The Ionization of Nitrous Oxide and Nitrous Dioxide by Electron Impact, with Smyth, H. D., *Physical Review*, vol. 36 (1930), pp. 478–481.

[12] A Theory of Collision Processes Involving no Radiation of Energy, with Morse, P. M., *Physical Review*, vol. 37 (1931) p. 449.

[13] Strahlungslose Stossprozesse bei kleinen Geschwindigkeiten, with Morse, P. M., *Annalen der Physik*, vol. 9 (1931), pp. 579–606.

[14] Die spezifische Wärme von quasi-freien Elektronen, with Morse, P. M., *Zeitschrift für Physik*, vol. 9 (1931), pp. 666–667.

[15] Unelastische Stösse zwischen Molekülen, with Morse, P. M., *Helvetica Physica Acta*, vol. 4 (1931), pp. 136–137.

[16] Lösung des Eigenwertproblems eines Potentialfeldes mit zwei Minima, with Morse, P. M., *Helvetica Physica Acta*, vol. 4 (1931), pp. 337–354.

[17] Theory of Inelastic Collisions, *Physical Review*, vol. 40 (1932), p. 1036.

[18] Theory of Continuous Absorption of Oxygen at 1450 Å, *Physical Review*, vol. 42 (1932), pp. 518–524.; Erratum *Physical Review*, 44 (1933), p. 234.

[19] Theorie der unelastischen Stösse zwischen Atomen, Habilitationsschrift, *Helvetica Physica Acta*, vol. 5 (1932), pp. 369–422.

[20] Relativistisch invariante Störungstheorie des Diracschen Elektrons. I. Teil: Streustrahlung und Bremsstrahlung, *Annalen der Physik*, 5. Folge, vol. 21 (1934), pp. 367–389. Berichtigung, p. 744.

[21] Remarque de discussion à propos de la chaleur spécifique des électrons dans les métaux, *Zeitschrift für technische Physik*, vol. 15 (1934), p. 520.

[22] Probleme der modernen Physik, Report on the "International Conference of Pure and Applied Phvsics", London, October 1–6, 1934; reprinted in *Basler Nachrichten*, Nr. 276 (1934), 6 p.

[23] Bemerkungen zur Intensität der Streustrahlung bewegter freier Elektronen, *Helvetica Physica Acta*, vol. 8 (1935), pp. 197–204.

[24] Remarques sur la production des paires d'électrons, Tagung der Schweizerischen Physikalischen Gesellschaft, Lausanne, May 4–5, 1935 *Helvetica Physica Acta*, 8 (1935), pp. 324–326.

[25] Remarque à propos des temps multiples dans la théorie d'interaction des charges entre elles, *C.R. de la Société de Physique et d'Histoire naturelle de Genève*, 52 (1935), pp. 98–101.

[26] Austauschkräfte zwischen Elementarteilchen & Fermische Theorie des β-Zerfalls als Konsequenzen einer möglichen Feldtheorie der Materie, *Helvetica Physica Acta*, 9 (1936), pp. 389–404.

[27] Radioactive β-Decay and Nuclear Exchange Force as a Consequence of a Unitary Field Theory, *Nature*, 137 (1936) p. 1032.

[28] Radioactivité γ avec un spectre continu. Essai d'une nouvelle théorie du champ, *C.R. de la Société de Physique et d'Histoire naturelle de Genève*, 53 (1936), pp. 64–69.

[29] Invariante Störungstheorie des Elektron-Neutrino-Teilchens unter dem Einfluss von elektromagnetischem Feld und Kernkraftfeld (Feldtheorie der Materie II), *Helvetica Physica Acta*, 9 (1936), pp. 533–554.

[30] Artificial Radioactivity giving Continuous γ-Radiation, *Nature*, vol. 137 (1936), pp. 1070–1071.

[31] Un grand savant à Genève: Max Planck et la Physique contemporaine, *Journal de Genève*, 2 juin 1936, 1 p.

[32] Fragen aus der Physik der Atomkerne und des Elektrons, Zur Vortragswoche des Physikalischen Institutes der ETH vom 30.06. bis 04.07.36 *Schweizer. Archiv für Angewandte Wissenschaft und Technik*, 1936, pp. 219–223.

[33] Über die Methode der physikalischen Naturbeschreibung, *Verhandlungen der Naturforschenden Gesellschaft*, vol. 47 (1936), pp. 181–205.

[34] Neutrino Theory of Light, *Nature*, vol. 139 (1937), pp. 198–199.

[35] La correspondance entre les potentiels retardés de la physique classique et de la physique quantique, *C.R. Société de Physique et d'Histoire Naturelle de Genève*, vol. 54 (1937), pp. 44–47.

[36] Etablissement de la formule des potentiels retardés dans la physique quantique, *C.R. Société de Physique et d'Histoire Naturelle de Genève*, vol. 54 (1937), pp. 47–50.

[37] Der heutige Stand der exakten Naturwissenschaften, *Feuille centrale de la Société de Zofingue*, vol. 8 (1937), pp. 503–514.

[38] On the Existence of heavy Electrons, *Physical Review*, vol. 52 (1937), pp. 41–42.

[39] Die Wechselwirkungskräfte in der Elektrodynamik und in der Feldtheorie der Kernkräfte (Teil I), *Helvetica Physica Acta*, vol. 11 (1938), pp. 225–244.

[40] Die Wechselwirkungskräfte in der Elektrodynamik und in der Feldtheorie der Kernkräfte (Teil II und III), *Helvetica Physica Acta*, vol. 11 (1938), pp. 299–328.

[41] Über die Energieverluste von Elementarteilchen mit ganzzahligem Spin, *Helvetica Physica Acta*, vol. 11 (1938) pp. 378–380.

[42] A propos de l'interaction entre les particules élémentaires, *C.R. Académie des Sciences de Paris*, vol. 207 (1938), pp. 387–389.

[43] Rigorous Theory of Interaction between Nuclear Particles, *Physical Review*, vol. 54 (1938), pp. 889–892.

[44] Theory of Mesons and Nuclear forces, *Nature*, vol. 143 (1939), p. 560–561.

[45] Sur l'intégration de l'équation $(\sum_1^4 \partial_{x_j}^2 - l^2)Q = -\rho$ en utilisant la méthode de Sommerfeld, *C.R. Société de Physique et d'Histoire naturelle de Genève*, 56 (1939), pp. 43–45.

[46] Sur l'interaction entre les particules nucléaires, *Nature*, vol. 143 (1939), p. 560.

[47] Sur l'interaction entre particules nucléaires, with Patry, J.F.C., *Helvetica Physica Acta*, vol. 12 (1939), pp. 300–303.

[48] A New Model of the Point Charge Electron and of Other Elementary Particles, *Nature*, 144, (1939), p.118.

[49] Théorie classique des forces d'échanges, with Patry, J.F.C., *Helvetica Physica Acta*, vol. 13 (1940), pp. 167–192.

[50] Influence du champ pseudoscalaire sur la théorie classique des forces d'échange, *Helvetica Physica Acta*, vol. 13 (1940), pp. 347–354.

[51] Die Schwierigkeiten in der Feldtheorie der Austauschkräfte, *Zeitschrift für technische Physik*, vol. 21 (1940), pp. 275–276.

[52] Un nouveau modèle de l'électron ponctuel en théorie classique, *Helvetica Physica Acta*, vol. 14 (1941), pp. 51–80.

[53] La signification du temps propre en mécanique ondulatoire, *Helvetica Physica Acta*, vol. 14 (1941), pp. 322–323.

[54] Remarque à propos de la création de paires de particules en théorie de la relativité, *Helvetica Physica Acta*, vol. 14 (1941), pp. 588–594.

[55] Le rôle de l'invariance spinorielle et l'invariance de jauge dans un nouveau principe fondamental, *Actes Société helvétique Sciences Naturelles*, 1941, pp. 83–84.

[56] La mécanique du point matériel en théorie de relativité et en théorie des quanta, *Helvetica Physica Acta*, vol. 15 (1942), pp. 23–37.

[57] La notion du temps, Conférence prononcée le 22 novembre 1941 à Bucarest, *Disquisit. Math. et Phys. Bucarest*, vol. 2 (1942), pp. 301–317.

[58] Solutions invariantes $D_{\chi^2}(x, y)$ de l'équation $(\Box - \chi^2)D = 0$ dans l'espace pseudo-euclidien, *C.R. Société de Physique et d'Histoire naturelle de Genève*, vol. 59 (1942), pp. 49–52.

[59] Solutions invariantes $D_{\chi^2}(x, y)$ de l'équation de Schrödinger relativiste, *C.R. Société de Physique et d'Histoire naturelle de Genève*, vol. 59 (1942), pp. 53–55.

[60] Remarques à propos de la relation entre spin et statistique, *Helvetica Physica Acta*, vol. 15 (1942), pp. 327–329.

[61] Le rôle de l'invariance spinorielle et l'invariance de jauge dans un nouveau principe fondamental, *Helvetica Physica Acta*, vol. 15 (1942), pp. 513–515.

[62] Une méthode nouvelle de la quantification des champs, *Archives des Sciences Physiques et Naturelles*, vol. 24 (1942), pp. 193–222.

[63] Une méthode nouvelle de la quantification des champs (Suite), *Archives des Sciences Physiques et Naturelles*, vol. 24 (1942), pp. 261–271.

[64] Une méthode nouvelle de la quantification des champs (Suite et fin), *Archives des Sciences Physiques et Naturelles*, vol. 25 (1942), pp. 5–34.

[65] Un principe qui relie la théorie de la relativité et la théorie des quanta, *Helvetica Physica Acta*, vol. 16 (1943), pp. 173–202.

[66] Le freinage du rayonnement en théorie des quanta, *Helvetica Physica Acta*, vol. 16 (1943), pp. 427–428.

[67] An unambiguous method of avoiding divergence difficulties in quantum theory, *Nature*, vol. 153 (1944), pp. 143–144.

[68] Principe de correspondance d'une mécanique asymptotique classique, *C.R. Société de Physique et d'Histoire naturelle de Genève*, vol. 61 (1944), pp. 156–158.

[69] Principe de correspondance d'une mécanique asymptotique quantifiée, *C.R. Société de Physique et d'Histoire naturelle de Genève*, vol. 61 (1944), pp. 159–161.

[70] Le freinage de radiation de l'électron de Dirac en mécanique asymptotique, with Bouvier, P., *C.R. Société de Physique et d'Histoire naturelle de Genève*, vol. 61 (1944), pp. 162–165.

[71] Un modèle de l'électron ponctuel, II, *Helvetica Physica Acta*, vol. 17 (1944), pp. 3–26.

[72] La charge gravifique et le spin de l'électron classique, III, *Helvetica Physica Acta*, vol. 18 (1945), pp. 21–44.

[73] Mécanique fonctionnelle, *Helvetica Physica Acta*, vol. 18 (1945), pp. 195–220.

[74] L'état actuel de la théorie des particules élémentaires et des quanta, *Experientia*, vol. 1 (1945), pp. 33–36.

[75] La découverte des rayons X et la nature du rayonnement, *Bulletin de l'Institut National Genevois*, vol. 52 (1946), pp. 3–13.

[76] Rayonnement d'accélération d'un électron dans l'effet Compton, with Bouvier, P., *Helvetica Physica Acta*, vol. 19 (1946), pp. 237–239.

[77] Opérateurs non linéaires en théorie des quanta, with Rivier, D., *Helvetica Physica Acta*, vol. 19 (1946), pp. 240–242.

[78] Une propriété de l'opérateur S en mécanique asymptotique, *Helvetica Physica Acta*, vol. 19 (1946), pp. 242–243.

[79] The present state of the S-operator theory, Report on an International Conference on Fundamenral Particles and Low Temperature, Cavendish Laboratory, 22–27 July 1946, Physical Society Cambridge Conference Report, (1947), pp. 199–200 (London: Taylor and Francis, Ltd. 1947).

[80] A possible new type of spin-spin interaction, *Physical Review*, vol. 73 (1948), p. 808.

[81] A convergent expression for the magnetic moment of the neutron, with Rivier, D., *Physical Review*, vol. 74 (1948), p. 218. Erratum: idem, p. 986.

[82] Causalité et structure de la matrice S, with Rivier, D., *Helvetica Physica Acta*, vol. 23 (1950), pp. 215–222.

[83] A propos des divergences en théorie des champs quantifiés, with Rivier, D., Internationaler Kongress über Kernphysik und Quantenelektrodynamik, Basel, 5–9 September 1949, *Helvetica Physica Acta*, vol. 23, suppl. 3 (1950), pp. 236–239.

[84] Relativistic Quantum Theory for Finite Time Intervals, *Physical Review*, vol. 81 (1951), pp. 130–133.

[85] Elimination des constantes arbitraires dans la théorie relativiste des quanta, with Green, T.A., *Helvetica Physica Acta*, vol. 24 (1951), pp.153–174.

[86] Restriction of Possible Interaction in Quantum Electrodynamics, with Petermann, A., *Physical Review*, vol. 82 (1951), pp. 548–549.

[87] The normalization group in quantum theory, with Petermann, A., *Helvetica Physica Acta*, vol. 24 (1951), pp. 317–318.

[88] Théorème H et unitarité de S, *Helvetica Physica Acta*, vol. 25 (1952), pp. 577–580.

[89] Thermodynamique en Relativité Générale, with Wanders, G., *Helvetica Physica Acta*, vol. 26 (1953), pp. 307–316.

[90] Thermodynamique dans un continu, Riemannien par domaines, et théorème sur le nombre de dimensions ($d < 3$) de l'espace, *Helvetica Physica Acta*, vol. 26 (1953), pp. 417–420.

[91] La normalisation des constantes dans la théorie des quanta, with Petermann, A., *Helvetica Physica Acta*, vol. 26 (1953), pp. 499–520.

[92] Acausalité de l'interaction non locale, with Wanders, G., *Helvetica Physica Acta*, vol. 27 (1954), pp. 667–682.

[93] Zur Deutung der relativistischen Wellenfunktionen, with Wanders, G., *Helvetica Physica Acta*, vol. 28 (1955), pp. 352–355.

[94] Particule élémentaire et particule composée, with Wanders, G., *Archives des Sciences*, Genève, vol. 8 (1955), pp. 71–79.

[95] Violation of Parity Conservation and General Relativity, *Physical Review*, vol. 106 (1957), pp. 388–389.

[96] Théorie de la radiation de photon de masse arbitrairement petite, *Helvetica Physica Acta*, vol. 30 (1957), pp. 209–215.

[97] Transformation de jauge et conservation de la charge leptonique, *Archives des Sciences*, Genève, vol. 10 (1957), pp. 243–247.

[98] Le théorème CPT, *Industries Atomiques*, vol. 2 (1958), pp. 17–26.

[99] Field Ouantization and Time Reversal in Real Hilbert Space, *Helvetica Physica Acta*, vol. 32 (1959), pp. 254–256.

[100] The sign of Absolute Temperature T in Phenomenological Thermodynamics, *Helvetica Physica Acta*, vol. 33 (1960), pp. 605–608.

[101] Quantum Theory in Real Hilbert Space, *Helvetica Physica Acta*, vol. 33 (1960), pp. 727–752.

[102] Antilinear Field Operators (Fields of the 2nd kind), with Guenin, M., *Helvetica Physica Acta*, vol. 34 (1961), pp. 506–508.

[103] Quantum theory in real Hilbert space II (Addenda and Errata), with Guenin, M., *Helvetica Physica Acta*, vol. 34 (1961), pp. 621–628.

[104] Quantum theory in real Hilbert space III: field of the first kind (linear field operators), with Guenin, M., Piron, C. et Ruegg, H., *Helvetica Physica Acta*, vol. 34 (1961), pp. 675–698.

[105] Relativistic Thermodynamics III: Velocity of Elastic Waves and Related Problems, *Helvetica Physica Acta*, vol. 35 (1962), pp. 568–591.

[106] Théorie des Quanta dans l'espace de Hilbert réel IV: champs de 2e espèce (opérateurs de champs antilinéaires, T et CP-covariance), with Guenin, M., *Helvetica Physica Acta*, vol. 35 (1962), pp. 673–695.

[107] Non-relativistic Thermodynamics IV (Sign Questions and Onsager Symmetry Relations in Phenomenological Theory), *Helvetica Physica Acta*, vol. 36 (1963), pp. 875–885.

[108] A Generalization of the Principle of Detailed Balancing in μ - Space, *Helvetica Physica Acta*, vol. 37 (1964), pp. 521–531.

[109] Phenomenological Thermodynamics V: The 2nd Law Applied to Extensive Function with the Use of Lagrange Multipliers, with Scheurer, P. B., *Helvetica Physica Acta*, vol. 40 (1967), pp. 887–906.

[110] Théorème de Noether pour la fonction entropie, with Scheurer, P. B., *Helvetica Physica Acta*, vol. 42 (1969), pp. 618–619.

[111] Charge Conjugation C as Proper Time Reversal, with Scheurer, P. B., *Helvetica Physica Acta*, vol. 43 (1970), p. 738.

[112] Sur une question de calcul des variations sous contraintes, with Poncet, J. C. and Scheurer, P. B., *Helvetica Physica Acta*, vol. 44 (1971), pp. 522–529.

[113] On a Covariant Expression of Energy-Momentum in the Relativistic Theory of Gravitation, with Chevalier, J., *Helvetica Physica Acta*, vol. 45 (1972), pp. 587–592.

[114] Minimum Radius of Particles with Spin, *Helvetica Physica Acta*, vol. 45 (1972), pp. 616–618.

[115] Thermocinétique phénoménologique du fluide, Cours de 3e cycle, 3e cycle de Physique en Suisse Romande (CICP), 162 pp.

[116] *Thermocinétique phénoménologique galiléenne,* with Scheurer, P. B., Basel und Stuttgart: Birkhäuser Verlag, 1974, 253 pp.

[117] Is Anti-Gravitation Possible?, in Enz, Ch. and Mehra, J. (eds), *Physical Reality and Mathematical Description.* Dordrecht: D. Reidel Publishing Comp., 1974, pp. 448–452.

Index